CHEMISTRY AND PHYSICS OF MODERN MATERIALS

Processing, Production and Applications

CHEMISTRY AND PHYSICS OF MODERN MATERIALS

Processing, Production and Applications

Edited by

Jimsher N. Aneli, DSc, Alfonso Jiménez, PhD, and Stefan Kubica, PhD

Gennady E. Zaikov, DSc, A. K. Haghi, PhD, and Alberto D'Amore, DSc

Reviewers and Advisory Board Members

Apple Academic Press

TORONTO NEW JERSEY

Apple Academic Press Inc.
3333 Mistwell Crescent
Oakville, ON L6L 0A2
Canada

Apple Academic Press Inc.
9 Spinnaker Way
Waretown, NJ 08758
USA

Printed in the United States of America on acid-free paper

International Standard Book Number-13: 978-1-926895-45-1 (Hardcover)

Library of Congress Control Number: 2013942333

Library and Archives Canada Cataloguing in Publication

Chemistry and physics of modern materials : processing, production and applications/edited by Jimsher N. Aneli, DSc, Alfonso Jiménez, PhD, and Stefan Kubica, PhD.

Includes bibliographical references and index.
ISBN 978-1-926895-45-1
1. Nanotechnology. 2. Nanostructured materials. I. Aneli, J. N., editor of compilation II. Jiménez, Alfonso, 1965- editor of compilation III. Kubica, Stefan Jakub, editor of compilation

T174.7.C44 2013   620'.5   C2013-903906-6

Apple Academic Press also publishes its books in a variety of electronic formats. Some content that appears in print may not be available in electronic format. For information about Apple Academic Press products, visit our website at **www.appleacademicpress.com** and the CRC Press website at **www.crcpress.com**

ABOUT THE EDITORS

Jimsher N. Aneli, DSc

Jimsher N. Aneli, DSc, is a professor at the Institute of Machine Mechanics at Georgian Technical University of Tbilisi State University in Tbilisi, Georgia. He is a scientist in the field of chemistry and the physics of oligomers, polymers, composites, and nanocomposites. He is the author of over 250 scientific articles and inventions. His research interests include the technology of polymer composites with special designations, such as composites with electrical conducting and high mechanical properties; the creation of polymer chemically stable sensors with high sensitivity to mechanical deformations, temperature, electric, and magnetic fields; and hydrogen electric.

Alfonso Jiménez, PhD

Alfonso Jiménez, PhD, is currently a professor of analytical chemistry and food science and technology at the University of Alicante, in Spain since 2001. He is the Head of the Polymer and Nanomaterials Analysis Group. He has edited 13 books on polymer degradation, stabilization biodegradable, and sustainable composites and has written 67 research papers published in journals in the area of analytical chemistry and polymer science. He has chaired organizing committees for several international conferences on biopolymers and polymer chemistry, in particular BIOPOL-2007 and BIOPOL-2009 held in Alicante, Spain. His main research areas include chemical characterization of polymers and biopolymers, environmentally friendly additives in polymers, characterization of biodegradable polymers and sustainable composites, modification of PLA for flexible films manufacturing, natural antioxidants for active packaging, and TPUs obtained from vegetable oils.

Stefan Kubica, PhD

Stefan Kubica, PhD, is head of the research center at the Institute for Engineering of Polymer Materials & Dyes, which is a research institute with its headquarters in Torun, Poland. It has divisions at Gliwice, Piastow (formerly Rubber Research Institute "STOMIL") and Zgierz, Poland. Professor Stefan Kubica is world well-known specialist in the field of polymer investigation of properties and applications of low-molecular compounds, oligomers, composites and nanocomposites. Professor Kubica is a contributor and co-contributor in several monograph and has written about 150 original papers.

REVIEWERS AND ADVISORY BOARD MEMBERS

Gennady E. Zaikov, DSc

Gennady E. Zaikov, DSc, is Head of the Polymer Division at the N. M. Emanuel Institute of Biochemical Physics, Russian Academy of Sciences, Moscow, Russia, and professor at Moscow State Academy of Fine Chemical Technology, Russia, as well as professor at Kazan National Research Technological University, Kazan, Russia. He is also a prolific author, researcher, and lecturer. He has received several awards for his work, including the the Russian Federation Scholarship for Outstanding Scientists. He has been a member of many professional organizations and on the editorial boards of many international science journals.

A. K. Haghi, PhD

A. K. Haghi, PhD, holds a BSc in urban and environmental engineering from University of North Carolina (USA); a MSc in mechanical engineering from North Carolina A&T State University (USA); a DEA in applied mechanics, acoustics and materials from Université de Technologie de Compiègne (France); and a PhD in engineering sciences from Université de Franche-Comté (France). He is the author and editor of 65 books as well as 1000 published papers in various journals and conference proceedings. Dr. Haghi has received several grants, consulted for a number of major corporations, and is a frequent speaker to national and international audiences. Since 1983, he served as a professor at several universities. He is currently Editor-in-Chief of the International Journal of Chemoinformatics and Chemical Engineering and Polymers Research Journal and on the editorial boards of many international journals. He is also a faculty member of University of Guilan (Iran) and a member of the Canadian Research and Development Center of Sciences and Cultures (CRDCSC), Montreal, Quebec, Canada.

Alberto D'Amore, DSc

Alberto D'Amore, DSc, is currently Associate Professor of Materials Science and Technology at Second University of Naples-SUN in Rome, Italy. He has authored more than one hundred scientific papers published in international journals and books. He is a member of the scientific committees of many international conferences and is on the editorial boards of several international journals. Currently he is the Chairman of the International-Times of Polymers (TOP) and Composites conferences.

CONTENTS

LIST OF CONTRIBUTORS

Alekseeva, O. M.
Institute of Biochemical Physics N. M. Emanuel Academy of Sciences, Russia, 119334, Moscow, st. Kosygin d.4. (495) 939-74-09, Email: olgavek@yandex.ru

Alinkina, E. S.
Emanuel Institute of Biochemical Physics, Russian Academy of Sciences, Moscow, 119334 Russia

Aloev, V. Z.
Kabardino-Balkarian State Agricultural Academy, Nal'chik – 360030, Tarchokov st., 1 a, Russian Federation

Altshuler, M. L.
FGBU I. Mechnikov Research Institute for Vaccines and Sera of RAMS. Moscow. 105064 Maliy Kazyonniy side. Fax: 8 (495) 916-11-52. e-mail: labpitsred@yandex.ru

Antonov, Y. A.
N. M. Emanuel Institute of Biochemical Physics, Russian Academy of Sciences, Kosygin Stir. 4. 119334 Moscow, Russia, Tel. +7(495)939-74-02; Fax:+7(495)137-41-01. e-mail:chehonter@yandex.ru

Aphashagova, Z. Kh.
Kabardino-Balkarian State University, Chernyshevskii st., 173, Nal'chik 360004, Russian Federation

Babkin, V. A.
403343 SF VolgSABU, c. Mikhailovka, region Volgograd, S.Michurina 21. E-mail:sfi@reg.avtlg.ru

Bazylyak, L. I.
Chemistry of Oxidizing Processes Division; Physical Chemistry of Combustible Minerals Department; Institute of Physical–Organic Chemistry & Coal Chemistry named after L. M. Lytvynenko; National Academy of Science of Ukraine, 3[a]Naukova Str., Lviv–53, 79053, UKRAINE; e–mail: hop_vfh@ukr.net; bazyljak.L.I@nas.gov.ua

Bekusarova, S. A.
Gorsky State Agrarian University, Kirov str., 37, Vladikavkaz, Republic of North Ossetia Alania, Russia, ggau@globalalania.ru;

Belostotskaja, I. S.
Emanuel Institute of Biochemical Physics of Russian Academy of Sciences, 4 Kosygin Street. 119334 Moscow, Russia

Blinkova, L. P.
FGBU I. Mechnikov Research Institute for Vaccines and Sera of RAMS. Moscow. 105064 Maliy Kazyonniy side. Fax: 8 (495) 916-11-52. e-mail: labpitsred@yandex.ru

Bome, N. A.
Tyumen State University, Semakova str., 1, 625003, Tyumen, Russia, +7(3452) 46-40-61, 46-81-69. president@utmn.r

Burlakova, E. B.
Emanuel Institute of Biochemical Physics of Russian Academy of Sciences, 4 Kosygin Street. 119334 Moscow, Russia

Bychkova, A. V.
Emanuel Institute of Biochemical Physics, Russian Academy of Sciences, Kosygina str., 4, Moscow, 119334, Russia, E-mail: annb0005@yandex.ru

Chalykh, A. E.
Institute of physical chemistry and electrochemistry named by Frumkin, Leninsky prospect, 31, GSP-1, Moscow, 119991

Deberdeev, R. Ya.
Kazan State Technological University, ul. Karla Marksa 68, Kazan, Republic of Tatarstan, 420015 Russia

Derkach, S. A.
Mechnicov Institute of Microbiology and Immunology of Academy of Medical Sciences of Ukraine

Durlakova, T. B.
Emanual Institute of Biochemical Physics of the Russian Academy of Science

Evlashkina, V. F.
FGBU I. Mechnikov Research Institute for Vaccines and Sera of RAMS. Moscow. 105064 Maliy Kazyonniy side. Fax: 8 (495) 916-11-52. e-mail: labpitsred@yandex.ru

Evteeva, N. M.
Institute of biochemical physics of N. M.Emanuelja of the Russian academy, Sciences, the Russian Federation, 119991 Moscow, Kosygina str. 4, Fax: 095 1374101.

Fatianova, E. A.
Katarína Valachová, Mária Baňasová, Ľubica Machová, Ivo Juránek, Štefan Bezek, Ladislav Šoltés Department "General and Inorganic chemistry", South-West State University, Kursk, Russia

Fatkullina, L. D.
Emanuel Institute of Biochemical Physics, Russian Academy of Sciences, Kosigina 4, Moscow, 119334 Russia

Fattakhov, S. G.
Arbuzov Institute of Organic and Physical Chemistry of the Russian Academy of Science

Generozova, I. P.
Timiryazev Institute of Plant Physiology of the Russian Academy of Science

Gerasimov, V. K.
Institute of physical chemistry and electrochemistry named by Frumkin, Leninsky prospect, 31, GSP-1, Moscow, 119991

Gumargalieva, K. Z.
N. N. Semenov Institute of Chemical Physics, RAS, Moscow

Gurchenkova, Y. A.
Sochi Institute of Russian people`s friendship university, Kuibyshev str. 32, E-mail: sfrudn@rambler.ru, Fax +7 (8622) 411043.

Hadavi Moghadam, B.
Department of Textile Engineering, University of Guilan, Rasht, Iran

Haghi, A. K.
University of Guilan, Rasht, Iran, E-mail: Haghi@Guilan.ac.ir

Hasanzadeh, M.
Department of Textile Engineering, University of Guilan, Rasht, Iran, Tel.: +98-21-33516875; fax: +98-182-3228375, Email: m_hasanzadeh@aut.ac.ir

Iordanskii, A. L.
N. N. Semenov Institute of Chemical Physics, RAS, 4 street Kosygina, 4119991, Moscow

Ivanova, N. N.
Institute of Dermatology and Venerology of Academy of Medical Sciences of Ukrainestr. Chernishevskaya 7/9, 61057 Kharkov, Ukraine, e-mail: jet-74@mail.ru, fax +38-057- 706-32-03

Ivo Juránek
Institute of Experimental Pharmacology and Toxicology, Slovak Academy of Sciences, Bratislava 84104, Slovakia

Katarína Valachová
Institute of Experimental Pharmacology and Toxicology, Slovak Academy of Sciences, Bratislava 84104, Slovakia, Tel.: +421902303680. E-mail address: katarina.valachova@savba.sk

Kazakova, L. I.
Institute of Theoretical and Experimental Biophysics Russian Academy of Science, 142290 Pushchino, Moscow region 142290, Russia

Kharuta, L. G.
Sochi Institute of Russian people`s friendship university, Kuibyshev str. 32, E-mail: sfrudn@rambler.ru, Fax +7 (8622) 411043.

Khokhriakov, N. V.
Izhevsk State Agricultural Academy, Basic Research and Educational Center of Chemical Physics and Mesoscopy, Udmurt Scientific Center, Ural Division, Russian Academy of Science, Russia, Izhevsk, 426000, e-mail: korablev@udm.net.

Khrapova, N. G.
Emanuel Institute of Biochemical Physics of Russian Academy of Sciences, 4 Kosygin Street. 119334 Moscow, Russia

Khrustova, N. V.
N. M. Emanuel Institute of Biochemical Physics, Russian Academy of Sciences, 4 Kosygin Street, 119334 Moscow, Russia

Khuzakhanov, R. M.
Kazan State Technological University, K.Marx, 68, Kazan, Tatarstan, Russia, 420015

Kirsh, I. A.
Moscow State University of Food Production, 33 Talalihin st. Moscow Russia 109316, kaf.vms@rambler.ru

Klodzinska, E.
Institute for Engineering of Polymer Materials and Dyes, 55 M. Sklodowskiej-Curie str., 87-100 Torun, Poland, E-mail: Ewa.Klodzinska@impib.pl

Kokov, L. S.
Sklifosovsky Research Institute for Emergency Medicine, Moscow

Korablev, G. A.
Izhevsk State Agricultural Academy, Basic Research and Educational Center of Chemical Physics and Mesoscopy, Udmurt Scientific Center, Ural Division, Russian Academy of Science, Russia, Izhevsk, 426000, E-mail: korablevga@mail.ru,biakaa@mail.ru

Kosenko, R. Yu.
N. N. Semenov Institute of Chemical Physics, RAS, 4 street Kosygina, 4119991, Moscow

Kotsar, E. V.
Institute of Dermatology and Venerology of Academy of Medical Sciences of Ukrainestr. Chernishevskaya 7/9, 61057 Kharkov, Ukraine, e-mail: jet-74@mail.ru, fax +38-057- 706-32-03

Kovarski, A. L.
Emanuel Institute of Biochemical Physics, Russian Academy of Sciences, Kosygina str., 4, Moscow, 119334, Russia

Koverzanova, E. V.
N.N. Semenov Institute of Chemical Physics, RAS, Moscow

Kozlov, G. V.
Kabardino-Balkarian State Agricultural Academy, Nal'chik – 360030, Tarchokov st., 1 a, Russian Federation

Kozlova, Z. G.
Emanuel Institute of Biochemical Physics of Russian Academy of Sciences, 119334, Moscow, 4 Kosygin St. E-mail: yevgeniya-s@inbox.ru, fax: (495) 137-41-01

Kozlowa, N. I.
Sochi Institute of Russian people`s friendship university, Kuibyshev str. 32, E-mail: sfrudn@rambler.ru, Fax +7 (8622) 411043.

Krikunova, N. I.
Emanual Institute of Biochemical Physics of the Russian Academy of Science

Kubica, S.
Institute for Engineering of Polymer Materials and Dyes, 55 M. Sklodowskiej-Curie str., 87-100 Torun, Poland, Email: S.Kubica@impib.pl

Kushnireva, Ye. V.
N. M. Emanuel Institute of Biochemical Physics, Russian Academy of Sciences, 4 Kosygin Street, 119334 Moscow, Russia

Kuvardin, N. V.
Department "General and Inorganic chemistry", South-West State University, Kursk, Russia

Kytsya, A.R.
Chemistry of Oxidizing Processes Division; Physical Chemistry of Combustible Minerals Department; Institute of Physical–Organic Chemistry & Coal Chemistry named after L. M. Lytvynenko; National Academy of Science of Ukraine, 3[a]Naukova Str., Lviv–53, 79053, UKRAINE; e–mail: hop_vfh@ukr.net; bazyljak.L.I@nas.gov.ua

Ladislav Šoltés
Institute of Experimental Pharmacology and Toxicology, Slovak Academy of Sciences, Bratislava 84104, Slovakia

Leonova, V. B.
Emanuel Institute of Biochemical Physics, Russian Academy of Sciences, Kosygina str., 4, Moscow, 119334, Russia

Lomakin, S. M.
Establishment of the Russian Academy of Sciences Institute of biochemical physics of N.M.Emanuelja, 119991 Moscow, Kosygina str. 4, Fax: (095 1374101

Lubica Machová
Institute of Experimental Pharmacology and Toxicology, Slovak Academy of Sciences, Bratislava 84104, Slovakia

Luschenko, G. V.
North Caucasus Research Institute of mountain and foothill agriculture, Williams's str, 1, 391502, Mikhailovskoye, Republic of North Ossetia Alania, Russia, t/f +7(8672) 73-03-40

Maedeh Sajedi
Textile Engineering department, Faculty of Engineering, University of Guilan, Rasht P.O.BOX 3756, Guilan, Iran

Magomedov, G. M.
Dagestan State Pedagogical University, Yaragskii st., 57, Makhachkala 367003, Russian Federation

Mahmood Saberi Motlagh
Textile Engineering department, Faculty of Engineering, University of Guilan, Rasht P.O.BOX 3756, Guilan, Iran

Makhrova, N. V.
Kazan State Technological University, ul. Karla Marksa 68, Kazan, Republic of Tatarstan, 420015 Russia

Malamatov, A. Kh.
Kabardino-Balkarian State University, Chernyshevskii st., 173, Nal'chik 360004, Russian Federation

Mária Baňasová
Institute of Experimental Pharmacology and Toxicology, Slovak Academy of Sciences, Bratislava 84104, Slovakia

Mavrov, G. I.
Institute of Dermatology and Venerology of Academy of Medical Sciences of Ukrainestr. Cherni-shevskaya 7/9, 61057 Kharkov, Ukraine, e-mail: jet-74@mail.ru, fax +38-057- 706-32-03

Medvedeva, I. B.
Emanuel Institute of Biochemical Physics, Russian Academy of Sciences, ul. Kosigina 4, Moscow, 119334 Russia

Misharina, T. A.
Emanuel Institute of Biochemical Physics, Russian Academy of Sciences, ul. Kosigina 4, Moscow, 119334 Russia, Email: e-mail: tmish@rambler.ru

Misin, V. M.
Emanuel Institute of Biochemical Physics of Russian Academy of Sciences, 4 Kosygin Street. 119334 Moscow, Russia

Nikiforov, G. A.
Emanuel Institute of Biochemical Physics of Russian Academy of Sciences, 4 Kosygin Street. 119334 Moscow, Russia

Nikiforova, O. V.
FGBU I. Mechnikov Research Institute for Vaccines and Sera of RAMS. Moscow. 105064 Maliy Kazyonniy side. Fax: 8 (495) 916-11-52. e-mail: labpitsred@yandex.ru

Niyazy, F. F.
Department “General and Inorganic chemistry”, South-West State University, Kursk, Russia, Email: FarukhNiyazi@yandex.ru

Olkhov, A. A.
Moscow State Academy of Chemical Technology of M.V. Lomonosov, 119571 Moscow, prosp. Vernadskogo, 86, Email: aolkhov72@yandex.ru

Pakhomov, Y. D.
FGBU I. Mechnikov Research Institute for Vaccines and Sera of RAMS. Moscow. 105064 Maliy Kazyonniy side. Fax: 8 (495) 916-11-52. e-mail: labpitsred@yandex.ru

Pomogova, D. A.
Moscow State University of Food Production, 33 Talalihin st. Moscow Russia 109316

Remizov, A. B.
Kazan State Technological University, K.Marx, 68, Kazan, Tatarstan, Russia, 420015

Rosenfeld, M. A.
Emanuel Institute of Biochemical Physics, Russian Academy of Sciences, Kosygina str., 4, Moscow, 119334, Russia

Rusanova, S. N.
Kazan State Technological University, K.Marx, 68, Kazan, Tatarstan, Russia, 420015

Rybalko, A. E.
Sochi Institute of Russian people`s friendship university, Kuibyshev str. 32, E-mail: sfrudn@rambler.ru, Fax +7 (8622) 411043.

Sechko, E. V.
Kazan State Technological University, K.Marx, 68, Kazan, Tatarstan, Russia, 420015

Shabarchina, L. I.
Institute of Theoretical and Experimental Biophysics Russian Academy of Science, 142290 Pushchino, Moscow region 142290, Russia, E-mail: shabarchina@rambler.ru

Shishkina, L. N.
N. M. Emanuel Institute of Biochemical Physics, Russian Academy of Sciences, 4 Kosygin Street, 119334 Moscow, Russia. E-mail: shishkina@sky.chph.ras.ru

Shtol’ko, V. N.
Emanuel Institute of Biochemical Physics of Russian Academy of Sciences, 4 Kosygin Street. 119334 Moscow, Russia

Shugaev, A. P.
Timiryazev Institute of Plant Physiology of the Russian Academy of Science

Simonova, Yu. S.
Moscow State Academy of Chemical Technology of M.V. Lomonosov, 119571 Moscow, prosp. Vernadskogo, 86, Email: aolkhov72@yandex.ru

Skipina, K. P.
Sochi Institute of Russian people`s friendship university, Kuibyshev str. 32, E-mail: sfrudn@rambler.ru, Fax +7 (8622) 411043.

Sofina, S. Ju.
Kazan State Technological University, K.Marx, 68, Kazan, Tatarstan, Russia, 420015

Sogrina, D. A.
Moscow State University of Food Production, 33 Talalihin st. Moscow Russia 109316

Sorokina, O. N.
Emanuel Institute of Biochemical Physics, Russian Academy of Sciences, Kosygina str., 4, Moscow, 119334, Russia

Starostina, I. A.
Kazan State Technological University, K.Marx, 68, Kazan, Tatarstan, Russia, 420015

Štefan Bezek
Institute of Experimental Pharmacology and Toxicology, Slovak Academy of Sciences, Bratislava 84104, Slovakia

Stoyanov, O. V.
Kazan State Technological University, K.Marx, 68, Kazan, Tatarstan, Russia, 420015, Email: ov_stoyanov@mail.ru

Stoyanova, L. F.
Kazan State Technological University, K.Marx, 68, Kazan, Tatarstan, Russia, 420015, Email: ov_stoyanov@mail.ru

Sukhorukov, G. B.
School of Engineering & Materials Science, Queen Mary University of London, London, UK.

Terenina, M. B.
Emanual Institute of Biochemical Physics of the Russian Academy of Science

Titova, M. I.
A. V. Vishnevsky Institute of Surgery, Moscow

Turovsky, A. A.
Chemistry & Biotechnology of Combustible Minerals Division; Physical Chemistry of Combustible Minerals Department; Institute of Physical–Organic Chemistry & Coal Chemistry named after L. M. Lytvynenko; National Academy of Science of Ukraine, 3[a]Naukova Str., Lviv–53, 79053, UKRAINE e–mail: hop_vfh@ukr.net

Tzomatova, F. T.
North Caucasus Research Institute of mountain and foothill agriculture, Williams's str, 1, 391502, Mikhailovskoye, Republic of North Ossetia Alania, Russia, t/f +7(8672) 73-03-40

Usachev, S. V.
N. N. Semenov Institute of Chemical Physics, RAS, Moscow

Usmanova, R. R.
Ufa State technical university of aviation; 12 Karl Marks str., Ufa 450000, Bashkortostan, Russia, E-mail: Usmanovarr@mail.ru

Vahid Mottaghitalab
Textile Engineering department, Faculty of Engineering, University of Guilan, Rasht P.O.BOX 3756, Guilan, Iran

Vasiliev, Yu. G.
Izhevsk State Agricultural Academy, Basic Research and Educational Center of Chemical Physics and Mesoscopy, Udmurt Scientific Center, Ural Division, Russian Academy of Science, Russia, Izhevsk, 426000, E-mail: korablev@udm.net.

Volodkin, A. A.
Establishment of the Russian Academy of Sciences Institute of biochemical physics of N.M.Emanuelja, 119991 Moscow, Kosygina str. 4, Fax: (095 1374101

Vorobyeva, A. K.
Emanuel Institute of Biochemical Physics, Russian Academy of Sciences, ul. Kosigina 4, Moscow, 119334 Russia

Weisfeld, L. I.
N.M. Emanuel Institute of Biochemical Physics of Russian Academy of Sciences, Kosygina str., 4, 119334, Moscow, Russia, chembio@sky.chph.ras.ru;

Yakh'yaeva, Kh. Sh.
Dagestan State Pedagogical University, Yaragskii st., 57, Makhachkala 367003, Russian Federation

Yanovsky, Yu. G.
Institute of Applied Mechanics of Russian Academy of Sciences, Leninskii pr., 32 A, Moscow-119991, Russian Federation

Zaikov, G. E.
Kinetics of Chemical & Biological Processes Division; Institute of Biochemical Physics named after N. N. Emanuel; Russian Academy of Sciences, 4 Kosygin Str., Moscow, 119991, Russia; e-mail: chembio@sky.chph.ras.ru

Zakharov, D. S.
403343 SF VolgSABU, c. Mikhailovka, region Volgograd, S.Michurina 21. E-mail:sfi@reg.avtlg.ru

Zhigacheva, I. V.
Emanual Institute of Biochemical Physics of the Russian Academy of Science, Email: zhigacheva@mail.ru

Zhirikova, Z. M.
Kabardino-Balkarian State Agricultural Academy, Nal'chik – 360030, Tarchokov st., 1 a, Russian Federation

Zhuravleva, I. L.
N. M. Emanuel Institute of Biochemical Physics, Russian Academy of Sciences, Kosygin Stir. 4. 119334 Moscow, Russia

LIST OF ABBREVIATIONS

3D	three-dimensional
aPP	atactic polypropylene
BAP	6-benzylaminopurine
BM-MSCs	bone marrow mesenchymal stem cells
CN	carbon nanotubes
Co/C	cobalt/carbon
CTS	chitosan
Cu/C	copper/carbon
DLTP	dilaurylthiodipropionate
DMF	dimethylformamide
DNA	deoxyribonucleic acid
ECM	extracellular matrix
ECs	endothelial cells
ED	electron diffraction
EDR	epoxy diane resin
EGF	epidermal growth factor
EHD	electrohydrodynamic
EMD	electron microdiffraction
EPR	electron paramagnetic resonance
ER	epoxy resin
FS	fine suspensions
FTIR	Fourier transform infrared spectroscopy
HA	hyaluronan
HA	hydroxyl apatite
HA/CTS	hydroxyaptite/chitosan
HFIP	hexafluoroisopropanol
HSCs	hematopoietic stem cells
hSF	human skin fibroblasts
IAA	indole-3-acetic acid
IR	infrared

LBM	lattice boltzmann method
LDPE	low density polyethylene
MEK	methylethylketone
MS	Murashige-Skoog
MSC	mesenchymal stem cells
MWCNT	multi-walled carbon nanotube
NAA	naphtaleneacetic acid
NC	nanocomposites
nc-Si	nanocrystalline silicon
NGF	nerve growth factor
Ni/C	nickel/carbon
NS	nanostructures
NSCs	neural stem cells
ODE	ordinary differential equation
PAA	poly(acrylic acid)
PBS	phosphate buffered saline
PCL	policaprolactone
PCM	polymeric composite materials
PE	polyethylene
PEO	polyethylenoxide
PEPA	polyethylene polyamine
PET	polyethyelene terephthalate
PGA	poly(glycolic acid)
PGA-PHA	polyglycolic acid-polyhydroxyalkanote
PGR	plant growth regulator
PLA	poly(lactic acid)
PLGA	poly(lactic-co-glycolic acid)
PLLA	poly(L-lactic acid)
PLLA-CL	poly(L-lactic acid)-co-poly(ε-caprolactone)
PMMA	poly(methyl methacrylate)
PNR	peripheral nerve regeneration
PP	polypropylene
PS	polystyrene
PU	polyurethane
RA	rheumatoid arthritis
RHR	heat release rate

ROS	reactive oxygen species
SCs	Schwann cells
SEM	scanning electron microscope
SF	synovial fluid
SMCs	smooth muscle cells
TE	tissue engineering
TEM	transmission electron microscopy
TFA	trifluoroacetic acid
TGA	thermogravimetric analysis
THF	tetrahydrofuran
TPa	terapascal
UCB	umbilical cord blood
UCM	upper-convected maxwell
UV	ultraviolet
WBOS	Weissberger's biogenic oxidative system

PREFACE

Technical and technological development demands the creation of new materials, which are stronger, more reliable, and more durable, i.e. materials with new properties. Technologically advanced projects in the creation of new materials go along the way of nanotechnology.

Modern materials can also be referred to as a qualitatively new round in human progress. This is a wide concept, which embraces many areas, including information technologies, medicine, military equipment, robotics, and so on. We may narrow the concept of chemistry and physics of modern materials and consider it with the reference to polymeric materials as well as composites on their basis as well. For instance in the first chapter of this book, practical and theoretical hints on production and application of nanofibers, nanotubes, nanofillers, and nanocomposites are reviewed in detail. The prefix nano- means that in the context of these concepts the technologies based on the materials, elements of constructions and objects are considered, whose sizes make 10^{-9} meters. Though it sounds fantastic, science reached the nano-level long ago. Unfortunately, everything connected with such developments and technologies for now is impossible to apply to mass production because of low productivity and high cost. That means nanotechnology and nanomaterials are only accessible in research laboratories for now, but it is only a matter of time before can be more widely applied.

What sort of benefits and advantages will manufacturers have after the implementation of nanotechnologies in their manufactures and use of nanomaterials? Nanoparticles of any material have absolutely different properties rather than micro- or macroparticles. This results from the fact that along side the reduction of particles' sizes of the materials to nanometric sizes, physical properties of a substance change too. For example, the transition of palladium to nanocrystals leads to the increase in its thermal capacity in more than 1.5 times; it causes the increase of solubility of bismuth in copper in 4000 times, and the increase of self-diffusion coefficient of copper at a room temperature on 21 orders. Such changes in properties of substances are explained by the quantitative change of surface and volume atoms' ratio of individual particles, i.e. by the high-surface area. Insertion of such nanoparticles in a polymeric matrix while using the apparently old and known materials gives a chance of receiving the qualitatively and quantitatively new possibilities in their use.

Chemistry and Physics of Modern Materials: Processing, Production and Applications is a comprehensive anthology covering many of the major themes of modern materials, addressing many of the major issues, from concept to technology to implementation. It is an important reference publication that provides new research and updates on a variety of modern materials uses through case studies and supporting technologies, and it also explains the conceptual thinking behind current uses and potential uses not yet implemented. International experts with countless years of experience lend this volume credibility. As an example, drying of nanostructured fabrics is presented in Chapter 2. The term "nanocomposites" based on thermoplastic matrix and containing natural laminated inorganic structures are referred to as laminated nanocomposites. Such materials are produced on the basis of ceramics and polymers, however, with the use of natural laminated inorganic structures such as montmorillonite or vermiculite that are present, for example, in clays. A layer of filler ~1nm thick is saturated with monomer solution and later polymerized. The laminated nanocomposites in comparison with initial polymeric matrix possess much smaller permeability for liquids and gases. These properties allow applying them to medical and food-processing industries. Such materials can be used in the manufacture of pipes and containers for the carbonated beverages. These topics are reviewed in Chapters 15, 16, 18, 19, 20 and 28. These composite materials could also be eco-friendly, absolutely harmless to the person, and possess fire-resistant properties. The derived thermoplastic laboratory samples have been tested and really confirmed those statements.

It should be noted that manufacturing technique of thermoplastic materials causes difficulties for today, notably dispersion of silicate nanoparticles in monomer solution. To solve this problem, it is necessary to develop the dispersion technique, which could be transferred from laboratory conditions into the industrial ones.

What advantages the manufacturers can have, if they decide to reorganize their manufacture for the use of such materials, can be predicted even today. As these materials possess more mechanical and gas-barrier potential in comparison with the initial thermoplastic materials, then their application in manufacture of plastic containers or pipes will lead to raw materials saving by means of reduction of product thickness.

On the other hand, the improvement of physical and mechanical properties allows applying nanocomposite products under higher pressures and temperatures. For example, the problem of thermal treatment of plastic containers can be solved. Another example of the application of the valuable properties of laminated nanocomposites concerns the motor industry.

As mentioned earlier, another group of materials is metal containing nanocomposites. Thanks to the ability of metal particles to create the ordered structures (clusters), metal containing nanocomposites can possess a complex of valuable properties. The typical sizes of metal clusters from 1 to 10 nanometers correspond to their huge specific surface area. Such nanocomposites demonstrate the superparamagnetism and catalytic properties; therefore, they can be used while manufacturing semiconductors, catalysts, and optical and luminescent devices, etc.

Such valuable materials can be produced in several ways, for example, by means of chemical or electrochemical reactions of isolation of metal particles from solutions. In this case, the major problem is not so much the problem of metal restoration but the preservation of its particles, i.e. the prevention of agglutination and formation of large metal pieces.

For example under laboratory conditions metal is deposited in such a way on the thin polymeric films capable to catch nano-sized particles. Metal can be evaporated by means of high energy, and nano-sized particles can be produced, which then should be preserved. Metal can be evaporated while using explosive energy, high-voltage electric discharge or simply high temperatures in special furnaces.

The practical application of modern materials (not going into details about high technologies) can involve the creation of polymers possessing some valuable properties of metals. For example, the polyethylene plate with the tenth fractions of palladium possesses the catalytical properties similar to the plate made of pure palladium.

An example of applying metallic composite is the production of packing materials containing silver and possessing bactericidal properties. By the way, some countries have already been applying the paints and the polymeric coverings with silver nanoparticles. Owing to their bactericidal properties, they are applied in public facilities (painting of walls, coating of handrails, etc.). The technology of polymeric nanocomposites' manufacture forges ahead; its development is directed to simplification and cheapening the production processes of composite materials with nanoparticles in their structure. However, the nanotechnologies develop high rates; what seemed impossible yesterday will be accessible to the introduction on a commercial scale tomorrow.

The prospects in the field of modern materials upgrading are retained by nanotechnologies as well. Ever-increasing manufacturers' demand for new and superior materials stimulates scientists to find new ways of solving tasks on the qualitatively new nanolevel.

The desired event of fast implementation of modern materials in mass production depends on the efficiency of cooperation between the scientists and the manufacturers in many respects. Today's high technology problems of applied character are successfully solved in close consolidation of scientific and business worlds. The microstructure of modern materials has inhomogeneities in the scale range of nano to micrometers. Modern materials cover the range between inorganic glasses and organic polymers. Fillers of polymers have been used for a long time with the goal of enhanced performance of polymers and especially of rubber.

In this new book, top scientists offer an up-to-date global perspective on the latest developments in Chemistry and Physics of Modern Materials. This cuts across every scientific and engineering discipline to provide important presentations of current accomplishments in material sciences. This state-of-the art book provides empirical and theoretical research concerning the chemistry and physics of modern materials.

— Jimsher N. Aneli, DSc, Alfonso Jiménez, PhD, and Stefan Kubica, PhD

CHAPTER 1

PRACTICAL AND THEORETICAL HINTS ON PRODUCTION AND APPLICATION OF NANOFIBERS, NANOTUBES, NANOFILLERS AND NANOCOMPOSITES

A. K. HAGHI

CONTENTS

1.1 INTRODUCTION

The term "nano" like many other prefixes used in conjunction with Systeme International d'unites (SI units), "nano" comes from a language other than English. Originating from the Greek word "nannas" for "uncle," the Greek word "nanos" or "nannos" refer to "little old man" or "dwarf.' Before the term nanotechnology was coined and became popular, the prefixes "nanno-" or "nano-" were used in equal frequency, although not always technically correct.

The use of nanostructure materials is not a recently discovered era. Back in the 4th century AD the Romans used nano sized metal particles to decorate cups. The first known and the most famous example is the Lycurgus cup. In this cup as well as the famous stained glasses of the tenth, eleventh and twelfth centuries, metal nanoparticles account for the visual appearance. This property is used in other ways in addition to stained glass. For example particles of titanium dioxide (TiO_2) have been used for a long time as the sun-blocking agent in sunscreens.

Today, nanotechnology refers to technological study and application involving nanoparticles. In general, nanotechnology can be understood as a technology of design, fabrication and applications of nanostructures and nanomaterials. Nanotechnology also includes fundamental understanding of physical properties and phenomena of nanometerials and nanostructures. Nanostructure is the study of objects having at least one dimension within the nano-scale. A nanoparticle can be considered as a zero-dimensional nano-element, which is the simplest form of nanostructure.

1.2 NANOTECHNOLOGY AND BIOMIMETICS

Since the turn of the century, the research and development on the nature-inspired manufacturing technology, generally referred to as "biomimetics," have been coming to the fore in Europe and the United States. The term "biomimetics" (in Japan, the term is translated literally as "Imitation of Living Things") was proposed by German-American neurophysiologist Otto Schmitt in the latter half of the 1950s.

Scientists found that, on the surface of lotus leaves, the surface microstructure and secretion of wax-like compounds have a synergetic effect that produces super-hydrophobic property and is self-cleaning. The discovery of the lotus effect induced surface-structure research on living things that dwell inaquatic environments. The petals of roses, clivias, and sunflowers show superhydrophilicity, but at the same time they have so strong an adsorption power to water that they hold water drops even if they are held upside down.

Other researches in biomimetics are: Adhesive materials inspired by gecko legs, swimsuit material inspired by sharkskin riblet, Structural color materials inspired by butterflies and jewel beetles, anti-reflective materials inspired by the Moth-eye structure, low friction materials inspired by sandfish, sensor materials inspired by the sensing ability of insects.

1.3 NANOPROCESSING

There are two approaches to the synthesis of nanomaterials and the fabrication of nanostructures: top-down and bottom-up. Which depend upon the final size and shape of a nanostructure or nanodevice. Attrition or milling is a typical top-down method in making nanoparticles, whereas the colloidal dispersion is a good example of bottom-up approach in the synthesis of nanoparticles. Bottom-up approach is often emphasized in nanotechnology literature. Examples include the production of salt and nitrate in chemical industry, the growth of single crystals and deposition of films in electronic industry. The bottom-up approach refers to the approach to build a material up from the bottom: atom-by-atom, molecular-by-molecular, or cluster-by-cluster. Although the bottom-up approach is nothing new, it plays an important role in the fabrication and processing of nanostructures and nanomaterials. When structures fall into a nanometer scale, there is little choice for a top-down approach. All the tools we have possessed are too big to deal with such tiny objects.

1.4 NANOFIBERS AND SYNTHESIS

Human beings have used fibers for centuries. In 5000 BC, our ancestors used natural fibers such as wool, cotton silk and animal fur for clothing. Mass production of fibers dates back to the early stages of the industrial revolution. The first man-made fiber –viscose – was presented in 1889 at the World Exhibition in Paris. Traditional methods for polymer fiber production include melt spinning, dry spinning, wet spinning and gel-state spinning. These methods rely on mechanical forces to produce fibers by extruding a polymer melt or solution through a spinneret and subsequently drawing the resulting filaments as they solidify or coagulate. These methods allow the production of fiber diameters typically in the range of 5 to 500 microns. At variance, electrospinning technology allows the production of fibers of much smaller dimensions. The fibers are produced by using an electrostatic field.

A nanofiber is a nanomaterial in view of its diameter, and can be considered a nanostructured material if filled with nanoparticles to form composite nanofibers. Also Nanofibers are defined as fibers with diameters less than 1000 nm nanometers. There are various techniques for polymeric nanofiber production such as Drawing, Template Synthesis, Phase Separation, Self-Assembly and Electrospinning.

1.4.1 DRAWING

Drawing is a process whereby nanofibers are produced when a micropipette with a tip few micrometers in diameter is dipped into a droplet near the contact line using a micromanipulator and then withdrawn from the liquid at about 1×10^{-4} m s^{-1}. When the end of the micropipette touches the surface, the drawn nanofiber is then deposited (Fig. 1). To draw fibers a viscoelastic material is required so that it can endure the strong deformation and at the same time it has to be cohesive enough to support the stress that build up in the pulling process. The drawing process can be considered as dry spinning at a molecular level.

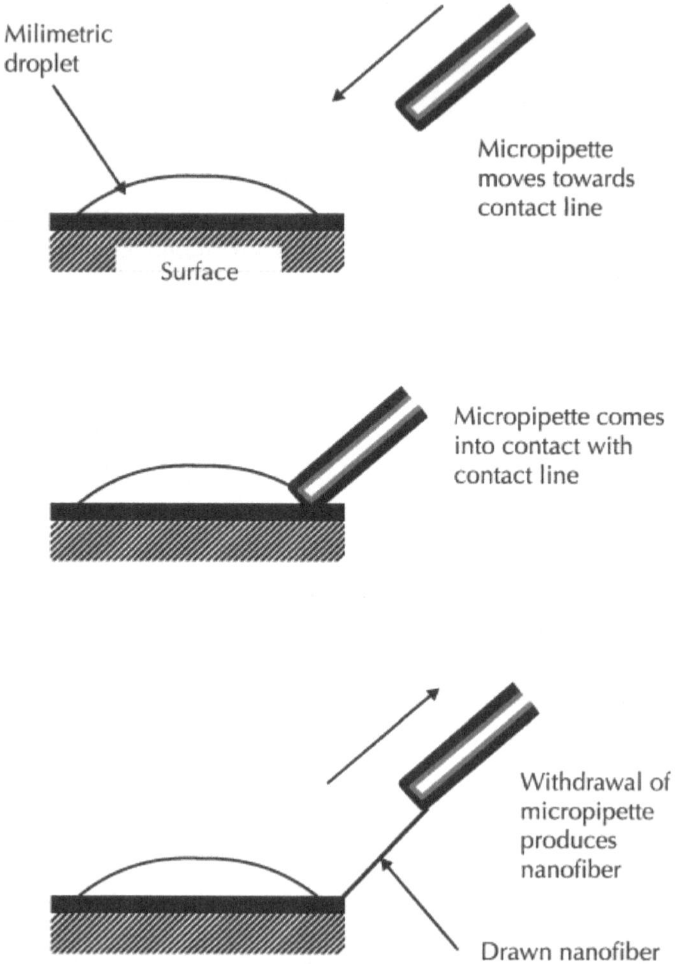

FIGURE 1 Obtaining nanofiber by drawing.

This method can be used to make one-by-one and very long single nanofibers.

1.4.2 TEMPLATE SYNTHESIS

Template synthesis implies the use of a template or mold to obtain a desired material or structure. A nanoporous membrane is used as template to make nanofibers of solid (i.e., fibril) or hollow (i.e., tubule) shape. Using this method nanometer tubules and fibrils of various raw materials such as electronically conducting polymers, metals, semiconductors and carbons can be fabricated. This method cannot, however, make one-by-one continuous fibers.

1.4.3 PHASE SEPARATION

In this technique porous polymer scaffolds are produced by removal of the solovent through freez-drying or extraction. This process takes a longer time to transfer the solid polymer into the nanoporous foam. Phase separation can be induced by changing the temperature, called the thermal induced phase separation or by adding non-solvent-induced-phase separation.

1.4.4 SELF-ASSEMBLY

The self-assembly refers to the process in which individual, pre-existing components organize themselves into desired patterns and functions. However, this method like phase separation method is time consuming. Finally it seems that the electrospinning is the only method, which can be further, developed to produce large quantity of one-by-one continuous nanofibers from synthetic polymers or natural polymeric materials.

However, it should be pointed out that the development of functional nanofibers via self-assembly method is an increasing interested area. Many kinds of polymer nanostructures have been developed successfully by self-assembly.

1.5 ELECTROSPINNING

The term "electrospinning" has been derived from "electrostatic spinning." The innovative setup to produce fibers dates back to 1934 when Formhals published a patent describing an experimental setup for the production of polymer filaments.

The richness of fiber structures and also of nonwoven architectures that can be accessed via electrospinning is extremely broad and it has become apparent that highly fine fibers with diameters down to just a few nanometers can be prepared via electrospinning.

A basic electrospinning setup, as shown in Fig. 2a, consists of a container for polymer solution, a high-voltage power supply, spinneret (needle) and an electrode collector. During electrospinning, a high electric voltage is applied to the polymer solution and the electrode collector leading to the formation of a cone-shaped solution droplet at the tip of the spinneret, so called "Taylor cone." A solution jet is created when the voltage reaches a critical value, typically 5–20 kV, at which the electrical forces overcome the surface tension of the polymer solution. Under the action of the high electric field, the polymer jet starts bending or whipping around stretching it thinner. Solvent evaporation from the jet results in dry/semidry fibers, which randomly deposit onto the collector forming a nonwoven nanofiber web in most cases (Fig. 2b).

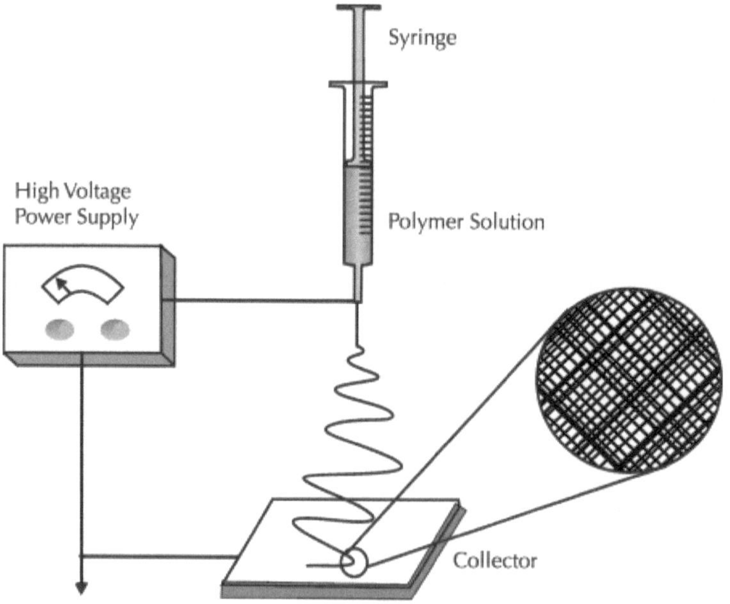

(a)

FIGURE 2(a) Basic apparatus for electrospinning.

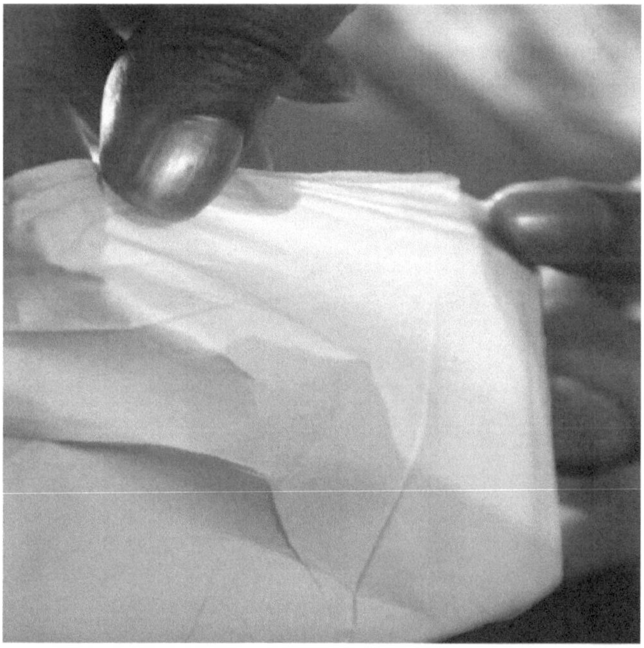

(b)

FIGURE 2(b) A photo of typical elctrospun nanofiber mat.

1.6 DEFINITION AND CLASSIFICATION OF POLYMERS

Polymers (or macromolecules) are very large molecules made up of smaller units, called monomers or repeating units, covalently bonded together (Fig. 3). This specific molecular structure (chainlike structure) of polymeric materials is responsible for their intriguing mechanical properties.

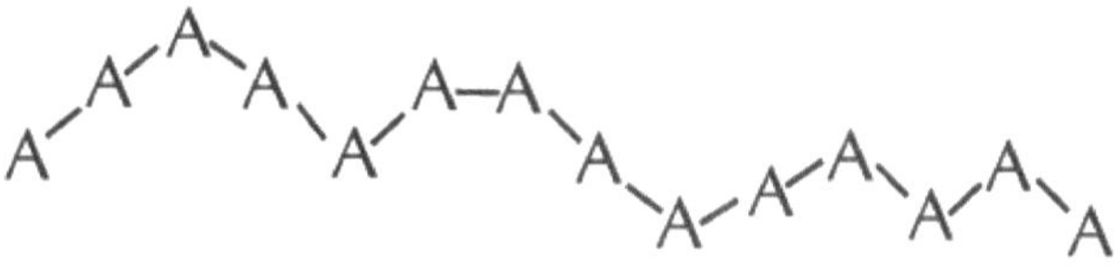

FIGURE 3 Polymer chain.

A *linear polymer* consists of a long chain of monomers. A *branched polymer* has branches covalently attached to the main chain. *Cross-linked polymers* have monomers of one chain covalently bonded with monomers of another chain. Cross-linking results in a three-dimensional network; the whole polymer is a giant macromolecule (Fig. 4).

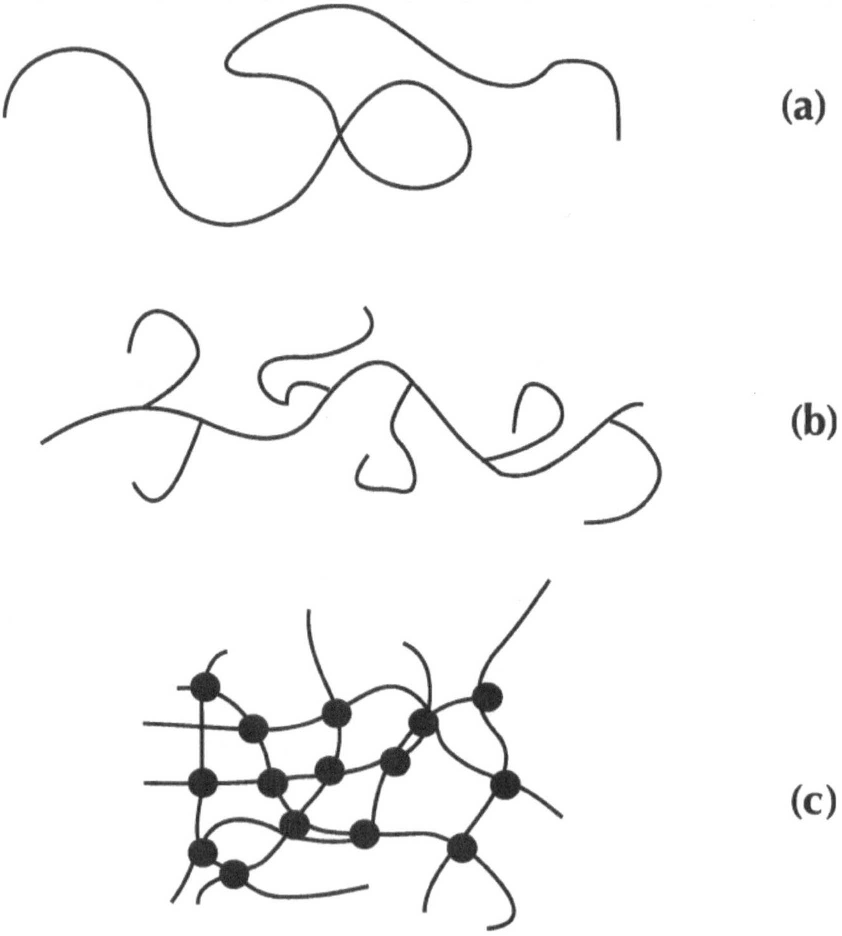

FIGURE 4 Types of molecular configuration: (a) Linear chain. (b) Branched molecule. (c) Cross-linked network: molecules are linked through covalent bonds; the network extends over the whole sample, which is a giant macromolecule.

1.6.1 HOMOPOLYMERS

The formal definition of a homopolymer is a polymer derived from one species of polymer. However, the word homopolymer often is used broadly to describe polymers whose structure can be represented by multiple repetition of a single type of repeat unit which may contain one or more species of monomer unit. The latter is sometimes referred to as a structural unit.

1.6.2 COPOLYMERS

The formal definition of a copolymer is a polymer derived from more than one species of monomer. Copolymers have different repeating units. Furthermore, depending on the arrangement of the types of monomers in the polymer chain, we have the following classification:

- In *random* copolymers two or more different repeating units are distributed randomly.
- *Alternating* copolymers are made of alternating sequences of the different monomers.
- In *block* copolymers long sequences of a monomer are followed by long sequences of another monomer.
- *Graft* copolymers consist of a chain made from one type of monomers with branches of another type (Fig. 5).

Another way to classify polymers is to adopt the approach of using their response to thermal treatment and to divide them into Thermoplastics and Thermosets. Thermoplastics are polymers which melt when heated and resolidify when cooled, while Thermosets are those which do not melt when heated but, at sufficiently high temperatures, decompose irreversibly.

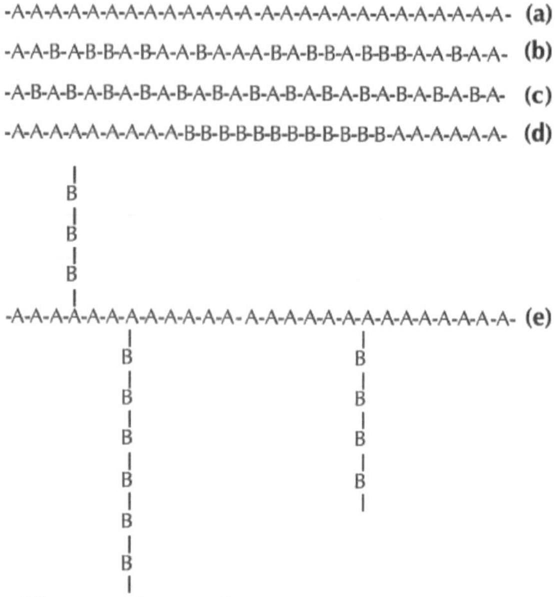

FIGURE 5 a) Homopolymer b) Random copolymer c) Alternating copolymer d) Block copolymer e) Graft copolymer.

In 1929 scientists proposed a generally useful differentiation between two broad classes of polymers: *condensation polymers,* in which the molecular formula of the structural unit (or units) lacks certain atoms present in the monomer from which it is formed, or to which it may be degraded by chemical means, and *addition polymers*, in which the molecular formula of the structural unit (or units) is identical with that of the monomer from which the polymer is derived. Condensation polymers may be formed from monomers bearing two or more reactive groups of such a character that they may condense inter-molecularly with the elimination of a by-product, often water. For example, polyesters, which are formed by condensation, are shown in this reaction.

n HO-R-OH + HOOC-R^1-COOH HO[-R-COO-R^1-COO-$]_n$H + (n-1)H_2O

The most important class of addition polymers consists of those derived from unsaturated monomers, such as vinyl compounds.

$$CH_2{=}CHX\ CH_2\text{-}CHX\text{-}CH_2\text{–}CHX\text{-},$$

In Tables 1 and 2 are some examples of addition and condensation polymers.

TABLE 1 Addition of Polymers.

Polymer name	Monomer(S)	Polymer	Use
Polyethylene	$CH_2{=}CH_2$ (ethene)	$-CH_2-CH_2-$	Most common polymer. Used in bags, wire insulation, and squeeze bottles
Polypropylene	$CH_2{=}CH$ with CH_3 (1-propene)	$-CH_2-CH_2-$ with CH_3	Fibers, indoor-outdoor carpets, bottles
Polystyrene	$CH_2{=}CH$ 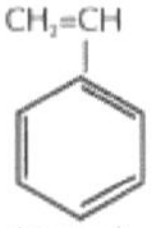 (styrene)	$-CH_2-CH-$ 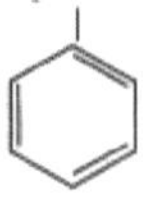	Styrofoam, molded objects such as tableware (forks, knives and spoons), trays, videocassette cases.
Poly(vinyl chloride) (PVC)	$CH_2{=}CH$ with Cl (vinyl chloride)	$-CH_2-CH-$ with Cl	Clear food wrap, bottles, floor covering, synthetic leather, water and drain pipe
Polytetrafluoroethylene (Teflon)	$CF_2{=}CF_2$ (tetraflouroethene)	$-CF_2-CF_2-$	Nonstic surface, plumbing tape, chemical resistant containers and films
Poly (methyl methacrylate) (Lucite, Plexiglas)	$CO_2{=}CH_3$ / $CH_2{=}C$ / CH_3 (methyl methacrylate)	CO_2CH_3 / $-CH_2-C-$ / CH_3	Glass replacement, paints, and household products
Polyacrylonitrile (Acrilan, Orlon, Creslan)	$CH_2{=}CH$ with CN (acrylonitrile)	$-CH_2-CH-$ with CN	Fibers used in knit shirts, sweaters, blankets, and carpets
Poly (vinyl acetate) (PVA)	$CH_2{=}CH$ with $OOCCH_3$ (vinyl acetate)	$-CH_2-CH-$ with $OOCCH_3$	Adhesives (Elmer's glue), paints, texture coatings, and chewing gum
Natural rubber	CH_3 / $CH_2{=}C-CH{=}CH_2$ (2-methyl-1,3-butadiene)	CH_3 / $-CH_2-C{=}CH-CH_2-$	Rubber bands, gloves, tires, conveyor belts, and household materials
Polychlorprene (neoprene rubber)	Cl / $CH_2{=}C-CH{=}CH_2$ (2-methyl-1,3-butadiene)	Cl / $-CH_2-C{=}CH-CH_2-$	Oil and gasoline resistant rubber
Styrene butadiene rubber (SBR)	$CH_2{=}CH$ 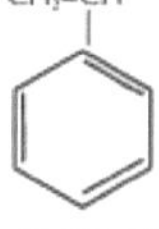 $CH_2{=}C-CH{=}CH_2$	$-CH_2-CH-CH_2-CH-CH-CH_2-$ 	Non-bounce rubber used in tires

TABLE 2 Condensation of Polymers.

Polymer name	Monomers	Polymer	Use
Polyamides (nylon)	$HO\overset{O}{\overset{\|}{C}}(CH_2)_n\overset{O}{\overset{\|}{C}}OH$ $H2N(CH_2)_nNH_2$	$-\overset{O}{\overset{\|}{C}}(CH_2)_n\overset{O}{\overset{\|}{C}}-NH(CH_2)_nNH-$	Fibers, molded objects
Polyesters (Dacron, Mylar, Fortrel)	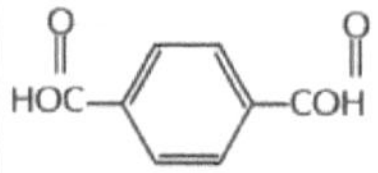 $HO(CH_2)_nOH$	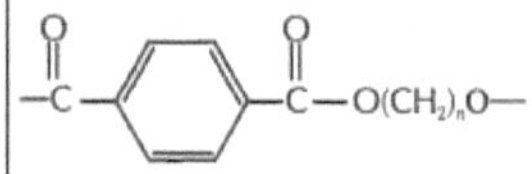	Linear polyesters, fibers, recording tape
Polyesters (Glyptal resin)	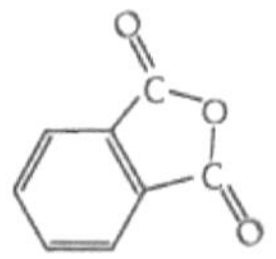 $HOCH_2\underset{OH}{\underset{\|}{C}}HOH_2OH$	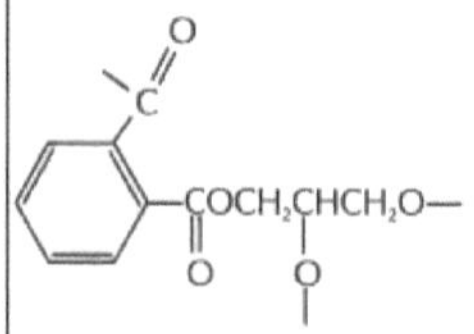	Cross-linked polyester, paints
Polyesters (Casting resin)	$HO\overset{O}{\overset{\|}{C}}CH{=}CH\overset{O}{\overset{\|}{C}}OH$ $HO(CH_2)_nOH$	$-\overset{\|}{C}CH{=}CH\overset{O}{\overset{\|}{C}}-O(CH_2)_nO-$	Cross-linked with styrene and benzoyl peroxide, fiberglass boat resin, casting resin
Phenol-formaldehyde (Bakelite)	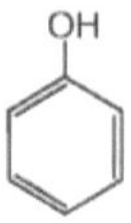 $CH_2{=}O$	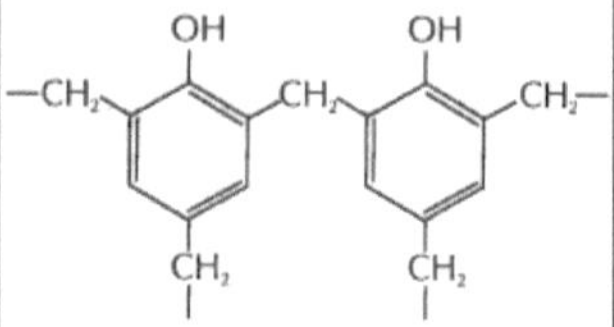	Mixed with fillers, molded electrical cases, adhesives, laminates, varnishes
Cellulose acetate (cellulose is a polymer of glucose)	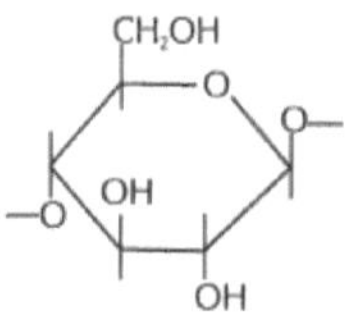 CH_3COOH	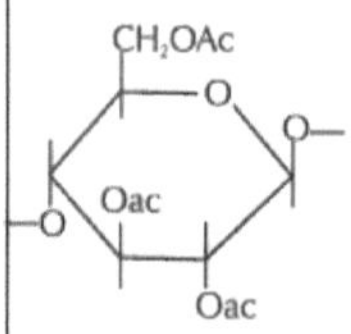	Photographic film
Silicones	$C1-\overset{CH_3}{\underset{CH_3}{Si}}-C1$ H_2O	$-O-\overset{CH_3}{\underset{CH_3}{Si}}-O-$	Water-repellent coatings, temperature-resistant fluids and rubber
Polyurethanes	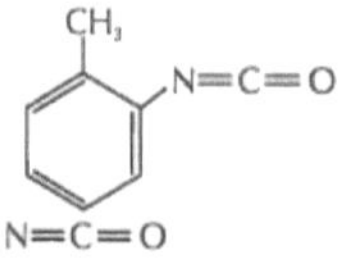 $HO(CH_2)_nOH$	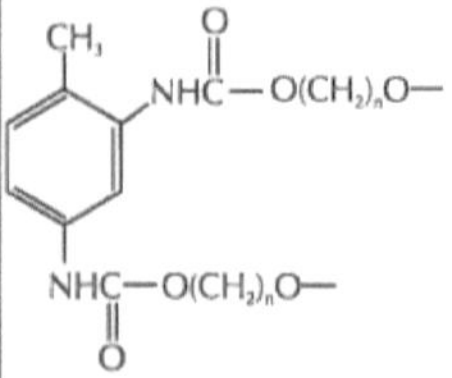	Foams, rigid and flexible, fibers

1.6.3 SYNTHETIC POLYMERS

In our everyday lives a seemingly endless range of synthetic polymers surrounds us. Synthetic polymers include materials that we customarily call plastics and rubbers. These include such commonplace and inexpensive items as polyehthlene grocery sacks, polyethylene terphthalate soda bottles, and nylon backpacks. At the end of the scale are less common and much more expensive polymers that exhibit specialized properties, including Kevlar bullet proof vests, polyacetal gears in office equipment, and high-temperature resistant fluorinated polymer seals in jet engines.

The first completely synthetic polymer (Bakelite) was not made until 1905 and the industry did not really take off until after the Second World War, with developments in methods for polymerizing polyethylene and polypropylene as well as in the production of synthetic fibers. The basic chemistry principles involved in polymer synthesis have not changed much since the beginning of polymer production. Major changes in the last 70 yr have occurred in the catalyst field and in process development. Synthetic polymers can be divided into three broad based divisions:

1. *Fibers* the most well known examples are nylon and terylene. They have high strength in one dimension that can be processed into long strands. The range of strength for fibers is truly impressive, from those used for textiles (clothing, carpets; relatively weak), to commercial monofilament fishing line up to 130 km long, to materials that can be woven into bulletproof vests (aramids).
2. *Synthetic rubbers*: these are not produced merely as substitutes for natural rubber. Some of these synthetic rubbers have properties superior to those of the natural product and are thus better suited for specific purposes, e.g., Neoprene rubber.
3. *Plastics* were originally produced as substitutes for wood and metals. Now they are important materials in their own right and tailor-made molecules are synthesized for a particular use, e.g., heat resistant polymers which are used in rocketry.

1.6.4 NATURAL POLYMERS

As their name implies, natural polymers occur in nature. Also referred to as biopolymers, they are synthesized in the cells of all organisms. It is interesting to note that two of the most prevalent types, polysaccharides and proteins, each contain diverse compounds with extremely different properties, structures, and uses. For example, the protein in egg white (albumin) serves a much different function (nutrition) from that in silk or wool (structural). Likewise, the properties of starch and cellulose could hardly be more different. Although each is made up of polymers based on the condensation of glucose, the final molecular structure differ dramatically. Both sustain life, but in completely different ways.

The utilization of natural polymers for non-food uses can be traced back for to ancient times. Skin and bone parts of animals, plant fibers, starch, silk, etc. are typical examples of the natural polymers used in different periods of the human history. Modern technologies provide new insights of the synthesis, structures, and properties of the natural polymers.

1.6.5 STARCH POLYMER

Starch (Fig. 6) consists of two major components: amylase, a mostly linear α-D(1-4)-glucan and amylopectin, an α-D(1-4)-glucan and amylopectin, an α-D(1-6) linkages at the branch point.

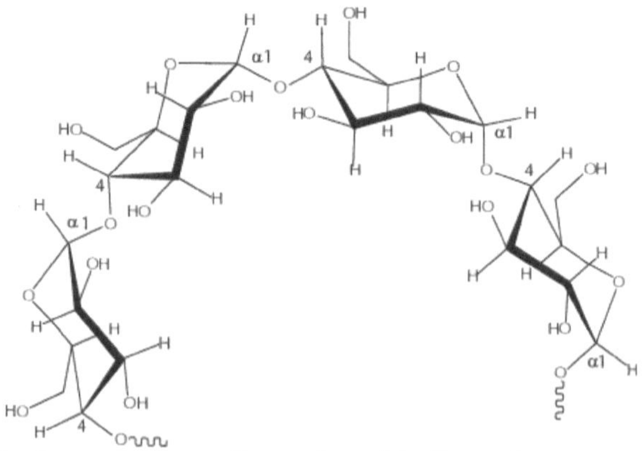

FIGURE 6 A short section of starch, with four glucose molecules.

Starch is unique among carbohydrates because it occurs naturally as discrete granules. This is because the short-branched amylopectin chains are able to form helical structures with crystallise. Starch granules exhibit hydrophilic properties and strong inter-molecular association via hydrogen bonding due to the hydroxyl groups on the granule surface. The melting point of native starch is higher than the thermal decomposition temperature: hence the poor thermal stability of native starch and the need for conversion to starch based materials with a much improved property profile. Also starch has been intensively studied in order to process it into a thermo-plastic in the hope of partially replacing some petrochemical polymers. In its natural form, starch is not meltable and therefore cannot be processed as a thermoplastic. However, starch granules can be thermoplasticized through a gelatinization process.

1.6.6 CELLULOSE

Cellulose is an organic compound with the formula $(C_6H_{10}O_5)_n$, a polysaccharide consisting of a linear chain of several hundred to over ten thousand β(1→4) linked D-glucose units.

Cellulose is the structural component of the primary cell wall of green plants. Some species of bacteria secrete it to form biofilms. Cellulose is the most common organic compound on Earth. About 33% of all plant matter is cellulose (the cellulose content of cotton is 90% and that of wood is 40–50%).

For industrial use, cellulose is mainly obtained from wood pulp and cotton. It is mainly used to produce paperboard and paper; to a smaller extent it is converted into a wide variety of derivative products such as cellophane and rayon. Converting cellulose from energy crops into biofuels such as cellulosic ethanol is under investigation as an alternative fuel source (Fig. 7a).

Cellulosic natural fibers (e.g., abaca, bamboo, jute, flax, and hemp) have been long used as load-bearing materials to reinforce polymer matrix. Compared to traditional reinforcement fibers, e.g., glass fibers and carbon fibers, cellulosic fibers show the advantages of low material cost, low environmental impact (renewability and carbon dioxide neutral, i.e., no excess carbon dioxide is returned to the environment when composted or combusted).

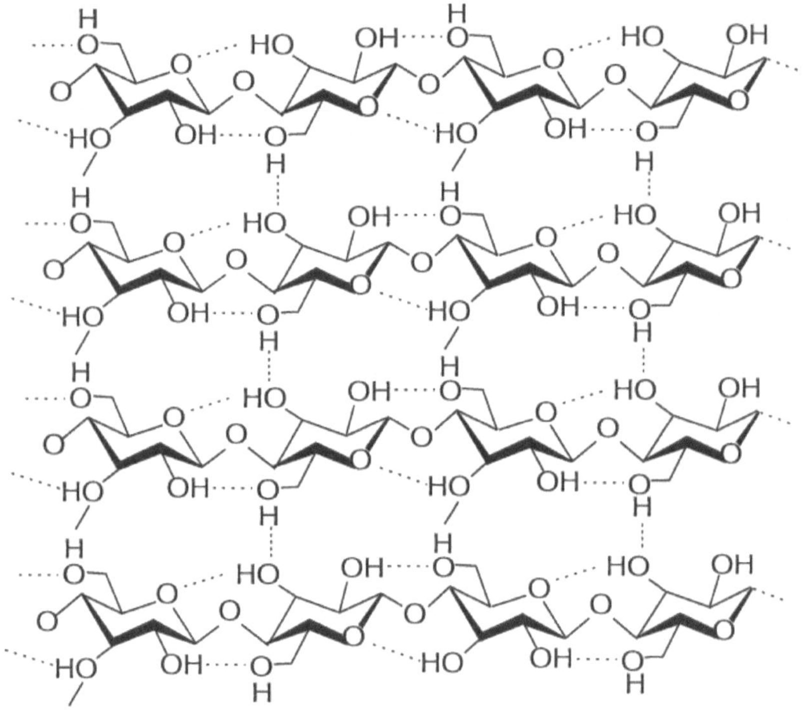

FIGURE 7(a) A strand of cellulose (conformation I_α), showing the hydrogen bonds (dashed) within and between cellulose molecules.

1.6.7 PROTEINS

The word "protein" is derived from the Greek word prôtos, meaning "primary" or "first rank of importance." Proteins are composed of amino acids and although there are hundreds of different proteins they all consist of linear chains of the same twenty L-α-amino acids, which are linked though peptide bonds formed by a condensation reaction. Each amino acid contains an amino group and carboxylic acid group with the general formula:

$$NH_2 - CH\text{-}R - COOH$$

The R group differs for the various amino acids. The amino acid units are linked together through a peptide bond to form a polypeptide chain. Removing one amino acid or changing it from the protein sequence can be detrimental to its structure and in the same way its biological meaning (Fig. 7b).

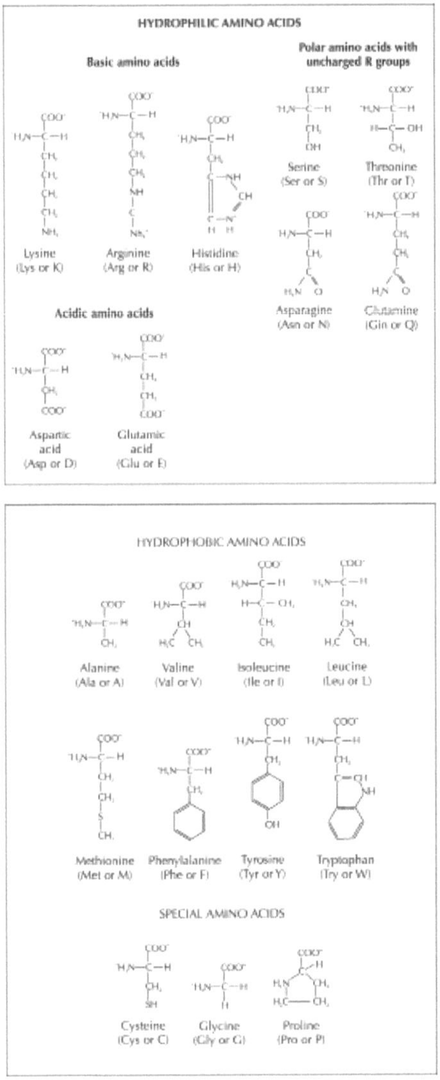

FIGURE 7(b) Molecular structure of the twenty essential amino acids that are present in all living organisms. A linear chain of amino acids linked by the amino group of one residue to the carboxyl group of the next makes up a protein.

Proteins can be divided into two main types, based on their overall shape. *Fibrous proteins*, as their name implies, have fiberlike structures and are used for structure or support. They are tough and insoluble macromolecules, often having several α-helical chains wound together into ropelike bundles. In sharp contrast, the shape of *globular* proteins is spherical or ellipsoidal. They are soluble in water and their solutions have low viscosities.

1.7 CARBON NANOTUBES

In 1960, Bacon of Union Carbide reported observing straight hollow tubes of carbon that appeared as grapheme layers of carbon. In 1970s, Oberlin et al., observed these tubes again by a catalysis-enhanced chemical vapor deposition (CVD) process. In 1985, random events led to the discovery of a new molecule made entirely of carbon, sixty carbons arranged in a soccer ball shape. In fact, what had been discovered was an infinite number of molecules: the fullerenes, C_{60}, C_{70}, C_{84}, etc., every molecule with the characteristic of being a pure carbon cage. These molecules were mostly seen in a spherical shape. However, it is until 1991 that Iijima of NEC observed a tubular shape in the form of coaxial tubes of graphitic sheets, ranging from two shells to approximately 50. Later this structure was called multi-walled carbon nanotube (MWNT). Two year later, Bethune et al., and Iijima and Ichihashi managed to observe the same tubular structure, but with only a single atomic layer of graphene, which became known as a single-walled carbon nanotube (SWNT).

CNTs have typical diameters in the range of ~1–50 nm and lengths of many microns (even centimeters in special cases). They can consist of one or more concentric graphitic cylinders. In contrast, commercial (PAN and pitch) carbon fibers are typically in the 7–20 μm diameter range, while vapor-grown carbon fibers (VGCFs) have intermediate diameters ranging from a few hundred nanometers up to around a millimeter. The variation in diameter of fibrous graphitic materials is summarized in Fig. 8. Crucially, conventional carbon fibers do not have the same potential for structural perfection that can be observed in CNTs. Indeed, there is a general question as to whether the smallest CNTs should be regarded as very small fibers or heavy molecules, since the diameters of the smallest nanotubes are similar to those of common polymer molecules. This ambiguity is characteristic of nanomaterials, and it is not yet clear to what extent conventional fiber composite understanding can be extended to CNT composites (Fig. 8).

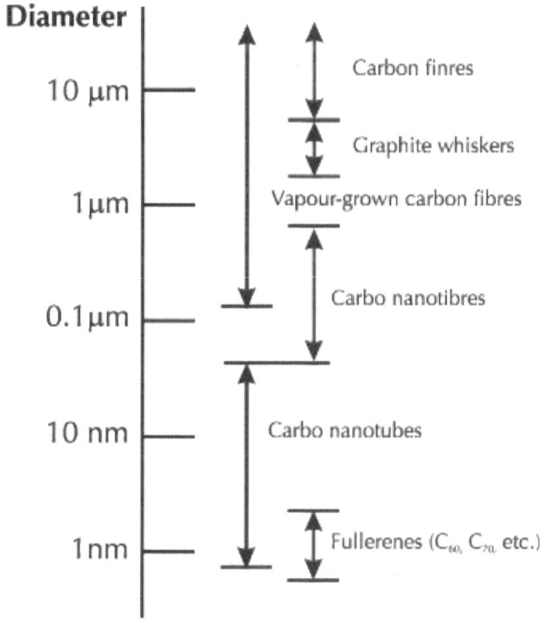

FIGURE 8 Comparison of diameters of various fibrous carbon-based materials.

1.7.1 PROPERTIES OF CARBON NANOTUBES

In 1991, Japanese researcher Idzhima was studying the sediments formed at the cathode during the spray of graphite in an electric arc. His attention was attracted by the unusual structure of the sediment consisting of microscopic fibers and filaments. Measurements made with an electron microscope showed that the diameter of these filaments does not exceed a few nanometers and a length of one to several microns.

Having managed to cut a thin tube along the longitudinal axis, the researchers found that it consists of one or more layers, each representing a hexagonal grid of graphite, which is based on hexagon with vertices located at the corners of the carbon atoms. In all cases, the distance between the layers is equal to 0.34 nm that is the same as that between the layers in crystalline graphite.

Typically, the upper ends of tubes are closed by multilayer hemispherical caps, each layer is composed of hexagons and pentagons, reminiscent of the structure of half a fullerene molecule.

The extended structure consisting of rolled hexagonal grids with carbon atoms at the nodes are called nanotubes.

Lattice structure of diamond and graphite are shown in Fig. 9. Graphite crystals are built of planes parallel to each other, in which carbon atoms are arranged at the corners of regular hexagons. The distance between adjacent carbon atoms (each side of the hexagon) $d_0 = 0{,}141 nm$, between adjacent planes – 0.335 nm.

Each intermediate plane is shifted somewhat toward the neighboring planes, as shown in the figure.

The elementary cell of the diamond crystal represents a tetrahedron, with carbon atoms in its center and four vertices. Atoms located at the vertices of a tetrahedron form a center of the new tetrahedron, and thus, are also surrounded by four atoms each, etc. All the carbon atoms in the crystal lattice are located at equal distance (0.154 nm) from each other.

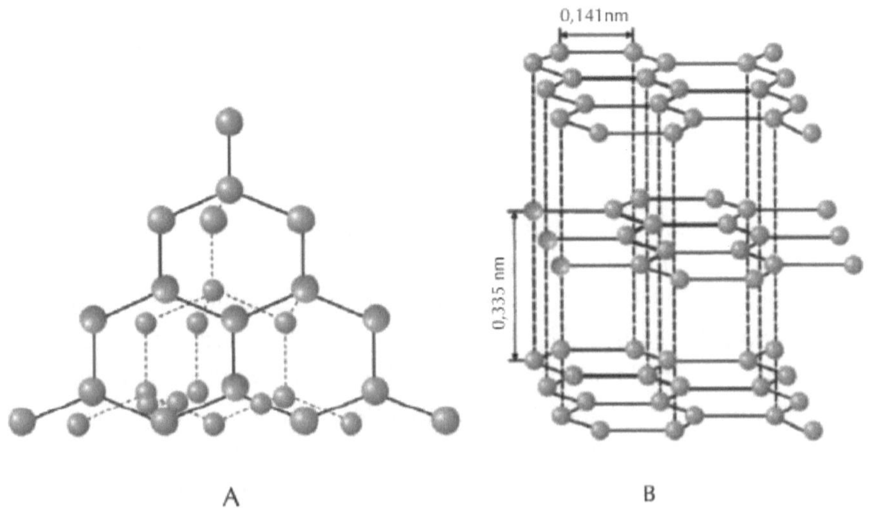

FIGURE 9 The structure of the diamond lattice (a) and graphite (b).

Nanotubes are rolled into a cylinder (hollow tube) graphite plane, which is lined with regular hexagons with carbon atoms at the vertices of a diameter of several nanometers. Nanotubes can consist of one layer of atoms – single-wall nanotubes SWNT and represent a number of "nested" one into another layer pipes – multi-walled nanotubes – MWNT.

Nanostructures can be built not only from individual atoms or single molecules, but the molecular blocks. Such blocks or elements to create nanostructures are graphene, carbon nanotubes and fullerenes.

1.7.2 GRAPHENE

Graphene is a single flat sheet, consisting of carbon atoms linked together and forming a grid, each cell is like a bee's honeycombs (Fig. 10). The distance between adjacent carbon atoms in graphene is about 0.14 nm.

Graphite, from which slates of usual pencils are made, is a pile of graphene sheets (Figs. 10–11). Graphenes in graphite is very poorly connected and can slide relative to each other. So, if you conduct the graphite on paper, then after separating graphene from sheet the graphite remains on paper. This explains why graphite can write.

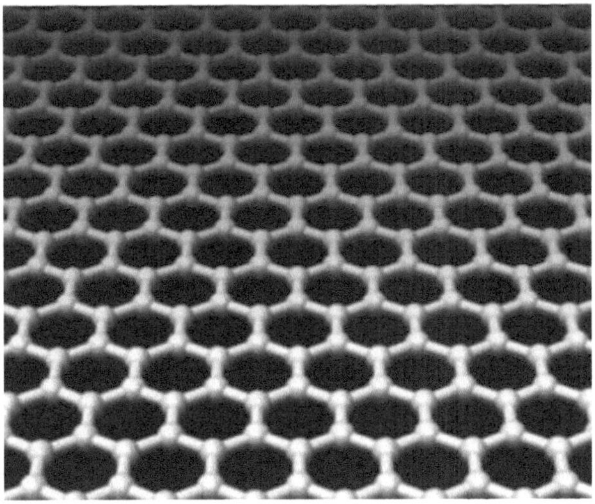

FIGURE 10 Schematic illustration of the graphene. Light balls – the carbon atoms, and the rods between them – the connections that hold the atoms in the graphene sheet.

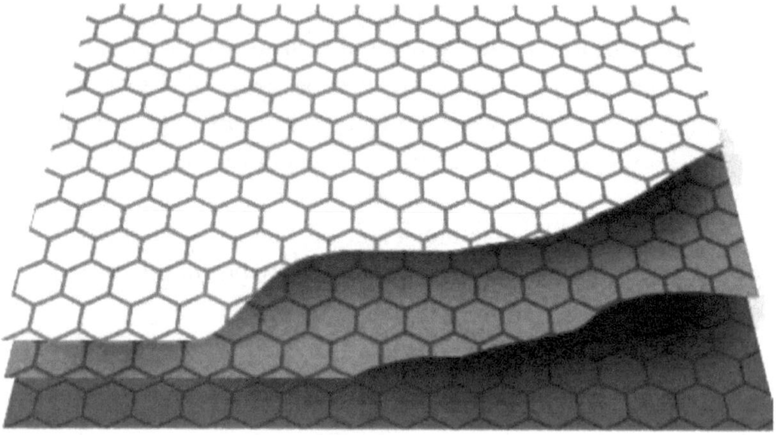

FIGURE 11 Schematic illustration of the three sheets of graphene, which are one above the other in graphite.

Many perspective directions in nanotechnology are associated with carbon nanotubes.

Carbon nanotubes are a carcass structure or a giant molecule consisting only of carbon atoms.

Carbon nanotube is easy to imagine, if we imagine that we fold up one of the molecular layers of graphite – graphene (Fig. 12).

FIGURE 12 Carbon nanotubes.

The way of folding nanotubes – the angle between the direction of nanotube axis relative to the axis of symmetry of graphene (the folding angle) – largely determines its properties.

Of course, no one produces nanotubes, folding it from a graphite sheet. Nanotubes formed themselves, for example, on the surface of carbon electrodes during arc discharge between them. At discharge, the carbon atoms evaporate from the surface, and connect with each other to form nanotubes of all kinds – single, multi-layered and with different angles of twist (Fig. 13).

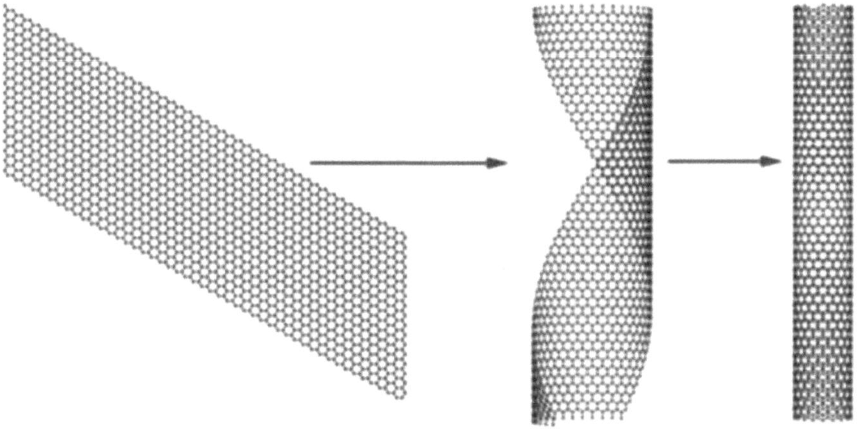

FIGURE 13 One way of imaginary making nanotube (right) from the molecular layer of graphite (left).

The diameter of nanotubes is usually about 1 nm and their length is a thousand times more, amounting to about 40 microns. They grow on the cathode in perpendicular direction to surface of the butt. The so-called self-assembly of carbon nanotubes from carbon atoms occures. Depending on the angle of folding of the nanotube they can have conductivity as high as that of metals , and they can have properties of semiconductors (Fig. 14).

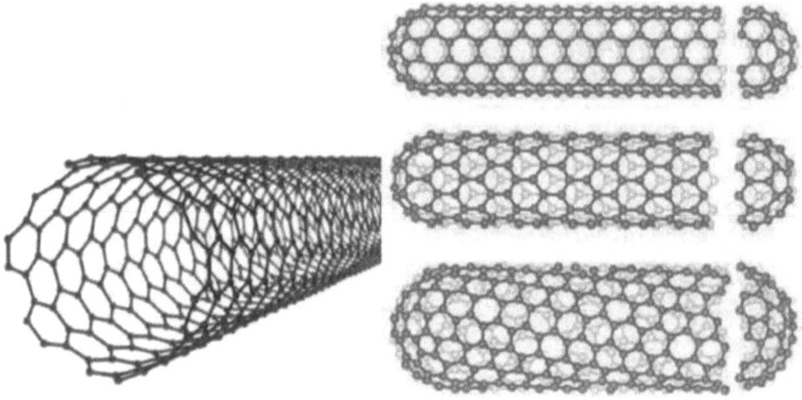

FIGURE 14 Left – schematic representation of a single-layer carbon nanotubes, on the right (top to bottom) – two-ply, straight and spiral nanotubes.

Carbon nanotubes are stronger than graphite, although made of the same carbon atoms, because carbon atoms in graphite are located in the sheets. And everyone knows that sheet of paper folded into a tube is much more difficult to bend and break than a regular sheet. That's why carbon nanotubes are strong. Nanotubes can be used as a very strong microscopic rods and filaments, as Young's modulus of single-walled nanotube reaches values of the order of 1–5 TPa, which is much more than steel! Therefore, the thread made of nanotubes, the thickness of a human hair is capable to hold down hundreds of kilos of cargo.

It is true that at present the maximum length of nanotubes is usually about a hundred microns – which is certainly too small for everyday use. However, the length of the nanotubes obtained in the laboratory is gradually increasing – now scientists have come close to the millimeter border. So there is every reason to hope that in the near future, scientists will learn how to grow a nanotube length in centimeters and even meters!

1.7.3 FULLERENES

The carbon atoms, evaporated from a heated graphite surface, connecting with each other, can form not only nanotube, but also other molecules, which are closed convex polyhedra, for example, in the form of a sphere or ellipsoid. In these molecules, the carbon atoms are located at the vertices of regular hexagons and pentagons, which make up the surface of a sphere or ellipsoid.

The molecules of the symmetrical and the most studied fullerene consisting of 60 carbon atoms (C_{60}), form a polyhedron consisting of 20 hexagons and 12 pentagons and resembles a soccer ball (Figure 15). The diameter of the fullerene C_{60} is about 1 nm. The image of the fullerene C_{60} many consider as a symbol of nanotechnology.

FIGURE 15 Schematic representation of the fullerene C_{60}.

1.7.4 CLASSIFICATION OF NANOTUBES

The main classification of nanotubes is conducted by the number of constituent layers.

Single-walled nanotubes (Fig. 16) – the simplest form of nanotubes. Most of them have a diameter of about 1 nm in length, which can be many thousands of times more. The structure of the nanotubes can be represented as a "wrap" a hexagonal network of graphite (graphene), which is based on hexagon with vertices located at the corners of the carbon atoms in a seamless cylinder. The upper ends of the tubes are closed by hemispherical caps, each layer is composed of hexa – and pentagons, reminiscent of the structure of half of a fullerene molecule. The distance d between adjacent carbon atoms in the nanotube is approximately equal to $d = 0.15$ nm.

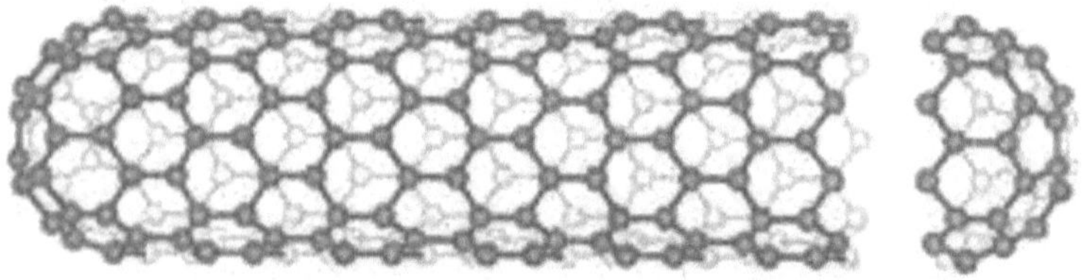

FIGURE 16 Graphical representation of single-walled nanotube.

Multi-walled nanotubes consist of several layers of graphene stacked in the shape of the tube. The distance between the layers is equal to 0.34 nm, that is the same as that between the layers in crystalline graphite (Fig. 17).

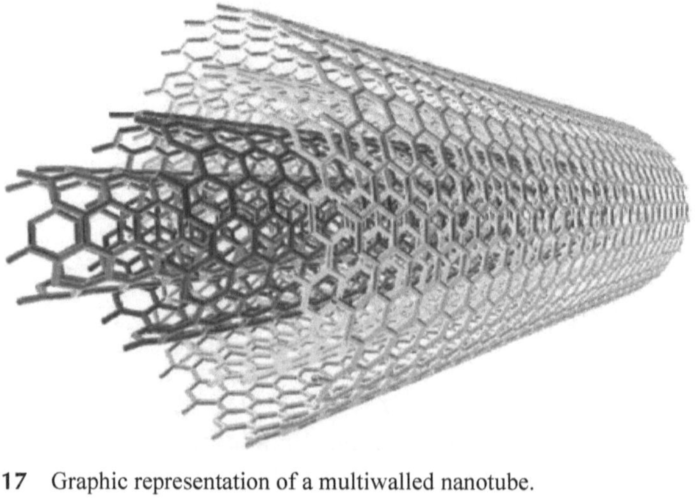

FIGURE 17 Graphic representation of a multiwalled nanotube.

Due to its unique properties (high fastness (63 GPa), superconductivity, capillary, optical, magnetic properties, etc.), carbon nanotubes could find applications in numerous areas:

- additives in polymers;
- catalysts (autoelectronic emission for cathode ray lighting elements, planar panel of displays, gas discharge tubes in telecom networks);
- absorption and screening of electromagnetic waves;
- transformation of energy;
- anodes in lithium batteries;
- keeping of hydrogen;
- composites (filler or coating);
- nanosondes;
- sensors;
- strengthening of composites;
- supercapacitors.

Chirality – a set of two integer positive indices (*n, m*), which determines how the graphite plane folds and how many elementary cells of graphite at the same time fold to obtain the nanotube.

From the value of parameters (*n, m*) are distinguished:

- direct (achiral) high-symmetry carbon nanotubes
    - armchair $n = m$
    - zigzag $m = 0$ or $n = 0$
- helical (chiral) nanotube

In Fig. 18a is shown a schematic image of the atomic structure of graphite plane – graphene, and shown how a nanotube can be obtained from it. The nanotube is fold up with the vector connecting two atoms on a graphite sheet. The cylinder is obtained by folding this sheet so that were combined the beginning and end of the vector. That is, to obtain a carbon nanotube from a graphene sheet, it should turn so that the lattice vector $\overline{R}$ has a circumference of the nanotube in Fig. 18b. This vector can be expressed in terms of the basis vectors of the elementary cell graphene sheet $R = nr_1 + mr_2$. Vector $\overline{R}$, which is often referred to simply by a pair of indices (*n, m*), called the chiral vector. It is assumed that $n > m$. Each pair of numbers (*n, m*) represents the possible structure of the nanotube.

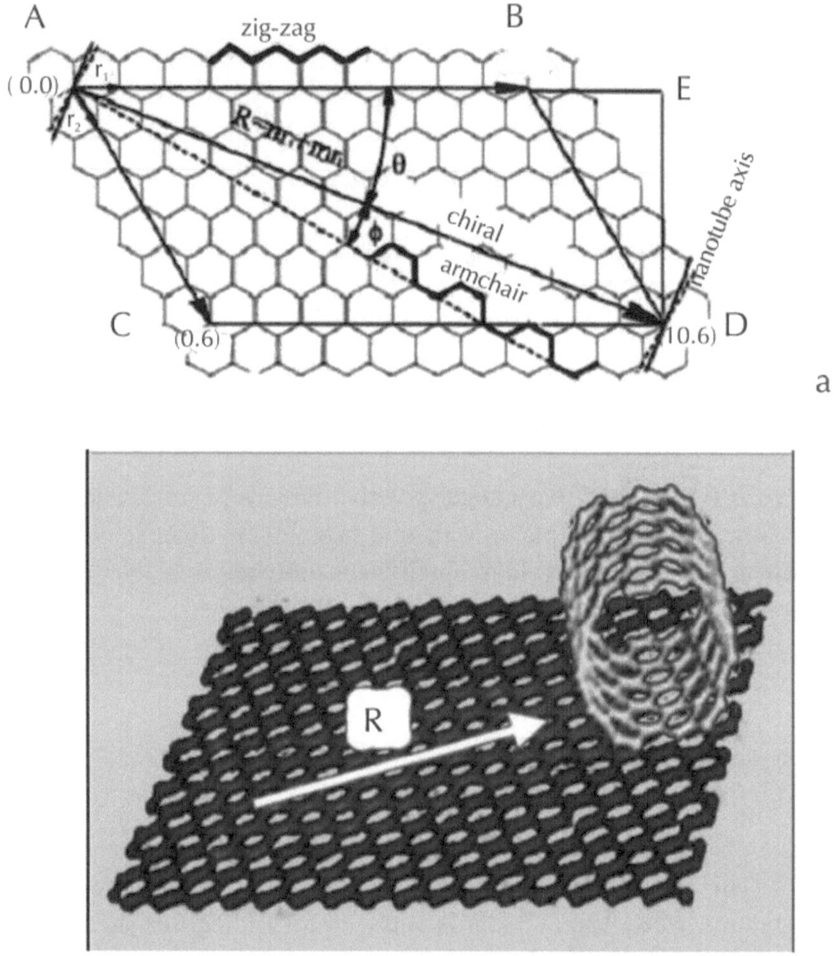

FIGURE 18 Schematic representation of the atomic structure of graphite plane.

In other words the chirality of the nanotubes (*n, m*) indicates the coordinates of the hexagon, which as a result of folding the plane has to be coincide with a hexagon, located at the the beginning of coordinates (Fig. 19).

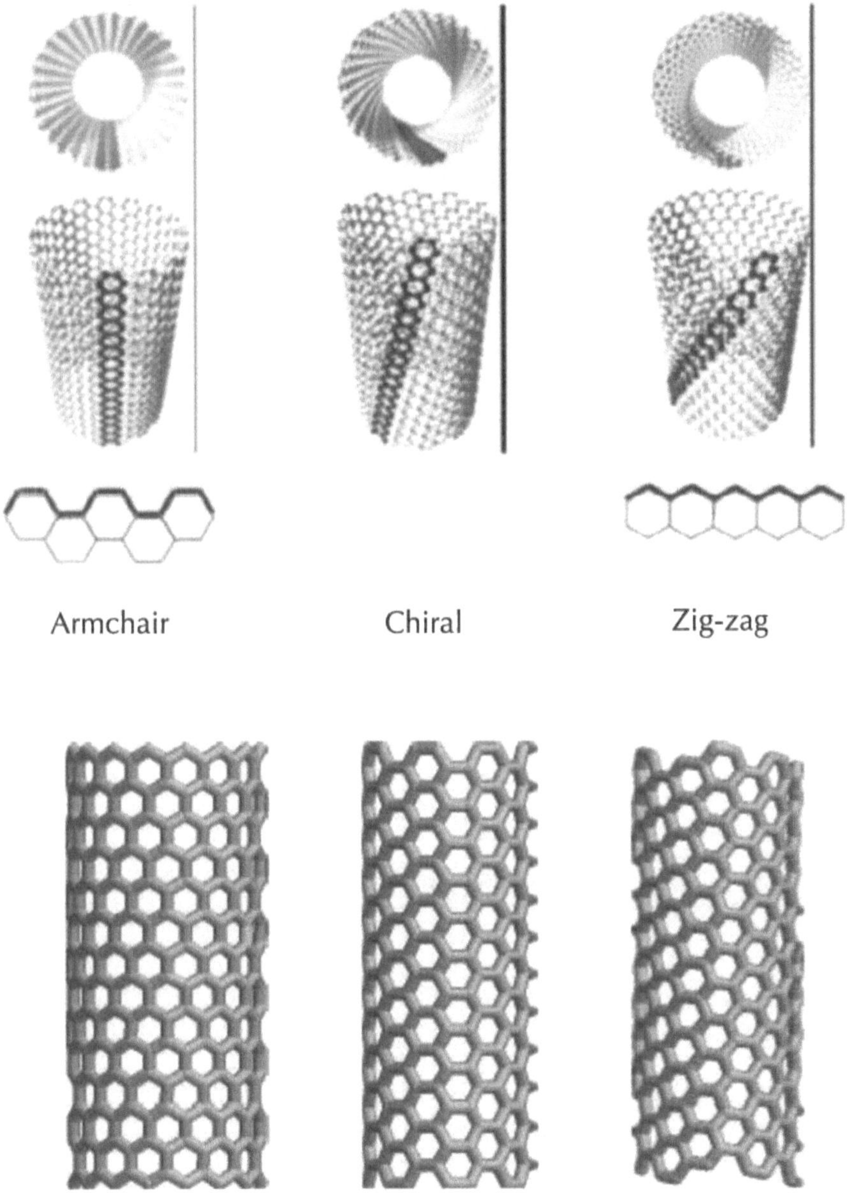

FIGURE 19 Single-walled carbon nanotubes of different chirality (in the direction of convolution). Left to right: the zigzag (16.0), armchair (8.8) and chiral (10.6) carbon nanotubes.

Many of the properties of nanotubes (for example, zonal structure or space group of symmetry) strongly depend on the value of the chiral vector. Chirality indicates what property has a nanotube – a semiconductor or metallicheskm. For example, a nanotube (10.10) in the elementary cell contains 40 atoms and is the type of metal, whereas the nanotube (10.9) has already in 1084 and is a semiconductor (Fig. 19).

If the difference $n-m$ is divisible by 3, then these CNTs have metallic properties. Semimetals are all achiral tubes such as "chair." In other cases, the CNTs show semiconducting properties. Just type chair CNTs ($n = m$) are strictly metal (Fig. 20).

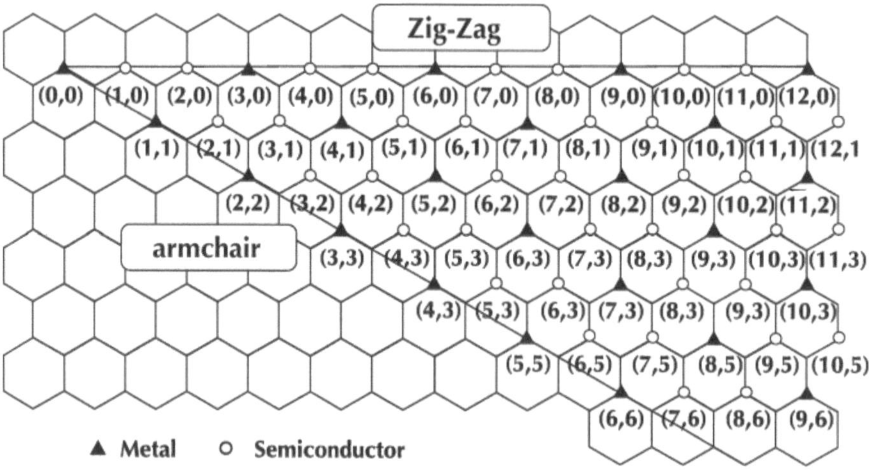

FIGURE 20 The scheme of indices (n, m) of lattice vector $\overline{R}$ tubes having semiconductor and metallic properties.

1.8 NANOCOMPOSITES

Materials are selected for a given application based principally on the material's *properties*. Most engineering structures are required to bear loads, so the material property of greatest interest is very often its *strength*. Strength alone is not always enough, however, as in aircraft or many other structures a great penalty accompanies weight. It is obvious an aircraft must be as light as possible, since it must be able to fly.

Although the terms nanomaterial and nanocomposite represent new and exciting fields in materials science, such materials have actually been used for centuries and have always existed in nature. A nanocomposite is defined as a composite material where at least one of the dimensions of one of its constituents is on the nanometer size scale. The term usually also implies the combination of two (or more) distinct materials, such as a ceramic and a polymer, rather than spontaneously phase-segregated structures. The constituent materials will remain separate and distinct at macroscopic level within the finished structure. Generally, two categories of constituent materials, matrix and reinforcement, exist in the nanocomposite. The matrix materials maintain the relative positions of the reinforcement materials by surrounding and supporting them, and conversely the reinforcements impart their special mechanical or physical properties to enhance the matrix properties. Thus the composite will have the properties of both matrix and reinforcement, but the properties of a composite are distinct from those of the constituent materials. Thousands of years ago, people used straw to reinforce mud in brick making to increase the strength of the brick.

1.8.1 NANOCOMPOSITE NANOFIBERS

A composite nanofiber filter media consisting of nanofibers from electrospinning in combination with a wet-laid substrate material has been successfully used in Ultra-Web® filter cartridges in a wide range of industrial, consumer and defense filtration applications since 1981. Nature has been the inspiration for many material designs and architectures and the strongest materials are often made out of composites. Closer examination of these composites reveals an intricate hierarchical organization from the nano-scale level, which gives it superior properties to meet the functional requirement. Hierarchical organization has been shown to play an important role in maximizing the performance of the structure using abundant elements. The organic matrix, which all natural materials are made of, is mainly carbon. Researchers are currently exploring additional nanofiber composite designs for several new uses including providing a highly permeable aerosol barrier in protective gear such as facemasks, medical gowns and drapes, and protective clothing applications. Nanofibers are a natural fit for these applications, as high air permeability is desired to improve user comfort, and high aerosol efficiency is needed to provide adequate protection from aerosolized threats. The potential to include active chemistry in the nanofiber layer provides additional opportunities for functionality.

Another interesting material is VGCF (vapor grown carbon fibers). It was more than a hundred years ago that Hughes and Chambers explored this type of extremely small carbon fibers. It was shown that the carbon fibers can be synthesized from vapor phase carbons. Because of the lower production costs, there is significant commercial potential for VGCF. Also, VGCF has excellent mechanical property, very surprisingly thermal conductivity, and superior electrical properties. Since VGCF can be produced economically for around 10 dollars per kilogram, it is commercially attractive compared to conventional carbon fibers.

1.9 TISSUE ENGINEERING

In tissue engineering, nanofibers, nanotubes or composite nanomaterials have been used as porous 3D scaffolds for engineering various tissues such as skin, blood vessels, nerve, tendon, bone and cartilage. Nanofibers have emerged as biocompatible, biodegradable, and drug delivery vehicles. They are used for the replacement of tissues with structural or physiological deficiencies. The use of nanofibers in tissue restoration can potentially produce an efficient, compact organ with a rapid recovery process. This is because of the large surface area that polymer and protein-based nanofibers possess. Additionally, the large surface area helps to support cell growth on the scaffold material. These characteristics are also advantageous for wound healing, the construction of biocompatible prostheses and bone substitutes, implant apithelialization, and drug delivery applications.

Chitin, the second most abundant natural polysaccharide, is synthesized by a number of living organisms. Chitin occurs in nature as ordered microfibrils, and is the major structural component in the exoskeleton of arthropods and cell walls of fungi and yeast. The main commercial sources of chitin are crab and shrimp shells, which are abundantly supplied as waste products of the seafood industry. Because chitin is not readily dissolved in common solvents, it is often converted to its more deacetylated derivative, chitosan. Because of its solubility in acidic, neutral and alkaline solutions, chitosan is preferred over chitin for a wide range of applications. Chitin and chitosan are biocompatible, biodegradable, and nontoxic, and are antimicrobial and hydrating agents. Chitin and chitosan are easily processed into gels, membranes nanofibers, beads, microparticles, nanoparticles, scaffolds, and sponge forms. There are a number of promising applications of nanoscale thin films and fibers based on chitin/chitosan.

1.9.1 MUSCULOSKELETAL TISSUE ENGINEERING

Though nanofibers have been studied as scaffolds for several tissue types, musculoskeletal tissue is almost certainly the most well studied one.

Skeletal tissue engineering requires, essentially, a scaffold conducive to cell attachment and maintenance of cell function, together with a rich source of osteoprogenitor cells in combination with selected osteoinductive growth factors.

1.9.2 ELECTROSPINNING PRODUCTION FOR TISSUE ENGINEERING

In synthesizing a tissue engineering construct or scaffold, the most important design considerations are mechanical strength, degradation time, surface chemistry, and scaffold architecture. Electrospinning has the advantage of requiring a minimal amount of specialized laboratory equipment to modify three-dimensional scaffolds, therefore making it an attractive technology for the field of tissue engineering. The first two variables – mechanical strength and degradation time – are almost exclusively controlled by the choice of scaffold material. Fortunately, electrospinning is capable of generating polymer fibers from a wide range of materials. Commodity polymers such as polystyrene, poly(ethylene oxide), poly(methyl methacrylate), and poly(ε-caprolactone) are frequently cited in publications, but high-end materials are also frequently electrospun and published as proof-of-concept. This includes multiple types of collagen and even DNA. Water, chloroform, methanol, ethanol, tetrahydrofuran, dimethylformamide, hexafluoroisopropanol, and mixtures thereof are primarily used as the solvent phase, though other solvents like dimethylacetamide are not uncommon. Solvent choice is dictated by polymer solubility, boiling point, and dielectric constant. Macromolecules that are incapable of being electrospun on their own due to low molecular weight, high entanglement concentrations, or possibly cost can be electrospun either in conjunction with a second component via a blended solution (PEO is a popular choice) or core-shell electrospinning, as mentioned previously. Core-shell electrospinning generates fibers with an internal core and external shell of varying materials. By using an outer shell material that is both capable of being electrospun and degraded during post-processing (generally PEO for its solubility in water), fiber mats of the internal material can be generated. While researchers have conducted *in-vitro* cell viability tests on many materials that would serve as poor *in-vivo* implants due to their vast breakdown time or potentially toxic byproducts (for instance, polystyrene, polyetherimide, or poly(ethylene terephthalate) many other materials have shown promise for *in-vivo* work. These include purchasable materials such as poly(ε-caprolactone) (PCL), poly(L-lactic acid) (PLLA), poly(glycolic acid) (PGA), poly(lactide-co-glycolide) (PLGA). Specially-synthesized polymers are also of interest. Finally, naturally occurring materials such as collagen and chitosan are a desirable choice due to their biocompatibility. These materials are often expensive or difficult to electrospun, however, and may not automatically be the best choice. Specific material choice will likely vary from one tissue-specific construct to the next, and new materials are constantly being synthesized and electrospun for this purpose .

As mentioned above, highly toxic solvents are often employed in creating polymer solutions for electrospinning. Among these is hexafluoroisopropanol (HFIP), which is commonly used for solvating materials such as collagen that are largely insoluble in other solvents. Fortunately, these solvents evaporate during the whipping portion of the electrospinning process; in the event that they do not evaporate completely, highly toxic components could potentially be released during scaffold degradation. Synthesizing a scaffold without the use of toxic solvents would clearly be more ideal, either by using water as the solvent phase or by electrospinning from the melt phase. Attempts to electrospin from the melt phase have been largely unsuccessful due to the high melt viscosity and quick cooling of the jet in-flight, though a few researchers have succeeded. Using an aqueous solution to electrospun materials is much more promising, but the number of water-soluble materials is limited and these materials will likely solvate during future contact with any water-based solutions such as PBS. One group has discovered a means of inhibiting re-solvation by affecting the molecular organization of the material – silk – following the electrospinning process. This strategy may be unique to silk, but offers hope that materials capable of being electrospun from an inert solvent such as water can still be candidates for tissue engineering constructs.

Surface chemistry is another vital component to consider in tissue engineering constructs. As-spun mats are obviously limited to the material of construction, though it is possible to affect the surface chemistry through careful synthesis procedures and post-processing to graft on RGD peptides.

The final design variable – scaffold architecture – can be divided into three different categories: fiber diameter, pore diameter, and porosity. There are currently no effective means of independently varying these three critical variables. Of these variables, fiber diameter and porosity are the easiest to measure. Average fiber diameter is determined via Scanning Electron Microscope (SEM) measurements, and is readily controllable through processing variables such as flow rate, polymer concentration, solvent choice, applied voltage, and needle-to-target distance. Researchers have successfully developed a scaling argument to calculate the terminal diameter, or the minimum fiber diameter capable of being achieved under a certain set of fluid and electric field conditions. Porosity can be measured by multiple techniques, including simple gravimetric measurements of length, width, height, and mass. Efforts to control porosity have mainly focused on increasing the overall porosity through solvent choice, and adding dispersing agents to the fiber mat during deposition to control the distance between fiber layers. Pore diameter measurements are much more complicated, but can be conducted by mercury porosimetry or liquid extrusion porosimetry. Reporting pore diameters can be equally complicated, as measurements are often reported in multiple statistics. Three examples are median pore diameter by volume, the median pore diameter by area, and the average pore diameter. All three values have very different physical significance. The median pore diameter by volume represents the pore diameter at which half the available volume is filled with mercury, easily determined from a plot of cumulative intruded volume, V, vs. pore diameter, D. The median pore diameter by area represents the median diameter for cumulative surface area, A, as a function of pore diameter. The area is calculated by dividing the differential volume intrusion at a given pore diameter by D/4, since the volume, diameter, and outer wall area of a cylinder are connected by the relationship $D = 4V/A$. While the volume of a cylinder scales as the square of the diameter, the area of the outer walls of a cylinder scales only by diameter to the first power. The median pore diameter by volume is therefore weighted to larger pores than the median pore diameter by area. Finally, the average pore diameter is calculated as 4 times the total intrusion volume V divided by the cumulative surface area, per the relationship described earlier. These three values often vary by a factor of 10 or more and do not completely address the statistical information most useful in characterizing the electrospun mat. Despite this, pore diameter is of critical importance to the design of scaffold architecture as it is the only measurement that specifically addresses quantized units of void volume available for tissue growth. Due to the concise control over porosity and pore diameter that are required to successfully build tissue engineering scaffolds via the electrospinning process, future research will likely address these two areas of interest.

Though electrospinning creates randomly interconnected void spaces throughout a scaffold, there is currently little to no way of creating a uniform, controlled, three-dimensional pore structure that can be incorporated into an electrospun scaffold. Currently, few techniques are capable of drastically affecting the scaffold porosity in a homogeneous manner. The addition of dissolvable spacers to the electrospun mat during deposition is capable of increasing the scaffold porosity; however, if the size scale of the spacer is much greater than that of the fiber diameter, inhomogeneities in the scaffold structure occur. It has been found that the use of ice crystals generates the proper space between depositing fibers and can therefore greatly increase the scaffold porosity over traditionally-spun mats. Another area of interest is the reduction of scaffold fiber diameters. The diameters of electrospun fibers generally reside on the upper limits of the natural ECM's 50–500 nm range. Fibers diameters even end up as micron scale, depending on the material and electrospinning solution concentration.

Despite the simplicity, diversity, and control offered by electrospinning, it is by no means a perfect solution to the creation of nanofibrous ECM analogues. Like other technologies, trade-offs are often made with regard to mechanical strength of degradation time, or scaffold porosity and fiber diameter; however, the possibility of generating increasingly biomimetic scaffolds drives the field for the introduction of new electrospinning technologies.

Reconstruction of gray and white matter defects in the brain is especially difficult after large lesions or in the chronic situation, where an injury has occurred sometime previously. In addition to neural death, there is demyelination, rejection or aberrant sprouting of injured axons, glial/minengeal scarring, and often progressive tissue cavitations. In these circumstances, there is a need for cell replacement and some form of neural tissue engineering to develop scaffolds that facilitate reconstruction and restore continuity across the traumatized region. Among various scaffolds useful for sustained three-dimensional growth of neural cells, porous nanofiber composites have shown great potential to mimic the natural extracellular matrix (ECM) in terms of structure, porosity, and chemical composition. Thus, these nanofibers must be adequately processed to obtain a porous matrix of suitable morphology.Electrospinning is the most effective method which has recently established the reputation for its capability to produce nanofiber composite scaffolds, and is counted as a new addition to the conventional techniques (e.g., Phase separation, and self-assembly). In electrospinning process, a high voltage is applied to a polymer solution that is pumped to a spinneret facing an earthed target (collector). Upon reaching a critical voltage, the surface tension of the polymer at the spinning tip is counterbalanced by localized charges generated by the electrostatic force, and the droplet elongates and stretches into a Taylor cone where a continuous jet is ejected.

Electrospun nanofibers can be made of natural and synthetic materials with a combination of reinforcing agent to form composite Nanomaterials. In the case of nanofiber-reinforced composite materials; a phase consisting of strong, stiff fibrous components is embedded in a more ductile matrix phase. Chitosan [β-(1,4)-2-amino-2-deoxy-D-glucose] is a unique polysaccharide obtained by deacetylation of chitin, a natural polymer having a primary amino groups at C_2 and hydroxyl groups at C_3 and C_6 positions. It is environmentally friendly, nontoxic, and biodegradable. To enhance the electrospinnability and structural properties of chitosan, blend systems with poly (vinyl alcohol) has been explored. PVA is a synthetic polymer with non-toxic, non-carcinogenic, and bio-adhesive properties. Applying the blend system of CS/PVA would combine the properties that are unique to each such as superior structural properties and biocompatibility. However, the main disadvantage of the obtained nanofibers is low volume porosity which is not suitable for some neural tissue engineering applications. In the contrary, porous scaffolds has the ability to promote initial cell attachment and subsequently migration into and through the matrix, mass transfer of nutrients and metabolites, vascularization, and permit sufficient space for development and later remodeling of organized tissue. For this purpose, Single-walled carbon nanotubes (SWNTs) were used as an additive in CS/PVA fiber-reinforced composites due to their high aspect ratio, porosity, and chemical orientation. SWNTs are essentially pure carbon polymers, with each carbon atoms in the lattic bound covalently to only three neighboring carbon atoms. The highly symmetric beam-and truss structure formed by the covalently bound carbon atoms gives high stability and strength, while retaining flexibility. In the present study, chitosan/poly(vinyl alcohol) reinforced SWNT (SWNT-CS/PVA) nanofibrous membranes have been developed as scaffold. It is suggested that, the combination of SWNTs with CS/PVA blends may improve the morphology and porosity of the scaffold. For this purpose, a number of properties of the generated nanofibers such as molecular identity, porosity, and morphology are characterized.

1.9.3 BIOMEDICAL APPLICATIONS

Electrospun fibers are generally collected as two-dimensional membranes with randomly arranged structures, and this has greatly limited their application. In fact, in order to make use of the electrospinning technique in biomedical applications, it is important to fabricate fibers with controllable 3D macro and microstructures. Such a nano/microfiber scaffold presents a high surface to volume ratio and porosity and has the potential to provide enhanced cell adhesion and, due to the similarity of their 3D structure to natural ECM, they supply a micro/nano environment for cells to grow and carry out their biological functions. In fact, cells present typically a diameter in the range of 6–20 μm and respond to stimuli from the macro environment down to the molecular level.

Hence, nanofibrous structures have been strongly pursued as scaffolds for tissue engineering applications and for a broad range of biomedical applications.

The natural ECM is a complex structure, it consists mainly of two classes of macromolecules: nanometer diameter fibrils (collagens) and polysaccharide chains of proteoglycans and glycosaminoglycan, and it may contain other important substances such as various minerals. Embedded fibrous collagens are organized in a 3D fiber network, which provides structural and mechanical stability. The ECM of natural tissue is characterized by fiber networks with wide range of pore diameter distribution, high porosity, and effective mechanical properties. High porosity provides more structural space for cell accommodation and makes the exchange of nutrient and metabolic waste between a scaffold and environment more efficient. The fulfilment of all of these characteristics is fundamental criteria for the design of successful tissue-engineered scaffolds. The task to reproduce ECM is in fact challenging, since it meets the specific requirements of the tissue and organ in question. For instance, fibrils that compose the ECM of tendon are parallel and aligned, while those found on the skin are mesh-like.

Electrospinning not only is able to fabricate nanofibers, but moreover it is a technique that can use a wide range of materials to be electrospun, as we have seen before.

1.9.4 WOUND DRESSING

Wounds presenting large amounts of cell loss require immediate coverage with a dressing, primarily to protect the wound. An ideal dressing should mimic the functions of native skin, protecting the injury from loss of fluid and proteins, enabling the removal of exudates, inhibiting exogenous micro-organism invasion and improving aesthetic appearance of the wound site.

Post-surgical adhesion is the most important challenge that affects wound healing and occurs with the use of either conventional bandages or barrier devices. In order to prevent post-surgical adhesion, a study used PLAGA electrospun non-woven bioabsorbable nanofiber matrices as bandages in a rat model and showed excellent anti-adhesion effect and prevented complete surgical adhesions.

1.9.5 CONTROLLED DRUG DELIVERY

Polymeric drug delivery systems have numerous advantages compared to conventional dosage forms, such as improving therapeutic effects reducing toxicity, convenience, etc. Pharmaceutical Release dosage can be designed as rapid, immediate, delayed, pulsed, or modified dissolution depending on the polymer carriers used and other included additives.

Biodegradable polymers have been made into fibrous scaffolds as drug carriers. The main advantage of fibrous carriers is that they offer site-specific delivery of any number of drugs from the scaffold into the body. In addition, the drug can be capsulated directly into fibers with different sizes, and these systems have special properties and surprising results for drug release different from other formulations.

Drug delivery with polymer nanofibers is based on the principle that the dissolution rate of drug particulate increases with increased surface area of both the drug and the corresponding carrier if necessary. Furthermore, unlike common encapsulation involving some complicated preparation process, therapeutic compounds can be conveniently incorporated into the carrier polymers using electrospinning.

The resulting nanofibrous membrane containing drugs can be applied topically for skin and wound healing or post-processed for other kinds of drug release. Thus, electrospinning show potential as an alternative polymer fabrication technique to drug release systems from particles to fibers.

1.9.6 BONE

The requirement for new bone to replace or restore the function of traumatized, damaged or lost bone is a major clinical and socioeconomic need.

The natural bone is composed of a hierarchical distribution of collagen and vital minerals (mainly calcium phosphate). Especially at the bone-cartilage interface both the concentrations and the orientations of the collagen and calcium phosphate particles are precisely organized. The engineering of a fully functional bone tissue still remains elusive, despite the excellent progress achieved up to date.

The typical composite scaffold consists of a biodegradable polymer homogeneously incorporated with various additives including tricalcium phosphate, hydroxyapatite, calcium carbonate, carbon nanotubes, hydrogels and proteins.

Bone formation was assessed in a rat model, by seeding mesenchymal stem cells on a PCL electrospun scaffold presenting an ECM-like topography. The cell-polymer constructs were cultured with osteogenic supplements in a rotating bioareactor for 4 weeks and subsequently implanted in the omenta of rats for 4 weeks. After explantation, the constructs presented a rigid and bone-like appearance and mineralization and type I collagen were detected.

An electrospun silk fibroin membrane seeded with a mouse preosteoblast cell line was analysed *in vivo* in a rabbit model with a 8 mm bilateral full-thickness calvarial bone defect. The prosthesis was shown to possess good biocompatibility and to effectively enhance new bone formation in vivo within 12 weeks.

A hybrid twin-screw extrusion/electrospinning process, which generates continuous spatial gradations in composition and porosity of nanofibers, was recently developed for the fabrication of non woven meshes of PCL incorporated with ß-tricalcium phosphate, to be used in the area of bone tissue regeneration and especially towards the controlled formation of the bone-cartilage interface. The scaffolds were seeded with mouse preosteoblast cells and within 4 weeks the tissue constructs revealed the formation of continuous gradations in ECM with various markers including collagen synthesis and mineralization, with resemblance to the type of variations observed in the typical bone-cartilage interface in terms of the distributions of concentration of Ca^+ particles and of mechanical properties associated with this.

1.9.7 CARTILAGE

Unlike bone, which has shown some prowess for repair and even regeneration, cartilage is recalcitrant to repair, mostly due to its hierarchical organization and geometry. In fact, cartilage presents a very complex stratified tissue structure.

Cartilage is composed mainly of type II collagen, so that electrospinning of type II collagen was performed and scaffolds demonstrated to produce a suitable environment for chondrocyte growth, which potentially establishes the foundation for the development of articular cartilage repair.

1.9.8 SKIN TISSUE ENGINEERING

The complex nature of wound healing requires the migration and proliferation of keratinocytes, both phenomena temporally-regulated by numerous growth factors activating cell membrane receptors up-regulated in the wound environment.

Tissue engineering projects innovative scaffolds to promote the adhesion and proliferation of human skin fibroblasts and keratinocytes,

Electrospun poly(lactic acid-co-glycolic acid (PLGA) matrices with fiber diameters from 150 to 6000 nm were fabricated and tested for their efficacy as skin substitutes by seeding them with human skin fibroblasts (hSF). hSF acquired a well spread morphology and showed significant progressive growth on fiber matrices in the 350–1100 nm diameter range.

An electrospun fibrinogen scaffold was cross-linked with one of the three cross-linkers: glutaraldehyde vapor, 1-ethyl-3-(3-dimethylaminopropyl) carbodiimide hydrochloride (EDC) in ethanol and genipin in ethanol. All three cross-linked scaffolds were seeded with human foreskin fibroblasts. EDC and genipin in ethanol proved to be highly effective in enhancing scaffold mechanical properties and in retarding the rate of scaffold degradation, in respect to the non cross-linked scaffold. Yet, this study demonstrated also that these cross-linked scaffolds had a negative impact on the ability of fibroblasts to migrate below the surface of the scaffold and remodel it with collagen.

1.9.9 VASCULAR TISSUE ENGINEERING

The vascular tissue arrangement precedes and dictates the architecture of the new tissue to be engineered, so that we have to consider both the question of vascular tissue engineering *per se* and also as a condition for musculoskeletal, skin and neural tissue engineering.

Making a selection of materials to be electrospun for arterial blood vessels, the energy and shape recovery are critical parameters to be considered. Energy stored during the expansion of the blood vessel should be recoverable and used in the contraction of the vessel without any distortion to the vessel.

An attractive option of using electrospinning to fabricate vascular grafts is its ability to electrospin small diameter tubes of different sizes with uniform thickness and fiber distribution throughout the scaffold. In fact, many vascular graft scaffolds have been fabricated. An electrospun polymer blend of type I collagen, elastin, and PLGA was used to fabricate a tubular scaffold of 4.75 mm inner diameter. The scaffolds were shown to be biocompatible and to possess tissue composition and mechanical properties similar to native vessels. Also, it was found to support both smooth muscle cells and endothelial cells.

The constructed vascular grafts should moreover express anti-coagulant activity until the endothelial cell lining is fully achieved. A solution blend of PLCL and a tri-n-butylamine salt of heparin was electrospun. Its soaking in PBS determined a burst release of heparin in the first 12 h, after which relatively sustained release rate was observed for 4 weeks.

1.9.10 CARDIAC TISSUE ENGINEERING

In the context of heart valve engineering, it is emphasized the need for including the requirements derived from "adult biology" of tissue remodelling and establishing reliable early predictors of success or failure of tissue engineered implants.

In the fabrication of cardiac graft, electrospun PCL scaffolds were coated with purified type I collagen solution to promote cell attachment. Neonatal rat cardiomyocytes were cultured on the electrospun PCL scaffolds. The cardiomyocytes attached well to the scaffold and contraction of the cardiomyocytes was observed. Tight arrangement and intercellular contacts of the cardiomyocytes were formed throughout the entire mesh, although more cells were found on the surface. The electrospun scaffold was sufficiently soft such that contractions of the cardiomyocytes were not impeded and stable enough for handling. By suspending the mesh across a ring, the cardiomyocytes are allowed to contract at their natural frequency. Thus electrospun patches seeded with cardiomyocytes are gaining interest among the scientific community for the recovery of infarctuated myocradia.

The use of electrical stimulations has been shown to increase adsorption of serum proteins onto electrically conducting polymer, which leads to significantly enhanced neurite extension. In developing nano-fibrous scaffolds to modulate various cell functions such as proliferation, differentiation and migration through electrical simulation, a blend of polyaniline and gelatin was electrospun and it was found to be biocompatible, supporting attachment, migration and proliferation of H9c2 rat cardiac myoblasts.

1.9.11 NEURAL TISSUE ENGINEERING

Engineering the neural tissue would be important for a number of applications, ranging from neural probes for neurodegenerative diseases to guidance scaffolds for axonal regeneration in patients with traumatic nerve injuries. Neural injury may be treated more effectively using nerve guidance channels containing longitudinally aligned fibers and this is true for both the PNS and the spinal cord.

It is well demonstrated that aligned electrospun PLLA nanofibers direct NSC neurite outgrowth, being a good candidate to be used as a potential scaffold in neural tissue engineering. Also, PLLA electrospun nanofiber were shown to support the serum-free growth of primary motor and sensory neurons. Also PLC/chitosan electrospun nanofibers demonstrated good results on Schwann cell proliferation and maintenance of cell morphology, with spreading bipolar elongations to the nanofibrous substrates. Also, a copolymer of methyl methacrylate and acrylic acid was electrospinned and cultured with neural stem cells and it was demonstrated that when type I collagen was immobilized onto the nanofibers surface, cell attachment and viability was enhanced.

In vivo studies have been performed and recently an electrospun guidance channel made of a blend of PLGA and PCL was used to regenerate a 10-mm nerve gap in a rat model of sciatic nerve transaction, with no additional biological coating or drug loading treatment. Also, an electrospun bilayered chitosan tube was fabricated, comprising an outer layer of chitosan film and an inner layer of chitosan nonwoven nano and microfiber mesh and moreover, the inner layer of the tube was covalently bound with peptides with modified domains for laminin-1. This tube was grafted to bridge a rat-injured sciatic nerve and nervous regeneration obtained was similar to the control isograft. Moreover, it was demonstrated that chitosan nano and microfiber mesh tubes with a deacetylation rate of 93% present good nerve regeneration in a rat sciatic nerve injury model.

1.10 FILTER MEDIA

In industrial factory, working office and hygienic surgical operation room, air purification is essential requirement to protect people and precision equipment. Filter media is utilized to purify air which contains solid particles (virus, mine dust and anther dust etc.) and liquid particles (smog, evaporated water and chemical solvents etc.). So far, high-efficiency-particulate-air (HEPA) filter made of non-woven glass fiber mesh has been utilized to capture particles and 300 nm size particles can be excluded with 99.97% filter efficiency. Generally, the efficiency of nanofibers has been found much more effective than other commercial high efficiency air filter media. In most cases, the nanofiber webs are applied on a substrate chosen to provide mechanical properties, while the nanofiber dominates the filtration performance. Electrospinning can also be used to produce charged fibers for use in filtration media.

Obviously, the charge induction and charge retention characteristics are related to the polymer material used for electrospinning.

1.11 GAS SENSORS

Nanofibers fabricated via electrospinning have specific surface approximately one to two orders of the magnitude larger than flat films, making them excellent candidates for potential applications in sensors. A gas sensor is a device, which detects the presence of different gases in an area, especially those gases, which might be harmful to humans or animals. The development of gas sensor technology has received considerable attention in recent years for monitoring environmental pollution. It is well known that chemical gas sensor performance features such as sensitivity, selectivity, time response, stability, durability, reproducibility, and reversibility are largely influenced by the properties of the sensing materials used. Many kinds of materials such as polymers, semiconductors, carbon graphites, and organic/inorganic composites have been used as sensing materials to detect the targeted gases based on various sensing techniques and principles. It is worth noting that the sensitivity of chemical gas sensors is strongly affected by the specific surface of sensing materials. A higher specific surface of a sensing material leads to a higher sensor sensitivity, therefore many techniques have been adopted to increase the specific surface of sensing films with fine structures, especially to form the nanostructures, taking advantage of the large specific surface of nanostructured materials.

1.12 SMART TEXTILE

Textiles are used in everyday life, for example as garment to protect ourselves from heat or cold, fabrics covering the surfaces of floors, or the upholstery of car seats.

On the other hand electronic devices are spreading – but still some people do not have the knowledge to use them. If an electronic device is integrated in things we already use it is easier to accept.

Smart textiles could offer improved interfaces and make it easier for the user to accept electronic devices in everyday life. *Large area textiles* could be used outside for drainage and irrigation in agriculture, in buildings sensor systems within textile reinforced concrete could be used to detect cracks after earthquakes. Another possibility of use could be sensors that control air conditioning or electronics that are used as guiding systems in case of emergency. *Industrial textiles* with electronics may prevent breakdowns due to abrasion of conveyor bands. Another important category is *wearable electronics*, these could be used to survey workers in hazardous environments (sensors could provide information about the surrounding). Medicine could utilize vests with integrated pulse meters for monitoring of patients.

The desire for autonomous self-powered systems has led to considerable research into systems that harvest or scavenge energy from the garments wearer. Human energy is primarily stored as fat (other energy forms are very limited) and is first available for harvesting during metabolism when it is converted into heat and movement. On average, muscles convert just 25% of chemical energy to mechanical movement, while the remaining 75% is dissipated as heat. As a result, potential energy which can be harvested using piezo elec-tronic or mechanical technologies is extremely limited.

Despite this, the primary focus of research activities has been on technologies for harvesting mechanical energy. While design of entirely energy autonomous integrated systems is not likely with existing technologies, re-searchers expect emergence of technologies that will make this possible within the next 10 yr.

Also we have intelligent textiles, which are fibers and fabrics with a significant and reproducible automatic change of properties due to defined environmental influences. Other textiles that are more passive can be called high performance textiles. Microfibers are very passive, but waterproof, but at the same time permeable to water vapor.

One of the main reasons for the rapid development of intelligent textiles is the important investment made by the military industry. This is because they are used in different projects such as extreme winter condition jackets or uniforms that change color so as to improve camouflage effects. According to functional activity smart textiles can be classified in three categories:

1.12.1 PASSIVE SMART TEXTILES

The first generations of smart textiles, which can only sense the environmental conditions or stimulus, are called Passive Smart Textiles.

1.12.2 ACTIVE SMART TEXTILES

The second generation has both actuators and sensors. The actuators act upon the detected signal either directly or from a central control unit. Active smart textiles are shape memory, chameleonic, water-resistant and vapor permeable (hydrophilic/non porous), heat storage, thermo regulated, vapor absorbing, heat evolving fabric and electrically heated suits.

1.12.3 ULTRA SMART TEXTILES

Very smart textiles are the third generation of smart textiles, which can sense, react and adopt themselves to environmental conditions or stimuli. A very smart or intelligent textile essentially consists of a unit, which works like the brain, with cognition, reasoning and activating capacities. The production of very smart textiles is now a reality after a successful marriage of traditional textiles and clothing technology with other branches of science like material science, structural mechanics, sensor and actuator technology, advance processing technology, communication, artificial intelligence, biology, etc.

1.13 NANOPHENOMENON IN OIL PRODUCTION

Oil-saturated layers are porous materials with different pore sizes, pore channels and composition of rocks that define the features of interaction formation and injected fluids with the rock. Taking the mentioned into account we can conclude that the displacement of oil from oil fields in production wells is not a mechanical process of substitution of oil displacing it with water, but a complex physical-chemical process in which the decisive role is played by the phenomenon of ion exchange between reservoir and injected fluids with the rock, i.e., nanoscale phenomena.

The mechanism of displacement of oil in the reservoir and its recovery is largely determined by the molecular-surface processes occurring at phase interfaces (the rock-forming minerals – saturate the reservoir fluids and gases – displacing agents). Therefore the problem of wetability is one of the major problems in oil and gas field of nanoscience.

As the clay is an ultra-system, a huge amount of research on regulation of the position of clay minerals in porous media with good reason can be attributed to nanoscience. To it also should be included the study of gas hydrates, a number of processes regulating the properties of the pumped oil and gas trapped water, water – oil preparation.

Filtering oil in reservoirs at a depth of 13 km of hard rock is determined by the hydrodynamics of the Darcy law. Of the Navier-Stokes equation

$$\rho\left(\frac{v}{t}+(\nabla\cdot v)v\right)=-\nabla p+\mu v, \qquad (1)$$

At very low Reynolds numbers it follows that we can neglect the inertial forces and simplify viscous friction:

$$-\nabla p-\mu\frac{v}{d^2}=0, \qquad (2)$$

d – characteristic pore size. Equation (2) yields Darcy's law

$$v=-\frac{K}{\mu}\nabla p, \qquad (3)$$

where

$K=d^2$ — permeability of the medium, a μ — viscosity of the fluid.

Typical permeability values range from $5mD$ to $500mD$. The permeability of coarse-grained sandstone is $10^{-8}-10^{-9}sm^2$, the permeability of dense sandstone around $10^{-2} sm^2$. Medium with the permeability $1D$ passes of fluid flow with a viscosity 1 sP at a pressure gradient $1aym/sm$.

During filtration oil fills and moves in the pores with the size of **Error! Objects cannot be created from editing field codes.**. In order to displace oil from the reservoir one should have a medium with density and viscosity of oil and the size of $1mkm$.

It seems that it is necessary to conduct experimental and theoretical studies of possible ways to obtain microbubbles nanoscale environments, to make the calculations and estimates of energy costs upon receipt of such media in different ways and their application to problems of oil production.

Chemically, oil is a complex mixture of hydrocarbons (HC) and carbon compounds. It consists of the following elements: carbon (84–87%), hydrogen (12–14%), oxygen, nitrogen, sulfur (1–2%). The sulfur content can reach up to 3–5%. Oils can contain the following parts: a hydrocarbon, asvalto-resinous, porphyrins, sulfur and ash. Oil has a dissolved gas that is released when it comes to the earth's surface.

The main part of petroleum hydrocarbons are different in their composition, structure and properties, which may be in gaseous, liquid and solid state. Depending on the structure of the molecules they are classified into three classes – paraffinic, naphthenic and aromatic. But a considerable proportion of oil is hydrocarbons of mixed structure containing structural elements of all three above-mentioned classes. The structure of the molecules determines their chemical and physical properties.

Carbon is characterized by its ability to form chains in which the atoms are connected in series with each other. In remaining connections hydrogen atoms are attached to the carbon. The number of carbon atoms in the molecules of paraffinic hydrocarbons exceeds the number of hydrogen atoms twice, with some constant excess in all the molecules equal to 2. In other words, the general formula of this class of hydrocarbons is C_nH_{2n+2}. Paraffinic hydrocarbons chemically more stable and refer to the limiting HC.

Depending on the number of carbon atoms in the molecule hydrocarbons may be in one of the three states of aggregation. For example, if there are one to four carbon atoms in a molecule (CH_4–C_4H_{10}), the hydrocarbon is a gas, from 5 to 16 (C_5H_{16} –$C_{16}H_{34}$) – a liquid hydrocarbon, and if more than 16 ($C_{17}H_{36}$ и т.д.) – solid.

Thus, paraffin hydrocarbons in oil can be represented by gases, liquids and solid crystalline substances. They have different effects on the properties of oil: gas reduces viscosity and increases the vapor pressure.

Fluid paraffins dissolve well in oil only at elevated temperatures, forming a homogeneous mixture. Hard paraffins also dissolve well in oil forming the true molecular mixtures. Paraffin hydrocarbons (with the exception of ceresin) can be easily crystallized in the form of plates and plate strips.

Naphthenic (tsiklanovae or alicyclic) hydrocarbons have cyclic structure (C/C_nH_{2n}), to be exact, they are composed of several groups – CH_2 – interconnected in ringed system. Oil contains mainly naphthenes consisting of five or six groups of CH_2. All connections of carbon and hydrogen are saturated, so the naphthenic oil has stable properties. Compared with paraffin, naphthenes have a higher density and lower vapor pressure and have better solvent power.

Aromatic hydrocarbons (arena) are represented by the formula C_nH_n , are most poor by hydrogen. The molecule has a form of a ring with unsaturated carbon connections. The simplest representative of this class of hydrocarbons is benzene C_6H_6, which consists of six groups of CH. For aromatic hydrocarbons a large solubility, higher density and boiling point are typical.

Asphalt-resinous portion of oil is a substance of dark color, which is partially soluble in gasoline. They have the ability to swell in solvents, and then pass into mixture.

The solubility of asphaltenes in the resin-carbon systems increases with decreasing concentration of light hydrocarbons and increasing concentrations of aromatic hydrocarbons.

The resin does not dissolve in gasoline and is a polar substance with a relative molecular mass of 500–1200. They contain the bulk of oxygen, sulfur and nitrogen compounds of oil.

Asphaltic-resinous substances, and other polar components are surface-active compounds and natural oil-water emulsion stabilizers.

Special nitrogenous compounds of organic origin are called porphyrins. It is assumed that they were formed from animal hemoglobin and chlorophyll of plants. These compounds are destroyed at temperatures of 200–250°C.

Sulfur is prevalent in petroleum and hydrocarbon gas and is contained both in the eree State and in the form of compounds (hydrogen sulphide, mercaptans).

Ash is the residue, which is formed by burning oil. This is a different mineral compound, usually iron, nickel, vanadium, and sometimes sodium.

Properties of oil determine the direction reprocessing and affect the products derived from petroleum, so there are different types of classification, which reflect the chemical nature of oil and determine possible areas of processing.

For example, in the base of the classification, reflecting the chemical composition is laid the preference content of one or more classes of hydrocarbons in the oil.

Naphthene, paraffin, paraffin-naphthene, paraffin-naphthene-aromatic, naphthene-aromatic and aromatic hydrocarbons are being distinguished. Thus, all fractions in the paraffin oils contain a significant quantity of alkanes. In the paraffin-naphthene-aromatic hydrocarbons of all three classes are contained in approximately equal amounts. Naphthene-aromatic oil is characterized mainly by the content of cycloalkanes and arenes, especially in the heavy fractions.

Classification is also used by the content of asphaltenes and resins. In the technical classification oils are divided into classes according to the sulfur content; Types, by the output of factions at certain temperatures; Groups, by the potential content of base oils; Species, by content of solid alkanes (of papafins).

Figure 20 shows the components of the reservoir oil, which have different average integrated over the period of development and the entire volume of the reservoir values of physico-chemical properties. Here,

M_3 – reserves of reservoir oil at reservoir conditions;

m_B – Mass of water in the reservoir area drained by the end of development, t;

ω – watering at the end of development, the proportion of units;

m_{II} – Mass of mobile oil, t;

m_{u3} – Mass of extracted oil, t;

m_{H3} – Mass of the nonremovable movable oil, nonremovable t;
m_c – Mass of the adsorbed and structured oil t;
m_{II} – Mass of oil remaining in pillars, t.

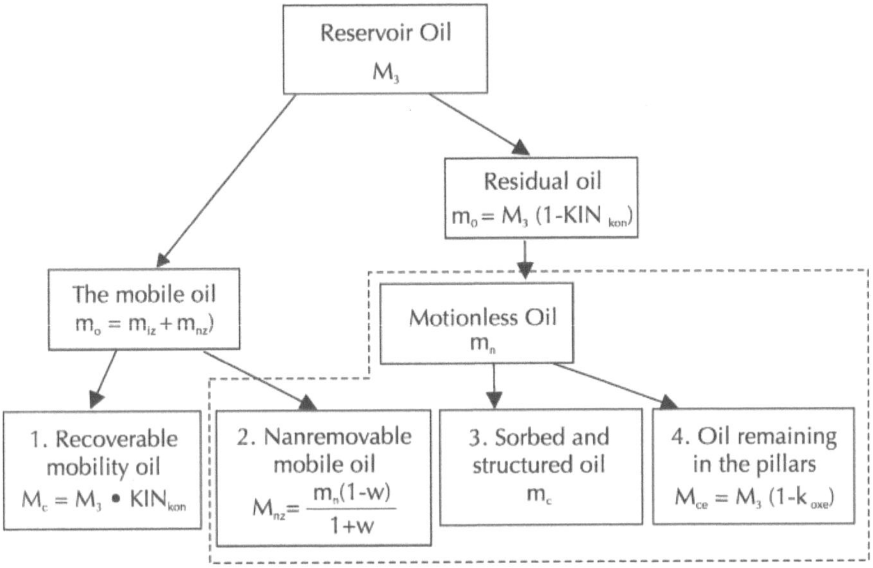

FIGURE 21 The constituents of reservoir oil.

The mobile oil (m_{II}) is part of the reservoir oil moving along layer due to the impact of external influences. Recoverable mobile oil (m_{u3}) is certain part of the mobile oil, which can be extracted from the reservoir as a result of industrial activity with taking into account of economic and technological limitations. Nonremovable mobile oil (m_{H3}) is part of the mobile oil, which will not be extracted from the reservoir using the technologies as a result of industrial activity on the economic and technological constraints.

The residual oil (m_0) is part of the reservoir oil, located in the reservoir at the end of the displacement. Motionless oil (m_H) is part of the reservoir oil, remaining motionless in the reservoir due to external influence. Sorbed and structured oil (m_c) is part of motionless oil, retained near the surface of the collector by the intermolecular interaction. Oil remaining in pillars (m_{II}) is part of motionless oil, is not involved in the process of drainage.

Based on the proposed separation of produced oil into separate components, the average integral value of the physico-chemical properties of produced oil $\overline{X_3}$ for counting of reserves must be found from the expression:

$$\overline{X}_3 = \frac{X_{u3} \times m_{u3} + X_{u3} \times m_{u3} + X_c \times m_c + X_{\text{Ц}} \times m_{\text{Ц}}}{M_3} \tag{4}$$

$$M_3 = m_{u3} + m_{u3+} m_c + m_{\text{Ц}} \tag{5}$$

or

$$X_3 = X_{u3} d_{u3} + X_{u3} + X_c d_c + X_{\text{Ц}} d_{\text{Ц}} \tag{6}$$

where, X_{u3}, X_{H3}, X_c, X_{II} – the mean of the integral value of the corresponding component of reservoir oil; d_{u3}, d_{H3}, d_c, d_{II} – the mass fraction of the corresponding component of reservoir oil $d_i = m_i/m_3$,
where;

m_i – The mass of i component of reservoir oil; m_3 – The mass of reservoir oil (geological reserves of oil reservoir).

Equations (4) and (6) allow us to calculate the average integral value of the physico-chemical properties of reservoir oil by using a different source (relative or absolute).

The average integral value of the physico-chemical properties of mobile oil X_n for use in the calculation of oil displacement processes must be found out of the expression:

$$\overline{X}_n = \frac{X_{u3} \times m_{u3}}{m_n} + \frac{X_{u3} \times m_{u3}}{m_n} \tag{7}$$

$$\overline{X}_n = X_{u3} c_{u3} + X_{u3} C_{u3} \tag{8}$$

or

where;

$X_{u3} c_{u3} X_{u3} = m_{H3} / m_n$ – property value and the mass fraction of component of recovered part of mobile oil, respectively; $X_{H3} c_{H3} = m_{H3} / m_n$ – Property value and the mass fraction of component of nonremovable part of mobile oil, respectively.

In practice, there are a number of contradictions in the calculation methods of geological reserves of hydrocarbons and the calculation of oil recovery processes. For example, an anachronism in the method of counting, which is based on the volume (sealer) ratio of produced oil, because the value of reserves is obtained depending on the conditions of oil, since the magnitude of the volume ratio is dependent on the parameters of technology of oil preparation (the number of stages of oil separation and the temperature and pressure conditions).

The density is determined for each formation zone in the case of inhomogeneity of the properties of reservoir oil in the layer. In this case, during the separation of the extracted products on the commodity oil and associated gas only their masses will be dependent on the properties of reservoir oil and the parameters of ground preparation.

1.13.1 NANOBUBBLES OF GAS, OIL IN THE PORE CHANNELS AND WATER

The average diameter of the channels of the porous medium d is easy to estimate using the well known relationship_

$$d = \frac{4m}{S} \tag{9}$$

where, m – Porosity of the medium, fraction of a unit; S – Specific surface (surface per a volume).

Clay Oil Gas Motherboard Thicknesses (OGMT) usually characterized by a porosity of 10-20%, and their specific surface of not less than $10^8 m^{-1}$. Consequently, the diameters of the pore channels of clay OGMT not exceed an average of 3 nm.

The minimum diameter of a gas bubble in a water medium can be calculated from the following assumptions:

The gas pressure p_g in a bubble caused by the action of two components: of the deposit pressure $p_{n\pi}$ and the pressure of surface tension p_σ.

$$p_g = p_{n\pi} \tag{10}$$

or in accordance with the law of Laplace

$$p_g = p_{na} + \frac{2\sigma}{R} \tag{11}$$

where, σ – Coefficient of the surface tension of the liquid, in which formed a gas bubble; R – the bubble radius.

Gas bubbles can be considered an ensemble of hydrocarbon molecules with a mass equal to or less than the buoyant force acting on it in the liquid, which gives a formal record

$$NMg = \rho Wg \tag{12}$$

where, N – Number of molecules in the ensemble; M – Mass of each molecule; g – Acceleration of free fall; ρ – Density the liquid; W – Volume of the bubble.

The molecular energy of the ensemble, keeping a bottle from collapse (i.e., from the dissolution of gas), is defined by the van der Waals forces, in which force of intermolecular interaction can be ignored, since the gas has virtually no internal pressure. In this case, the magnitude of this energy per unit volume of a gas bubble is equal to the pressure p_0 in the bubble:

$$p_0 = \frac{NkT}{W - W_0} \tag{13}$$

where;

T – Absolute temperature of the gas ^0K; k – Boltzmann's constant; $W_0 = Nw_0$; $w_0 = 12w$; Volume per one molecule of gas at the critical temperature and pressure; w ; the real volume of a gas molecule.

From (5.13) after simple transformations, with taking into account expression for w_0 we have

$$N = \frac{p_0 W}{kT + p_0 w_0} \tag{14}$$

which after substitution in Eq. (12) gives

$$p_0 = \frac{kT}{M/\rho - w_0} \tag{15}$$

The condition of the bubbles floating necessarily implies the existence of the phase boundary, i.e. surface bounding the volume of the bubble. A necessary and sufficient condition for such a boundary is the equality of pressures in gas and liquid phases:

$$p = -p_0 \tag{16}$$

Substituting the values of pressures from Eqs. (11) and (15) gives the desired diameter of the bubble in the form of:

$$d_n = 2R = \frac{4\sigma(Vp_0 - M/\rho)}{kT + p_{n\pi}(w_0 - M/\rho)} \tag{17}$$

Calculation shows that the diameter of the molecule the simplest hydrocarbon gas – methane is equal to 0.38 nm, $w_0 = 0{,}345 \cdot 1^{-27} m^3$.

If this take into account, that $M = 27{,}2 \cdot 10^{-27} kg$, $\rho \sim 10^3 kg/m^3$, $T = 360^0 K$, $\sigma = 62{,}3 \cdot 10^{-3} n/m$, then from Eq. (17) is easy to determine the value of the minimum diameter of the gas bubble.

Under hydrostatic pressure of petroleum $P_{n\pi} = 20$ Mpa, it is 7 nm. This is more than twice the diameter of the pore channels. Consequently, in the cramped conditions of the pore space of OGMT formation of a gas bubble is impossible, because the capillary pressure in the ducts with a diameter of 3 nm to more than 5 times higher than the pressure p_0 at a given temperature. This violates the necessary condition for the existence of a gas bubble in a liquid (5.16) and prevents phase separation, since the external pressure leads to a collapse of gas bubbles, i.e., to its "dissolution" of the pore fluid, if it occurs at all.

It is clear that the required number of methane molecules to form a bubble is equal to

$$N = \frac{W_i}{w_0} = \frac{179{,}6 \cdot 10^{-27}}{0{,}345 \cdot 10^{-27}} = 520$$

where, W_n; Volume of the bubble with diameter of 7 nm.

The minimum diameter of a bubble of oil in water is calculated based on approximately similar considerations.

A bubble of oil in an aqueous medium can be considered as an ensemble of molecules, which are able to form an interface between the liquid phases. The above (average) radius of the volume of such an ensemble can be obtained from Eq. (11):

$$R = \frac{2\sigma}{p_g - p_{n\pi}} \tag{18}$$

where, σ – Border tension coefficient in the water- oil, equal to approximately $45 \cdot 10^{-3} n/m — p_g = p_{n\pi} + p$,

where, p – additional molecular pressure in the equation of van der Waals forces, which for a liquid at condition of it incompressibility it is possible to express from the of the same equation of van der Waals forces:

$$p = -\left(\frac{kT}{w} + p_{n\tau}\right) \tag{19}$$

where, w – The real volume of one molecule of the liquid, and "minus" sign due to the fact that the pressure force directed toward the center of the bubble of oil.

Then from Eqs. (18) and (19) we have finally,

$$d_n = 2R = \frac{4\sigma w}{kT + wp_{n\pi}} \tag{20}$$

The resulting formula determines the minimum value of the diameter of the bubble of oil in a water medium. In this case the real volume of the hydrocarbon molecule with one carbon atom – methane, $w \sim 3{,}5 \cdot 10^{-28} m^3$.

The total mass of hydrocarbons that are brought by failution flow per unit area of contact OGMT – rock reservoir is defined as

$$G = \int_0^t V_G dt \tag{21}$$

where, V_G – Velocity of hydrocarbon generation Organic Materials (OM) from OGMT; t – Time; G – the productivity of generation.

The formation of hydrocarbons from the Dispersed Organic Material (DOM) can be regarded as an elimination process in which the initial component is the reactive part of the DOM, and the final product (hydrocarbon molecules). Then, the reaction velocity of elimination is written well-known formula

$$V = \varepsilon(\Gamma - x) = \varepsilon\Gamma e^{-\varepsilon t} \tag{22}$$

since $x = \Gamma(1 - e^{-\varepsilon t})$.

In such a formulation, V – The velocity of formation hydrocarbon from DOM; ε; Integrated constant of reaction velocity; $\tilde{A}$ – The initial concentration of the reaction capable of part of the DOM in the breed; x – the mass of hydrocarbons that has developed in time t; ε – is usually determined experimentally from the velocity of formation of hydrocarbon out from the given DOM at different temperatures.

Equation (22) is valid for open systems in which the products of the reaction easily draw off the center of the reaction. In conditions of porous medium of natural OGMT with a limited pore volume it is necessary to introduce a factor considering the difficulty of derivatives removal (reaction products) into the Eq. (22).

As noted above, gas (or oil) cannot exist in the pore channels of clay medium in the form of separate phase, the elimination reaction can continue only until the capacity of the pore space is depleted relative to the hydrocarbon material.

The magnitude of this capacity is determined by the solubility of hydrocarbon in the pore water under specified pressure and temperature. The limiting value of the concentration of derivatives in the layer place express the maximum capacity of the reaction volume. The velocity of reaction considering this case, is limited by the introduction of Eq. (22) factor $\alpha = 1 - C / C_0$, where C – the current concentration of hydrocarbon in the pore water, C_0 – its maximum value.

Thus, the velocity of generation of hydrocarbon per unit area of the roof of OGMT is (actually generation process is implemented in OGMT thick h, but all products of this generation pass through the surface of contact generating thickness with the collector. Therefore, the amount of generation is convenient to normalize the surface area of the roof OGMT.)

$$V_G = \varepsilon h\Gamma(1 - C / C_0)e^{-\varepsilon t} \tag{23}$$

where, h – Thicknsse of OGMT.

The magnitude of the current concentration of hydrocarbons in the flow of pore water, squeezed out from OGMT, by definition, is equal

$$C = \frac{x}{W} = \frac{h\Gamma}{W}\left(1 - e^{-\varepsilon t}\right) \tag{24}$$

where, W – the volume of pore water, released as a result of compaction OGMT and/or passed through it over time t; x – the mass of the produced hydrocarbon substances.

Substituting Eqs. (23) and (24) (21), we have

$$G = \varepsilon h\Gamma \int_0^t \left[1 - \frac{h\Gamma\left(1 - e^{-\varepsilon t}\right)}{C_0 W}\right] e^{-\varepsilon t} dt \tag{25}$$

which after integration and simple transformations gives

$$G = h\Gamma\left(1 - e^{-\varepsilon t}\right)\left[1 - \frac{h\Gamma}{2WC_0}\left(1 - e^{-\varepsilon t}\right)\right] \tag{26}$$

According to Eq. (23) obtained formula remains valid until $C < C_0$. When $C = C_0$ then the hydrocarbon generation ceases. If the current hydrocarbon concentration exceeds a specified limit (i.e., hydrocarbon products of degradation are beginning to stand out in a separate phase, forming gas bubbles or droplets of oil) then the formula loses physical meaning.

Equation (26) is valid during the primary migration of hydrocarbons and loses its physical meaning in the transition to the process of secondary migration. From this it follows that the inequality $G > 0$ is always performed, $G = 0$ at $t = 0$ and when $t \to \infty$, G tends to $G_0 = h\Gamma$, which is quite realistic.

Since $W = w_0 + W_\phi$, where, W_ϕ – volume of failuation, which flow through unit area of OGMT by thickness h, and w_0 – Volume of pore fluid contained in a block of OGMT with a single base and height h, value $h\Gamma / 2C_0 W$ is always less than 1. However, w_0 is the volume in which occurs in the reaction of degradation of DOM. It is equal to the volume of voids of OGMT, including the volume occupied by the DOM. This can be written as $W_0 = hn + h\Gamma / \rho$, which in turn suggests that the inequality

$$\frac{h\Gamma}{2C_0 W} = \frac{h\Gamma}{2C_0\left(hn + h\Gamma / \rho + W_\delta\right)} < 1$$

It should be noted that the density of DOM does not exceed the $2/t/m^3$, and is often significantly less; $h/n > h\Gamma$ in most cases is of practical interest, and $2C_0$ at a depth of petroleum is always more than $1kg/m^3$.

As formulated in the task, the movement of pore fluid in the clay thickness in the inviolate natural environment under the principles of geofluid dynamics of slow flow is realized in thefiylation regime. It is absolutely clear that the alien pore water molecules of hydrocarbon – derivatives will be forced to fall on the axis of the pore channels with greater frequency than the water molecules. This is due to the fact that near the walls of the pore channels molecules of water are additionally bonded by surface forces of mineral skeleton, whereas in the center of the channel the resultant of these forces is equal to zero. Therefore, the concentration of hydrocarbons in foliation flow is greater than C_0. However, the cramped conditions of the pore space in OGMT, as noted, do not allow such molecules to combine into a separate phase. Consequently, they are forced to migrate in a homogeneous unstable mixture with molecules of the pore water.

Based on the above, one can calculate the concentration of hydrocarbons in failation stream, if their concentration in the volume of pore space is equal to C. The ideology of this calculation is fairly obvious.

If the channel pore with radius R and the length l contains N molecules of a solvent such as water, we can write down

$$R^2 l = N \cdot \frac{4}{3} \pi r^3$$

where, r – radius of molecules of the solvent, the total number of which is n_t at the fixed moment of time are located on the channel axis (i.e., there are always vacancies). Then $l = 2rn_t$, which after substituting into the initial equation gives,

$$n_t = \frac{2}{3} N \frac{r^2}{R^2}$$

In other words, the number of molecules that fall at the same time on the axis of the pore channel, in R^2/r^2 time smaller than two-thirds of their total number N in the channel. And because the frequency of contact of the hydrocarbon molecules with the axis of the pore channel is priority compared with the water molecules, their concentration in this part of the pore space is equal to:

$$C_{\delta} = C\frac{N}{n_{t}} = \frac{3}{2}C\frac{R^{2}}{R^{2}} \tag{27}$$

The resulting formula characterizes the concentration of hydrocarbon-substances in filiations flow [Eq. (27)] holds for the reaction volumes commensurate with the volume occupied by the molecules of derivatives, which is characteristic of the pore space of clay OGMT).

A capillary pressure force directed against the forces of buoyancy occurs at the hydrocarbon – cluster contact with manifold overlapping tight layer and therefore;

$$K_{np} \leq \frac{n_{\gamma}}{2}\left[\frac{gH}{\sigma}(\rho_{0} - \rho) + \sqrt{\frac{n}{2K}}\right]^{2} \tag{28}$$

where, K_{np} and n_{n} – coefficients of permeability and porosity of the proposed tires, respectively; K and n – coefficients of permeability and porosity of the manifold, respectively; σ– border tension between the hydrocarbon-phase and the reservoir water; H– Depth of reservoir; g– Acceleration of free fall; ρ_{0} – Density of reservoir water; ρ – density of the hydrocarbon phase.

From the known Arrhenius equation the coefficient of the reaction velocity of degradation is equal to:

$$\varepsilon = Ae^{-E/RT} \tag{29}$$

where, E – The activation energy of degradation; R – Universal gas constant; T – Absolute temperature of the system.

APPENDIX A

Nanocomposites nowadays are considered a reliable alternative for either anterior or posterior restorations. This became possible due to the latter developments in composites microstructures and mechanical properties, and to factors such as the aesthetic appeal and the adhesiveness to the dental structure.

Adhesion of such restorations is driven by the adequate penetration of a fluid resin into the collagen network. The role of the adhesive layer is to withstand the stress caused by the shrinkage of the composite during setting and all the mechanical challenges imposed by the oral dynamics to the restoration, keeping it in place 1. Add to that, it is also responsible for the marginal sealing of the restoration.

Consequently, the breakage of the adhesive bonds results in poor marginal sealing, inducing tooth sensitivity and pulpal damages.

Within the adhesive zone, the adhesive layer located above the hybrid layer is considered the weakest spot of an adhesive restoration and, thus, is more likely to fail.

Microcracks may propagate within the adhesive layer, leading to interfacial opening. Besides, adhesives also undergo hydrolytic degradation through time, resulting in elution of their components to the wet oral environment and weakening even more the interface. These situations are highly related to the nature of the polymer network that constitutes the adhesives.

Attempts to improve the mechanical properties of polymer- based materials, especially of dental composites, have been made by adding a percentage of inorganic filler particles. Previous reports on composites have shown considerable improvement of properties such as elastic modulus, fracture toughness, flexural strength and hardness with the increase of the filler volume. The presence of closely spaced filler particles has been suggested also to increase the strength and the fracture energy of brittle polymers through creation of crack stopping steps.

Several recent commercial adhesive systems present filler particles in their constitution. However, whether the addition of filler particles improves the mechanical behavior of these adhesives still remains unclear, since their mechanical properties rely on other factors that cannot be studied isolated using commercial adhesive systems.

Only a few studies have systematically studied the effect of the filler content on the strength of the adhesives.

Scientists observed a significant increase in flexural strength of an ethanol-based one bottle adhesive by adding up to 1% of filler.

The strength of the adhesives should be sufficient to resist the polymerization shrinkage of the composite without creating microcracks, voids or any other discontinuity in the tooth/restoration interface. It is expostulated whether the addition of filler particles could not only improve the strength of adhesives but also reduce the number of structural flaws that could lead to catastrophic failure. The Weibull statistics have been employed to verify the effect of the flaw distribution on the mechanical properties of several materials, and is perfectly suitable to determine the effect of microstructural alterations on the distribution of failure stress. Besides, parameters others than the Weibull modulus and the characteristic strength (which help to visualize the extremes of the stress distribution) are provided by such approach.

Theoretical treatment of polymer nanocomposites behavior specific features was performed with the help of fractal and percolation models. It has been shown that these models explain both qualitatively and quantitatively experimentally obtained nanocomposites specific features as a function of nanofiller particles size. These estimations allow selecting nanocomposites behavior three ranges and determining dimensional boundaries of the definition "nanoparticle."

Polymer nanocomposites of different classes have a specific features number, which are well-known from the experimental point of view, but studied theoretically weakly enough. The strong growth of reinforcement degree E_n/E_m (where E_n and E_m are elasticity moduli of nanocomposite and matrix polymer, respectively) at particles diameter D_p of the initial disperse nanofiller decreasing and the following sharp decay of E_n/E_m at D_p<10 nm can be adduced. Such character of the dependence (E_n/E_m) (D_p) defined nanofiller particles classification, which looks schematically as follows: the nanofiller with particles size of D_p=1000–5000 nm is defined as "diluent," with D_p=100–1000 nm as "semireinforcing" and with D_p<100 nm as "reinforcing." Besides, the nanofillers with D_p<35 nm are often called "superreinforcing." The notion of "nanoparticle" definition boundary, in itself which is accepted conditionally enough equal to 100 nm, is not quite clear, either.

Beginning from the appearance of nanomaterials conception itself, the main emphasis in their study was made on numerous division surfaces in their decisive role as the basis of essential properties change.

It is obvious enough, that the main cause of various effects kinds (mainly, dimensional ones) insufficient theoretical description is deficiency of theoretical models, allowing the indicated effects theoretical treatment. These models take into consideration an important factors number, which were not paid proper attention earlier, namely, nanofiller particles surface structure and polymer matrix molecular characteristics of nanocomposites. Let us consider some typical dependence of particulate-filled polymer nanocomposites properties on nanoparticles size (diameter) D_p.

The following parameters will be accepted further as constant for estimations simplicity. The characteristic ratio C_∞, which is polymer matrix chain statistical flexibility indicator, is accepted equal to 6.5, which is a mean value of this parameter for polymers large number. Then the statistical segment length l_{st} can be estimated according to the equation:

$$l_{st} = l_0 C_\infty \tag{A.1}$$

where, l_0 is main chain skeletal bond length, which is equal to ~ 0.154 nm. In this case it was accepted further l_{st}=1 nm.

The volume filling degree φ_n is accepted constant and equal to 0.15. One from the nanoparticles specific features is the strong dependence of their density ρ_n on the size D_p, expressed by the following formula:

$$\rho_n = 0.188\left(D_p\right)^{1/3} \tag{A.2}$$

Then the nanoparticles specific surface S_u can be determined according to the Eq. (A.3):

$$S_u = \frac{6}{\rho_n D_p} \tag{A.3}$$

The nanoparticles surface structure is characterized by its fractal dimension d_{surf}, which is determined with the help of the Eq. (A.4):

$$S_u = 410\left(\frac{D_p}{2}\right)^{d_{surf}-d} \tag{A.4}$$

where, d is dimension of Euclidean space, in which a fractal is considered (it is obvious, that in our case d=3).

Within the frameworks of fractal analysis an interfacial layer in nanocomposites is considered as a result of two fractal objects (filler particles surface and polymer matrix) interaction, for which there is the only linear scale l, defining these objects interpenetration distance. Since the filler elasticity modulus is considerably higher than the corresponding parameter for polymer matrix, then the indicated interaction comes to filler indentation in polymer matrix and then $l=l_{if}$, where l_{if} is interfacial layer thickness. In this case it can be written as:

$$l_{if} \sim a\left(\frac{R_p}{a}\right)^{2(d-d_{surf})/d} \tag{A.5}$$

where, a is a lower linear scale of polymer matrix fractal behavior, which is accepted equal to l_{st},
R_p is a nanofiller particle radius.

Further an interfacial regions relative fraction φ_{if} can be determined according to the Eq. (A.6):

$$\varphi_{if} = \varphi_n\left[\left(\frac{R_p + l_{if}}{R_p}\right)^3 - 1\right] \tag{A.6}$$

The nanocomposites reinforcement degree (E_n/E_m) (at φ_n=const) is connected with nanoparticles surface structure, characterized by fractal dimension d_{surf}, as follows:

$$\frac{E_n}{E_m} = 15.2\left[1-(d-d_{surf})^{1/t}\right] \tag{A.7}$$

where, t is percolation index, equal to 1.7.

One more variant of parameter (E_n/E_m) calculation gives the percolation relationship:

$$\frac{E_n}{E_m} = 1+11(c\varphi_n b_\alpha)^t \tag{A.8}$$

where, c is proportionality coefficient between φ_{if} and φ_n, determined from the Eq. (A.6),

b_α is parameter, characterizing an interfacial adhesion level nanofiller-polymer matrix.

The Eqs. (A.1)–(A.8) combination allows to determine parameter b_α as a function of nanofiller particles size D_p, that is shown in diagram form in Fig. A.1. As one can see, the dependence $b_\alpha(D_p)$ qualitatively repeats the mentioned above dependence $(E_n/E_m)(D_p)$ – first D_p decreasing results to strong enough (especially at $D_p \leq 100$ nm) an interfacial adhesion level growth and then (at $D_p \leq 15$ nm) this parameter sharp decay is observed. Such analogy supposes that in case $D_p \leq 10$ nm the sharp decay is due to corresponding reduction of an interfacial adhesion nanofiller-polymer matrix level, characterized by parameter b_α. This supposition can be verified by plotting the dependence $(E_n/E_m)(D_p)$, calculated according to the Eq. (A.9):

$$\frac{E_n}{E_m} = 1 + \frac{0.19 W_n l_{st} b_\alpha}{D_p^{1/2}} \tag{A.9}$$

Where;
W_n is nanofiller mass contents, which is equal to:

$$W_n = \rho_n \varphi_n \times 100 \text{, mass\%.} \tag{A.10}$$

The dependence $(E_n/E_m)(D_p)$, calculated according to the Eqs. (A.9) and (A.10), was shown in Fig. A.2. In its turn, the performed above quantitative analysis allows to determine the dependence $(E_n/E_m)(D_p)$ such behavior causes, in particular, reinforcement degree sharp decay at $D_p \leq 10$ nm causes. This decay is due to two factors: an interfacial adhesion level reduction, characterized by parameter b_α, at $D_p \leq 10$ nm (see Fig. A.1) and nanofiller mass contents W_n decreasing, necessary for the condition φ_n=const realization, at D_p decrease and, consequently, ρ_n reduction according to the Eq. (A.2). The last effect is typical for nanoparticles exactly, which are fractal objects with structure dimension $d_f \approx 2.5$. The fractal object density is defined as follows:

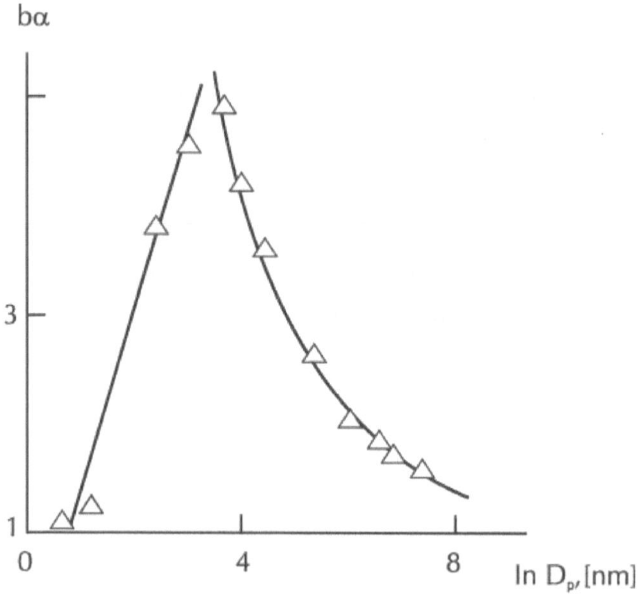

FIGURE A-1 The dependence of parameter b_α on nanofiller particles diameter D_p in logarithmic coordinates.

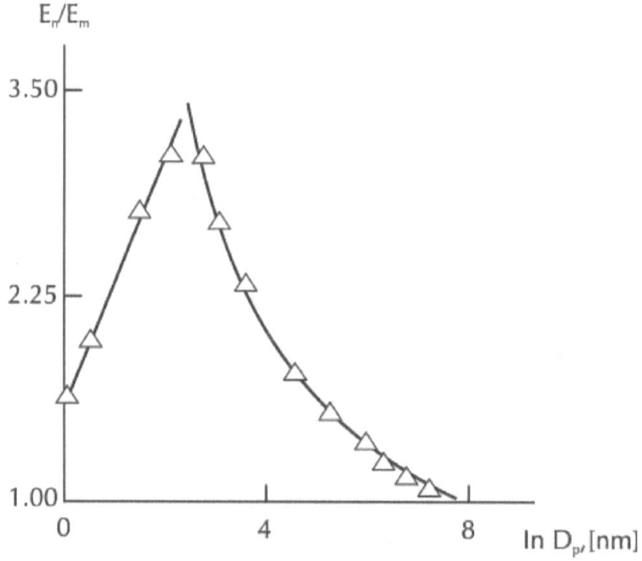

FIGURE A-2 The dependence of reinforcement degree E_n/E_m on nanofiller particles diameter in logarithmic coordinates.

$$\rho_n = \rho_{dens}\left(\frac{R_p}{a}\right)^{d_f-d} \tag{A.11}$$

where, ρ_{dens} is the density of nanofiller material with Euclidean structure.

From the Eq. (A.11) it follows, that for samples with Euclidean structure d_f=d and for any R_p ρ_n=ρ_{dens}. For fractal objects the principally different picture is observed. So, for R_p=10 nm, a=1 nm and d_f=2.5 the value ρ_n is in three times smaller than ρ_{dens}. Therefore D_p increase from 1 up to 750 nm results to W_n growth from 3.5 up to 32.3 mass%, i.e. by one order, at the condition φ_n=0.15=const conservation.

Within the frameworks of thermodynamical approach the interfacial layer thickness for two incompatible polymers l_{if} can be determined according to the following equation:

$$l_{if} = \frac{2\mathrm{A}}{\left(\chi_{\mathrm{AB}}c\right)^{1/2}}, \tag{A.12}$$

where, A is Kuhn segment length, χ_{AB} is Flory-Huggins interaction parameter, c is constant, which is equal to 6 in the limit l_{if}>>R_g (where R_g is macromolecular coil gyration radius) and c=9 for l_{if}>>R_g . Since for the considered hypothetical nanocomposites the value l_{if} varies within the limits of 1.0–36 nm and R_g mean value makes up ~10 nm, then as c the indicated above values average magnitude, i.e. c=7.5, can be considered.

Assuming, that nanofiller-polymer matrix interaction obeys to the same laws, as the incompatible polymers pair interaction, we calculated the value χ_{AB} according to the Eq. (A.12) at A=1.5 nm. In Fig. A.3 the comparison of parameters b_α and χ_{AB}, characterizing nanofiller-polymer matrix interaction (or an interfacial adhesion level) is adduced in logarithmic coordinates. As one expected, for parameters, characterizing the same effect, χ_{AB} increase resulted to b_α growth.

In Fig. A.4 the dependence of Flory-Huggins interaction parameter χ_{AB} on D_p is adduced in double logarithmic coordinates. Besides, in the same Figure the average value χ_{AB} for polymers pair polystyrene-poly-p-methylstyrene is shown by a shaded line. As it follows from the data of Fig. A.4, the intensification of interaction nanofiller-polymer matrix, characterized by Flory-Huggins parameter χ_{AB}, at nanofiller particles size D_p reduction is observed. In Fig. A.5 the dependence of the interfacial layer thickness l_{if} on D_p is adduced in logarithmic coordinates, from which l_{if} growth at D_p

increase follows. As one can see, this dependence on condition can also be divided into two parts: the dependence $l_{if}(D_p)$ slow and fast growth. The value $D_p \approx 100$ nm, i.e., the nanoparticle size boundary magnitude, is boundary between the indicated parts. In other words, this dependence also confirms different behavior of filler nano- and microparticles.

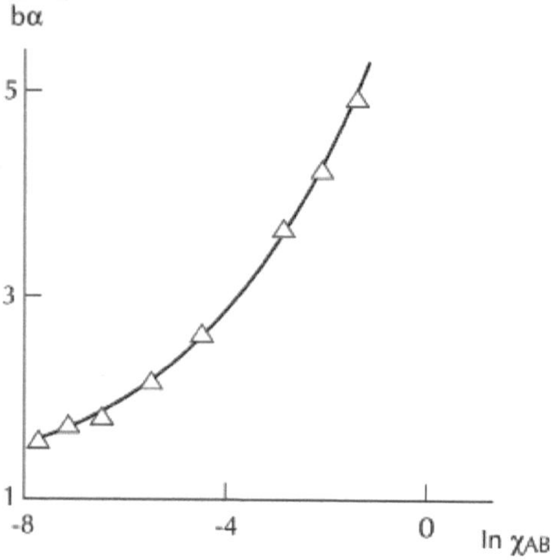

FIGURE A-3 The dependence of parameter b_α on Flory-Huggins interaction parameter χ_{AB} in logarithmic coordinates.

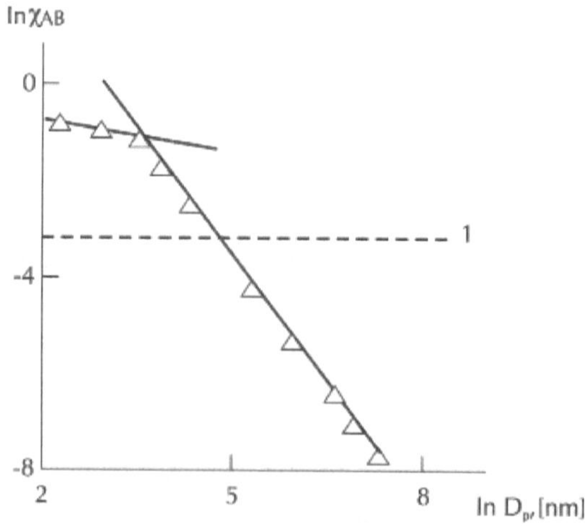

FIGURE A-4 The dependence of Flory-Huggins interaction parameter χ_{AB} on nanofiller particles diameter D_p in double logarithmic coordinates. The horizontal shaded line 1 indicates the average value χ_{AB} for pair of incompatible polymers polystyrene-poly-p-methylstyrene.

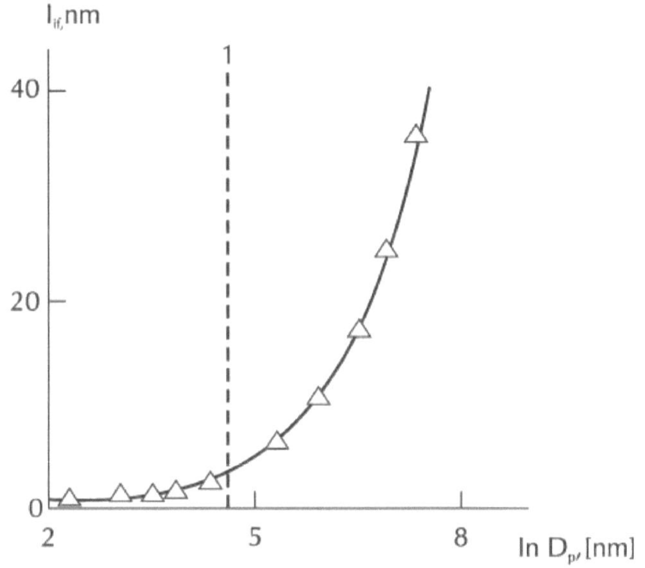

FIGURE A-5 The dependence of interfacial layer thickness l_{if} on on nanofiller particles diameter D_p in logarithmic coordinates. The vertical shaded line 1 indicates the nanoparticles dimensional boundary D_p=100 nm.

APPENDIX B

Polymer nanocomposites can be defined as polymers containing fillers with one dimension smaller than 100 nm. In contrast to traditional polymer composites with high loadings (60 vol.%) of micrometer-sized filler particles, polymer nanocomposites are being developed with very low loadings (less than 5 vol.%) of well-dispersed nanofillers. While elastomeric compositeswith nanoscale spherical fillers have been in use for more than 100 yr, one in the last 15 yr new fillers have emerged, providing an opportunity for the development of high-performance multifunctional nanocomposites. For example, transparent conducting polymer/nanotube composites are under developments solar cell electrodes, nanoparticle filled amorphous polymers are being used as scratch-resistant, transparent coatings in cell phone and compact-disc technology, and nanoparticles are being considered for enhancing matrix properties of traditional composites to increase out-of-plane properties and add conductivity and sensing capabilities.

The recent resurgence of interest in polymer nanocomposites has emerged for several reasons. First, nanoscale fillers often have properties that are different from the bulk properties of the same material. For example, as the size of silicon nanoparticles decreases, the band gaps changes, and the color of the particles changes. As another example, single-wall carbon nanotubes can exhibit stiffness, strength, and strain-to-failure that substantially exceeds that of traditional micrometer-diameter carbon fiber.

These features of nanoparticles provide an opportunity for creating polymer composites with unique properties. Second, nanoscale fillers are small defects. Micrometer-scale fillers are similar in size to the critical crack size causing early failure while nanofillers are an order of magnitude smaller. This can prevent early failure, leading to nanocomposites with enhanced ductility and toughness. Similarly, it has been shown that nanoparticles can increase the electrical breakdown strength and enduranceand are small optical scattering defects. Third, due to the large surface area of the fillers, nanocomposites have a large volume of interfacial matrix material with properties different from the bulk polymer.

One of the challenges in developing polymer nanocomposites for advanced technology applications is a limited ability to predict the properties. While the techniques exist to tailor the surface chemistry and structure of nanoparticle surfacesusing a myriad of methods, the impact of the nanoscale filler surface on the morphology, dynamics, and properties of the surrounding polymer chains cannot be quantitatively predicted. Therefore, the properties of a significant volume fraction of the polymer, the interfacial polymer, are unknown making it difficult to predict bulk properties.

Other challenges include predicting the impact of heterogeneous filler distribution and filler geometries (e.g., wavy fibers). The solutions to this predicament are focused experimental and multiscale modeling efforts, which are ongoing and too broad to address here. Instead, the goal of this paper is to briefly introduce polymer nanocomposites and describe the impact of the interfacial region on both composite properties and the ability to model behavior.

In Fig. B.1 the schematic diagrams load-time (P–t) are adduced for two cases of polymeric materials samples fracture in impact testing: by both instable (a) and stable cracks (b). As it is known, the area under P–t diagram, gives mechanical energy consumed with samples fracture. The polymeric materials macroscopic fracture process, defined by the magistral crack propagation, begins at the greatest load P.

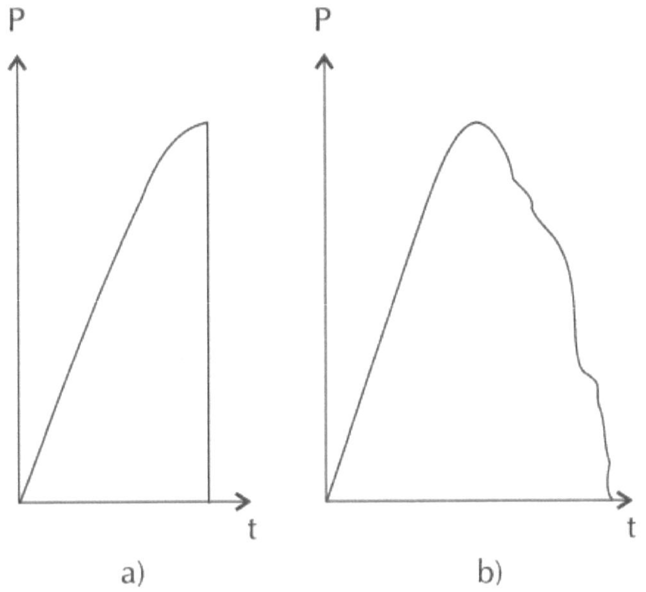

FIGURE B-1 Schematic diagrams load-time (P-t) in instrumented impact tests. The fracture by instable crack (a) and stable one (b).

The fractal dimension d_f is the most general informant of an object structure (in our case – polymeric material) and the true structural characteristic, describing structure elements distribution in space. The value d_f can be determined according to the following equation:

$$d_f = 3 - 6\left(\frac{\varphi_{cl}}{C_\infty S}\right)^{1/2}, \tag{B.1}$$

where, φ_{cl} is a relative fraction of local order domains (clusters) in polymeric material structure; C_∞ is characteristic ratio, which is equal to 7 for polyethylenes, S is macromolecule cross-sectional area, which is equal to 14.3 Å^2 for HDPE.

In its turn, the value φ_{cl} is determined according to the following percolation relationship:

$$\varphi_{cl} = 0.03(1-K)(T_m - T)^{0.55}, \tag{B.2}$$

where, K is cristallinity degree, equal to 0.48 and 0.55 for neat HDPE and nanocomposite HDPE/$CaCO_3$, respectively, T_m is melting temperature, equal to ~ 406 and 405 K for the mentioned materials, respectively, T is testing temperature.

Let us note, that d_f calculation according to the Eq. (B.1) gives values, corresponding to other methods of this parameter estimation. So the value d_f can be calculated alternatively according to the following equation:

$$d_f = (d-1)(1+\nu), \tag{B.3}$$

where, d is dimension of Euclidean space, in which fractal is considered (it is obvious, that in our case d=3), ν is Poisson's ratio, estimated with the aid of the relationship:

$$\frac{\sigma_Y}{E} = \frac{1-2\nu}{6(1+\nu)}, \tag{B.4}$$

where, σ_Y is yield stress, E is elasticity modulus.

The estimations according to the Eqs. (B.1) and (B.4) have given the following values d_f at testing temperature 293 K: for HDPE 2.73 and 2.68, for nanocomposite HDPE/$CaCO_3$ – 2.75 and 2.73, respectively. As one can see, the good enough correspondence is obtained (the discrepancy by d_f fractional part, which has the main information amount about structure, does not exceed 7%).

In Fig. B.2 the dependence $A_p(d_f)$ for the studied polymeric materials is adduced, which has turned out to be linear, common for the near HDPE and nanocomposite HDPE/$CaCO_3$ and is described by the following empirical correlation:

$$A_p = 13.5(d_f - 2.5), \text{ kJ/m}^2. \tag{B.5}$$

From the Eq. (B.5) it follows, that at d_f=2.5 the value A_p=0. The mentioned fractal dimension corresponds to the ideally brittle fracture condition, that defines the condition A_p=0. For real solids the greatest fractal dimension of their structure is equal to 2.95, that allows to determine the greatest value A_p according to the Eq. (5), which is equal to ~ 6.1 kJ/m^2.

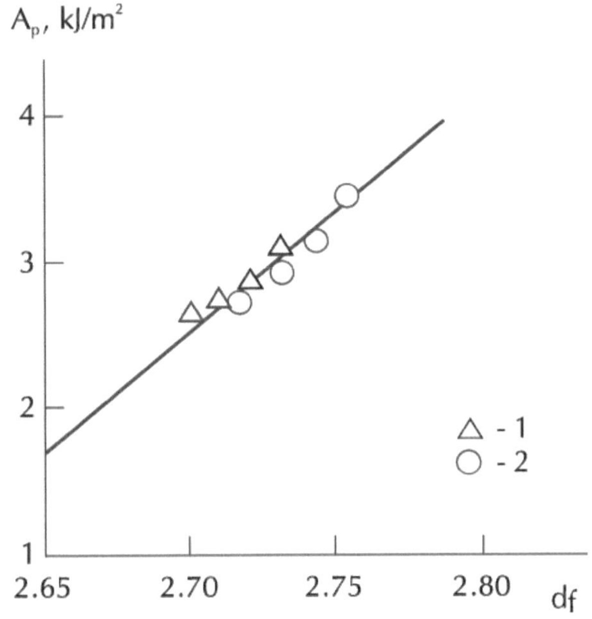

FIGURE B-2 The dependence of impact toughness A_p on structure fractal dimension d_f for HDPE (1) and nanocomposite HDPE/$CaCO_3$ (2).

In general, energy dissipation at an impact grows at polymeric materials molecular mobility level increase. Within the frameworks of fractal analysis this level can be characterized with the aid of the fractal dimension D_{ch} of a polymer chain part between its fixation points (chemical cross-links, physical entanglements nodes, clusters, etc.). Such analysis method can be applied successfully for the value A_p description it case of particulate-filled nanocomposites phenylone/β-sialone. The value D_{ch} can be determined with the aid of the following equation:

$$D_{ch} = \frac{\ln N_{cl}}{\ln\left(4-d_f\right)-\ln\left(3-d_f\right)}, \tag{B.6}$$

where, N_{cl} is a statistical segments number per chain part between clusters, which is determined as follows. Firstly the density of physical entanglements cluster network ν_{cl} is determined:

$$\nu_{cl} = \frac{\varphi_{cl}}{C_\infty l_0 S}, \tag{7}$$

where, l_0 is the main chain skeletal bond length, which is equal to 1.54 Å for polyethylenes.

Then the estimation of polymer chains total length per polymer volume unit L was carried out as follows:

$$L = S^{-1} \tag{8}$$

The chain part length between clusters L_{cl} is determined according to the equation:

$$L_{cl} = \frac{2L}{\nu_{cl}}. \tag{9}$$

The statistical segment length l_{st} is determined as follows [14]:

$$l_{st} = l_0 C_\infty . \tag{10}$$

And at last the value N_{cl} can be determined as ratio [6]:

$$N_{cl} = \frac{L_{cl}}{l_{st}} \tag{11}$$

In Fig. B.3 the dependence of impact toughness A_p on fractal dimension D_{ch} for the studied materials is adduced. As it should be expected, A_p growth at D_{ch} increase is observed, analytically described by the following relationship:

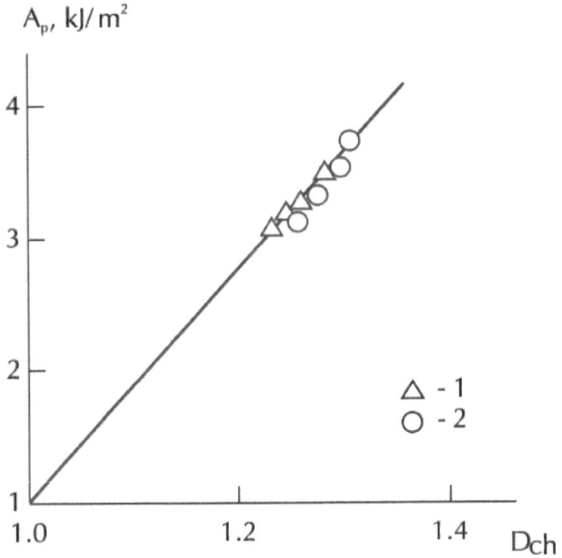

FIGURE B-3 The dependence of impact toughness A_p on fractal dimension D_{ch} of chain part between clusters for HDPE (1) and nanocomposite HDPE/$CaCO_3$ (2).

$$A_p = 6.75(D_{ch} - 1), \text{ kJ/m}^2. \quad (12)$$

The Eq. (12) allows to determine the greatest value A_p for the studied materials at the condition D_{ch}=2.0: this value is equal to 6.75 kJ/m^2, that is close to the cited above estimation according to the Eq. (5) – the average discrepancy makes less than 10%.

KEYWORDS

- **Biomimetics**
- **Lycurgus cup**
- **Moth-eye structure**
- **Taylor cone**

REFERENCES

1. Ziabari, M.; Mottaghitalab, V.; McGovern, S. T.; Haghi, A. K. *Chem. Phys. Lett.*, **2008**, *25*, 3071.
2. Ziabari, M.; Mottaghitalab, V.; McGovern, S. T.; Haghi, A. K. *Nanoscale Res. Lett.*, **2007**, *2*, 297.
3. Ziabari, M.; Mottaghitalab, V.; Haghi, A. K. *Korean J. Chem. Eng.*, **2008**, *25*, 905.
4. Ziabari, M.; Mottaghitalab, V.; Haghi, A. K. *Korean J. Chem. Eng.*, **2008**, *25*, 919.
5. Haghi, A. K.; Akbari, M.; *Phys. Status Solidi*, **2007**, *204*, 1830.
6. Kanafchian, M.; Valizadeh, M.; Haghi, A. K. *Korean J. Chem. Eng.*, **2011**, *28*, 428.
7. Kanafchian, M.; Valizadeh, M.; Haghi, A. K. *Korean J. Chem. Eng.*, **2011**, *28*, 763.
8. Kanafchian, M.; Valizadeh, M.; Haghi, A. K. *Korean J. Chem. Eng.*, **2011**, *28*, 751.
9. Kanafchian, M.; Valizadeh, M.; Haghi, A. K.; *Korean J. Chem. Eng.*, **2011**, *28*, 445.
10. Afzali, A.; Mottaghitalab, V.; Motlagh, M.; Haghi, A. K. *Korean J. Chem. Eng.*, **2010**, *27*, 1145.
11. Wan, Y. Q.; Guo, Q.; Pan, N. *Int. J. Nonlinear Sci. Numer. Simul.*, **2004**, *5*, 5.
12. Feng, J. J. *J. Non-Newtonian Fluid Mech.*, **2003**, *116*, 55.
13. He, J.; Wan, Y.; Yu, J.-Y. *Polymer*, **2005**, *46*, 2799.
14. Zussman, E.; Theron, A.; Yarin, A. L. *Appl. Phys. Lett.*, **2003**, *82*, 73.
15. Reneker, D. H.; Yarin, A. L.; Fong, H.; Koombhongse, S. *J. Appl. Phys.*, **2000**, *87*, 4531.
16. Theron, S. A.; Yarin, A. L.; Zussman, E.; Kroll, E. *Polymer*, **2005**, *46*, 2889.
17. Z-Huang, M.; Zhang, Y.-Z.; Kotak, M.; Ramakrishna, S. *Compos. Sci. Technol.*, **2003**, *63*, 2223.
18. Schreuder-Gibson, H. L.; Gibson, P.; Senecal, K.; Sennett, M.; Walker, J.; Yeomans, W.; Ziegler, D. *J. Adv. Mater.*, **2002**, *34*(3), 44.
19. Ma, Z.; Kotaki, M.; Inai, R.; Ramakrishna, S. *Tissue Eng.*, **2005**, *11*, 101.
20. Ma, Z.; Kotaki, M.; Yong, T.; He, W.; Ramakrishna, S. *Biomaterials*, **2005**, *26*, 2527.
21. Gao, Y.; Xu, S.; Wu, R.; Wang, J.; Wei, J. *J. Appl. Polym. Sci.*, **2008**, *107*, 391.
22. El-Hag Ali A.; Abd El-Rehim, H. A.; Hegazy, E. S. A.; Ghobashy, M. M. *Radiat. Phys. Chem.*, **2006**, *75*, 1041.
23. Yao, L.; Krause, S. *Macromolecules*, **2003**, *36*, 2055.
24. Kim, S. J.; Yoon, S. G.; Lee, Y. H.; Kim, S. I. "208 Stimuli Responsive Drug Delivery Systems: From Introduction to Application," *Polym. Int.*, **2004**, *53*, 1456.
25. Kim, S. J.; Kim, M. S.; Kim, S. I.; Spinks, G. M.; Kim, B. C.; Wallace, G. G. *Chem. Mater.*, **2006**, *18*, 5805.
26. Kim, S. J.; Yoon, S. G.; Lee, Y. M.; Kim, H. C.; Kim, S. I. *Biosens. Bioelectron.*, **2004**, *19*, 531.
27. Wu, Y.; Sasaki, T.; Irie, S.; Sakurai, K. *Polymer*, **2008**, *49*, 2321.
28. Phongying, S.; Aiba, S. I.; Chirachanchai, S. *Polymer*, **2007**, *48*, 393.
29. Fei, B.; Lu, H.; Xin, J. H. *Polymer*, **2006**, *47*, 947.
30. Berner, B.; Dinh, S. M., Eds.; *Electronically Controlled Drud Delivery*; CRC Press: Boca Raton, F. L., 1998, p. 3.
31. Kikuchi, A.; Okano, T. In *Polymeric Drug Delivery Systems*; Kwon, G. S., Ed. Taylor & Francis Group: Boca Raton, F. L., 2005, p. 275.

32. Li, H.; Yuan, Z.; Lam, K. Y.; Lee, H. P.; Chen, J.; Hanes, J.; Fu, J. *Biosens. Bioelectron.*, **2004**, *19*, 1097.
33. Homma, M.; Seida, Y.; Nakano, Y. *J. Appl. Polym. Sci.*, **2000**, *75*, 111.
34. Ricka, J.; Tanaka, T. *Macromolecules*, **1984**, *17*, 2916.
35. Gu, W. Y.; Lai, W. M.; Mow, V. C. *J. Biomech. Eng.*, **1998**, *120*, 169.
36. Lai, W. M.; Hou, J. S.; Mow, V. C. *J. Biomech. Eng.*, **1991**, *113*, 245.
37. Hodgkin, A. L.; Huxley, A. F. *J. Physiol.*, **1952**, *116*, 473.
38. Onsager, L. *Phys. Rev.*, **1931**, *37*, 405.
39. Kedem, O.; Katchalsky, A. *J. Gen. Physiol.*, **1961**, *45*, 143.
40. Helfferich, F. *Ion Exchange*; McGraw-Hill: New York, N. Y., 1962.
41. Frank, E. H.; Grodzinsky, A. J.; Phillips, S. L.; Grimshaw P. E. In *Biomechanics of Diarthrodial Joints*, Vol. 1, Mow, V. C.; Ratcliffe, A.; S.L-Woo, Y., Eds.; Springer-Verlag: New York, N. Y., 1990, pp. 261–209.
42. Comper, W. D., Ed. *Extracellular Matrix, Volume 1 – Tissue Function*;. Harwood Academic Publishers: Amsterdam, The Netherlands, 1996.
43. Li, H. *Int. J. Solids Struct.*, **2009**, *46*, 1326.
44. Li, H.; Chen, J.; Lam, K. Y. *Biosens. Bioelectron.*, **2007**, *22*, 1633.
45. Lam, K. Y.; Li, H.; Ng, T. Y.; Luo, R. *Eng. Anal. Boundary Elem.*, **2006**, *30*, 1011.
46. Chen, J.; Li, H.; Lam, K. Y. *Mater. Sci. Eng., C*, **2005**, *25*, 710.
47. Li, H.; Luo, R.; Lam, K. Y. *J. Membr. Sci.*, **2007**, *289*, 284.
48. Li, H.; Luo, R.; Lam, K. Y. *J. Biomech.*, **2007**, *40*, 1091.
49. Li, H.; Ng, T. Y.; Yew, Y. K.; Lam, K. Y. *Biomacromolecules*, **2005**, *6*, 109.
50. Li, H.; Ng, T. Y.; Cheng, J. Q.; Lam, K. Y. *Comput. Mech.*, **2003**, *33*, 30.
51. Li, H.; Cheng, J. Q.; Ng, T. Y.; Chen, J.; Lam, K. Y. *Eng. Struct.*, **2004**, *26*, 531.
52. Reddy, J. N.; *An Introduction to the Finite Element Method*, 2nd edition; McGraw-Hill: New York, N. Y., 1993.
53. Shang, J.; Shao, Z.; Chen, X. *Polymer*, **2008**, *49*, 5520.
54. Shang, J.; Shao, Z.; Chen, X. *Biomacromolecules*, **2008**, *9*, 1208.
55. Tan, S.-H.; Inai, R. *Polymer*, **2005**, *46*, 6128.
56. Deitzel, J. M. *Polymer*, **2001**, *42*, 8163.
57. Zarkoob, S. *Polymer*, **2004**, *45*, 3973.
58. Atheron, S. *Polymer*, **2004**, *45*, 2017.
59. He, J.; Wan, Y. Q. *Polymer*, **2004**, *45*(19), 6731.
60. Aouada, F. A.; Guilherme, M. R.; Campese, G. M.; Girotto, E. M.; Rubira, A. F.; Muniz, E. C. *Polym. Test.*, **2006**, *25*, 158.
61. Ramanathan, S.; Block, L. H. *J. Controlled Release*, **2001**, *70*, 109.
62. Jensen, M.; Birch Hansen, P.; Murdan, S.; Frokjaer, S.; Florence, A. T. *Eur. J. Pharm. Sci.*, **2002**, *15*, 139.
63. Kim, S. J.; Park, S. J.; Kim, I. Y.; Shin, M. S.; Kim, S. I. "210 Stimuli Responsive Drug Delivery Systems: From Introduction to Application," *J. Appl. Polym. Sci.*, **2002**, *86*, 2285.
64. Tanaka, T.; Nishio, I.; Sun, S. T.; Ueno-Nishio, S. *Science*, **1982**, *218*, 467.
65. Bajpai, K.; Bajpai, J.; Soni, S. N. *Express Polym. Lett.*, **2008**, *2*, 26.
66. Kishi, R.; Hara, M.; Sawahata, K.; Osada, Y. In *Polymer Gels: Fundamentals and Biomedical Applications*, DeRossi, D.; Kajiwara, K.; Osada, Y.; Yamauchi, A., Eds.; Plenum Press: New York, N. Y., 1991, p. 205.

67. Irvin, D. J.; Goods, S. H.; Whinnery, L. L. *Chem. Mater.*, **2001**, *13*, 1143.
68. Filipcsei, G.; Feher, J.; Zrinyi, M. *J. Mol. Struct.*, **2000**, *554*, 109.
69. Kwon, I. C.; Bae, Y. H.; Okano, T.; Berner, B.; Kim, S. W. *Die Macromolekulare Chemie, Macromol. Symp.*, **1990**, *33*, 265.
70. Kwon, I. C.; Bae, Y. H.; Kim, S. W. *Nature*, **1991**, *354*, 291.
71. Kwon, I. C.; Bae, Y. H.; Okano, T.; Kim, S. W. *J. Controlled Release*, **1991**, *17*, 149.
72. Kwon, I. C.; Bae, Y. H.; Kim, S. W. *J. Controlled Release*, **1994**, *30*, 155.
73. Hsu, C. S.; Block, L. H. *Pharm. Res.*, **1996**, *13*, 1865.
74. Kajiwara, K.; Ross-Murphy, S. B. *Nature*, **1992**, *355*, 208.
75. Osada, Y.; Okuzaki, H.; Hori, H. *Nature*, **1992**, *355*, 242.
76. Salehpoor, K.; Shahinpoor, M.; Mojarrad, M. In *Polymeric-Based Systems Conference, Proceedings of the SPIE*, Vol. 2716, 1996, p. 36. *Proceedings of the SPIE – The International Society for Optical Engineering*, San Diego, CA, USA, 1996, p. 36.
77. Hirai, T.; Zheng, J.; Watanabe, M.; Shirai, H. In *Smart Structures and Materials*, 2000. Electroactive Polymer Actuators and Devices (EAPAD), Rd., Y. Bar Cohen, *Proceedings of the SPIE*, Vol. 3987, 2000, p. 281, *Proceedings of the SPIE – The International Society for Optical Engineering*, Convention Center, San Jones, CA, USA, 2000, Vol. 3987, 281.
78. Sukigara, S.; Gandhi, M.; Ayutsede, J.; Micklus, M.; Ko, F. *Polymer*, **2003**, *44*, 5727.
79. Sukigara, S.; Gandhi, M.; Ayutsede, J.; Micklus, M.; Ko, F. *Polymer*, **2004**, 45, 3708.
80. Park, K. E.; Jung, S. Y.; Lee, S. J.; Min, B.-M.; Park, W. H. *Int. J. Biol. Macromol.*, **2006**, *38*, 165.
81. Yuan, X.; Zhang, Y.; Dong, C.; Sheng, J. *Polym. Int.*, **2004**, *53*, 1704.
82. Ki, C. S.; Baek, D. H.; Gang, K. D.; Lee, K. H.; Um, I. C.; Park, Y. H.; *Polymer*, **2005**, 46.
83. Shahinpoor, M. In *Smart Structures and Materials 2000: Electroactive Polymer Actuators and Devices* (EAPAD), Rd., Y. Bar Cohen, Proceedings of the SPIE Vol. 3987, **2000**, p.187, *Proceedings the of SPIE – The International Society for Optical Engineering*, Convention Center, San Jones, CA, USA, **2000**, *3987*, 187.
84. Sun, S.; Mak, A. F. T. *J. Polym. Sci., Part B: Polym. Phys. Ed.*, **2001**, *39*, 236.
85. Fei, J.; Zhang, Z.; Gu, L. *Polym. Int.*, **2002**, *51*, 502.
86. Kim, S. J.; Ryon Shin, S. U.; Kim, N. G.; Kim, S. I. *J. Macromol. Sci.- Pure Appl. Chem.*, **2005**, *42A*, 1073.
87. Kim, S. J.; Yoon, S. G.; Lee, Y. M.; Kim, S. I. *Sens. Actuators, B: Chem.*, **2003**, *88*, 286.
88. Homma, M.; Seida, Y.; Nakano, Y. *J. Appl. Polym. Sci.*, **2001**, *82*, 76.
89. Kim, S. Y.; Shin, H. S.; Lee, Y. M.; Jeong, C. N. *J. Appl. Polym. Sci.*, **1999**, *73*, 1675.

CHAPTER 2

ON THE DRYING OF NANOSTRUCTURED FABRICS

A. K. HAGHI

CONTENTS

2.1 INTRODUCTION

Drying wet nano-porous media are coupled in a complicated way. The structure of the solid matrix varies widely in shape. There is, in general, a distribution of void sizes, and the structures may also be locally irregular. Energy transport in such a medium occurs by conduction in all of the phases. Mass transport occurs within voids of the medium. In an unsaturated state these voids are partially filled with a liquid, whereas the rest of the voids contain some gas. It is a common misapprehension that non-hygroscopic fibers (i.e., those of low intrinsic for moisture vapor) will automatically produce a hydrophobic fabric. The major significance of the fine geometry of a textile structure in contributing to resistance to water penetration can be stated in the following manner:

The requirements of a water repellent fabric are (*a*) that the fibers shall be spaced uniformly and as far apart as possible and (*b*) that they should be held so as to prevent their ends drawing together. In the meantime, wetting takes place more readily on surfaces of high fiber density and in a fabric where there are regions of high fiber density such as yarns, the peripheries of the yarns will be the first areas to wet out and when the peripheries are wetted, water can pass unhindered through the fabric.

For thermal analysis of wet nanostructured fabrics, the liquid is water and the gas is air. Evaporation or condensation occurs at the interface between the water and air so that the air is mixed with water vapor. A flow of the mixture of air and vapor may be caused by external forces, for instance, by an imposed pressure difference. The vapor will also move relative to the gas by diffusion from regions where the partial pressure of the vapor is higher to those where it is lower.

Again, heat transfer by conduction, convection, and radiation and moisture transfer by vapor diffusion are the most important mechanisms in very cool or warm environments from the skin.

Meanwhile, Textile manufacturing involves a crucial energy-intensive drying stage at the end of the process to remove moisture left from dye setting. Determining drying characteristics for textiles, such as temperature levels, transition times, total drying times and evaporation rates, etc. is vitally important so as to optimize the drying stage. In general, drying means to make free or relatively free from a liquid. We define it more narrowly as the vaporization and removal of water from textiles.

2.2 HEAT

When a wet nano-porous fabric is subjected to thermal drying two processes occur simultaneously, namely:

a) Transfer of heat to raise the wet fabric temperature and to evaporate the moisture content.
b) Transfer of mass in the form of internal moisture to the surface of the fabric and its subsequent evaporation.

The rate at which drying is accomplished is governed by the rate at which these two processes proceed. Heat is a form of energy that can across the boundary of a system. Heat can, therefore, be defined as "the form of energy that is transferred between a system and its surroundings as a result of a temperature difference." There can only be a transfer of energy across the boundary in the form of heat if there is a temperature difference between the system and its surroundings. Conversely, if the system and surroundings are at the same temperature there is no heat transfer across the boundary.

Strictly speaking, the term "heat" is a name given to the particular form of energy crossing the boundary. However, heat is more usually referred to in thermodynamics through the term "heat transfer," which is consistent with the ability of heat to raise or lower the energy within a system.

There are three modes of heat transfer:

– Convection;
– Conduction; and
– Radiation.

All three are different. Convection relies on movement of a fluid. Conduction relies on transfer of energy between molecules within a solid or fluid. Radiation is a form of electromagnetic energy transmission and is independent of any substance between the emitter and receiver of such energy. However, all three modes of heat transfer rely on a temperature difference for the transfer of energy to take place.

The greater the temperature difference the more rapidly will the heat be transferred. Conversely, the lower the temperature difference, the slower will be the rate at which heat is transferred. When discussing the modes of heat transfer it is the rate of heat transfer Q that defines the characteristics rather than the quantity of heat.

As it was mentioned earlier, there are three modes of heat transfer, convection, conduction and radiation. Although two, or even all three, modes of heat transfer may be combined in any particular thermodynamic situation, the three are quite different and will be introduced separately.

The coupled heat and liquid moisture transport of nano-porous material has wide industrial applications in textile engineering and functional design of apparel products. Heat transfer mechanisms in nano-porous textiles include conduction by the solid material of fibers, conduction by intervening air, radiation, and convection. Meanwhile, liquid and moisture transfer mechanisms include vapor diffusion in the void space and moisture sorption by the fiber, evaporation, and capillary effects. Water vapor moves through textiles as a result of water vapor concentration differences. Fibers absorb water vapor due to their internal chemical compositions and structures. The flow of liquid moisture through the textiles is caused by fiber-liquid molecular attraction at the surface of fiber materials, which is determined mainly by surface tension and effective capillary pore distribution and pathways. Evaporation and/or condensation take place, depending on the temperature and moisture distributions. The heat transfer process is coupled with the moisture transfer processes with phase changes such as moisture sorption and evaporation.

Mass transfer in the drying of a wet fabric will depend on two mechanisms: movement of moisture within the fabric which will be a function of the internal physical nature of the solid and its moisture content; and the movement of water vapor from the material surface as a result of water vapor from the material surface as a result of external conditions of temperature, air humidity and flow, area of exposed surface and supernatant pressure.

2.3 CONVECTION HEAT TRANSFER IN NANOSTRUCTURED FABRICS

A very common method of removing water from textiles is convective drying. Convection is a mode of heat transfer that takes place as a result of motion within a fluid. If the fluid, starts at a constant temperature and the surface is suddenly increased in temperature to above that of the fluid, there will be convective heat transfer from the surface to the fluid as a result of the temperature difference. Under these conditions the temperature difference causing the heat transfer can be defined as:

$$\Delta T = \text{(surface temperature)} - \text{(mean fluid temperature)}$$

Using this definition of the temperature difference, the rate of heat transfer due to convection can be evaluated using Newton's law of cooling:

$$Q = h_c A \Delta T \tag{1}$$

where A is the heat transfer surface area and h_c is the coefficient of heat transfer from the surface to the fluid, referred to as the "convective heat transfer coefficient."

The units of the convective heat transfer coefficient can be determined from the units of other variables:

$$Q = h_c A \Delta T$$
$$W = (h_c) m^2 K$$

So the units of h_c are $W / m^2 K$.

The relationship given in Eq. (1) is also true for the situation where a surface is being heated due to the fluid having higher temperature than the surface. However, in this case the direction of heat transfer is from the fluid to the surface and the temperature difference will now be

$$\Delta T = \text{(Mean fluid temperature)} - \text{(Surface temperature)}$$

The relative temperatures of the surface and fluid determine the direction of heat transfer and the rate at which heat transfer take place.

As given in Eq. (1), the rate of heat transfer is not only determined by the temperature difference but also by the convective heat transfer coefficient h_c. This is not a constant but varies quite widely depending on the properties of the fluid and the behavior of the flow. The value of must depend on the thermal capacity of the fluid particle considered, i.e., mC_p for the particle. The higher the density and C_p of the fluid the better the convective heat transfer.

Two common heat transfer fluids are air and water, due to their widespread availability. Water is approximately 800 times denser than air and also has a higher value of C_p. If the argument given above is valid then water has a higher thermal capacity than air and should have a better convective heat transfer performance. This is borne out in practice because typical values of convective heat transfer coefficients are as follows:

Fluid	
Water	500–10,000
Air	5–100

The variation in the values reflects the variation in the behavior of the flow, particularly the flow velocity, with the higher values of h_p resulting from higher flow velocities over the surface.

2.4 CONDUCTION HEAT TRANSFER IN NANOSTRUCTURED FABRICS

If a fluid could be kept stationary there would be no convection taking place. However, it would still be possible to transfer heat by means of conduction. Conduction depends on the transfer of energy from one molecule to another within the heat transfer medium and, in this sense, thermal conduction is analogous to electrical conduction.

Conduction can occur within both solids and fluids. The rate of heat transfer depends on a physical property of the particular solid of fluid, termed its thermal conductivity k, and the temperature gradient across the medium. The thermal conductivity is defined as the measure of the rate of heat transfer across a unit width of material, for a unit cross–sectional area and for a unit difference in temperature.

From the definition of thermal conductivity k it can be shown that the rate of heat transfer is given by the relationship:

$$Q = \frac{kA\Delta T}{x} \tag{2}$$

ΔT is the temperature difference $T_1 - T_2$, defined by the temperature on the either side of the porous surface. The units of thermal conductivity can be determined from the units of the other variables:

$$Q = kA\Delta T / x$$
$$W = (k)m^2 K / m$$

The unit of k is $W / m^2 K / m$.

2.5 RADIATION HEAT TRANSFER IN NANOSTRUCTURED FABRICS

The third mode of heat transfer, radiation, does not depend on any medium for its transmission. In fact, it takes place most freely when there is a perfect vacuum between the emitter and the receiver of such energy. This is proved daily by the transfer of energy from the sun to the earth across the intervening space.

Radiation is a form of electromagnetic energy transmission and takes place between all matters providing that it is at a temperature above absolute zero. Infra-red radiation form just part of the overall electromagnetic spectrum. Radiation is energy emitted by the electrons vibrating in the molecules at the surface of a body. The amount of energy that can be transferred depends on the absolute temperature of the body and the radiant properties of the surface.

A body that has a surface that will absorb all the radiant energy it receives is an ideal radiator, termed a "black body." Such a body will not only absorb radiation at a maximum level but will also emit radiation at a maximum level. However, in practice, bodies do not have the surface characteristics of a black body and will always absorb, or emit, radiant energy at a lower level than a black body.

It is possible to define how much of the radiant energy will be absorbed, or emitted, by a particular surface by the use of a correction factor, known as the "emissivity" and given the symbol ε. The emissivity of a surface is the measure of the actual amount of radiant energy that can be absorbed, compared to a black body. Similarly, the emissivity defines the radiant energy emitted from a surface compared to a black body. A black body would, therefore, by definition, have an emissivity ε of 1. It should be noted that the value of emissivity is influenced more by the nature of texture of clothes, than its color. The practice of wearing white clothes in preference to dark clothes in order to keep cool on a hot summer's day is not necessarily valid. The amount of radiant energy absorbed is more a function of the texture of the clothes rather than the color.

Since World War II, there have been major developments in the use of microwaves for heating applications. After this time it was realized that microwaves had the potential to provide rapid, energy-efficient heating of materials. These main applications of microwave heating today include food processing, wood drying, plastic and rubber treating as well as curing and preheating of ceramics. Broadly speaking, microwave radiation is the term associated with any electromagnetic radiation in the microwave frequency range of 300 MHz–300 Ghz. Domestic and industrial microwave ovens generally operate at a frequency of 2.45 Ghz corresponding to a wavelength of 12.2 cm. However, not all materials can be heated rapidly by microwaves. Materials may be classified into three groups, i.e., conductors' insulators and absorbers. Materials that absorb microwave radiation are called dielectrics, thus, microwave heating is also referred to as dielectric heating. Dielectrics have two important properties:

(1) They have very few charge carriers. When an external electric field is applied there is very little change carried through the material matrix.
(2) The molecules or atoms comprising the dielectric exhibit a dipole movement distance. An example of this is the stereochemistry of covalent bonds in a water molecule, giving the water molecule a dipole movement. Water is the typical case of non-symmetric molecule. Dipoles may be a natural feature of the dielectric or they may be induced. Distortion of the electron cloud around non-polar molecules or atoms through the presence of an external electric field can induce a temporary dipole movement. This movement

generates friction inside the dielectric and the energy is dissipated subsequently as heat.

The interaction of dielectric materials with electromagnetic radiation in the microwave range results in energy absorbance. The ability of a material to absorb energy while in a microwave cavity is related to the loss tangent of the material.

This depends on the relaxation times of the molecules in the material, which, in turn, depends on the nature of the functional groups and the volume of the molecule. Generally, the dielectric properties of a material are related to temperature, moisture content, density and material geometry.

An important characteristic of microwave heating is the phenomenon of "hot spot" formation, whereby regions of very high temperature form due to non-uniform heating. This thermal instability arises because of the non-linear dependence of the electromagnetic and thermal properties of material on temperature. The formation of standing waves within the microwave cavity results in some regions being exposed to higher energy than others.

Microwave energy is extremely efficient in the selective heating of materials as no energy is wasted in "bulk heating" the sample. This is a clear advantage that microwave heating has over conventional methods. Microwave heating processes are currently undergoing investigation for application in a number of fields where the advantages of microwave energy may lead to significant savings in energy consumption, process time and environmental remediation.

Compared with conventional heating techniques, microwave heating has the following additional advantages:

- higher heating rates;
- no direct contact between the heating source and the heated material;
- selective heating may be achieved;
- greater control of the heating or drying process.

2.6 COMBINED HEAT TRANSFER COEFFICIENT

For most practical situations, heat transfer relies on two, or even all three modes occurring together. For such situations, it is inconvenient to analyze each mode separately. Therefore, it is useful to derive an overall heat transfer coefficient that will combine the effect of each mode within a general situation. The heat transfer in moist fabrics takes place through three modes, conduction, radiation, and the process of distillation. With a dry fabric, only conduction and radiation are present.

2.7 POROSITY AND PORE SIZE DISTRIBUTION IN FABRIC

The amount of porosity, i.e., the volume fraction of voids within the fabric, determines the capacity of a fabric to hold water; the greater the porosity, the more water the fabric can hold. Porosity is obtained by dividing the total volume of water extruded from fabric sample by the volume of the sample:

Porosity = volume of water/volume of fabric

= (volume of water per gram sample)(density of sample)

It should be noted that most of water is stored between the yarns rather than within them. In the other words, all the water can be accommodated by the pores within the yarns, and it seems likely that the water is chiefly located there. It should be noted that pores of different sizes are distributed within a fabric. By a porous medium we mean a material contained a solid matrix with an interconnected void. The interconnectedness of the pores allows the flow of fluid through the fabric. In the simple situation ("single phase flow") the pores is saturated by a single fluid. In "two-phase flow" a liquid and a gas share the pore space. In fabrics the distribution of pores with respect to shape and size is irregular. On the pore scale (the microscopic scale) the flow quantities (velocity, pressure, etc.) will clearly be irregular.

The usual way of driving the laws governing the macroscopic variables are to begin with standard equations obeyed by the fluid and to obtain the macroscopic equations by averaging over volumes or areas contained many pores.

In defining porosity we may assume that all the pore space is connected. If in fact we have to deal with a fabric in which some of the pore space is disconnected from the reminder, then we have to introduce an "effective porosity," defined as the ratio of the connected pore to total volume.

A further complication arises in forced convection in fabric, which is a porous medium. There may be significant thermal dispersion, i.e., heat transfer due to hydrodynamic mixing of the fluid at the pore scale. In addition to the molecular diffusion of heat, there is mixing due to the nature of the fabric.

KEYWORDS

- **Heat transfer**
- **Black body**
- **Emissivity**
- **Bulk heating**
- **Effective porosity**

REFERENCES

1. Haghi, A. K. *JTAC*, **2004**, *76*, 1035–1055.
2. Haghi, A. K. *JTAC*, **2003**, *74*, 827–842.
3. Haghi, A. K. *Some Aspects of Microwave Drying*; The Annals of Stefan cel Mare University, Year VII, No. 14, 22–25, 2000.
4. Haghi, A. K. *A Thermal Imaging Technique for Measuring Transient Temperature Field- An Experimental Approach*; The Annals of Stefan cel Mare University, Year VI, No. 12, 73–76, 2000.
5. Haghi, A. K. *Acta Polytech.*, **2001**, *41*(1), 55–57.
6. Haghi, A. K. *Acta Polytech.* **2001**, *41*(3), 20–23.
7. Haghi, A. K. *J. Comput. Appl. Mech.*, **2001**, *2*(2), 195–204.
8. Haghi, A. K. J. *Theor. Appl. Mech.*, **2002**, *32*(2), 47–62.
9. Haghi, A. K. *Acta Polytech.*, **2002**, *42*(2), 35–40.
10. Haghi, A. K. *J. Technol.*, **2002**, *35*(F), 1–16.
11. Haghi, A. K. *H. J.I. C.*, **2002**, *30*, 261–269.
12. Haghi, A. K. *J. Theor. Appl. Mech.*, **2003**, *33*, 83–94.
13. Haghi, A. K.; Mahfouzi, K.; Mohammadi, K. *JUCTM*, **2002**, *38*, 85–96.
14. Haghi, A. K. *Int. J. Appl. Mech. Eng.*, **2003**, *8*(2), 233–243.
15. Haghi, A. K.; Rondot, D. *IJC&Chem. Eng.*, **2004**, *23*, 25–34.
16. Haghi, A. K. *Heat and Mass Transport Through Moist Porous Materials*, 14th International Symposium on Transport Phenomena Proceedings, 209–214, 6-–9 July 2003, Indonesia.
17. Higdon, J.; Ford, G. *J. Fluid Mech.*, **1996**, *308*, 341–361.

18. Hong, K.; Hollies, N. R. S.; Spivak, S. M. *Textile Res. J.* **1988**, *58*(12), 697–706.
19. Hsieh, Y. L.; Yu, B.; Hartzell, M. *Textile Res. J.* **1992**, *62*(12), 697–704.
20. Huh, C.; Scriven, L. E. *J. Coll. Inter. Sci.*, **1971**, *35*, 85–101.
21. Incropera, F. P.; Dewitt, D. P. *Fundamentals of Heat and Mass Transfer*, 2nd ed.; Wiley: New York, 1985.
22. ISO 11092, *Measurement of Thermal and Water-vapour Resistance under Steady-state Conditions* (Sweating Guarded-hotplate Test), 1993.
23. Ito, H.; Muraoka, Y. *Textile Res. J.*, **1993**, *63*(7), 414–420.
24. Jackson, J.; James, D. *Can. J. Chem. Eng.*, **1986**, *64*, 364–374.
25. Jacquin, C. H.; Legait, B. *Phy.-Chem. Hydrol.* **1984**, *5*, 307–319.
26. Jirsak, O.; Gok, T.; Ozipek, B.; Pau, N. *Textile Res. J.* **1998**, *68*(1), 47–56.
27. Kaviany, M. *Principle of Heat Transfer in Porous Media;* Springer: New York, 1991.
28. Keey, R. B. *Drying: Principles and Practice*; Oxford: Pergamon, 1975.
29. Keey, R. B. *Introduction to Industrial Drying Operations*; Oxford: Pergamon, 1978.
30. Keey, R. B. *Rev. Prog. Coloration* **1993**; *23*, 57–72.
31. Kulichenko, A.; Langenhove, L., *J. Textile Inst.*, **1992**, *83*(1), 127–132.
32. Kyan, C.; Wasan, D.; Kintner, R. *Ind. Eng. Chem. Fundament*, **1970**, *9*(4), 596–603.
33. Le, C. V.; Ly, N. G., *Textile Res. J.*, **1995**, *65*(4), 203–212.
34. Le, C. V.; Tester, D. H.; Buckenham, P. *Textile Res. J.* **1995**, *65*(5), 265–272.
35. Lee, H. S.; Carr, W. W.; Becckham, H. W.; Wepfer, W. J. *Textile Res. J.*, **2000**, *70*, 876–885.
36. Lee, H. S. *Study of the Industrial Through-Air Drying Process For Tufted Carpet*, Doctoral thesis, Georgia Institute of Technology, Atalnta, GA, 2000.
37. Luikov, A. V. *Heat and Mass Transfer in Capillary Porous Bodies*; Pergamon Press: Oxford, 1966.
38. Luikov, A. V. *Int. J. Heat Mass Transfer*, **1975**, *18*, 1–14.
39. Metrax, A. C.; Meredith R. J. *Industrial Microwave Heating*; Peter Peregrinus Ltd: London, England, 1983.
40. Mitchell, D. R.; Tao, Y.; Besant, R. W. *Validation of Numerical Prediction for Heat Transfer with Airflow and Frosting in Fibrous Insulation*, paper 94-WA/HT-10 in Proceedings of ASME International Mechanical Engineering Congress, Chicago, Nov. 1994.
41. MOD Specification UK/SC/4778A SCRDE, Moisture Vapour Transmission Test Method.
42. Morton, W. E.; Hearle, J. W. S. *Physical Properties of Textile Fibers*, 3rd ed.; Textile Institute: Manchester, U. K., 1993.
43. Moyene, C.; Batsale, J. C. ; Degiovanni, A. *Int. J. Heat Mass Transfer*, **1988**, *31*(11), 2319–2330.
44. Mujumdar, A. S.; *Handbook of Industrial Drying*; Marcel Decker: New York, 1985.
45. Nasrallah, S. B.; Pere, P. *Int. J. Heat Mass Transfer*, **1988**, *31*(5), 957–967.
46. Nossar, M. S.; Chaikin, M.; Datyner, A. *J. Textile Inst*, **1973**, *64*, 594–600.
47. Patankar, S. V. *Numerical Heat Transfer and Fluid Flow*; Hemisphere Publishing: New York, 1980.
48. Penner, S.; Robertson, A. *Textile Res. J.*, **1951**, *21*, 775–788.
49. Provornyi, S.; Slobodov, E. *Theoret. Foundat. Chem. Eng.*, **1995**, *29*(1), 1–5.

50. Hollies, N. R. S. *Cotton Clothing Attributes in Subject Comfort*, 15th Textile Chemistry and Processing Conference, USDA New Orleans, 1975.
51. Hollies, N. R. S. In *Clothing Comfort*, Hollies, N. R. S.; Goldman, R. F. Eds.; Ann Arbor Science: Ann Arbor, 1977.
52. Scheurell, D. M.; Spivak, S. M.;, Hollies, N. R. S.; *Textile Res. J.*, 1985; 394–399.
53. De Vries, D. A. *Trans. Am. Geophys. Union*, **1958**, *39*(5), 909–916.
54. Philip, J. R.; de Vries, D. A. *Trans. Am. Geophys. Union*, **1957**, *38*, 222–232.
55. Eckert, E.; Faghri, M. *Int. J. Heat Mass Transfer*, **1980**, 23, 1613–1623.
56. Udell, K. S. *Int. J. Heat Mass Transfer*, **1985**, *28*(2), 485–495.
57. Udell, K. S. *J. Heat Transfer*, **1983**, *105*, 485–492.
58. Bouddour, A.; Auriault, J. L. ; Mhamadi, M.; Bloch, J. F. *Int. J. Heat Mass Transfer*, **1998**, *41*(15), 2263–2277.
59. Gibson, P. W. In *Computational Technologies for Fluid/Thermal/Structural/Chemical Systems with Industrial Applications*, Vol. II, pp 125–139, ASME, 1999.
60. Nordon, P.; David, H. G. *Int. J. Heat Mass Transfer*, **1967**, *10*, 853–866.
61. Farnworth, B. *Textile Res. J.*, **1983**, *56*, 581–587.
62. Farnworth, B. *Textile Res. J.*, **1986**, *56*, 653–665.
63. Farnworth, B.; Lotens, W. A.; Wittgen, P. *Tex. Res. J.*, **1990**, *60*(1), 50–53.
64. Osczevski, R. J.; Dolhan, P. A. *J. Coated Fabrics*, **1989**, *18*, 255–258.
65. Gretton, J. C.; Brook, D. B.; Dyson, H. M.; Harlock, S. C. *Textile Res. J.*, **1998**, *68*(12), 936–941.

CHAPTER 3

CONTENT OF UNSATURATED FATTY ACIDS CONTAINING 18 AND 20 CARBON ATOMS IN THE TOTAL LIPID MOIETY OF MITOCHONDRIAL MEMBRANES DETERMINES THE ACTIVITY COMPLEX I OF RESPIRATORY CHAIN

I. V. ZHIGACHEVA, T. B. DURLAKOVA, T. A. MISHARINA, M. B. TERENINA, N. I. KRIKUNOVA, I. P. GENEROZOVA, A. P. SHUGAEV, and S. G. FATTAKHOV

CONTENTS

3.1 INTRODUCTION

Cell membranes are among cell components damaged under the conditions of water deficit. Water deficit modifies the cell and organelle membranes and thus alters their functions and cell metabolism [1]. The alterations occur at the fatty acid composition of biological membranes. The content of saturated fatty acid increases while the content unsaturated fatty acid decreases [2]. These changes reflect on the activity of enzymes associated with chloroplast and mitochondrial membranes and result in decreasing the functional activity of these organelles [3]. As known from the literature, regulators of plant growth and development improve their tolerance to biotic and abiotic stresses, to water deficit in particular [4]. One of such growth regulators is melaphen – a melamine salt of bis(oxymetyl) phosphonic acid [5]:

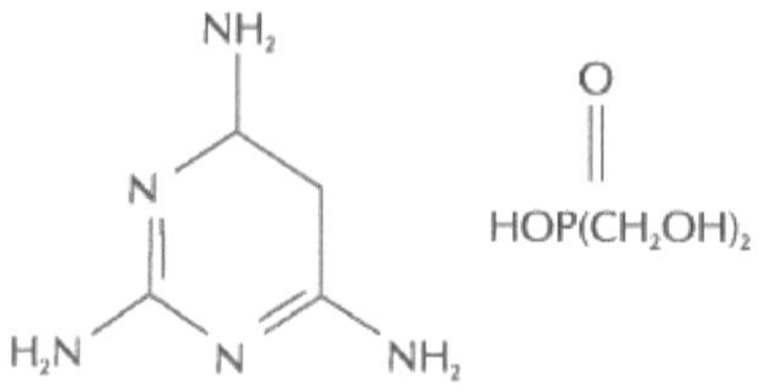

The aim of this work is study the effect of insufficient watering and the plant growth regulator – melaphen on the fatty acids composition lipid fraction of mitochondrial membranes and bioenergetical function of 5-day pea seedling mitochondria.

The development and, probability, survival of plant in any case is more dependent on availability of water than on any other environmental factors. At present, numerous data accumulated demonstrating that even weak water deficit affected plant metabolism and thus their growth and development [6]. Metabolism of plants survived even short-term strong drought could not be recovered [7]. Water deficit modified cell membranes, which affected their functions and disturbed cell metabolism [1]. The alterations occur at the level of glycolipids, monogalactosyl-diacyl-glycerol and digalactosyl-diacyl-glycerol [2, 8]. The content of unsaturated fatty acids decreases in these lipids which results in the decreasing the membrane "fluidity," alteration in the lipid – protein ratio, and eventually in the activity changes of the enzymes associated with membrane, first of all enzymes which enter into complex of electron-transport chain of mitochondria and chloroplasts [3]. The energy metabolism plays a significant role in adaptive response of the organism. Mitochondria play a key role in the energy, redox and metabolic processes in cell [9]. As the plant grows regulators enhance the resistance of plant to different kinds of stress and melaphen is such a preparation, it was interesting to find out if a preliminary treatment of seeds with melaphen would exert a protective effect under the conditions of water deficit.

3.2 MATERIALS AND METHODS

The study was carried out on mitochondria isolated from pea seedlings (*Pisum sativum*) obtained in standard conditions and in the conditions of insufficient watering.

Pea seeds germination. The seeds from the control group were washed with soap solution and 0.01% KMnO4 solution and left in water for 60 min. The seeds from the experimental group were placed in the 2×10^{-12} M melaphen solution for 60 min. After 1-day exposure, half of the seeds from the control group and half of the seeds treated with melaphen were placed onto a dry filter paper in open cuvettes. Two days later the seeds were placed into closed cuvettes with periodically watered filter paper and left for 2 days. On the 5th day the amount of germinated seeds was calculated and mitochondria isolated.

Isolation of mitochondria from 5-day sprouts epicotyls was performed by a method of [8] in our modification. The epicotyls having a length of 1.5 to 5 cm (20–25 g) were placed into a homogenizer cup, poured with an isolation medium in a ratio of 1:2, and then were rapidly disintegrated with scissors and homogenized with the aid of a press. The isolation medium comprised: 0.4 M sucrose, 5 mM EDTA, 20 mM KH_2PO_4 (pH 8.0), 10 mM KCl, 2 mM dithioerythritol, and 0.1% BSA (free of fatty acids). The homogenate was centrifugated at 25,000 g for 5 min. The precipitate was re-suspended in 8 ml of a rinsing medium and centrifugated at 3,000g for 3 min. The suspension medium comprised: 0.4 M sucrose, 20 mM

KH_2PO_4, 0.1%. BSA (free of fatty acids) (pH 7.4). *The supernatant was centrifuged for 10 min at 11,000 g for mitochondria sedimentation. The sediment was re-suspended in 2–3 ml of solution contained: 0.4 M sucrose, 20 mM KH_2PO_4 (pH 7.4), 0.1% BSA (without fatty acids) and mitochondria were precipitated by centrifugation at 11,000 g for 10 min. The suspension of mitochondria (about 6 mg of protein/ml) was stored in ice.*

The rate of mitochondria respiration was measured with the aid of Clarke oxygen electrodes and LP-7 polarograph (Czechia). Mitochondria were incubated in a medium containing 0.4 M sucrose, 20 mM HEPES-Tris buffer (pH 7.2), 5 mM KH_2PO_4, 4 mM $MgCl_2$, 5 mM EDTA and 0.1% BSA, 10 mM mlate + glutamate, pH 7.4. The rate of respiration was expressed in ng-atom O/(mg protein min).

Fatty acid methyl esters (FAMEs) were produced by acidic methanolysis of mitochondrial membrane lipids [11] or using one-step methylation of fatty acids, excluding the extraction of lipids [12]. MEFA were purified by the method of thin layer chromatography on the silica plates and hexanol elution. For a quantitative control of the methanolysis process. An internal standard – pentadecane was used.

FAME identification was performed by chromato-mass-spectrometry (GC–MS) using a Hewlett-Packard-6890 spectrophotometer with a HP-5972 mass-selective detector and after the retention times [13]. FAME were separated in the HP-5MS capillary column (30 m × 0.25 mm, phase layer – 0.25 μm) at programmed temperature increase from 60 to 285°C at the rate of 5°C /min. Evaporator temperature is 250°C, detector temperature is 289 C. Mass spectra were obtained in the regime of electron impact ionization at 70 eV and the scan rate of 1 sec. per mass decade in the scan mass range of 40–450 a.u.m.

FAME quantification was performed using a Kristall 2000M chromatograph (Russia) with flame-ionization detector and quarts capillary column SPB-1 (50 m × 0.32 mm, a nonpolar phase layer – 0.25 μm). FAME analysis was performed at programmed temperature increase from 120 to 270°C at the rate of 4°C/min. Temperature of injector and detector –270°C; the helium carrier gas rate was 1.5 ml/min. Each sample contained 2 μl of the hexane extract. The FAME content in samples was calculated as the ratio of peak area of a corresponding acid to the sum of peak areas of all found FAMEs.

Unsaturation index was calculated as a total percentage of unsaturated fatty acids with a certain number atoms multiplied by the number of double bonds, and divided by 100%. For example, for fatty acids with 18 carbon atoms unsaturation index is equal to $(18:2\omega6) \times 2 + (18:3\omega3) \times 3 + 18:1\omega9 + 18:1\omega7/100$

Lipid peroxidation (LPO) activity was assessed by fluorescent method [14]. Lipids were extracted by the mixture of chloroform and methanol (2:1). Lipids of mitochondrial membranes (3–5 mg of protein) were extracted in the glass homogenizer for 1 min at 10°C. Thereafter, equal volume of distilled water was added to the homogenate, and after rapid mixing the homogenate was transferred into 12-ml centrifuge tubes. Samples were centrifuged at 600 g for 5 min. The aliquot (3 ml) of the chloroform (lower) layer was taken, 0.3 ml of methanol was added, and fluorescence was recorded in 10-mm quartz cuvettes with a spectrofluorometer (FluoroMaxHoribaYvon, Germany). The excitation wavelength was 360 nm, the emission wavelength was 420–470 nm. The results were expressed in arbitrary units per mg protein. The using of this method permits recording both fluorescence of 4-hydroxynonenals and the fluorescence of MDA. The emission wavelength depends on the nature of the Schiff's bases: the Schiff's bases formed by 4-hydroxynonenals have fluorescence wavelength 430–435 nm; those formed by MDA, 460–470.

Statistics. Tables and figures present means and their standard deviations (M+m). In Figs. 2 and 3, correlations between unsaturation coefficients of C 18 and C 20 fatty acids and the rates of NAD-dependent substrate oxidation are presented; they were calculated using Statistica v. 6 software for Windows.

Reagents. The following reagents were used: potassium carbonate, methanol, chloroform (Merck, Germany), hexane (Panreac, Spain), acetyl chloride (Acros, Belgium), sucrose, Tris, EDTA, FCCP, malate, glutamate, FA-free BSA (Sigma, United States), Hepes (MB Biomedicals, Germany).

3.3 RESULTS AND DISCUSSIONS

Insufficient watering resulted in 3-fold increase in content of LPO products in pea seedling mitochondrial membranes (Fig. 1).

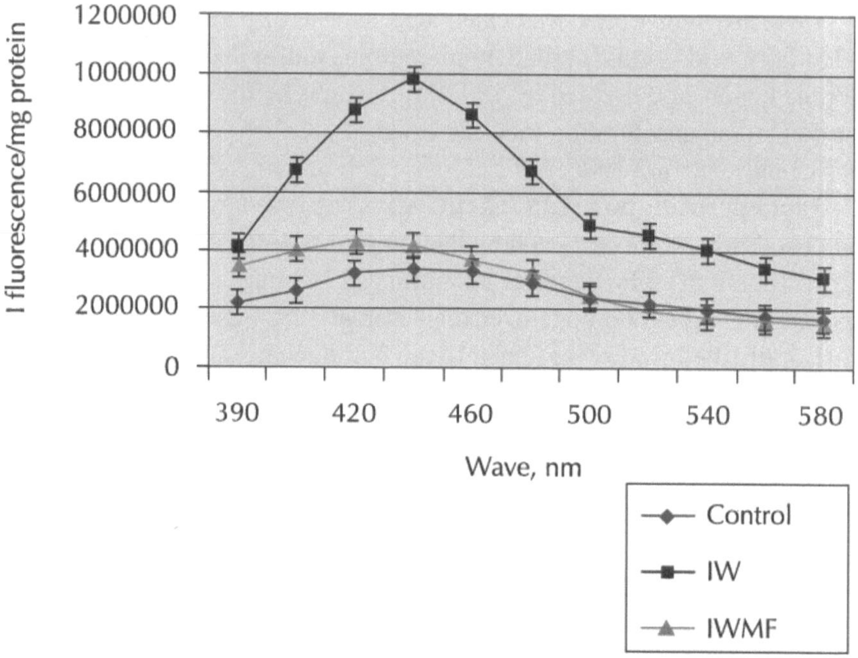

FIGURE 1 Fluorescence spectra of LPO products in mitochondrial membranes of 5-days pea seedlings under condition of insufficient watering (IW) and the treatment of seeds with melaphen (IW+MF).

The treatment of seeds with a 2×10^{-12} M melaphen solution decreased the content of LPO products to the control values. Insufficient watering promoted LPO accompanied by modification of the fatty acid composition of pea seedling mitochondrial membranes. Water deficit led to the increase in the relative content of saturated and a decrease in the content of unsaturated fatty acids in mitochondrial membranes of pea seedlings (Table 1).

TABLE 1 Effects of insufficient watering (IW) and melaphen (MF) on the relative fatty acids content in mitochondrial membrane lipids of pea seedlings,%.

Fatty acid	Control	Control+MF	IW	IW+MF
12:0	0.34 ± 0.03	0.34 ± 0.01	0.94 ± 0.30	0.34 ± 0.20
14:0	0.68 ± 0.03	0.64 ± 0.02	0.67 ± 0.20	0.69 ± 0.20
16:1ω7	0.36 ±0,03	0.36 ± 0.004	0.47 ± 0.13	.42 ± 0.005

TABLE 1 *(Continued)*

Fatty acid	Control	Control+MF	IW	IW+MF
16:0	18.64 ± 0.75	18.63 ± 0.05	20.74 ± 0.11	18.96 ± 0.50
17:0	0.45 ± 0.05	0.78 ± 0.12	0.66 ± 0.10	0.45 ± 0.16
18:2ω6	50.72 ± 0.80	50.74 ± 0.40	45.22 ± 0.10	50.65 ± 0.01
18:3ω3	11.3 ± 0.02	10.67 ± 0.01	9.18 ± 0.30	10.81 ± 0.09
18:1ω9	5.27 ± 0.40	5.25 ± 0.37	6.77 ± 0.20	5.22 ± 0.01
18:1ω7	0.81 ± 0.10	0.79 ± 0.24	0.61 ± 0.03	0.73 ± 0.05
18:0	4.10 ± 0.18	4.14 ± 0.32	5.83 ± 0.38	4.10 ± 0.15
20:2ω6	0.82 ± 0.01	0.80 ± 0.02	0.30 ±0.05	0.82 ± 0.01
20:1ω 9	2.22 ± 0.01	2.79 ± 0.01	1.57 ± 0.01	2.6 ± 0.03
20:1ω7	1.45 ± 0.01	1.14 ± 0.01	1.00 ± 0.01	1.52 ± 0.01
20:0	1.23 ± 0.03	1.20 ± 0.03	2.52 ± 0.20	1.30 ± 0.05
22:0	1.23 ± 0.11	1.20 ± 0.03	2.52 ± 0.20	1.04 ± 0.05
24:0	0.37 ± 0.02	0.55 ± 0.005	0.98 ± 0	0.35 ± 0.10

The relative content of linoleic acid was reduced by 11%, that of linolenic acid – by 29%. The content of stearic acid increased by 41%, which resulted in the decrease in the total content of C_{18} unsaturated fatty acids relative to the content of stearic acid from 16.61 ± 0.30 to 10.59 ± 0.20. Similar effect of the water deficit on the fatty acid composition of the mitochondrial membranes from maize, potato, and leaves of *Arabidopsis thaliana* and apricot was observed earlier [2, 15–18]. The authors detected a considerable decrease of the levels of linoleic and linolenic acids and an increase of the level of stearic acid in the membranes.

Substantial changes occurred also in the relative content of fatty acids with 20 carbon atoms. The pool of 20:2ω6 reduced by 2.7 times, 20:1ω 9 – 1.3 times. At the same time, the content of eicosanoic acid (20:0) increased more than twofold. As a result, the ratio of pool unsaturated fatty acids containing 20 carbon atoms (20:1ω7 + 20:1ω 9) + (20:2ω6) × 2 to eicosanic acid in mitochondrial membrane lipids decreased from 3.65 ± 0.03 to 1.20 ± 0.16.

The observed alterations possibly influence lipid–protein relation and thus alter the activity of the enzymes associated with the membrane. Indeed, insufficient watering results in a decrease of the maximal rates of NAD–dependent substrates oxidation. The rate of the pair glutamate + malate oxidation in the presence of uncoupling agent (FCCP) drops from 70.0 ± 4.6 down to 48.9 ± 3.2 ng oxygen atom/mg of protein min and the respiratory control rate (RCR) decreases from 2.27 ± 0.1 to 1.7 ± 0.2 (Table 2).

TABLE 2 Effects of insufficient watering (IW) and melaphen (MF) on the rate of NAD–dependent substrate oxidation by mitochondria isolated from pea seedlings, ng–atom/(mg protein min).

Treatment	V_0	V_3	V_4	V_3/V_4	FCCP
Control	20.0 ± 1.5	68.0 ± 4.1	30.0 ± 2.0	2.27 ± 0.1	60.0 ± 4.6
IW	14.0 ± 2.0	48.6 ± 3.0	40.2 ± 1.0	1.7 ± 0.2	41.9 ± 3.2
IW+MF	19.8 ± 3.0	66.0 ± 2.4	27.5 ± 1.3	2.4 ± 0.2	60.3 ± 5.2

Notes: Incubation medium contained 0.4 M sucrose, 20 mM Hepes–Tris, 5 mM KH_2PO_4, 2 mM $MgCl_2$, 5 mM EDTA, 10 mM malate + glutamate, pH 7.4. ADP (200 μM) and FCCP (0.5 μM) were added. V_0–the rate of oxygen decay in the presence of 10 mM malate + glutamate as substrate; V_3–the rate of substrate oxidation in the presence of ADP (State 3, 200 μM ADP); V_4–the rate of substrate oxidation after added ADP consumption (State 4); FCCP–the rate of substrate oxidation in the presence of FCCP (carbonylcyanide –*p*–trifluoromethoxyphenylhydrazone). The results of 10 experiments are presented, M±m.

The treatment of seeds with a 2×10^{-12} M melaphen solution before germination prevents the alteration of the oxidative phosphorylation efficiency caused by insufficient watering. Besides, the preliminary treatment with melaphen reduces the rates of NAD_dependent substrates oxidatio in the presence of ATP or FCCP to the control values (Table 2). Apparently, the described alterations are related with the physicochemical state of mitochondrial membranes.

Indeed the treatment with melaphen protects the unsaturated fatty acid from LPO and prevents thereby from changes in the fatty acid composition of seedling membranes in condition of insufficient watering (Table 1). In the group of seedlings subjected to insufficient watering combined with melaphen treatment (group IW + MF) concentrations of such saturated fatty acids as lauric, palmitic, and stearic acids are lower by 65%, 7.5%, and 30%, respectively, than in seedlings subjected to insufficient watering only (group IW). The level of C_{18}-unsturated fatty acids playing an important role in plant resistance to the effect of adverse factors of environment [19] increased. Thus, the level of linoleic acid increases by 12% and linolenic acid, by 15% (Table 1). The ratio of the sum of unsaturated C_{18} acids to saturated $C_{18:0}$ acids increases 1.5-fold in comparison with the group of insufficient watering. The level of C_{20} fatty acids also changed. The level of 20:2 ω 6 acid increases 2.73 times and that of 20:1ω9 acid, 2.28 times. The level of eicosanic acid decreases 1.92 times. As a result, the ratio of the sum of unsaturated C_{20} acids to saturated $C_{20:0}$ acids returns to the control values. The changes in the fatty acid composition of mitochondrial membranes were accompanied by changes in maximum rates of NAD-dependent substrates oxidation. A decrease in unsaturation coefficient of fatty acids in mitochondrial membranes led to decreasing the rates of NAD-dependent substrates oxidation and efficiency of oxidative phosphorylation.

On the basis of presented data, it may be supposed that a prevention of unsaturated fatty acids peroxidation, in particular C_{18} and C_{20} acids in membranes of plant tissues leads to enhancement of plant resistance to insufficient watering. In fact, a close correlation was observed between the unsaturation coefficient of C_{18} fatty acids in mitochondrial membranes ($\sum$unsaturated C_{18} fatty acids/$C_{18:0}$) and maximum rates of NAD-dependent substrate oxidation (the correlation coefficient r = 0.765) (Fig. 2).

An even greater correlation is observed between the unsaturation coefficient of C_{20} fatty acids (20:2 ω6) × 2+ 20:1 ω 9+ 20:1 ω 7/20:0) and maximum rates of NAD-dependent substrate oxidation (r = 0.964) (Fig. 3).

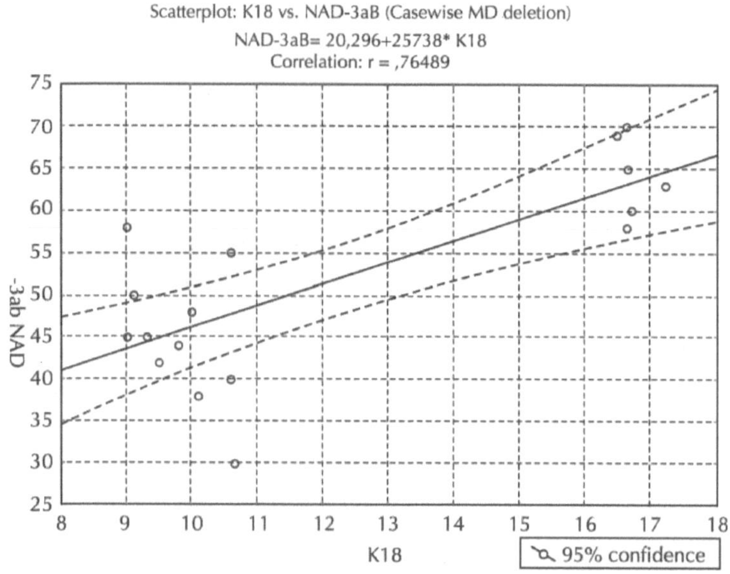

FIGURE 2 Correlation between the unsaturation coefficient of C_{20} fatty acids and maximum rates of NAD-dependent substrate oxidation. Y-axis shows the maximum rates of NAD-dependent substrate oxidation; X-axis-unsaturation coefficient of C_{18} fatty acids.

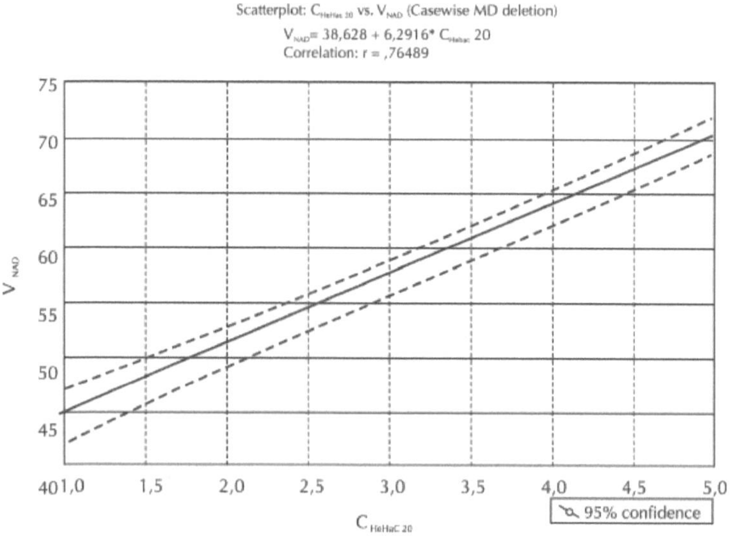

FIGURE 3 Correlation between the unsaturation coefficient of C_{20} fatty acids and maximum rates of NAD-dependent substrate oxidation. Y-axis shows the maximum rates of NAD-dependent substrate oxidation; X-axis-the ratio (20:2 ω6) × 2+ 20:1 ω 9+ 20:1 ω 7/20:0).

Changes in physical and chemical properties of mitochondrial membranes resulting in changes in the energy metabolism affected also physiological indices, e.g., seedling growth. As evident from Fig. 4, pea seed treatment with melaphen stimulated root growth by 5 times and sprouts growth by 3.5 times under conditions of water deficit. Observed stimulation of seedling root growth under insufficient watering has a great adaptive significance.

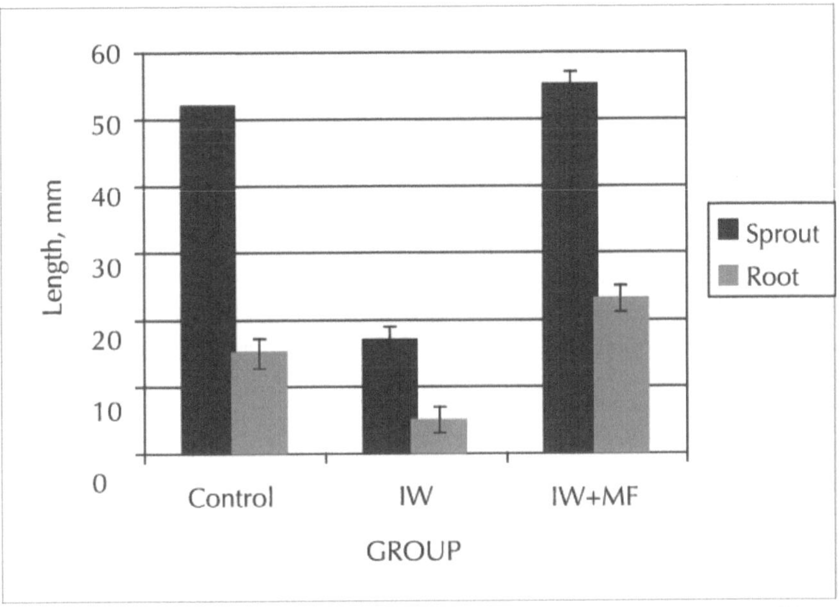

FIGURE 4 The lengths of sprouts and roots of 5-days pea seedling under condition of insufficient watering (IW) and treatment of pea seeds in this condition with melaphen (IW+MF).

Thus, under condition of insufficient watering, melaphen decreased the intensity of lipid peroxidation in mitochondrial membranes. As a result, the pool of unsaturated fatty acids containing 18 and 20 carbon atoms in the lipid phase of mitochondrial membranes remained unchanged. The prevention of changes in fatty acid composition of mitochondrial membranes affected the bioenergetic indices: there was maintained a high activity of the NADH-dehydrogenase complex of the respiratory chain of mitochondria.

KEYWORDS

- **Fluidity**
- **Melaphen**
- **NADH-dehydrogenase complex**
- **Pea seedlings**

REFERENCES

1. Kizis, D.; Lumbreras, V.; Pages, M. *FEBS Lett.*, **2001**. 498, 187–189.
2. Junior, R. R. M.; Oliveira, M. S.C.; Baccache, M. A.; de Paula, F. M. *Brazilian Arch. Biol. Technol.*, **2008**. 5, 361–367.
3. Shugaeva, N. A.; Vyskrebentseva, E. I.; Orekhova, S. O.; Shugaev, A. G. *Plant Physiol. (Russia)*, **2007**, *54*(3), 373–380.
4. Zhirmunskaya, N. M.; Shapovalov, A. A. *Agrokhimiya*, **1987**, *5*, 102–119.
5. Fattakhov, S. G.; Reznik, V. S.; Konovalov, A. I. *Proceedings of the 13th International Conference on Chemistry of Phosphorus Compounds;* Saint-Petersburg, 2002, p 80.
6. Ribas-Carbo, M.; Tailor, N. L.; Giles, L.; Busquets, S.; Finnegan, P. M.; Day, D. A.; et al. *Plant Physiol.*, **2005**, *139*, 466–473.
7. Boyer, J. S. *Science*, **1982**, *218*, 443–448.
8. Sahsah, Y.; Campos, P.; Gareil, M.; et al. *Plant. Physiol.*, **1988**, *104*, 577–586.
9. Atkin, O. K.; Macherel, D. *Ann. Bot.*, **2009**, *103*, 581–590.
10. Popov, V. N.; Ruuge, E. K.; Starkov, A. A. *Biohimiya*, **2003**, *68*(7), 910–916.
11. Carreau, J. P.; Dubacq, J. P. *J. Chromatogr.*, **1979**, *1516*, 384–390.
12. Wang, J.; Sunwoo, H.; Cherian, G.; Sim, I. S. *Science*, **2000**, *79*, 1168–1171.
13. Golovina, R. V.; Kuzmenko, T. E. *Chromatography*, **1977**, *10*, 545–546.
14. Fletcher, B. I.; Dillard, C. D.; Tappel, A. L. *Ann. Biochem.*, **1973**, *52*, 1–9.
15. Makarenko, S. P.; Konstanyinov, Y. M.; Khotimchenko, S. V.; Konenkina, T. A.; et al. *Plant Physiol. (Russia)*, **2003**, *50*, 487–492.
16. Gigon, A.; Matos, A. R.; Laffray, D.; Fodil, Y. Z. *Ann. Bot.*, **2004**, *94*, 345–351.
17. Leone, A.; Costa, A.; Grillo, S.; Tucci, M.; Horvarth, I.; Vigh L. *Plant Cell Environ.*, **1996**, *19*, 1103–1109.
18. Yun-Ping, G., Jia-Rui, L. *J. Zhejiang Univ (Agricalt. & Life Sci).*, **2002**, *28*, 513–517.
19. Demin, A. N.; Deryabin, A. N.; Sinkevich, M. S.; Trunova, T. I. *Plant Physiol. (Russia)*, **2000**, *55*, 710–720.
20. Torres-Franklin, M.-L.; Repellin, A.; Van-Biet, H.; d'Arcy-Lameta, A. *Environ. Exp. Bot.*, **2009**, 65, 162–169.

CHAPTER 4

UPDATES ON THE INFLUENCE OF MELAFEN-PLANT GROWTH REGULATOR

O. M. ALEKSEEVA

CONTENTS

4.1 INTRODUCTION

This investigation deals with the of melamine salt of bis (oximethyl) phosphinic acid – Melafen, that was synthesized by works of laboratory of A. E. Arbuzov Institution of organic and physical chemistry [1]. We tested the influence this hydrophilic substance at the wide concentration range (10^{-21}–10^{-3} M) on the functional properties of animal cells. The animal cells must have a number of the variable targets for hydrophilic substance action.

If we deal with the hydrophobic substance, we must notes the certain points. The hydrophobe substances go into the membrane directly or to the hydrophobe pockets into the protein molecules, and are incorporated there, and than the substances are dissipated among hydrophobic phase, or formed its own phase in the cellular compartments [2]. Its action has a negligible specificity without certain targets almost always. The hydrophobe substances have the possibility to show its specificity at protein hydrophobe pocket only. It can influence to the micro viscosity of lipid and protein membrane components [3]. Its actions to the protein structure and functions may be mediated by hydrophobic phase changing or immediate by hydrophobic targets bindings.

On the contrary the hydrophilic substance, as the Melafen, may have the contacts with any charged or polar surfaces. Thus the molecules of Melafen may have their interactions with a number nonspecific targets. But the incorporation to the membrane and dissipate among the lipid molecules doesn't happen. The targets are unknown, but the aftermath's actions of Melafen were known. Lower we will describe the overall results of Melafen actions.

From literary data it is known that the hydrophilic Melafen change the fatty acid composition and lipid and protein microviscosity of cellular microsome and mitochondrial membranes of vegetable [2]. Low concentrations (from 4×10^{-12} to 2×10^{-7} M) of Melafen changed the structural characteristics of plant and animal cell membranes. Melafen changes the microviscosity of free bilayer lipids and annular lipids bounded with protein clusters with different effective concentrations for plant and animal membranes. Melafen decreased the level of lipid peroxide oxidation (LPO) in biological membranes under bed environment conditions also.

The Melafen concentrations that affect to the microviscosity of free and annular lipids, decreases the intensity of LPO processes too [2]. On a foundation of the research conducted, authors assumed that high physiological activity of Melafen is linked to its actions to the physical and chemical state of biological membranes, resulting in change of lipid–protein interaction, influencing the activity by membrane–associated enzymes and channels. Melafen increases the effectiveness of energy metabolism of plant cells [3]. The exclusive influence of Melafen was shown on electron transport in respiratory chain of mitochondria [4], and stress-resistance of vegetables and cereal corn in bed environment conditions as result [5]. The seed treatment by the Melafen or Pyrafen (Melafen analog) reduces the intensity of lipid peroxidation and strengthens the mitochondria energy of six-day pea seedlings, which have undergone stress in conditions of moistening shortage and moderate cooling [6]. Melafen changes the fatty acid composition of mitochondria greatly in the presence of its effective concentrations [7]. These are the reason why the crop yields much rose [8–10]. The influences to the animal microsomes and mitochondrial were showed also [4–6].

Our laboratory investigated the possibility of Melafen influence to the first targets at animal objects: the protein in blood–vascular system, cellular membranes and its components. We found that Melafen under the large concentrations changes greatly (may be loosed) the quarter structure of bovine serum albumin (BSA) – the main soluble protein in blood–vascular system. The intrinsic fluorescence of two tryptophanils that are contained at the BSA molecule was quenched because the tryptophanils became access to the main quencher – H_2O [11].

But to the integral membrane–bounded proteins Melafen influences were negligible even under the large concentrations. Thus the organization of protein microdomains of erythrocyte ghost didn't change, that we registered with aid of differential scanning microcalorymetry (DSK). The thermo induced parameters of usual cytoskeleton protein components of cellular membrane: spektrin, ancyrin, actin, demantin, fragments of ion–channels and other, stood unchanged under the Melafen presence [12].

However the thermo induced parameters of lipids microdomains at membranes multulammelar liposomes, formed by individual neutral saturated phospholipids dimyristoilphosphatidylcholine (DMPC), were changed greatly under the Melafen influence at the wide concentration diapason. Having the using of the specific method of membrane extraction from steady state – the different rates of heat supplied to the cell with our liposomes suspension, we received the glaring picture of dependence of thermally induced parameters: enthalpies, maximum temperatures thermally induced transition and cooperatives transition, under Melafen concentrations in the range 10^{-17}–10^{-3} M. The main extreme was under the 10^{-14}–10^{-8} M of Melafen for all rates (1°C/min, 0.5°C/min; 0.25°C/min; 0.125°C/min) of heat supplied to the cell with our liposomes suspension [13].

But the reciprocal location and density of packaging of membranes at multulammelar liposomes didn't change under the Melafen influence at the wide concentration diapason. At this case the membranes of multulammelar liposomes were formed by egg lecithin that is the mix of natural saturated and no saturated, neutral and charged phospholipids [14]. At that type of membrane, where the belayer structure is reinforced by balancing on charge and by the location of length of fatty acids tails at nature phospholipids mix, the membrane thickness from data of x-ray diffraction method does not change under the Melafen presence [15]. Concerning the microdomains organizations in such mixture we say nothing. It is impossible, since the application of differential scanning calorimetry for lipids mixture have been hampered.

The structure (and functions and fate may be) of native cells were under Melafen influenced too. The small doses 10^{-11}–10^{-5} M Melafen influence *in vivo* on the morphology of the erythrocytes in mice. We obtained the decrease of height, area and volume of the AFM image of red blood cells that registered by AFM (atomic force microscopy) [16].

The low doses 10^{-12}–10^{-5} M Melafen influence to the fate of animal tumor cells *in vitro*. The content of protein "labels of apoptosis": protein–regulator p53 (increase) and antiapoptosis protein Bcl-2 (decrease) that were showed by immunoblotting methods at the EAC cells. This fact indicated that the apoptosis was developed under the 1.5 hour of Melafen action. And similar dozes of Melafen *in vivo* suppressed the growth of Luis carcinoma [17].

Thus, Melafen looses the structure of the soluble protein, changes the microdomaine organization of DMPC membrane and doesn't change the structural organizations of ghost and egg lecithin membranes. But Melafen global changes the erythrocyte morphology and delayed the rate of growth of solid Luis carcinoma.

Taking into account data obtained the A. E. Arbuzov Institution of organic and physical chemistry, about formation by the Melafen in aqueous solutions of supramolecular structures involving of water molecules [18] it can be assumed that just such the structures change the microdomaines organization in attackable delicately organized and labile structures. This assumes may be real only for interactions between biological objects with Melafen in aqua solutions. But in the cellular interior there is not the sufficient amount of free waters molecules for such supramolecular structures formation. It's hard to suppose, that the linkage with Melafen will find stronger, than with nature cellular chelating agents of water. To withdraw the water molecules from the coats of cell components in the presence of Melafen is not likely. On the contrary, to structure the bulk water and the neared membrane water layers it is quite likely.

However, at present chapter we are emphasized the attention to the Melafen aqua solutions actions at the animal cellular level *in vitro*.

This is why we tried to clear up the most possible specific targets for Melafen on animal's cells surface. And several types of easily emitted cells of different etiology or different origin were having picked up. Why this the main task of our work was the investigated the Melafen action to the three types of cellular plasmalemmal receptors – purinoreceptors P2Y, P2X, P2Z that present at the three types of cells: EAK cells, thymocytes and lymphocytes.

Melafen is a melamine salt of bis (hydroximethyl) phosphinic acid). It is a hydrophilic polifunction substance with multitargets for its actions (Fig. 1). There are the phosphoric, hydroximethyl groups and nitrogen contained structures at Melafen molecule, potentially pointed to the certain targets at the biological cells. The pre-treatment of crop seeds by aqueous solutions of Melafen at concentration 10^{-11}–$^{-9}$ M increased the yield of plant production by 11% or more due to the increasing of plants stress tolerance under the bed environment [10]. But the increasing of the concentration of Melafen to 10^{-7} M and higher inhibits the processes of plants growth completely [19]. Therefore, our studies were carried out at a wide range of concentrations (10^{-21} M–10^{-2} M). The main purpose was to determine how the aqueous solutions of Melafen in a wide range of concentrations influence to the function of animal cells *in vitro* only.

It can be assumed that some fragmental structural similarity of Melafen – organophosphorus plant growth regulator, and ATP molecules (Figs. 1 and 2) may define the binding with similar active sites. As result we may observe the both substances actions on purine receptors. But the vectors of consequences of Melafen and ATP molecules actions are opposed to each. Also we must note the global activating influence of ATP applications to the Ca^{2+}-signal transductions at the EAK cells at the 7–8 days of development that was described by Zamai [20] and Zinchenko [21]. The main metabolic pathways points are indicated and at this caze [22]. We deal with the testing of Melafen actions to the ATP-binding purinoreceptors at the base of some similarity of ATP and Melafen molecular construction.

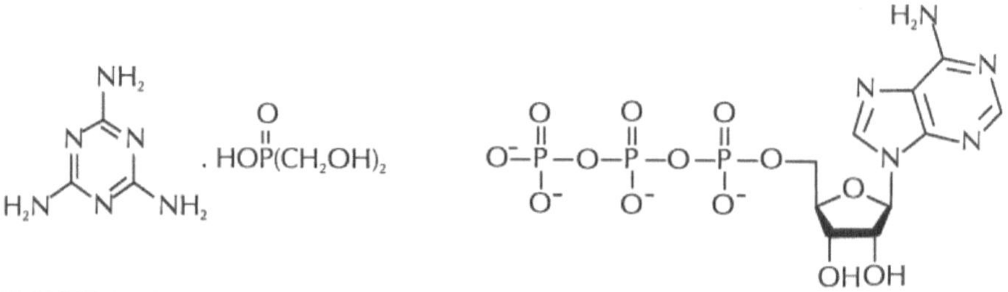

FIGURE 1 Structural formula of Melafen and ATP.

4.2 EXPERIMENTAL

4.2.1 MATERIALS

ATP ("Bochringer"Germany); HENKS (138 mM NaCl; 5.4 mM KCl; 1.2 mM $CaCl_2$; 0.4 mM KH_2PO_4; 0.8 mM $MgSO_4$; 0.3 mM Na_2HPO_4; 5.6 мM D-glucose и 10 мM HEPES pH 7.2), cytrat Na («PAN EKO» Russia); DMPC dimyristoilphosphatidylcholine ("Sigma"); NaCl, KCl, $CaCl_2$, KH_2PO_4, $MgSO_4$, Na_2HPO_4, HEPES ("Sigma), A23187 (Sigma).

4.2.2 METHODS

The cells EAK received by methods [21, 23]. EAK induced at pubescent white mice of males of NMRI introduction intraperitoneally on 10^6 cells of diploid strain of ascetic Ehrlich carcinoma. The cells EAK insulated from mice on 7 days after the transplantation.

Thymocytes insulated by the pulping through the capron net from thymus the white Wister rat. The cells were washed out 3 times centrifugation when 800 rate/min. A 10 min on medium of HENKS when 4°C. The pellet resuspended on medium of HENKS in concentration about $1–5 \times 10^8$ cells/ml.

The pooled fraction of lymphocytes and platelets received after spontaneous deposition erythrocytes from the blood of white Wister rat in the presence of medium of HENKS (1: 1) and 5% citrate Na. The cell viability was estimated on cells overtone 0.04% trypan blue and compiled in all experiences not less than 95%.

The registration of light diffusion by dilute suspension of EAK cell, of thymocytes, of lymphocytes and platelets, was held by the method Cornet [23], modified by Zinchenko [21] at a right angle under the wavelength 510 nm with aid of fluorescent spectrophotometer "MPF–44B" Perkin–Elmer.

4.3 RESULTS AND DISCUSSION

The using of light scattering method allowed us to investigate the overall cellular answers under the Melafen additions without any artificial messengers. Thanks to presented method we tested the action of Melafen to the animal cells and its components under a wide range of concentrations.

The three cellular objects were used, as the dilute cells cellular suspension: ascetic Ehrlich carcinoma (EAC) cells – transformed cells with uncontrolled growth, and normal cells thymocytes and lymphocytes (the white fraction of blood without the platelets and erythrocytes).

First – EAC cells, as a good model of cells with the complete cellular transduction system. The active P2Y purinoreceptors are presented at the cellular plasmalemma surface at the 7–8 days of carcinoma growth [20]. Thus the ATP or ADP additions initiated the Ca^{2+} – signal transduction (Fig. 2). ATP is the first messenger that deals with the extracellular signal transductions pathways. ATP is released to the extracellular space. At least two subtypes of receptors for extracellular ATP are currently known: the

G-protein-coupled P2Y receptors, which are methabotrophic receptors, and the ATP-gated cation channels classified as P2X receptors (and its subspecy – P2Z receptors). EAC cells gave a typical cellular response to a signal (ATP-addition). It sent the signal from the cell surface to inside to the endoplasmic reticulum (ER) $InsP_3$-receptor, and backward to the CRAC at the cell surface.

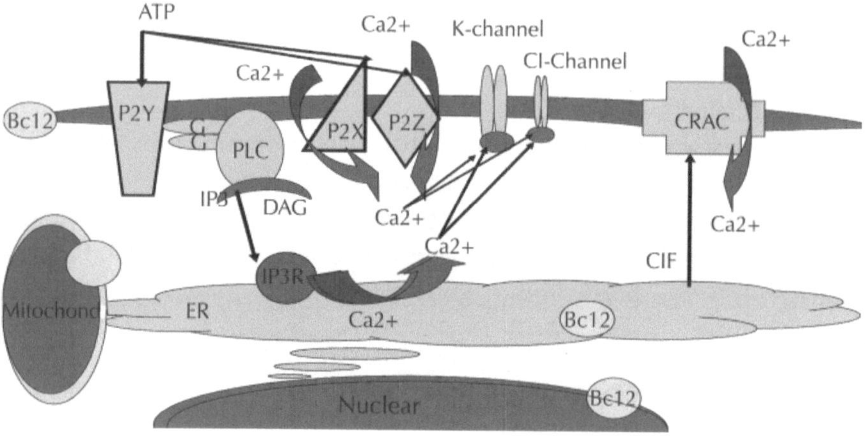

FIGURE 2 The total scheme of P2Y, P2X, P2z –related system of signal transduction (scheme was modified from Alekseeva O.M. 2010.).

At the Fig. 2 we show our cumulative scheme of the purine–dependent Ca^{2+}-transduction at the cells. The metabotropic purinoreceptors P2Y throughout the G-proteins activate PLC that produces $InsP_3$, then $InsP_3$ bounds with $InsP_3$-receptors that release Ca^{2+} from ER. And retrograde signal CIF (related to ER Ca^{2+}-store depletion) go to the cellular surface, and the large flow of Ca^{2+} introduced to the cell throughout the Ca^{2+}-release-activated Ca^{2+} channels (CRAC) [24].

So, after ATP addition occur two large increasing's of intracellular Ca^{2+}-concentrations. As result Ca^{2+}–dependent K^+ и Cl^- -channels are activated. The volume of cells changed, because these channels regulate the cell volume. The compensators H_2O-flows occurred to the cellular interiors. And cellular volumes are increased greatly and bitterly. Than Ca^{2+}-concentrations are decreased quickly, Ca^{2+} is attenuated by mitochondrial, or ER or it is removed from the cell. After than, Ca^{2+} concentration in the cell increases smoothly, again. Ca^{2+}-dependent K^{+} и Cl^{-}-channels are activated again. The cellular volume increase slowly, too. Thus the light scattering of dilute suspension of EAC cells were changed too twice. There are two large increasing of intracellular Ca^{2+}-concentrations, as the result – we obtain two maximums at the light scattering kinetic curves.

We recorded of right angle light scattering of dilute suspension of EAC cells with Perkin-Elmer-44B spectrophotometer at a wavelength of 510 nm. The control samples showed the bimodal cellular responses (Fig. 3). Urgently after the first addition of ATP to the cell suspension the big maximum appeared quickly on the kinetic curve (first peak). Then it bust momentary. After that the low plate is appeared. And than the slow rise up is developing to plate, that is equal or above to the reference levels (second peak). When we repeated the addition of ATP, the picture was repeated. Melafen inhibited the response development by the doze-dependent manner (Fig. 3).

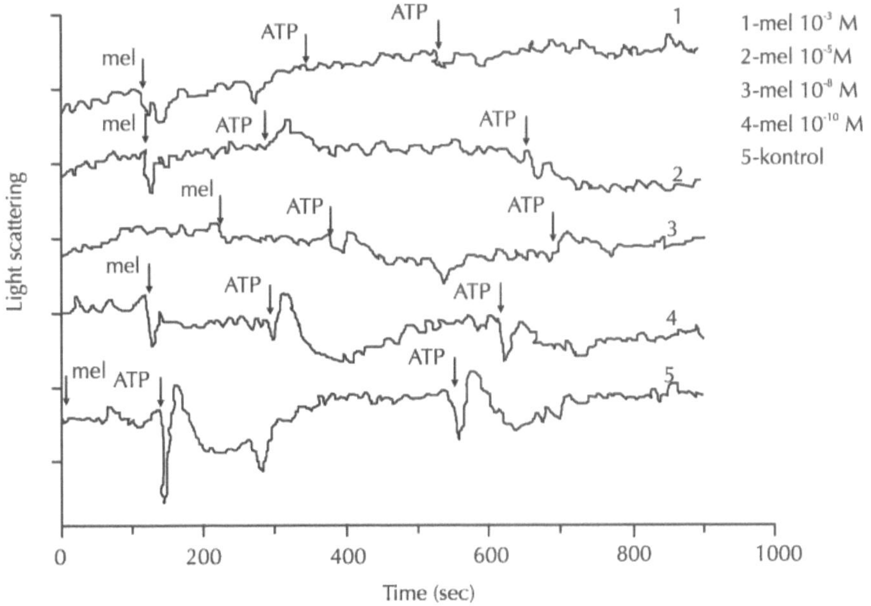

FIGURE 3 The influences of Melafen to the first and second cellular answer of EAC cells. Kinetic curves of light scattering of the cell suspension under the wide concentrations region of Melafen.

But Melafen has a bidirectional effect to the EAC cells (Table 1). We founded that at super low concentrations Melafen (10^{-12}, 10^{-13} M) stimulated the signal transduction, increasing the Ca^{2+}-releasing from intracellular Ca^{2+}-store (the first peak). But under the bigger concentrations the Melafen actions changed its vector and the depressing of the overall cell responses were began. At the case, when the Melafen concentrations were increased, the amplitudes of first and second extremes were decreased. The second cell response – the Ca^{2+}–entering through the plasma membrane (second peak) was not activated by Melafen (under the anywhere concentrations), and it shown the bigger sensitivity to the Melafen action. It was inhibited by the Melafen at the smaller concentrations. Thus, Melafen 10^{-7} M decreased by 50% the first extreme and by 70% – the second extreme. Melafen 10^{-4} M really eliminated the overall cell response fully; it inhibited the purine-dependent Ca^{2+}-transduction – both peaks. Hence, the carcinoma cells that characterized by uncontrolled cell division, can be depressed by such doses of Melafen, which are harmless for erythrocytes that we obtain earlier [16].

TABLE 1 The influences of Melafen to the first (A) and second (B) cellular answer of EAC cells at the wide concentrations diapason.

The melafen influences to EAC cell bimodal responses (1 peak and 2 peak) under the ATP adding

Sample + melafen	1 peac (rel.un.)	Δ (%)	2 Peac (rel.un.)	Δ (%)
Control	**27,5±0,1**	–	**20±0,1**	–
+10-13M	35±0,1	27±0,01	19±0,1	–5±0,01
+10-12M	33±0,1	20±0,01	19±0,1	–5±0,01
+10-11M	27,5±0,1	0	13±0,1	–35±0,01
+10-10M	25±0,1	–9±0,01	13±0,1	–35±0,01
+10-9M	22±0,1	–20±0,01	10±0,1	–50±0,01
+10-8M	16±0,1	–42±0,01	7±0,1	–65±0,01
+10-7M	13±0,1	–53±0,01	6±0,1	–70±0,01
+10-6M	12±0,1	–56±0,01	3±0,1	–85±0,01
+10-5M	11±0,1	–60±0,01	2±0,1	–90±0,01
+10-4M	7±0,1	–74,5±0,01	0	–100
+10-3M	0	–100	0	–100

Thus Melafen influence on two targets surfaces of cells simultaneously – on purinoreceptors PY2 and on CRAC, considerably reducing their activity in plants–growth stimulated doses. Really Melafen changes the functions of surface receptors and intracellular signal transduction.

It will be interesting to test the Melafen influence to the overall cellular answer of the normal cells that may be activated by ATP additions. Because that we recorded of right angle of light scattering of dilute suspension thymocytes and leucocytes. These cells have ranking among P2Y receptors have another types of purinoreceptors nonmethabotrophic channel formers P2X (at leukocytes) [25] and its modification P2Z (at thymocytes) [26]. The compositions of groups of P2 receptors at thymocytes are showing both P2Z and P2X receptor activation characteristics are in depending on the stage of cellular growth. The additions of ATP to the thymocytes suspension caused the two-phase of change of cell volume (Fig. 4).

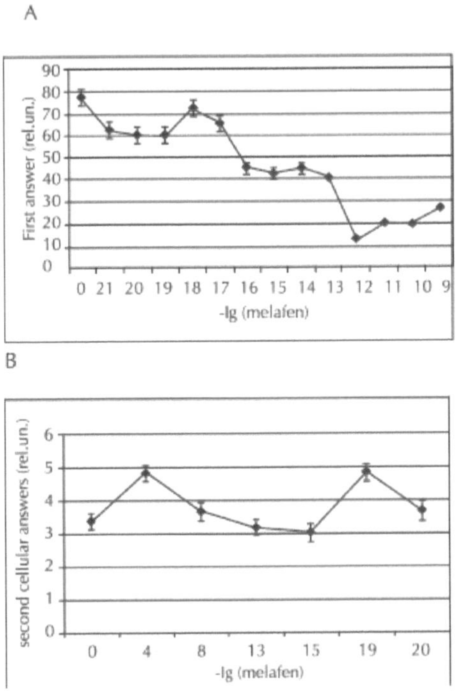

FIGURE 4 The influence of Melafen to the first (A) and second (B) cellular answers of cell suspension of rat thymocytes under the wide concentrations region of Melafen. The amplitude of light scattering risings after the ATP additions to the cell suspension in the dependence of Melafen concentrations.

Melafen influences on both phases. The questions, what certain points of Ca^{2+}-transduction at rat thymocytes are involved to Melafen effects, don't were closed. We may conclude that P2X and P2Z Ca^{2+}-channel formed receptors are susceptible to Melafen influence under a wide region of concentrations.

Additions of ATP to leucocytes suspension caused releasing of Ca^{2+} from intracellular stores through activating of metabotropic P2Y purine receptors (at the medium of measurement Ca^{2+} don't present). Thus we eliminated the possibility of introducing to the cell interior the extra cellular Ca^{2+}. We used the measurements medium without the Ca^{2+} ions. At this case the non-methabotrophic P2X- and P2Z- channel formers were silent structure. And as itself will lead the P2Y methabotrophic receptor? What activate the Ca^{2+}-ions releasing from intracellular Ca^{2+}-stores (endoplasmic reticulum). The Melafen attendance shortens the time the phase lag up to cellular answer development. The EAC cells behave analogously. Its go out to the stable behavior in the presence of Melafen faster. Impact on Melafen to the channel-former P2X the leucocytes receptors are coming clear to up (Fig. 5).

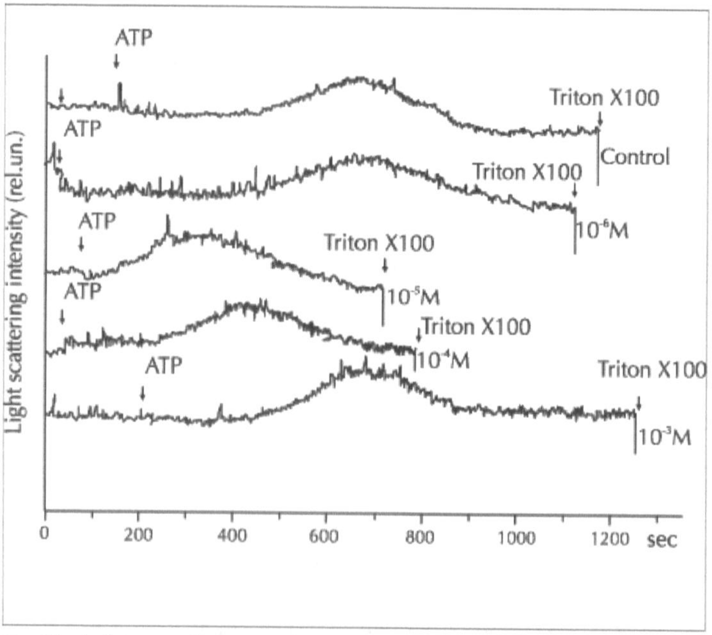

FIGURE 5 The influences of Melafen to the lag-phase of cellular answers of leucocytes. Kinetic curves of light scattering of the cell suspension under the wide concentrations region of Melafen.

KEYWORDS

- **Leukocytes**
- **Melafen**
- **Perkin-Elmer-44B spectrophotometer**
- **Thymocytes**

REFERENCES

1. Fattachov, S. G.; Reznik, V. S.; Konovalov, A. I. In Set of articles. *Melamine Salt of Bis(hydroxymethyl)phosphinic Acid. (Melaphene) As a New Generation Regulator of Plant growth regulator.* Reports of 13-th International Conference on Chemistry of Phosphorus Compounds; Saint Petersburg, 2002.
2. Alekseeva, O. M.; Narimanova, R. A.; Yagolnik, E. A.; Kim, Yu. A. *Phenosan Influence and its Hybrid Derivative on Membrane Components.* Bashkir State University All-Russian Conference Biostimulators in Medicine and Agriculture March 15 2011 Ufa pp 15–19 article in collection.
3. Vekshina, O. M.; Fatkullina, L. D.; Kim, Yu. A.; Burlakova, E. B. "The changes of structure and functions of erythrocyte membranes and cells of ascetic Ehrlich carcinoma when action of hybrid antioxidant of rising generation ICHFAN–10" *Bull. Exp. Biol. Med.*, **2007**. *4*, 402–406.
4. Zhigacheva, I. V.; Fatkullina, L. D.; Burlakova, E. B.; Shugaev, A. G.; Generozova, I. P.; Fattakhov, S. G.; A. I. Konovalov. "Effects of the Organophosphorus Compound "Melaphen" – Plant Growth Regulator – on Structural Characteristics of Membranes of Plant and Animal Origin" *Biol. Membrane* **2008**, *25*(2), 128–134.
5. Zhigacheva, I. V.; Fatkullina, L. D. ; Shugaev, A. G.; Fattakhov, S. G.; Reznik, V. S.; Konovalov, A. I. "Melafen and energy status of cells of plant and animal origin." *Dokl. Biochem. Biophys.*, **2006**, *409*, 200.
6. Zhigacheva, I. V.; Evseenko, L. S.; E. B. Burlakova; Fattakhov, S. G; Konovalov, A. I. "Influence of organophosphorus plant growth regulator on electron transport in respiratory chain of mitochondria" *Dokl. Akademii. Nauk.*, **2009**, *427*(5), 693–695.
7. Zhigacheva, I. V.; Fatkullina, L. D.; Rusina, I. F.; Shugaev, A. G.; Generozova, I. P.; Fattakhov, S. G.; Konovalov, A. I. "Antistress properties of preparation Melaphen" *Dokl. Akademii. Nauk.*, **2007**, *414*(2), 263–265.
8. Zhigacheva, I. V.; et al. "Fatty acids membrane composition of mitochondria of pea seedlings in conditions of moistening shortage and moderate cooling" *Dokl. Akademii. Nauk.*, **2011**, *437*(4), 558–560.
9. Zhigacheva, I. V.; et al. "Insufficient moistening and Melafen change the fatty–acid membrane composition of mitochondria from pea seedlings" *Dokl. Akademii. Nauk.*, **2010**, *432*(1), 124–126.

10. Kostin, V. I.; Kostin, O. V.; Isaichev, V. A. *Research Results Concerning the Application of Melafen When Cropping*. Investigation State and Utilizing Prospect of Growth Regulator "Melafen" in Agriculture and Biotechnology. Kazan, 2006, pp 27–37.
11. Alekseeva, O. M.; Yagolnik, E. A.; Yu.Kim, A.; Albantova, A. A.; Mil, E. M.; Binyukov, V. I.; Goloshchapov, A. N.; Burlakova, E. B. *Melafen Action on Some Links of Intracell Signalling of Animal Cells.* International Conference "Receptors and the Intracell Signalling", 24-26 May, 2011 Puschino, pp 435–438.
12. Alekseeva, O. M.; Krivandin, A. V.; Shatalova, O. V.; Yu.Kim, A.; Burlakova, E. B.; Goloshapov, A. N.; Fattakhov, S. G. *No Lipid Microdomains Destruction, But Stabilization by Melafen Treatment of Dimyristoilphosphatidylcholine Liposomes.* International Symposium "Biological Motility: from Fundamental Achievements to Nanotechnologies" Pushchino, Moscow Region, Russia, May 11–15, 2010, pp 8–12.
13. Alekseeva, O. M.; Shibryaeva, L. S.; Krementsova, A. V.; Yagolnik, E. A.; Yu.Kim, A.; Golochapov, A. N.; Burlakova, E. B.; Fattakhov, S. G.; Konovalov, A. I. "The aqueous melafen solutions influence to the microdomains structure of lipid membranes at the wide concentration diapason." *Dokl. Akademii. Nauk.*, **2011**, *439*(4), 548–550.
14. Tarakhovsky, Yu. S.; Kuznetsova, S. M.; Vasilyeva, N. A.; Egorochkin, M. A.; Kim, Yu. A. "Taxifolin interaction (digidroquercitine) with multilamellar liposomes from dimitristoyl phosphatidylcholine" *Biophysicist*, **2008**, *53*(1), 78–84.
15. Alekseeva, O. M.; Krivandin, A. V.; Shatalova, O. V.; Rykov, V. A.; Fattakhov, S. G.; Burlakova, E. B.; Konovalov, A. I. "The Melafen–Lipid– Interrelationship Determination in phospholipid membranes" *Dokl. Akademii. Nauk.*, **2009**, 427(6), 218–220.
16. Binyukov, V. I.; Alekseeva, O. M.; Mil, E. M.; Albantova, A. A.; Fattachov, S. G.; Goloshchapov, A. N.; Burlakova, E. B.; Konovalov, A. I. "The investigation of melafen influence on the erythrocytes in vivo by AFM method." *Dokl. Biochem. Biophys.*, **2011**, 441, 245–247.
17. Alekseeva, O. M.; Erokhin, V. N.; Krementsova, A. V.; Mil, E. M.; Binyukov, V. I.; Fattachov, S. G.; Yu.Kim, A.; Semenov, V. A.; Goloshchapov, A. N.; Burlakova, E. B.; Konovalov, A. I. "The investigation of melafen low dozes influence to the animal malignant neoplasms *in vivo* and *in vitro*"*Dokl. Akademii. Nauk.*, **2010**, 431(3), 408–410.
18. Rizkina, I. S.; Murtazina, L. I.; Kiselyov, J. V.; Konovalov, A. I. "Property of supramolecular nanoassociates, formed in aqueous solutions low and ultra–low concentrations of biologically active substance"*DAN*, **2009**, 428(4), 487–491.
19. Osipenkova, O. V.; Ermokhina, O. V.; Belkina, G. G.; Oleskina, Yu. P.; Fattakhov, S. G.; Yurina, N. P. "Effect of Melafen on Expression of *Elip1* and *Elip2* Genes Encoding Chloroplast Light–Induced Stress Proteins in Barley" *Pract. Biochem. Microbiol.*, **2008**, *44*(6), 701–708.
20. Zamai, A. C.; et al. *The ATP Influence to the Tumor Ascetic Cells at Different Stages of Cellular Growth.* Conference "Reception and Intracellular Signalization", "The ATP influence to the tumor ascetic cells at different stages of cellular growth" Puschino 2005, pp 48–51.
21. Zinchenko, V. P.; Kasimov, V. A.; Li, V. V.; Kaimachnikov, N. P. "Calmoduline inhibitor of R24571 induces the short–time entry Ca^{2+} and the pulsed secretion ATP in cells of ascetic Ehrlich carcinoma" *Biophysicist*, **2005**, *50*(6), 1055–1069.

22. Pedersen, S. F.; Pedersen, S. I.; Lambert, H.; Hoffmann, E. K. "P2 receptor–mediated signal transduction in Ehrlich ascites tumor cells." *Biochim. Biophys. Acta.*, **1998**, *1374*(1-2), 94–106.
23. Cornet, M.; Lambert, I. H.; Hoffman, E. K. «Relation between cytosceletal hypoosmotic treatment and volume regulation in Erlich ascites tumor cells». *J. Membr. Biol.* **1993**, *131*, 55–66.
24. Artalej, O. A.; Garcia-Sancho, J. "Mobilization of intracellular calcium by extracellular ATP and by calcium ionophores in the Ehrlich ascities tumor cells." *Biochem. Biophys. Acta.,* **1988**, *941.9*, 48–54.
25. Di Virgilio, F.; et al. *Blood*, **2001**, *97*, 587–600.
26. Nagy, P. V.; Fehér, T.; Morga, S.; Matkó, J. "Apoptosis of murine thymocytes induced by extracellular ATP is dose- and cytosolic pH-dependent." "Nucleotide receptors: an emerging family of regulatory molecules in blood cells." *Immunol. Lett.*, **2000**, *72*(1), 23–30.

CHAPTER 5

ON PHYSICO-CHEMICAL PROPERTIES OF ASCORBIC ACID AND PARACETAMOL HIGH-DILUTED SOLUTIONS

F. F. NIYAZY, N. V. KUVARDIN, E. A. FATIANOVA, and S. KUBICA

CONTENTS

5.1 INTRODUCTION

During last decades there is a tendency of the growing interest to the study of high-diluted solutions of bioactive substances. Besides, concentration ranges under study are related to the category of supersmall or, in other words, 'illusory' concentrations. Such solutions, unlike more saturated ones, but with pretherapeutic content of active substance, may possess high biological activity.

Use of bioobjects to reveal supersmall doses effect (SSD) in the substances is the most exact method today allowing not only to define the effect existence and to find out how it shows itself, but to determine concentration ranges of its action. However, the use of this method will entail great difficulties. In this connection it is necessary to search for other methods, including physico-chemical ones, allowing to define presence of SSD effect in the compounds vif only at the stage of preliminary tests. Study of physico-chemical bases of SSD effect display is one of the most interesting questions in the given sphere of research and attracts attention of many scientists [1–4].

Antineoplastic and antitumorous agents, radioprotectors, neutropic preparations, neupeptides, hormones, adaptogenes, immunomodulators, antioxidants, detoxicants, stimulants and inhibitors of plants growth and so on are included into the group of bioactive substances possessing SSD effect. Study of high-diluted solutions of bioactive compounds was carried out on one-component solutions, that are ones containing only one solute. But now, mainly multicomponent medical preparations, possessing several therapeutic actions, are used in medicine. So, preparations of analgesic-antipyretic action are possibly used at sharp respiratory illnesses, accompanied by muscular pain and rise of temperature. It is possible that effects of multicomponent medical preparation in supersmall concentrations and its separate components will differ.

We studied some physico-chemical properties of high-diluted aqueous solutions of paracetamol and ascorbic acid with the purpose of finding out peculiarities of their change at solutions dilution and also of definition of possible concentration ranges of SSD effect action. Paracetamol and ascorbic acid are the components of combined analgesic-antipyretic and antiinflammatory preparations [5]. Ascorbic acid is used as fortifying and stimulating remedy for immune system. Paracetamol (acetominophene) has anaesthetic and febrifugal effect.

There have been prepared one-component solutions of ascorbic acid and paracetamol, two-component solutions of paracetamol with ascorbic acid (relation of dissoluted compounds in solutions is 1:1), in the following concentrations of dissoluted substances (mol/l): 10^{-1}, 10^{-3}, 10^{-5}, 10^{-7}, 10^{-9}, 10^{-11}, 10^{-13}, 10^{-15}, 10^{-17}, 10^{-19}, 10^{-21}, 10^{-23}. Water cleansed by reverse osmosic was used as solvent. Solutions were prepared by successive dilution by 100 times using classical methods. Initial solution was 0,1M one. Before choosing of solution portion for the following dilution the sample was subjected to taking antilogs.

Prepared solutions were studied by cathetometric method of substances screening, the ones acting in supersmall concentrations, and also by method of electronic spestroscopy.

Cathetometric method of substances screening is based on the study of the change of solution meniscus height in capillary [6]. Results of measuring meniscus height of ascorbic acid, paracetamol, paracetamol with ascorbic acid solutions are given in Figs. 1, 2, and 3, accordingly.

Meniscus height of dilution with ascorbic acid concentration of 10^{-3} mol/l was 0.8 mm (Fig. 1). During further dilution value of meniscus height in the capillary reduces, but changes are not uniquely defined. The most lowering of meniscus height is observed in samples in which content of ascorbic acid is 10^{-9}, 10^{-13}, 10^{-15}, 10^{-17} mol/l.

Meniscus height reduces on an average by 13.75%. Lowering of meniscus height by 23.7%, in comparison with more concentrated solution is also observed for the sample with ascorbic acid content of 10^{-23} mol/l.

Equivalent lowering of meniscus in the capillary has been stated for solutions with ascorbic acid concentration of 10^{-9} mol/l and 10^{-13}–10^{-17} mol/l. Between these concentration ranges there is concentration range in which there are no essential changes. Being based on literature data and also on cathetometric method for screening of substances activity in supersmall doses, we can assume that dilutions of ascorbic acid in concentrations of 10^{-9} mol/l and 10^{-13}–10^{-17} mol/l show biological activity regarding bioobjects.

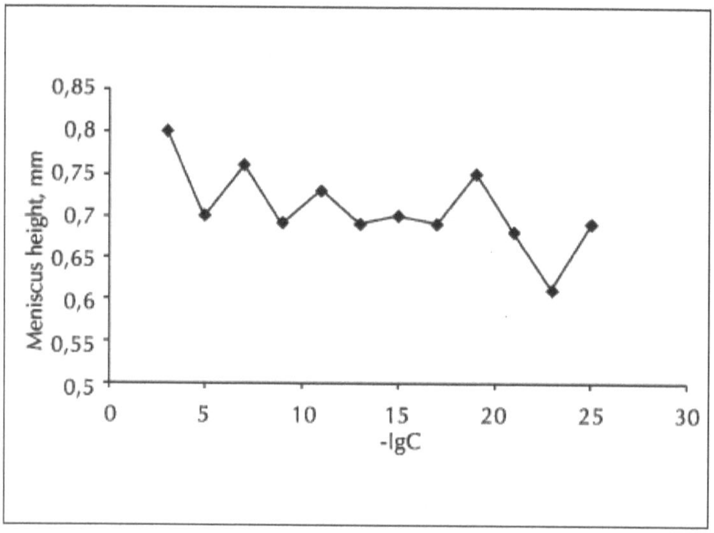

FIGURE 1 Values of meniscus height in the capillary of ascorbic acid solutions (concentration, mol/l).

While studying solutions of paracetamol by cathetometric method it can be observed that in first dilutions by 100 times (paracetamol concentrations being 10^{-3}–10^{-7} mol/l) height of meniscus reduces slightly, maximum 5.7%, regarding meniscus height of the first dilution. This slight lowering of meniscus height in the capillary is caused by rather large dose of active substance in these dilutions. However, sudden lowering of meniscus height in the capillary up to 0.71 mm, which is by 19.3% lower than meniscus height of the first dilution, is observed at diluting paracetamol solution up to the concentration of 10^{-9} mol/l (Fig. 2).

The same dependence is observed for dilution of paracetamol solution with concentration of 10^{-15} mol/l. So, at this concentration meniscus height is 0.7 mm.

Growth of meniscus height in the capillary is observed further for solutions with the following dilution. This process is motivated by very high dilution that is the solution, according to its composition and properties, tries to attain the state of pure solvent.

These changes are polymodal dependence effect – concentration, described for different substances and different properties in domestic and world scientific literature. From the figure and its description it is clearly seen that there are concentrations of paracetamol solution of 10^{-9} mol/l and 10^{-15} mol/l, for which there has been stated essential lowering of meniscus in the capillary regarding other concentrations. Between these concentration ranges there is a concentration range in which there are no essential changes, this range being the so-called "dead zone."

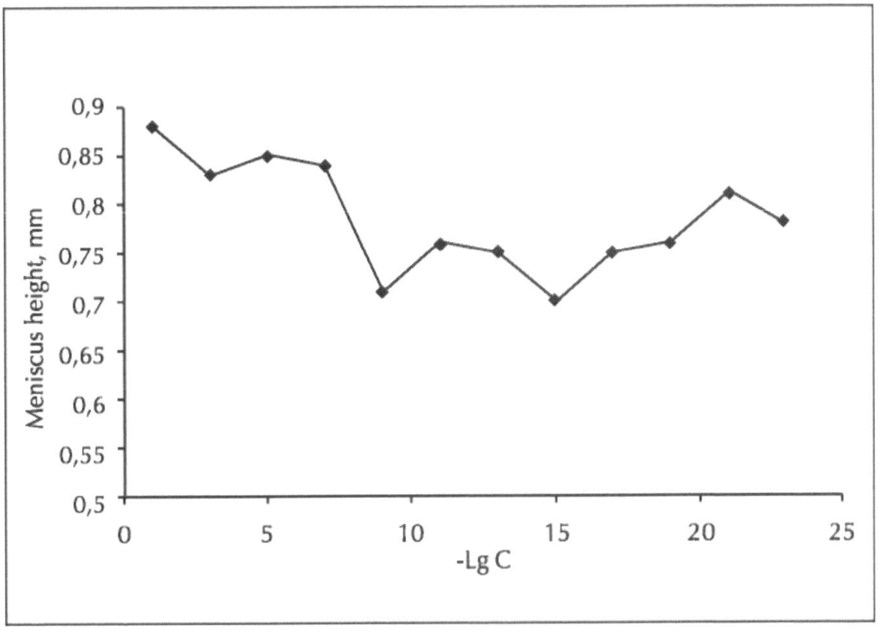

FIGURE 2 Values of meniscus height in the capillary of paracetamol solution (concentration, mol/l).

We have studied solutions of ascorbic acid and paracetamol mixture. It is necessary to note that not uniquely defined change of meniscus height is observed in solutions containing simultaneously two active substances. Meniscus of one -component solution is narrower than that of water, but in two-component solutions both reduction and increase of meniscus height values are possible in comparison with water (Fig. 3).

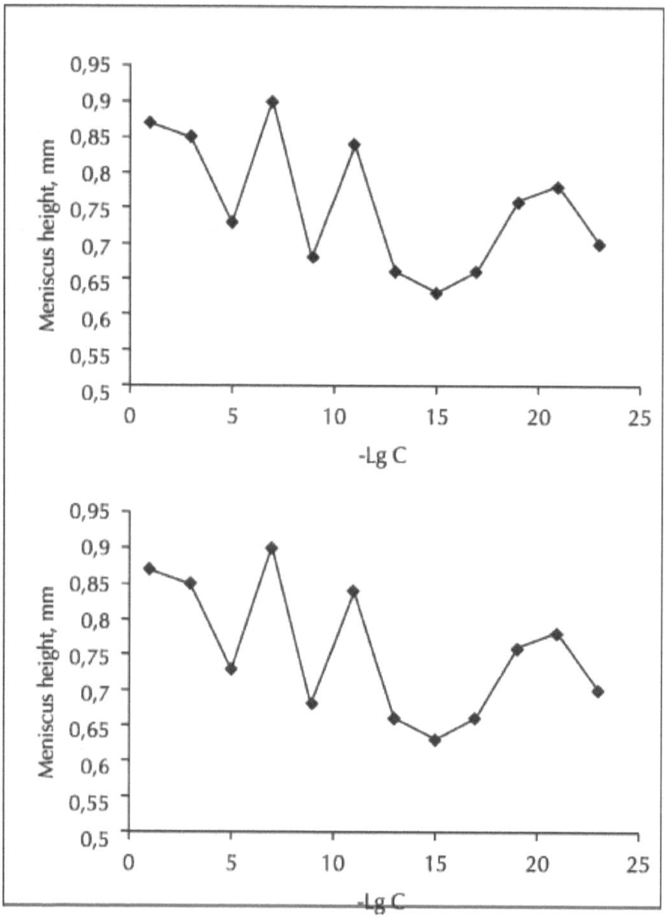

FIGURE 3 Values of meniscus height in the capillary of paracetamol solutions with ascorbic acid.

Lowering of meniscus height is observed in solutions with concentration of paracetamol with ascorbic acid 10^{-5}, 10^{-9}, 10^{-13}, 10^{-15}, 10^{-17} mol/l on 11.5%, 17.5%, 20%, 23.6% and 20% correspondingly. It is possible to distinguish two concentration ranges in the field of supersmall concentrations where reduction of meniscus height in the capillary takes place: this is -10^{-9} mol/l and 10^{-13}–10^{-17} mol/l. Received concentration ranges coincide with data of cathetometric studies of ascorbic acid and paracetamol solutions (Table 1).

TABLE 1 Results of study of ascorbic acid, paracetamol, paracetamol with ascorbic acid solutions of wide concentration range by method of electronic spectroscopy and cathetometric method.

Method of study	Concentration ranges, mol/l		
	Ascorbic acid solutions	**Paracetamol solutions**	**Solutions of paracetamol with ascorbic acid**
Cathetometric method of screening	10^{-9}, 10^{-13} - 10^{-17}, 10^{-23}	10^{-9}, 10^{-15}	10^{-9}, 10^{-13} - 10^{-17}
Electronic spectroscopy	10^{-9}, 10^{-15}, 10^{-19}	10^{-9}, 10^{-13}, 10^{-15}, 10^{-21}	10^{-9}, 10^{-13}, 10^{-15}, 10^{-21}

Increase of meniscus height in comparison with the solvent is observed in solutions with concentrations of diluted compounds of 10^{-1}, 10^{-3}, 10^{-7}, 10^{-11} mol/l. Growth of meniscus height above such value of the solvent can be explained by the change in SSD effect display.

We have analyzed ultra-violet spectra of solutions of ascorbic acid, paracetamol, paracetamol with ascorbic acid. All the spectra were read from spectrophotometer Cary 100, UV-Visible Specrtophotometer in the range of 200–350 nm.

The most absorption in all solutions takes place in short waves part of the effective wave band. More concentrated solutions (10^{-1}, 10^{-3} mol/l) of ascorbic acid have maximum absorption in the length interval of 220–280 nm, of paracetamol – 220–310 nm, paracetamol with ascorbic acid – 200–320 nm. Tendency to narrowing of absorption field, reduction of optical density value and quantity of tops up to one or two is registered in all solutions, under study, as content of diluted substance reduces.

Solutions spectra with ascorbic acid concentration of 10^{-15} and 10^{-19} mol/l are equal in form and differ from spectra of other dilutions. Lowering of ascorbic acid concentration is not accompanied by uninterrupted reduction of optical density value. Irregular growth of absorption in comparison with more concentrated solutions is observed in solutions with ascorbic acid concentrations of 10^{-9}, 10^{-15}, 10^{-19} mol/l (Table 2).

TABLE 2 Maximum values of absorption of ascorbic acid solutions in ultra-violet field.

Concentration, mol/l	Absorption maximum	
	Wave length, nm	A
10^{-3}	230–270	0.79
10^{-5}	265	0.04
10^{-7}	220–230	0.012
10^{-9}	225	0.03
10^{-11}	235	0.01
10^{-13}	220	0.015
10^{-15}	220	0.04
10^{-17}	220–240	0.005
10^{-19}	220	0.03
10^{-21}	220	0.005
10^{-23}	220	0

Community in given structures may be assumed taking into account coincidence of waves length and values of optical density for solutions with concentrations 10^{-9}, 10^{-15}, 10^{-19} mol/l.

Gradual reduction of paracetamol concentration in solutions is not accompanied by the same reduction of optical density value. Growth of absorption in comparison with more concentrated solutions is observed for solutions with concentrations 10^{-9}, 10^{-13}, 10^{-15}, 10^{-21} mol/l (Table 3). All this allows assuming the rise of structural changes in these solutions.

TABLE 3 Maximum values of absorption of paracetamol solutions in ultra-violet field.

Concentration, mol/l	Absorption maximum	
	Wave length, nm	A
10^{-1}	230	0.58
10^{-3}	270	0.53
10^{-5}	245	0.15

TABLE 3 *(Continued)*

Concentration, mol/l	Absorption maximum	
	Wave length, nm	A
10^{-7}	240	0.02
10^{-9}	235	0.015
10^{-11}	240	0.007
10^{-13}	245	0.007
10^{-15}	220–245	0.01
10^{-17}	235	0.003
10^{-19}	290	0.001
10^{-21}	225	0.02
10^{-23}	230	0.005

Spectrum of ascorbic acid and paracetamol solution in concentrations of 10^{-1} mol/l is characterized by wide absorption band in the field of 200–320 nm. Dilution of 0.1 M solution by 100 times is accompanied by reduction of absorption field width up to 200–280 nm without changing of spectrum form and intensity of absorption on peaks.

Changing of spectrum form accompanied by essential reduction of absorption from 4 to 0.095 with the maximum on the length 243 nm takes place while diluting solution of ascorbic acid with paracetamol up to the concentration 10^{-5} mol/l (Table 4).

TABLE 4 Maximum values of absorption of ascorbic acid with paracetamol solutions (1:1) in ultra-violet field.

Concentration, mol/l	Absorption maximum	
	Wave length, nm	A
10^{-1}	237	4.651
10^{-3}	235	4.208
10^{-5}	243	0.095
10^{-7}	205	0.242

TABLE 4 *(Continued)*

Concentration, mol/l	Absorption maximum	
	Wave length, nm	A
10^{-9}	205	0.293
10^{-11}	207	0.084
10^{-13}	206	0.367
10^{-15}	205	0.186
10^{-17}	207	0.086
10^{-19}	207	0.089
10^{-21}	206	0.261
10^{-23}	207	0.105

Two peaks on lengths 205–206 nm and 270–273nm are shown on spectra of solutions with paracetamol and ascorbic acid concentrations 10^{-7}, 10^{-9}, 10^{-13}, 10^{-15}, 10^{-21} mol/l. One peak on the length 207 nm is shown on spectra of solutions with paracetamol and ascorbic acid concentrations 10^{-11}, 10^{-17}, 10^{-19}, 10^{-23} mol/l. Display of maximum on the length 270 nm occurred under conditions that absorption on maximum 205 nm was not less than 0.1.

Increase of optical density is observed in spectra of solutions with paracetamol and ascorbic acid concentrations 10^{-7}, 10^{-9}, 10^{-13}, 10^{-19}, 10^{-21} mol/l.

KEYWORDS

- **Cathetometric**
- **Dead zone**
- **Illusory**
- **Paracetamol**

REFERENCES

1. Konovalov, A. I. Physico-chemical mystery of super-small doses. Chemistry and life. **2009**, *2*, 5–9.
2. Kuznetsov, P. E.; Zlobin, V. A.; Nazarov, G. V. On the question about physical nature of superlow concentrations action. Heads of reports at III International Symposium "Mechanisms of super-small doses action," Moscow, December 3–6, 2002; p. 229.
3. Chernikov, F. R. Method for evaluation of homoeopathic preparations and its physico-chemical foundations. Materials of Congress of homoeopathists of Russia. Novosibirsk, 1999; p. 73.
4. Pal'mina, N. P. Mechanisms of super-small doses action. Chemistry and life. **2009**, *2*, 10.
5. Maslikovsky, M. D. Combined analgesic-febrifugal and anti-inflammatory preparations. M., 1995, p.208
6. Niyazi, F. F.; Kuvardin, N. V. Method for determining abilities of bioactive substances to display "super-small doses" effect. Patent N 2346260 of 10.02.2009.

CHAPTER 6

PRACTICAL HINTS ON TESTING VARIOUS HEXAHYDROPYRIDOINDOLES TO ACT AS ANTIOXIDANTS

KATARHNA VALACHOVБ, MÁRIA BAŇASOVÁ, ĽUBICA MACHOVÁ, IVO JURÁNEK, ŠTEFAN BEZEK, and LADISLAV ŠOLTÉS

CONTENTS

6.1 INTRODUCTION

Inflammation of synovial joints is accompanied by a decrease in the viscosity of synovial fluid (SF), in which hyaluronan (HA, called also hyaluronic acid; Fig. 1) is the macromolecular component, which imparts the SF viscosity. It has been hypothesized that reactive oxygen species (ROS), produced by (infiltrated) neutrophils, may be responsible for the degradation of HA macromolecules within the SF of patients suffering from rheumatoid arthritis (RA). *In vitro* studies underline that of the various individual ROS, hydroxyl radicals – •OH are the most degradative against the HA chain [1].

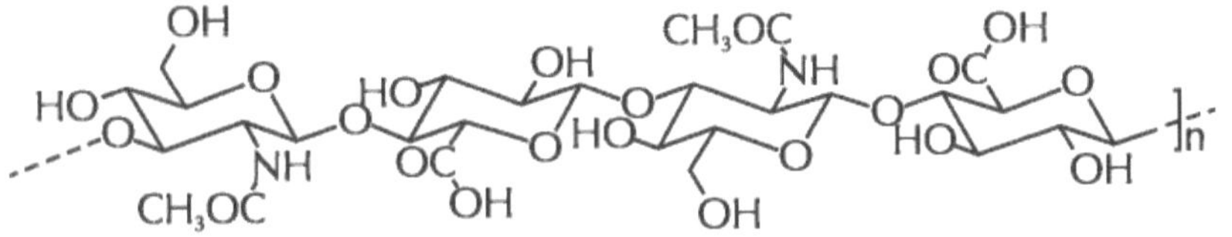

FIGURE 1 Hyaluronan – acid form.

Several *in vitro* generators of •OH radicals investigated, exploit primarily the system comprising hydrogen peroxide and reduced transition metal ions, mostly Fe(II) – the Fenton reactants. One fact has however been criticized on applying the Fenton's generator of •OH radicals, namely that the experiments at which H_2O_2 solution is applied like a "bolus" are really far from pathophysiological conditions, which are actually involved also in inflammation of synovial joints. *In vivo*, the process of generation of hydrogen peroxide as a pre-cursor of further ROS should run continually and the *in situ* "born" H_2O_2 molecules should be stepwise converted to •OH radicals. Of such continual generators of hydrogen peroxide and/or hydroxyl radicals we would like to call attention to the so called Weissberger's biogenic oxidative system – WBOS (cf. Scheme 1 [2–7].

AscH$^-$ and DHA denote ascorbate anion and dehydroascorbate.

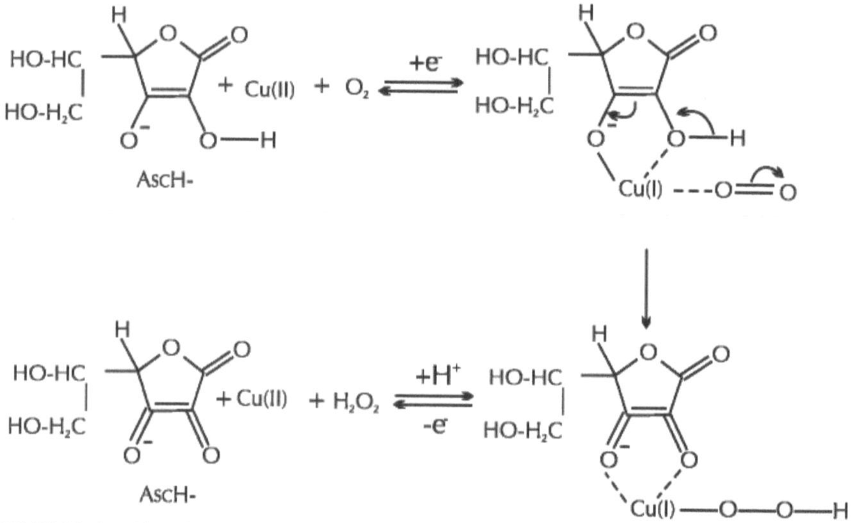

SCHEME 1 Chemistry of Weissberger's biogenic oxidative system: Hydrogen peroxide is generated by oxidation of ascorbate at catalytic action of Cu(II) ions (adapted from Ref. [8]).

6.1.1 WBOS – CONDITION SETTINGS

Taking into account the well known fact that in most human tissues, including that of SF, the concentrations of ascorbate never exceed the value of 200 μM [9] (in mean ≈ 100 μM), settings of ascorbate concentration in the WBOS to 100 μM is comprehensible. The second variable in the WBOS, namely the level of Cu(II) ions, can fall to a few μM /l (4.33 μM – as determined in *post mortem* collected SF from subjects without evidence of connective tissue disease [10]. Yet, as reported by Naughton et al. [11], the level of copper ions in the SF ultrafiltrates of RA patients equals 0.125 ± 0.095 μM. Therefore, the application of 0.1 μM of cupric ions (in the form of e.g., $CuCl_2$) as the second variable in the WBOS could well model pathophysiological conditions. Thus it can be claimed that the Cu(II) concentration equaling 0.1 μM along with 100 μM of ascorbate is a proper setting, especially valuable to model the situation within SF during the early stage of acute phase inflammation of synovial joints [12]. Under such [Cu(II)]:[ascorbate] setting, it is evident (cf. Scheme 1) that within one single reaction cycle 0.1 μmol H_2O_2 is yielded, which by action of e.g. the intermediate Cu(I)-complex is altered to 0.1 μmol of $^{\bullet}OH$ radicals. One fraction of these radicals reacts *in statu nascendi* with the present (target) HA macromolecules, while another fraction is scavenged by the ascorbate excess according to the reaction.

$$AscH^- + {}^{\bullet}OH \rightarrow Asc^{\bullet -} + H_2O \quad (1)$$

where $Asc^{\bullet -}$ denotes ascorbyl anion radicals, which disproportionate immediately, yielding ascorbate and DHA.

6.1.2 WBOS UTILITY

For testing the efficiency of a substance in function as preventive antioxidant and/or scavenger of the generated $^{\bullet}OH$ radicals (the substance H atom donating property) one should take into account that on applying a potentially "perfect/absolute inhibitor" of HA degradation induced by the $^{\bullet}OH$ radical, the experimental curve should copy the gray one (cf. Appendix, Fig. 1). On the contrary, the substance with nil $^{\bullet}OH$ radical trapping properties (donor of no H atom) should yield a curve, which superimposes that of the black one (cf. Appendix, Fig. 1). Results of measurements of time dependencies on the HA solution dynamic viscosity falling within the region between the gray and black curves relate to greater or lower efficiency of the test substance to act as a preventive antioxidant.

Since the primary/initiation step of the HA reaction with $^{\bullet}OH$ radical should yield a reactive *C*-type macroradical (hereafter denoted as $A^{\bullet}$), under aerobic conditions the $A^{\bullet}$ macroradicals should react with dioxygen yielding a peroxyl-type macroradical (hereafter denoted as $AOO^{\bullet}$).

$$HA + {}^{\bullet}OH \rightarrow A^{\bullet} + H_2O$$

$$A^{\bullet} + O{=}O \rightarrow AOO^{\bullet}$$

and the $AOO^{\bullet}$ macroradical immediately starts the propagation phase of the free-radical chain degradation of HA macromolecules. This fact could, however, be exploited for testing substances acting as chain-breaking antioxidants.

For testing the efficiency of a substance in function as a chain-breaking antioxidant and/or a scavenger of the propagated $AOO^{\bullet}$

radicals, one should take into account that a "perfect/absolute chain-breaker" of the propagation phase of the free-radical HA degradation should result in an experimental curve, as represented by the gray one (cf. Appendix, Fig. 2). On the other hand, the substance with nil chain-breaking properties (donor of no H atom) should yield a curve, which superimposes that of the black one (cf. Appendix, Fig. 2). Results of time-dependent measurements of dynamic viscosity of the HA solution falling within the region between the gray and black curves relate to a greater or lower efficiency of the test substance to act as a chain-breaking antioxidant.

Both spectrophotometric methods, namely the ABTS and DPPH decolorization assays, cannot be classified as absolute methods [13]. By the reduction action of the test substance, the color indicator ($ABTS^{\bullet+}$ or $DPPH^{\bullet}$) is converted to the final compound ABTS or $DPPH^{-}$ according to the reactions

$$ABTS^{\bullet+} + e^{-} \rightarrow ABTS \quad \text{(one electron reduction)}$$

$$DPPH^{\bullet} + e^{-} \rightarrow DPPH^{-} \quad \text{(one electron reduction)}$$

By measuring the reaction kinetics, one can classify the test substance as either fast or slowly acting reductant. Measurements at a pre-selected time interval after the reaction onset show that the substance reduction property can be coined, e.g., as an IC_{50} value. Although both assays are still often exploited, their impact is questionable since neither the $ABTS^{\bullet+}$ nor $DPPH^{\bullet}$ indicator represents any biomolecule, and is not even found in any biological system.

Vast amount of hexahydropyridoindoles with different physico-chemical properties have been synthesized so far [14–16]. Of them, all five derivatives, namely the dihydrochlorides of – stobadine (**1**) and SM1dM9dM10 (**2**) and the monohydrochlorides of – SME1i-ProC2 (**3**), SM1M3EC2 (**4**), and SMe1EC2 (**5**) (see Appendix, Fig. 3), which undergone pharmacodynamic studies are examined and reported in this paper from the point of view of their H atom donating as well as reductive properties.

6.2 RESULTS AND DISCUSSION

6.2.1 *WEISSBERGER'S BIOGENIC OXIDATIVE SYSTEM*

Scheme 1 implicates the statement that, e.g., at the ratio of the reactants [Cu(II)]:[ascorbate] = 0.1/100 the reaction cycle will be repeated

1000-times and at 100% efficacy of all elementary reaction steps the products will be DHA and H_2O_2 – both in the amount of 100 μmol. This statement is naturally incorrect since the product generated, i.e., hydrogen peroxide is decomposed yielding •OH radicals due to the presence of the reactant Cu(II) reduced to Cu(I)-intermediate [17]. According to reaction (1), however, during the early stage of the reaction cycles, there is a high molar surplus of ascorbate within the system, and thus most of the generated •OH radicals will be immediately scavenged and the exclusively detectable "radical product" in the reaction mixture will be $Asc^{\bullet-}$ – the ascorbyl anion radicals. This implicit conclusion was proved by EPR measurements of the aqueous system comprising $CuCl_2$ (0.1 μmol), ascorbic acid (100 μmol), and the spin-trapping agent 5,5-dimethyl-1-pyrroline-*N*-oxide (DMPO; 250 mmol) [18]: During the first approximately 60 min of the reaction of WBOS components, the exclusive EPR signal detected was that belonging to ascorbyl anion radical ($Asc^{\bullet-}$; Fig. 2a). The •DMPO-OH adduct was detectable as late as 1 h after the reaction initiation, i.e., after disappearance of the EPR signal of ascorbyl anion radical, pointing to the depletion of ascorbate in the reaction mixture monitored. Figure 2b shows an explanatory chart of the time courses of the integral EPR signals of $Asc^{\bullet-}$ anion radical and the •DMPO-OH adduct.

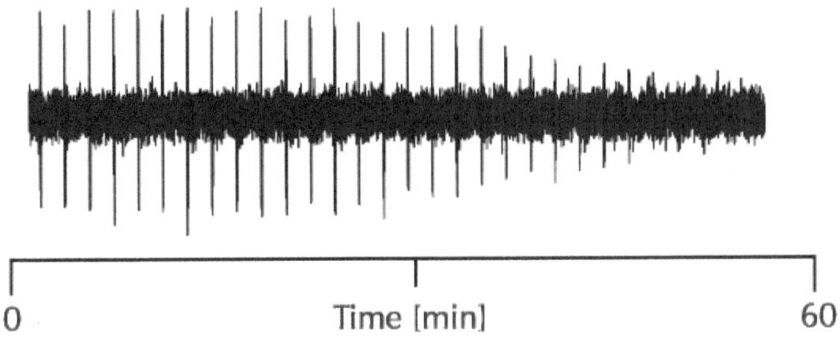

FIGURE 2(A) Time course of EPR spectra of the aqueous mixture containing $CuCl_2$ (0.1 μmol), ascorbic acid (100 μmol), and spin trapper DMPO (250 mmol) at room temperature – adapted from Šoltés et al. [18].

The record illustrates the scans of the $Asc^{\bullet-}$ anion radical evidenced in time from 0.5 to 56 min.

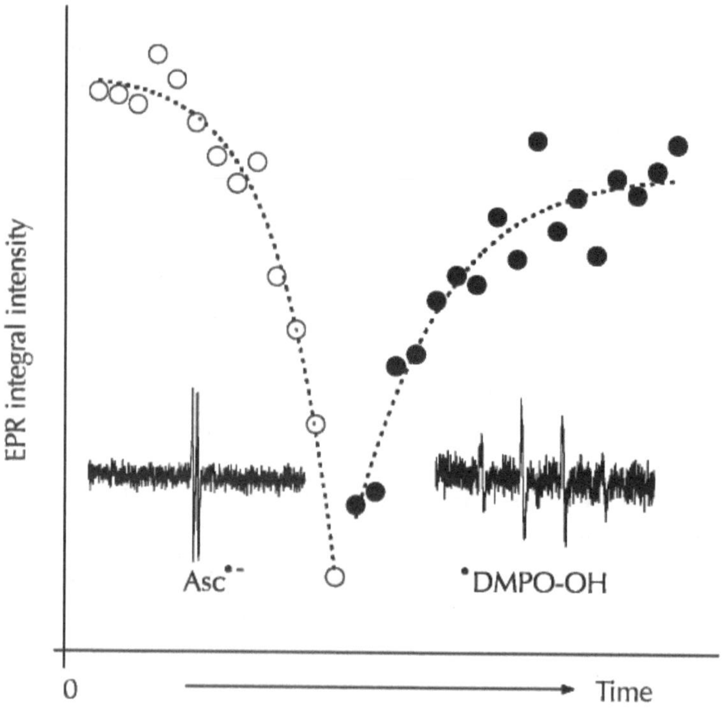

FIGURE 2(B) Illustrative representation of the time dependences of the integral EPR signals of Asc•- anion radical (○) and the •DMPO-OH adduct (•) – adapted from Šoltés *et al.* [18].

In the figure, both the EPR spectrum of the ascorbyl anion radical Asc•- and that of the •DMPO-OH aduct are depicted.

6.2.2 HYALURONAN FREE-RADICAL DEGRADATION

As can be deduced from the above-mentioned observations (cf. Figs. 2a and 2b), by applying WBOS, •OH radicals should be generated after a certain time delay, i.e., after the "consumption" of a certain amount of ascorbate, which acted as a scavenger of hydroxyl radicals. Yet, in spite of such an expectation, after ascorbic acid addition to the HA solution containing copper ions, the degradation of the biopolymer macromolecules starts practically immediately (cf. Appendix, Figs. 1 and 2, black curves). Dynamic viscosity *vs*. time relationship of the solution monitored decreases gradually and the initial dynamic viscosity (η) value = 9.93 mPa×s (corresponding to M_w = 808.7 kDa) decreases in time and in the 5th h its value equals 6.30 mPa×s (corresponding to $M_w \approx 420$ kDa [19]).

To elucidate the above-mentioned observations it is necessary to note here the following: Under the experimental conditions used within the solution mixture monitored (pH ≈ 6.0–6.5), the HA macromolecules (2.5 mg/ml) are highly ionized since the pK_a values of the D-glucuronic acid residues equal 3.12 [20]. The anions of the D-glucuronic acids (6.25 mmol) – the structural elements of HA macromolecule – naturally form salts with the (counter) cations of copper ions (0.1 μmol). Moreover, as reported [21], hyaluronic acid binds reversibly cupric ions (the binding constant = 3.0×10^3 l/mol [22]). Yet plausibly, due to the really high molar ratio of [–COO^-]:[Cu(II)] = 62,500, the copper cations will be dispersed throughout the chain of HA macromolecule, randomly forming a relatively sporadic population of Cu(II)-bond micronuclei. Under aerobic conditions the pre-formed micronuclei of the HA-Cu(II) complex may react with ascorbate generating thus *in situ* hydrogen peroxide, which decomposes and forms the highly reactive •OH radical. The latter, *in statu nascendi*, reacts with the chain of the HA macromolecule, yielding the A• macroradical. Hence, the substance whose role is to (preventively) inhibit the generation of •OH radicals, must diffuse as close as possible to the micronucleus of the HA-Cu(II) complex and either decomposes the molecules of H_2O_2 to inert components [23] or effectively donates the atom H from its molecule. In case that the given substance is an efficient H atom donor, the radical formed from the substance should be ineffective to re-initiate the reactions' cascade of the free-radical chain degradation of HA macromolecules. The latter condition mentioned is very well fulfilled by applying L-glutathione (GSH) [7,24,25], the endobiotic substance, which donates H atom really freely yielding a weakly reactive glutathiyl radical (GS•). Within the organism, G.S• radicals recombine rapidly to glutathione disulfide – GSSG (called also “oxidized” glutathione).

6.2.3 THE POTENCY OF SUBSTANCES 1, 2, 3, 4 OR 5 TO ACT AS PREVENTIVE ANTIOXIDANTS

On inspecting the experimental results represented in Fig. 3, panel A, one can state that from all five substances tested substance **1** is classifiable as a really efficient preventive antioxidant. In the concentrations used (100, 400, and 1000 μM) substance **1** inhibited the free-radical chain degradation of HA macromolecules almost completely (85.0, 82.2, and 78.1%) during the first hour. Yet even at the high potency of substance **1** to act as a preventive antioxidant, it was not efficient enough to inhibit the degradation of HAs (Fig. 3, panel B). The percentage of the inhibition of HA degradation at 5 h was practically independent of substance **1** concentration used and ranged between the values 24.4–31.3% (cf. Fig. 3, panel B, black line). From the point of view of the potency of the substances tested to act as preventive antioxidants, exclusively the action of substance **3** is valuable enough to be taken into account (cf. Fig. 3, panel A): Substance **3**, although less effective than **1** at 1 h, demonstrated a concentration dependent inhibitory action at 5 h with values of 23.2, 39.4, and up to 53.8% at the concentrations 100, 400, and 1000 μM (cf. Fig. 3, panel B). The remaining three substances, i.e., **2**, **4**, and **5**, concerning their potency to prevent free-radical chain degradation of HA macromolecules, were much less effective as compared to the efficiency of substances **1** and **3**. The registered negative values of the percentages of the inhibition of degradation of the HA macromolecules, evidenced markedly on testing the action of substance **4** (cf. Fig. 3, panels A and B, concentrations 100 and 400 μM), indicate that its radical formed within the reaction mixture might, most plausibly, initiate by itself a cascade of free-radical chain reactions. According to the above-mentioned facts, it can be concluded that substances **1** and **3** are much more relevant for application as preventive scavengers of HA degradation induced by WBOS.

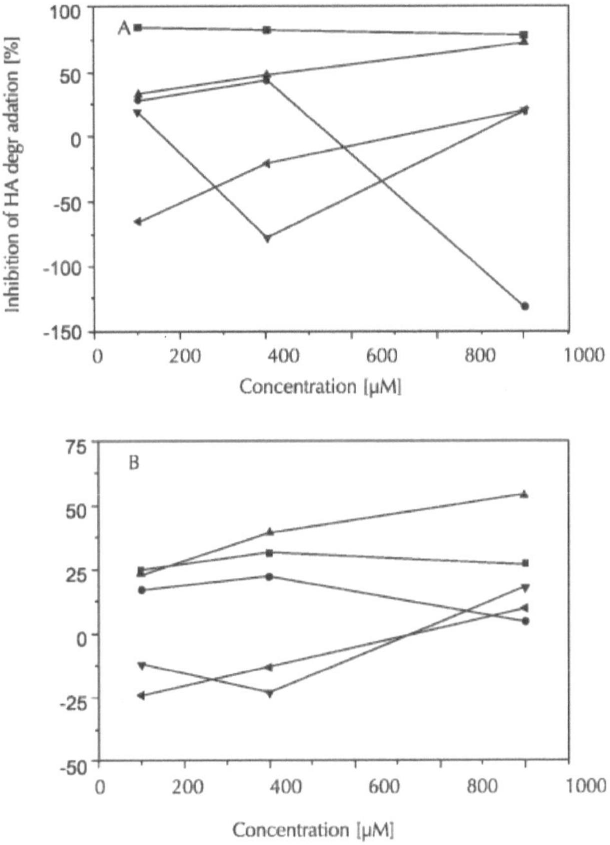

FIGURE 3 Percentage of HA degradation inhibition at 1 h (panel A) or 5 h (panel B) after the reaction onset inhibited by the substance tested to function as a preventive antioxidant.

6.2.4 SUBSTANCE 1, 2, 3, 4 OR 5 POTENCY TO ACT AS CHAIN-BREAKING ANTIOXIDANTS

Figure 4 shows the percentage of the inhibition of HA degradation at 5 h by the substance applied into the vessel 1 h after the reaction onset. At such an experimental setting, again substance **1** demonstrated the "greatest" efficacy, namely 23.2% at 100 μM. Simultaneously, the percentage of the inhibition of degradation of HAs by applying substance **1** indicates a weak positive concentration dependency with the value of inhibition equaling 38.3% at 1000 μM. The efficacy of substance **1** at 1000 μM was slightly exceeded on applying substance **5** (51.6%). The latter substance (**5**) was the only one, which unambiguously demonstrated a significant concentration-dependent inhibitory action (cf. Fig. 4, line **5**).

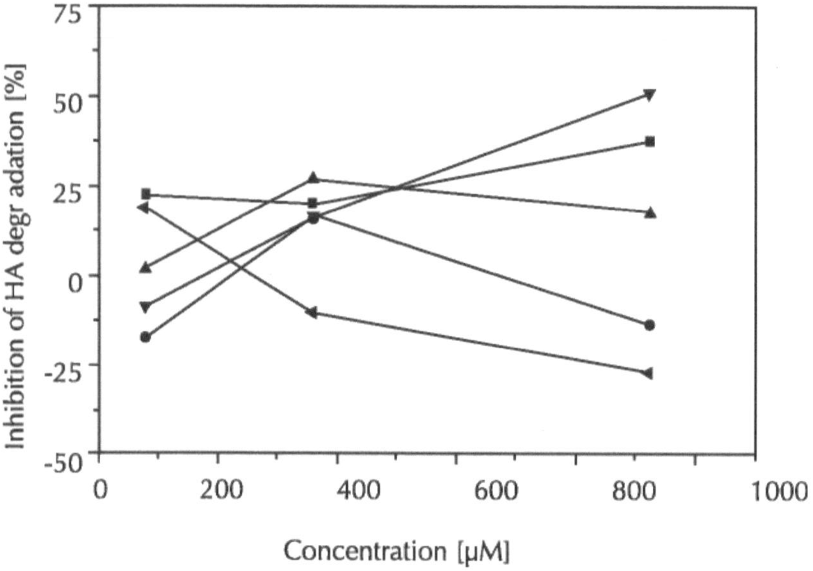

FIGURE 4 Percentage of inhibition of HA degradation at 5 h after the reaction onset inhibited by the substance tested to function as a chain–breaking antioxidant.

6.2.5 REDUCTIVE PROPERTY OF SUBSTANCES 1, 2, 3, 4 OR 5

A simple order of substances **1–5** based on the determined values of IC_{50} recorded in Table 1 shows that the order of the efficacy of the substances tested by both ABTS and DPPH assays is not identical. While the data obtained by ABTS decolorization assay resulted in the efficacy of the substances in the order: **4** > **1** > **5** > **3** > **2**, the order of the substances determined by the DPPH assay was: **3** > **5** > **4** > **2** > **1**. The most disputable result was the IC_{50} value = 122±5.2 μM determined for stobadine by the DPPH assay. The most appropriate explanation is the fact that both stobadine and substance **2** used in the DPPH assay were dihydrochlorides. Since this assay necessitates work in an absolutely non-aqueous environment – in absolute methanol – it is questionable if ever and to what extent the reduction capability of dihydrochlorides (**1** and **2**) or monohydrochlorides (**3**, **4**, and **5**) of hexahydropyridoindoles was influenced by the absence of H^+ protons, i.e., by H_3O^+ ions. Due to this fact, it can be stated that for correct determination of the IC_{50} values of the substances investigated, the ABTS assay is more relevant compared to the DPPH assay. The latter method has been already applied to some pyridoindole derivatives [26], yet a free base of these substances was used in this assay. The results of determining the IC_{50} values by the latter assay can be therefore significantly influenced by inefficient or practically nil ionization of substances **1**, **2**, **3**, **4**, and **5** in the $[H^+]/[H_3O^+]$-deficient milieu, i.e., in non-aqueous methanol.

TABLE 1 IC_{50} values of five hexahydropyridoindoles determined by ABTS and DPPH assays.

Substance	ABTS assay [M]	DPPH assay [M]
1, Stobadine	12.6±0.24	122±5.2
2, SM1dM9dM10	155±2.2	29.4±0.85
3, SME1i–ProC2	27.6±0.46	10.6±0.51
4, SM1M3EC2	10.8±0.38	21±1.4
5, SMe1EC2	17±2.5	16.9±0.67

Values are means ± SEM; n = 4.

The range of the IC_{50} values of substances **1**, **3**, **4**, and **5** from 10.8±0.38 to 27.6±0.46 μM determined by the ABTS assay indicates that these structural derivatives have similar reduction properties against $ABTS^{\bullet+}$ cation radical. The most effective reductants of $ABTS^{\bullet+}$ cation radical were substances **4** and **1**. Substance **2** cannot be included into the set of effective reductants for its high IC_{50} value (155±2.2 μM). A recent detailed examination showed that substance **2**, i.e., SM1dM9dM10, has to be tested especially carefully from the viewpoint of its reduction properties. This substance, in the process of a high-molar-mass hyaluronan degradation induced by ROS in WBOS, showed a significant pro-oxidative effect, which was especially evident in the concentration of 1000 μM (cf. Fig. 5, panels

A and B, gray curves). The pro-oxidative effect observed in substance **2** leads implicitely to two conclusions: (i) substance **2** is the least efficient reductant of $ABTS^{\bullet+}$ cation radical of the set of hexahydropyridoindoles tested and (ii) although the given substance **2** is an H atom donor, yet the radical formed from this substance might be effective in re-initiating the reactions cascade of the free-radical chain degradation of HA macromolecules. However, the latter fact is to be confirmed by some complementary experimental techniques.

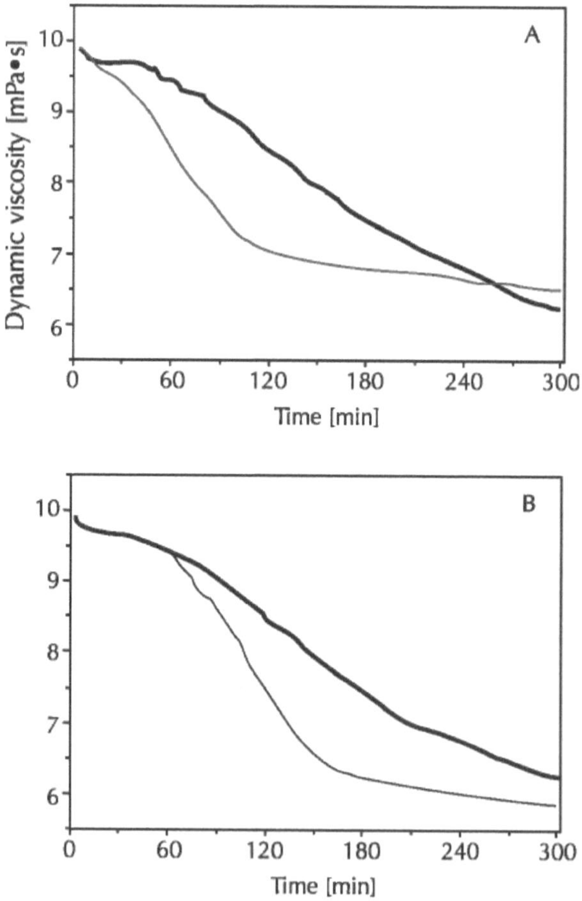

FIGURE 5 Pro-oxidative effect of substance **2** on HA degradation induced by WBOS. The substance was added to the oxidative system before the start of HA degradation (A) or after 1 h (B) in μM concentrations: 0 (black curve) and 1000 (gray curve).

6.3 CONCLUSIONS

The primary goal of the modifications of the hexahydropyridoindole chemical structure of stobadine {cis-(–)-2,3,4,4a,5,9b-hexahydro-2,8dimethyl-1-*H*-pyrido[4,3-b]indole} [27] (an enantiomer of the racemic drug carbidine [28, 29]) was to diminish adverse hypotensive effects of substance **1**, which are due to its α-adrenolytic activity [30, 31]. Since the acute toxicity of substances **2**, **3**, **4** or **5** has been much lower compared to substance **1** [15], diminishing adverse effects along with an increase of the antioxidative efficacy of the newly synthesized hexahydropyridoindoles would better meet the claims for introducing these prospective drugs into clinical practice. However, according to the above-presented observations, it is evident that any gain, if any, found on applying substances **2**, **3**, **4** or **5** did not exceed remarkably the well established antioxidative properties of substance **1** [31–33], i.e., stobadine – a cardioprotective drug [31, 34–37].

Stobadine, an amphiphilic substance with $pK_{a1} = 2.9$ and $pK_{a2} = 7.2$, can freely reach both lipoidal and hydrophilic environments in the human organism including those of (synovial) joints [38]. Due to its redox potential at neutral pH of +0.58 V, lying between that of ascorbate ($Asc^{•-}$, $H^+/AscH^- = +0.282$ V) and glutathione ($GS^•/GS^- = +0.920$ V) [39, 40], stobadine is a really proper reductant and H atom donor.

To avoid inappropriate applications and misinterpretation of the observations resulting from exploiting one single assay, the usage of a "battery" of assays measuring different aspects of the behavior of antioxidants has been recommended (for review see Ref. [13]). Although the ABTS and DPPH decolorization assays are the most frequently utilized, development of assays where more than one oxidative species is present in the reaction medium simultaneously should be considered inevitable. Efforts towards this direction can be appreciated by establishing the design of a standardized analytical method [41]. Within the latter method, a "cocktail" of ROS – H_2O_2, $^•OH$, and $AOO^•$ – acting practically simultaneously can be stated. This ROS cocktail damages the probe, i.e., the high-molar-mass hyaluronan, a process, which resembles that within the inflamed (synovial) joints.

Application of the DPPH decolorization assay in case of testing the reductive properties of salts of organic substances can lead to results, which should be reevaluated in the context of non-disociability of these salts in the non-aqueous environment. The ABTS assay, which operates in a partially aqueous, i.e., ionic milieu, is most plausibly a proper choice to broaden the insight into the reductive (antioxidative) properties of salts of organic substances.

6.4 EXPERIMENTAL PROCEDURES

6.4.1 *BIOPOLYMER AND CHEMICALS*

The high-molar-mass hyaluronan sample P9710-2A used with a weight–average of the molar masses M_w = 808.7 kDa and polymolecularity value M_w/M_n = 1.63, where the M_n is the number-average of the polymer molar masses, was the product of Lifecore Biomedical Inc., Chaska, MN, U.S.A. Analytical purity grade NaCl, $CuCl_2 \cdot 2H_2O$, ethanol 96%, and methanol were purchased from Slavus Ltd., Bratislava, Slovakia; L-ascorbic acid and potassium persulfate ($K_2S_2O_8$; *p.a.* purity, max. 0.001% nitrogen) were the products of Merck KGaA, Darmstadt, Germany; 2,2′-azino-bis-(3-ethylbenzothiazoline-6-sulfonic acid) (ABTS; *purum* >99%) was from Fluka, Steinheim, Germany; 2,2-diphenyl-1-picrylhydrazyl (DPPH) were the products of Sigma–Aldrich, Steinheim, Germany. The hexahydropyridoindoles were prepared at the Institute of Experimental Pharmacology and Toxicology, Bratislava, Slovakia. Deionized high-purity grade water, with conductivity of ≤0.055 μS/cm, was produced by using a water purification system of Thermo Scientific TKA, Niederelbert, Germany.

6.4.2 *SOLUTIONS*

The HA sample solutions (2.5 mg/ml) were prepared in the dark at room temperature in 0.15 M aqueous NaCl in two steps. First, 4.0 ml of the solvent was added to 20 mg HA, and 3.90, 3.85, 3.70 or 3.40 ml of the solvent was added after 6 h. All stock solutions, including those of each hexahydropyridoindole (**1**, **2**, **3**, **4** or **5**; 16 mM), L-ascorbic acid (16 mM), and cupric chloride (16 mM diluted to a 16 μM solution) were also prepared in 0.15 M aqueous NaCl.

6.4.3 HYALURONAN OXIDATIVE DEGRADATION

HA degradation was induced by the WBOS comprising L-ascorbic acid (100 μmol) and $CuCl_2$ (0.1 μmol). The procedure was as follows: a volume of 50 μl of $CuCl_2$ solution (16 μM) was added to the HA solution (7.90 ml) and after 30 s stirring the reaction mixture was left to stand for 7.5 min at room temperature. Then 50 μl of L-ascorbic acid solution (16 mM) were added to the reaction mixture and stirred again for 30 s. The solution mixture (8.0 ml) was then immediately transferred into the viscometer Teflon® cup reservoir.

Procedures to investigate the H atom donating property of the substances (**1**, **2**, **3**, **4** or **5**) were as follows:

(i) The solution of $CuCl_2$ (16 μM) in the volume of 50 μl was added to the HA solution (7.85, 7.70 or 7.40 ml), which was left to stand for 7.5 min at room temperature after stirring for 30 s. Then, 50, 200 or 500 μl of the substance solution (16 mM) were added to the solution mixture and stirred again for 30 s. Finally, 50 μl of the L-ascorbic acid solution (16 mM) were added to the solution mixture and stirred for 30 s. The reaction mixture (8.0 ml) was then immediately transferred into the viscometer Teflon® cup reservoir. By adding the substance in time 0 min, i.e., before adding ascorbic acid, we investigated the capability of the substance tested to scavenge •OH radicals, i.e., to act as a preventive antioxidant [7, 18].

(ii) In the second experimental setting, a similar procedure as that described in (i) was applied. However after leaving the solution mixture (7.90, 7.75, or 7.45 ml) for 7.5 min at room temperature, 50 μl of the L-ascorbic acid solution (16 mM) were added. After 1 h stirring of the reaction mixture, finally 50, 200 or 500 μl of the substance solution (16 mM) were added and stirred for further 30 s. The reaction mixture (8.0 ml) was then immediately transferred into the viscometer Teflon® cup reservoir. By adding the substance 1 h after admixing ascorbic acid, we investigated the capability of the substance tested to scavenge peroxy-type radicals, i.e., to act as a chain-breaking antioxidant [7, 25, 42, 43].

6.4.4 ROTATIONAL VISCOMETRY

The resulting reaction mixture (8.0 ml) was transferred into the Teflon cup reservoir of a Brookfield LVDV–II+PRO digital rotational viscometer (Brookfield Engineering Labs, Inc., Middleboro, MA, U.S.A.). The recording of viscometer output parameters started 2 min after the experiment onset. The changes of the η values of the reaction mixture were recorded at 25.0±0.1°C in 3-min intervals for up to 5 h. The viscometer Teflon® spindle rotated at 180 rpm, i.e., at a shear rate of 237.6 s^{-1}.

6.4.5 ABTS AND DPPH ASSAYS

The standard ABTS decolorization assay was applied as already reported [44−46]. Briefly, the aqueous solution of $ABTS^{\bullet+}$ cation radical was prepared 24 h before the measurements at room temperature as follows: ABTS aqueous stock solution (7 mM) was mixed with $K_2S_2O_8$ aqueous solution (2.45 mM) in equivolume ratio. The following day, 1.1 ml of the resulting solution was diluted with 96% ethanol to the final volume of 50 ml. The ethanol-aqueous reagent in the volume of 250 μl was added to 2.5 μl of the ethanolic solution of the substances **1**, **2**, **3**, **4**, or **5**. The concentration of each substance solution was 101−0.808 mM. The light absorbance of the sample mixture was recorded at 734 nm in the 6th min after mixing the reactants.

In DPPH decolorization assay, 2,2-diphenyl-1-picrylhydrazyl (1.1 mg) was dissolved in 50 ml of distilled methanol to generate $DPPH^{\bullet}$. The $DPPH^{\bullet}$ radical solution in the volume of 225 μl was added to 25 μl of the methanolic solution of the substances **1**, **2**, **3**, **4**, or **5** (in the concentration range of 10–0.078 mM) and in the 30th min the absorbance of the sample was measured at 517 nm. All measurements by both assays were performed quadruplicately in 96-well Greiner UV-Star microplates (Greiner-Bio-One GmbH, Germany) by using the Tecan Infinite M 200 reader (Tecan AG, Austria). The calculated values of IC_{50} are expressed as mean ± SEM.

APPENDIX

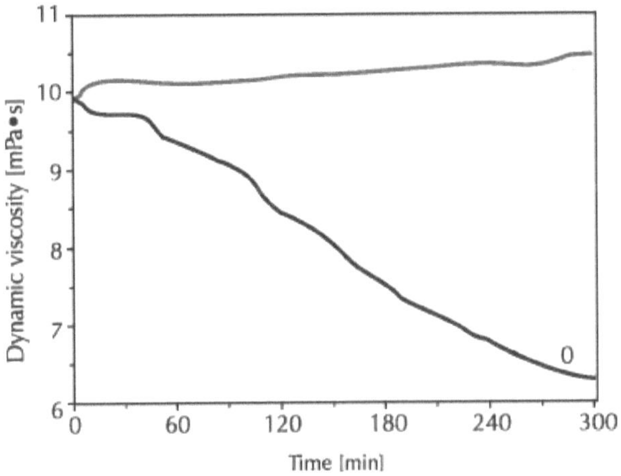

FIGURE 1 Time–dependent changes of dynamic viscosity values of the test HA solution (2.5 mg/ml).

The gray curve simulates the situation when no HA degradation occurs and the solution dynamic viscosity value rises slightly in time due to the phenomenon called rheopexy. The black curve (0) represents the real degradation of the biopolymer chains (here induced by 0.1 μmol Cu(II) *plus* 100 μmol ascorbate).

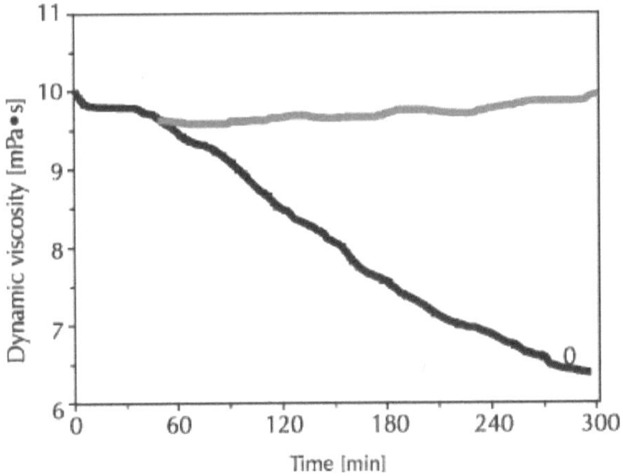

FIGURE 2 Time–dependent changes of dynamic viscosity values of the test HA solution (2.5 mg/ml).

The black curve (0) represents the real degradation of the biopolymer chains (here induced by 0.1 μmol Cu(II) *plus* 100 μmol ascorbate). The gray curve simulates the situation when the degradation of HAs initiated by the WBOS is broken–down (in the 60th minute) by addition of a "perfect/absolute chain–breaking" antioxidative substance and the solution dynamic viscosity value rises slightly in time due to the phenomenon called rheopexy.

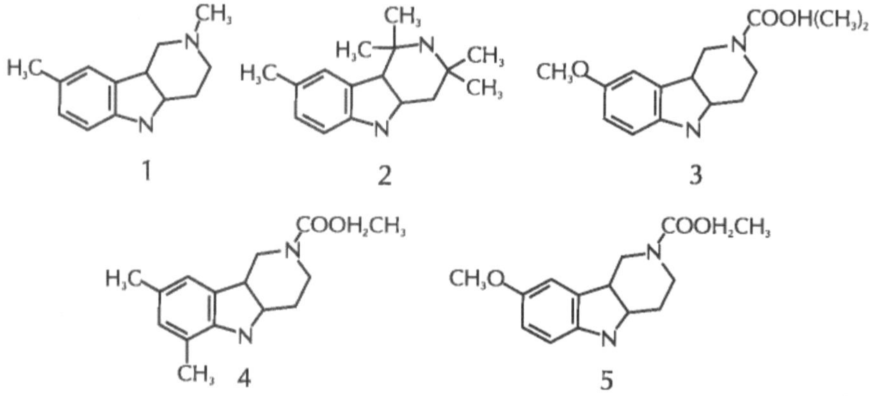

FIGURE 3 Structural formalae of stobadine (**1**), SM1dM9dM10 (**2**), SME1i-ProC2 (**3**), SM1M3EC2 (**4**), and SMe1EC2 (**5**).

KEYWORDS

- **Battery**
- **Bolus**
- **Born**
- **Cocktail**
- **Post mortem**

REFERENCES

1. Yamazaki, K.; Fukuda, K.; Matsukawa, M.; Hara, F.; Yoshida, K.; Akagi, M.; Munakata, H.; Hamanish, C. *Pathophysiology*, **2003**, *9*, 215–220.

2. Weissberger, A.; Luvalle, J. E. *J. Am. Chem. Soc.*, 1944, *66*, 700–705.
3. Butt, V. S.; Hallaway, M.; *Arch. Biochem. Biophys.* **1961**, *92*, 24–32.
4. Fisher, A. E. O.; Naughton, D. P. *Med. Hypotheses*, **2003**, *61*, 657–660.
5. Fisher, A. E. O.; Naughton, D. P. *Nutrition J.*, **2004**, *3*, 2.
6. Fisher, A. E. O.; Naughton, D. P. *Curr. Drug Del.*, 2005, *2*, 261–268.
7. Hrabárová, E.; Valachová, K.; Juránek, I.; Šoltés, L. *Chem. Biodiv.*, **2012**, *9*, 309–317.
8. Hrabárová, E.; *Voľnoradikálová degradácia hyaluronanu účinkom reaktívnych foriem kyslíka. Hodnotenie antioxidačných vlastností endogénnych a exogénnych látok s obsahom tiolovej skupiny.* PhD Thesis (in Slovak), Bratislava, 2012.
9. Halliwell, B.; Gutteridge, J. M. *Arch. Biochem. Biophys.*, **1990**, *280*, 1–8.
10. Niedermeier, W.; Griggs, J. H. *J. Chronic Dis.*, 1971, *23*, 527–536.
11. Naughton, D. P.; Knappitt, J.; Fairburn, K.; Gaffney, K.; Blake, D. R.; Grootveld, M. *FEBS Lett.*, 1995, *361*, 167–172.
12. Šoltés L. **In** *New Steps in Chemical and Biochemical Physics. Pure and Applied Science*, Pearce, E. M.; Kirshenbaum, G.; Zaikov, G. E., Eds.; Nova Science Publishers: New York, 2010, pp 133–148.
13. Magalhaes, L. M.; Segundo, M. A.; Reis, S.; Lima, J. L. F. C. *Anal. Chim. Acta*, **2008**, *613*, 1–19.
14. Horáková, Ľ.; Štolc, S. *Gen. Pharmac.*, **1998**, *30*, 627–638.
15. Štolc, S.; Šnirc, V.; Májeková, M.; Gásparová, Z.; Gajdošíková, A.; Štvrtina, S. L. *Cell. Mol. Neurobiol.*, **2006**, *26*, 1493–1502.
16. Štolc, S.; Považanec, F.; Bauer, V.; Májeková, M.; Wilcox, A. L.; Šnirc, V.; Račková, L.; Sotníková, R.; Štefek, M.; Gaspárová-Kvaltínová, Z.; Gajdošová, A.; Mihalová, D. *Pyridoindolové deriváty s antioxidačnými vlastnosťami, spôsob ich prípravy a použitia v liečebnej praxi a farmaceutické prostriedky.* Slovak Patent No. 287650, 2010.
17. Šoltés, L.; Brezová, V.; Stankovská, M.; Kogan, G.; Gemeiner, P. *Carbohydr. Res.*, **2006**, *341*, 639–644.
18. Šoltés, L.; Stankovská, M.; Brezová, V.; Schiller, J.; Arnhold, J.; Kogan, G.; Gemeiner, P. *Carbohydr. Res.*, **2006**, *341*, 2826–2834.
19. Stankovská, M. *Degradácia hyaluronanu rôznymi oxidačnými systémami: Návrh relevantných modelov testovania inhibičných účinkov látok v úlohe antioxidantov.* PhD Thesis (in Slovak), Bratislava, 2006.
20. Park, J. W.; Chakrabarti, B. *Biopolymers,* **1978**, *17*, 1323–1333.
21. Nagy, L.; Yamashita, S.; Yamaguchi, T.; Sipos, P.; Wakita, H.; Nomura, M. *J. Inorg. Biochem.*, **1998**, *72*, 49–55.
22. Figueroa, N.; Nagy, B.; Chakrabarti, B. *Biochem. Biophys. Res. Comm.*, **1977**, *74*, 460–465.
23. Miguel, M. G. *Flavour Fragr. J.*, **2010**, *25*, 291–312.
24. Dráfi, F.; Bauerová, K.; Valachová, K.; Poništ, S.; Mihálová, D.; Juránek, I.; Boldyrev, A.; Hrabárova, E.; Šoltés, L. *Neuroendocrinol. Lett.*, **2010**, *31*, 96–100.
25. Valachová, K.; Hrabárová, E.; Dráfi, F.; Juránek, I.; Bauerová, K.; Priesolová, E.; Nagy, M.; Šoltés, L. *Neuroendocrinol. Lett.*, **2010**, *31*, 101–104.
26. Račková, L.; Šnirc, V.; Májeková, M.; Májek, P.; Štefek, M. *J. Med. Chem.*, 2006, *49*, 2543–2548.
27. Beneš, L.; Štolc, S. *Drugs Future*, **1989**, *14*, 135–137.

28. Barkov, N. K. *Farmakologija i Toksikologija,* **1971**, *34*, 647–650.
29. Barkov, N. K. *Farmakologija i Toksikologija,* **1973**, *36,* 154–157.
30. Štolc, S.; Považanec, F.; Bauer, V.; Májeková, M.; Wilcox, A.; Šnirc, V.; Račková, L.; Sotníková R.; Štefek, M.; Gaspárová, Z.; Gajdošíková, A.; Mihálová, D.; Alföldi, J. *Pyridoindole Derivatives with Antioxidant Properties, the Way of their Production and Use in Medicinal Practice.* Slovak Patent No. 1321, 2003.
31. Juránek, I.; Horáková, Ľ.; Račková, L.; Štefek, M. *Curr. Med. Chem.*, **2010**, *17,* 552–570.
32. Vincenzi, F. F.; Hinds, T. R. *Life Sci.,* **1999**, *65*, 1857–1864.
33. Štefek, M.; Kyseľová, Z.; Račková, L.; Križanová, Ľ. *Biochim. Biophys. Acta*, **2005**, *1741,* 183–190.
34. Horáková, Ľ.; Giessauf, A.; Raber, G.; Esterbauer, H. *Biochem. Pharmacol.*, **1996**, *51,* 1277–1282.
35. Jančinová, V.; Nosáľ, R.; Danihelová, E. *Life Sci.*, 1999, *65*, 1983–1986.
36. Šoltés, L.; Bezek, Š.; Ujházy, E.; Bauer, V. *Biomed. Chromatogr.*, **2000**, *14,* 188–201.
37. Gallová, J.; Szalayová, S.; *Gen. Physiol. Biophys.*, 2004, *23,* 297–306.
38. Šoltés, L.; Kállay, Z.; Bezek, Š.; Fedelešová, V. *Biopharm. Drug Dispos.*, 1991, *12,* 29–35.
39. Buettner, G. R. *Arch. Biochem. Biophys.*, **1993**, *300*, 535–543.
40. Buettner, G. R. *Radiat. Res.*, **1996**, *145*, 532–541.
41. Orviský, E.; Šoltés, L.; Stančíková, M.; Vyletelová, Z.; Juránek, I. *Assessment of Antioxidative Properties of Hydrophilic Xenobiotics on the Basis of Inhibition of the Radical Degradation of Hyaluronan by Reactive Oxygen Species.* Slovak Patent No. 2764, 1994.
42. Valachová, K.; Rapta, P.; Kogan, G.; Hrabarová, E.; Gemeiner, P.; Šoltes, L. *Chem. Biodiv.*, **2009**, *6*, 389–395.
43. Valachová, K.; Vargová, A.; Rapta, P.; Hrabárová, E.; Dráfi, F.; Bauerová, K.; Juránek, I.; Šoltés, L. *Chem. Biodiv*., **2011**, *8*, 1274–1283.
44. Re, R. N.; Pellegrini, A.; Proteggente, A.; Pannala, M.; Yang, M.; Rice-Evans, C. *Free Radic. Biol. Med.*, **1999**, *26*, 1231–1237.
45. Cheng, Z.; Moore, J.; Yu, L. *J. Agric. Food Chem.*, **2006**, *54*, 7429–7436.
46. Hrabarova, E.; Valachova, K.; Rapta, P.; Soltes, L. *Chem. Biodiv.*, **2010**, *7,* 2191–2200.

CHAPTER 7

LAWS OF THE TRANSESTERIFICATION OF METHYL ESTER 3-(3′, 5′-DI-TERT. BUTYL-4′-HYDROXYPHENYL)-PROPIONIC ACID BY PENTAERYTHRITOL AND THE RESULT OF ITS REACTION

A. A. VOLODKIN, G. E. ZAIKOV, N. M. EVTEEVA, S. M. LOMAKIN, and E. KLODZINSKA

CONTENTS

7.1 INTRODUCTION

The results of the transesterification methyl ester 3-(3', 5'-di-tert.butyl-4'-hydroxyphenyl)-propionic acid, are known and generalised by polyols in reviews [1]. However, the information bound to pentaerythrityl-tetrakis-{3-(3', 5'-di-tert.butyl-4'-hydroxyphenyl)-propionate} is discordant and mainly based on declaring of the patent data [2–7]. The reaction pentaerythritol is step transesterification in which four bonds were formed, a yield and which interrelations changes in time. Pentaerythrityl-tetrakis-{3-(3', 5'-di-tert.butyl-4'-hydroxyphenyl)-propionate} (Phenosan-23) [1, 7] is an effective antioxidant and it was applied in technology of the reception of a products from polymers. In this connection interest to ester exchange is bound to workings out of the methods of synthesis "Phenosan-23." In the course of the reaction there is a viscosity augmentation that negatively influenced results and there is a necessity of rise in temperature > 120°C. At high temperatures in the conditions of alkaline catalysis oxidation and destruction processes that complicates the technology of allocation and finished the product purification become more active. Till now there are no data on the mechanism of step the transesterification and influence of intermediate products on "consumer" properties "Phenosan-23."

In the present work laws of the transesterification were positioned, evolved pentaerythrityl-di-{3-(3',5'-di-tert.buthyl-4'-hydroxyphenyl)propionate and pentaerythrityl-tris-{3-(3',5'-di-tert.buthyl-4'-hydroxyphenyl)-propionate. Their antioxidative properties in the modeling reaction inhibiting oxidations cumene were defined. On the basis of quantum-chemical calculations in approach PM6 initial, mediate and the transesterification finished products calculated thermodynamic equilibrium constants and energy H-O of communications of phenolic hydroxyl $D_{(OH)}$. Constitutions of the received compounds were confirmed with the yielded NMR ^{1}H and IR-spectrum. The technology of synthesis pentaerythrityl-tetrakis-{3-(3',5'-di-tert.buthyl-4'-hydroxyphenyl)-propionate depends on conditions of branch of methanol at last stage of the transesterification which was characterized anomalously by low value an equilibrium constant. In the conditions of the transesterification methanol allocation in a liquid-gas system changes in time that defines speeds of reactions.

7.2 EXPERIMENTAL

NMR ^1H spectrums wrote down on the device "Bruker WM-400" concerning a signal of residual protons solvent ($CDCl_3$). IR-spectra wrote down on a spectrometer "PERKIN-ELMER 1725-X in crystals. A chromatograph" Bruker LC-31 "column IBM Cyano, eluent: hexane-propanol-2 (9:1). Breakage coefficient (f) with participation **1- 5** defined on method **[8]** in conditions oxidations cumene at 50°C in the presence of the initiator of oxidation-azodiisobutyronitrile. Semiempirical method PM6 in allows counting energy of local minima of geometrical frames at gradients (gnorm) less than 0.1 with use of parameter EF. At calculation energy compound **3–5** and its radicals using the program Mopac2009 with an unlimited method of Hartrii-Foka (UHF) **[9]**.

Methyl Ester 3-(3′, 5′-Di-Tert.Butyl-4′-Hydroxyphenyl)-Propionic Acid *(1).*

Method A. To solution of 4.88 g (0.01 mol) 2,6-di-tert.butylphenolat potassium in 4 ml ДМCO at 115°C have added 2.5 ml (0.03 mol) methyl acrylate. In 3 h after refrigerating have neutralised 10%-s' HC1 and after crystallization from hexane have received 5.16 g (88%) **1**, m.p. 66°C (compare Ref. [10]: m.p. 66°C).

Method B. Of 11.78 g (0.01 mol) **5** and 4 ml of solution of MeONa of 5% in MeOH maintained 4 h. at 20°C, further to reactionary mass have added 20 ml of hexane. After crystallization have evolved 11.1 g (95%) **1**; m.p.66°C.

Pentaerythrityl-Di-{3-(3′,5′-Di-Tert.Buthyl-4′-Hydroxyphenyl)-Propionate} *(3).*

Of 14.6 g (0.05 mol) **1**, 6.8 g (0.05 mol) pentaerythritol in the presence of 3 g of accelerator (6% the GAME on clay) maintained an admixture at 190°C in an argon current in flow of 40 mines, 35 ml of hexane further have added. The low layer keeping 85% **3,** solvent have evaporated, the residual has chromatographied on silica gel: a yield of 8.4 g, m.p. 105–106°C. A NMR ^{1}H spectrum ($CDCl_3$, δ.): 1.43 (s, 36 H, ^{t}Bu); 2.67 (t, 4 H, Ar-$\underline{CH_2}$–CH_2, J = 8.21 Hz); 2.86 (t, 4H, Ar-CH_2-$\underline{CH_2}$, J=8.13 Hz); 3.28 (s. 4 H, CH_2OH) ;); 4.04 (s. 4 H, $COO\underline{CH_2}$); 5.01 (s., 2H, OH); 6.98 (s, 4 H, Ar). The IR-spectrum, ν / cm^{-1}: 3637 (OH); 3270 (br) (OH); 2951 (CH); 1744 (C=O).

Pentaerythrityl-Tris-{3-(3′,5′-Di-Tert.Buthyl-4′-Hydroxyphenyl)-Propionate} *(4).*

Mixture of 29.2 g (0.1 mol) **1**, 4.53 g (0.033 mol) pentaerythritol, 0.2 g (0.005 mol)) NaOH maintained 4 h at 160°C in the conditions of vacuum blowing at residual pressure $1.3.10^4$ Pases. According to LC the maintenance **4** in reactionary mass of 82%. Have received 22 g ArAlk **4**: m.p. 64–65°C. NMR ^{1}H spectrum ($CDCl_3$, δ): 1.43 (s, 54 H, ^{t}Bu); 2.62 (t, 6 H, Ar-CH_2-CH_2, J=8.63 Hz); 2.85 (t, 6H, Ar-CH_2-CH_2, J=8.68 Hz); 3.18 (s. 2 H, CH_2OH) ;); 4.01 (s. 6 H, $COOCH_2$); 5.09 (s., 3H. OH); 6.98 (s, 6 H, Ar). The IR-spectrum, ν / cm^{-1}: 3645 (OH); 3517 br. (OH); 2912 (CH); 1742 (C=O).

Pentaerythrityl-Tetrakis-{3-(3′,5′-Di-Tert.Buthyl-4′-Hydroxyphenyl)-Propionate} *(5).*

An admixture of 116.9 g (0.4 mol) **1**, 13, 6 g (0.1 mol) pentaerythritol, 1.5 g (~ 0.01 mol) K_2CO_3, (~ 0.04 mol) Na_2 CO_3 and 4.7 g (0.05 mol) phenol have heat up to 140-145°C and maintained in the conditions of vacuum blowing 2 h at residual pressure $1.3.10^4$ Pas., further 8 h at residual pressure $2.6.10^3$ Pas. A reaction mixture have dissolved in 300 ml cyclohexane, have cooled to 4–5°C, a deposit have separateed. A yield **of 5** 61.2 g (52%);: m.p.116-117 °C (compare. [7]: m.p. 116-117 °C)**.** NMR ^{1}H spectrum ($CDCl_3$, δ): 1.43 (s, 72 H, ^{t}Bu); 2.61 (t, 8 H Ar-CH_2-CH_2, J = 8.99 Hz); 2.86 (t, 8H, Ar-CH_2-$CH_{2,}$ J=7.72 Hz); 4.04 (s. 8H, $COOCH_2$); 5.01 (s.4H, OH); 6.98 (s 8 H, Ar). The IR-spectrum, ν / cm^{-1}: 3646 (OH); 2958 (CH); 1744 (C=O).

7.3 DISCUSSION OF RESULTS

The reaction of methyl ester 3- (3', 5'-di-tert.butyl-4'-hydroxyphenyl)-propionic acid (**1**) with pentaerythritol at temperature 140–200°C was accompanied by methanol allocation. In these conditions the admixture initial **1** and products by re-esterification **2–5** under the schema 1 were formed:

The Schema 1

$$ArAlkCOOMe + C(CH_2OH)_4 \underset{}{\overset{K_1}{\rightleftharpoons}} ArAlk\ COOCH2 + C(CH_2OH)3\ ,MeOH\ (1)$$

$$ArAlk\ COOCH_2\text{-}C(CH_2OH)3 + ArAlkCOOMe \overset{K_2}{\rightleftharpoons} (ArAlkCOOCH_2)_2 - C(CH_2COH)_2 + MeOH(2)$$

$$(ArAlkCOOCH_2)_2 - C(CH_2COH)_2 + ArAlkCOOMe \overset{K_3}{\rightleftharpoons} (ArAlkCOOCH_2)_3 - C(CH_2COH)_2 + MeOH(3)$$

$$(ArAlkCOOCH_2)_3 - C(CH_2COH)_2 + ArAlkCOOMe \overset{K_4}{\rightleftharpoons} (ArAlkCOOCH_2)_4C + MeOH(4)$$

$ArAlk{=}3,5\text{-}^tBu\text{-}4\text{-}HO\text{-}C_6H_2\text{-}CH_2\text{-}CH_2\text{-}$

The composition of reaction masses were analyzed by methods chromatography *LC* and NMR - spectroscopy. Compound **3–5** have evolved in an individual state and their constitutions confirmed NMR ^{1}H and IR-spectra. Interrelations of compounds **1–5** in reactionary masses were defined from spectrums on integrated intensity of protons methyl bunches of substituent $COOCH_3$ ($\delta = 3.70$, **1**), the methylene bunch of substituent $COOCH_2$ ($\delta = 4.04$, **3**), the methylene bunch of substituent $COOCH_2$ ($\delta = 4.01$, **4**) and the methylene bunch of substituent $COOCH_2$ ($\delta = 4.09$, **5**). At NMR ^{1}H spectrums **3** and **4** there were signals from protons CH_2OH of bunches. At an interrelation **1** – pentaerythritol, equal 4:1, concentration **2** in the conditions of re-esterification did not exceed 3–4%. Results of the analysis of reactionary masses method LC have coincided with the data of definition of composition of products of re-esterification on NMR ^{1}H to spectrums. From the kinetic data follows that in the conditions of methanol allocation at atmospheric pressure and temperature of 190°C balance in a fluid phase was positioned through 40–50 mins after the beginning of the reaction and further, within the next hour concentration of the compound **4** passed through a maximum, and concentration **5** increased (Fig. 1).

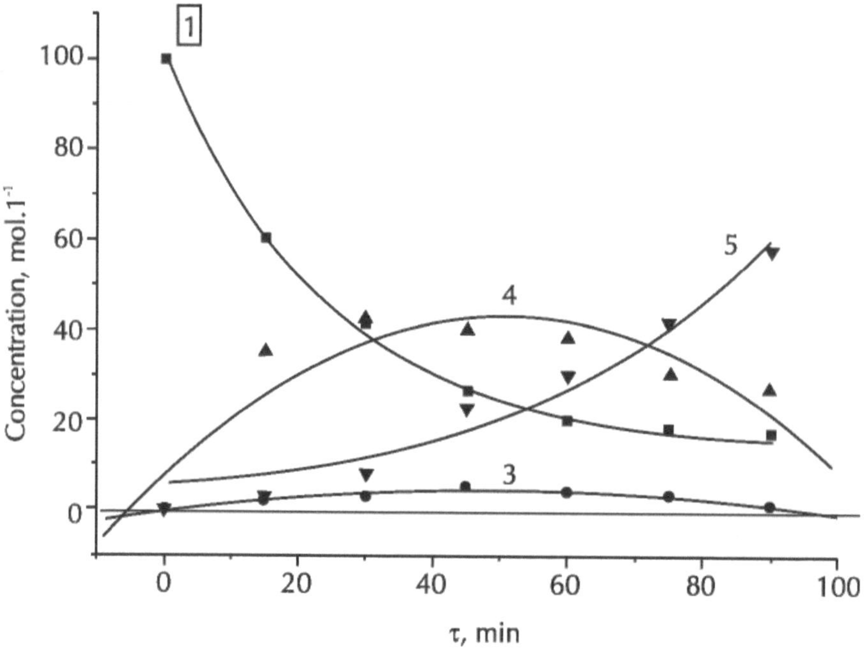

FIGURE 1 Changes of concentration of reacting components in the course of re-esterification **1** with pentaerythritol; an initial interrelation of reagents = 4:1 (mole), 190°C.

The change of the conditions of branch of methanol by vacuum blowing does not lead to change of an interrelation of compounds **5** and **4**. The received result it is possible interpretation as consequence of influence of a equilibrium constants of reversible reactions on a net result of step re-esterification. Results quantum – chemical calculations have specified that a thermodynamic equilibrium constant of reaction 4 $K_4 = 1.6.10^{-4}$ (298°C). It is logical to assume that in an admixture: **5–4**-methanol, balance it is moved towards formation of compound **4**. Anatropic re-esterifications at interaction **5** about a methanol-breakage proceeded at ambient temperature and led to formation initial **1**.

The energy of formation (E_f^o), enthalpies (E^o) and entropies (S^o) compounds were calculated in approach PM6 (program MOPAC 2009 [11]) that has allowed to find free energy change (ΔG^o) reversible reactions 1–4

(Tables 1 and 2) and to define energy of homolytic cleavage O–H of the communication.

TABLE 1 Enthalpies, entropies of compounds **1–5,** methanol and pentaerythritol (**6**), 298°C.

Compounds	***E°*** kJ.mol^{-1}	***S°*** J/grad/mol
6	33.0	437
1	67.9	713
2	89.7	869
3	147.1	1299.
4	206.6	1773
5	261.9	2170
MeOH	11.5	240

TABLE 2 Change of enthalpy, entropy and a energy of Gibbs (ΔG^{o}) in reactions 1–4, 298°C.

Reactions	**Δ *H°***	***ΔS°***	***ΔG°***	**Constant**
	kJ.mol^{-1}	**J/grad/mol**	**kJ.mol^{-1}**	**Equilibriums**
1 + 0.3 – 41 +12.5 K_1=6.4·10^{-3}				
2 + 1.0 – 43 +13.8 K_2=3.8·10^{-3}				
3 + 3.1 + 1 +2.8 K_3=3.2·10^{-1}				
4 – 1.1 – 76 +21.6 K_4=1.6·10^{-4}				

The formation of compound **5** in reaction 4 is bound to entropy reduction that, apparently, defines high reactivity **5** in reactions inhibiting oxidations. The value of constants balance of reactions 1–4 were calculated from the equation of Boltsmana: K=e – $^{\Delta Go/kT}$ where k –constant of Boltsmana. The sign on sizes Δ *H°*, Δ*S°* depends on the settlement data of enthalpy and entropy of frames of the cjompounds participating in reversible reactions 1–4. And the sign on value of energy of Gibbs is found from the equation of Gibbs.

The energy of homolytic cleavage O–H of communication is one of efficiency factors of derivative phenols as antioxidants. Calculation $D_{(OH)}$

was based on results of calculations energy formations of phenols and phenolic radicals unlimited a method of Hartrii-Foka (UHF) in approach PM6 (Table 3), allowing to calculate minima of energy of formation on gradient reduction energy binding orbitals of frames of phenol ($-E^{o}_{f\ InH}$), corresponding phenolic radical-$E^{o}_{f\ (In^{\cdot})}$. Also it is expressed by an interrelation [12]:

$$D_{(OH)} = -E^{o}_{f\ (In^{\cdot})} + E^{o}_{f\ (H)} - (-E^{o}_{f\ \{InH\}}),$$

where $E^{o}_{f\ (H)}$ – energy of hydrogen atom.

TABLE 3 The energy of formation of compounds **1–5**, its phenolic radicals and energy a homolytic cleavage of the communication O–H in approach PM6.

Compound	$-E^{o}_{f\ (InH)}$	$-E^{o}_{f\ (In^{\cdot})}$	$D_{(OH)}$	
			calculation	Ref. [13]
	kJ.mol^{-1}			
1	667.4	569.4	316.0	
2	1287.5	1186.4	319.1	
3	1741.3	1641.0	318.3	
4	2208.3	2079.0	347.3	
5	2682.0	2579.0	321.0	341

Results of calculation $D_{(OH)}$ frames **5** and experimental data [13] differ on 20 kJ.mol^{-1} that in our opinion, is bound as to a design procedure of experimental results, and calculations in approach PM6. On an extreme measure, results of definition of energy of homolytic cleavage O–H of communication in the molecula 2,4,6-tri-tert.butyl-phenol from the data [13–15] differ on 10 kJ.mol^{-1}. Also it is known [16] that results of calculations by quantum-chemical methods depend on programs of calculations.

The value of energy of homolytic cleavage H–O of communication of the molecula **4** essentially above among the presented compounds that should affect on antioxidative properties. If compound **5** is the most effective antioxidant in the conditions of thermoageing of polymers, **4** – it is ineffective. Electrical dipole moment is an index of symmetry of frame and on this index the frame **4** is more symmetric (μ =0.97 Db) in comparison with frame **5** (μ =5.6 Db). Geometry of frames of compounds based on calculation in approach PM6, visualisation of mopac-files made in program ChemBio3d and further files with dilating «mop» transformed to files with dilating «sk2» (Figs. 2 and 3).

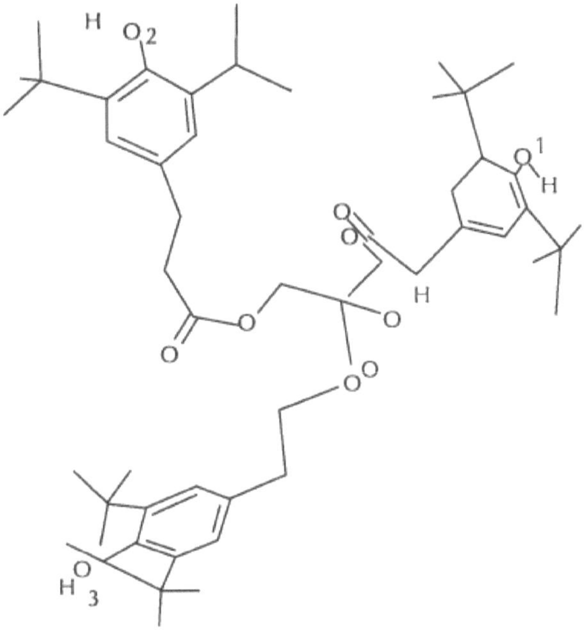

FIGURE 2 Frame **4**.

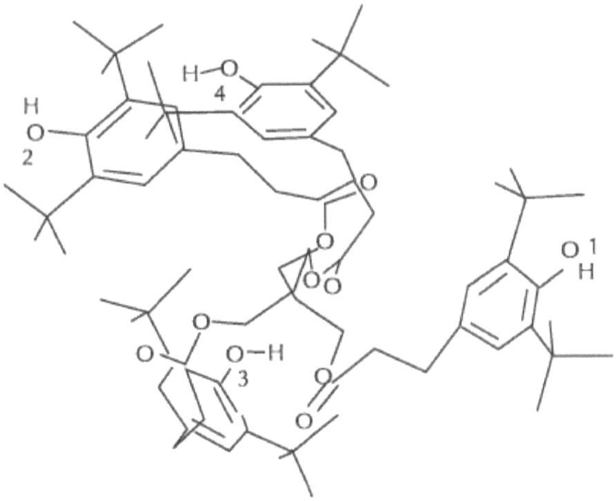

FIGURE 3 Frame **5**.

In frame **5** nonsymmetry it is invoked nonequivalent substituents (in comparison with **4)** that it leads to change of the value energy a homolytic cleavage of communication. X-ray diffraction study pentaerythrityl-tetrakis-{3-(3',5'-di-tert.buthyl-4'-hydroxyphenyl)-propionate specifies on nonequivalent substituents in frame of the molecula [17] that it is possible to treat, how coincidence of the experimental and settlement data. .

Cumene-soluble compounds **3–5** were used in the estimation of the influence of the intermolecular interaction on the antioxidation parameters: the period of oxidation inhibition involving the antioxidant (τ) and coefficient of stopping of chain (f) in reaction a peroxy radicals with antioxidant **3–5** at constant speed initiation (W_i=1.5·10^{-8} mol. l^{-1}·s^{-1}) defined a gasometric method [8] (Fig. 4).

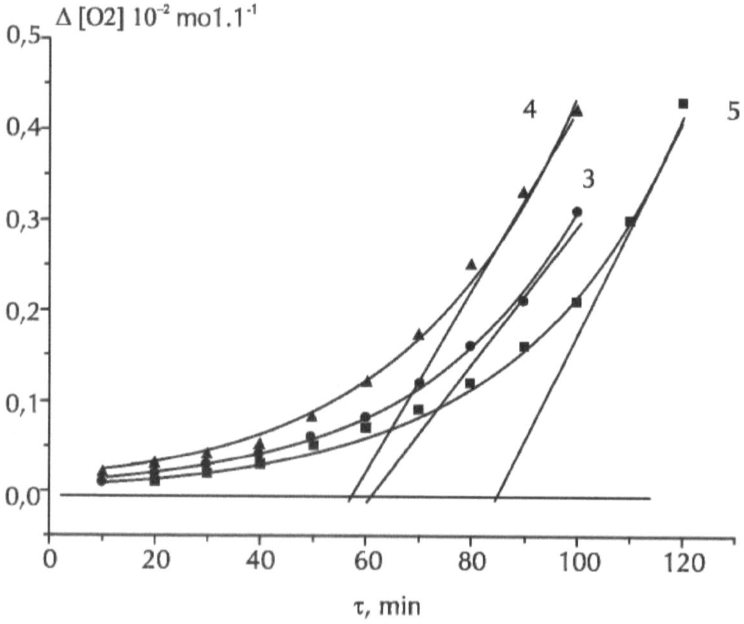

FIGURE 4 Kinetic dependences of oxygen uptake in reaction of the initiated oxidation cumene at 50°C in the presence of bonds **3**, **4**, **5** and azodiisobutyronitrile in co-ordinates Δ [O_2] – τ; Wi = 1.5·10^{-8} mol.l^{-1}·c^{-1}; **[3]** $_o$ = 1.34·10^{-5} **[4]** $_o$ =2.55.10^{-5}; **[5]** $_o$ =9.1.10^{-7} mol.l^{-1}.

The specified parameters are bound by expression: f = τ· W· $[InH]^{-1}$, where InH - compounds **4–5.** The coefficient of stopping of chain (the inhibiting container of an antioxidant) characterizes number of chains, which are broken by one molecula. From this data follows that in the presence of **5** coefficient f ≈ 8, whereas in reaction inhibiting oxidations in the presence of an antioxidant **4** f ≈ 2. Value of coefficient of breakage with participation **3** f ≈4. Hence, on the value of coefficient of stopping of chain the antioxidant **5** is more effective in comparison with **4**. Specific reaction rate constant (k_7) defined from dependence $\Delta [O_2] / [RH] = -k_2/k_7 \cdot \ln (1-t \cdot \tau^{-1})$, where k_2 – a kinetic constant of initiation ROO$^{\cdot}$ from cumene (according to [18, 19] k_2 = 1.75 l.mol^{-1}. c^{-1}), [RH] – concentration cumene (7.18 mol.l^{-1}), $\Delta [O_2]$ – quantity of absorbed oxygen. For the antioxidant **3** k_7 = 2.5±0.2·10^4, the antioxidant **4** k_7 = ±1.8 0.2·10^4, the antioxidant **5** k_7 = 3.1±0.2·10^4 l.mol-1.s^{-1}. On data [20] for compound **1** k_7 =2.3·10^4, by data [21] for **5** k_7 = 1.5·10^4 l.mol^{-1}.s^{-1}. It is necessary to notice that in Ref. [21] the industrial stabilizer "Irganox1010" which composition has not been analyzed, apparently, has been used.

KEYWORDS

- **Consumer**
- **Gasometric**
- **Hartrii-Foka**
- **Phenosan-23**

REFERENCES

1. Zaikov, G. E.; Volodkin, A. A. *Not Coloured Stabilizers of Polymers on the Basis of Methyl Ester 3-(3', 5'-di-tert.butyl-4'-hydroxyphenyl) Propionic Acid.* Designs from Composite Stuffs, Moscow, 2003, 11 [in Russian].
2. Burton, L. *4-Hydroxy-3,5-dialkylphenylpropionic Acid Ester Derivation.* U.S. Patent 4659863, 1987, Chem. Abstrs, 1989, Vol. 107, pp 77439.
3. Haeberli, J.; Park, K.; Vellturo, A. *Nurnberg G., 2,6-Dialkilphenols.* Ger. Patent 2364126, 1972, Chem. Abstrs, 1974, Vol. 81, pp 120260.
4. Orban, I. *Antioxydant of Polypropylene.* Eur. Patent 206789, 1987, Chem. Abstrs., 1987, Vol. 107, pp 58657.
5. Nachitani, M. *2,6-Dialkylphenols Ester Derivation.* Jap. Patent 12341, 1981, Chem. Abstrs., 1985, Vol. 95, pp 42686.

6. Dexter, M.; Meier, E. *Alkylphenols*. U.S. Patent 3364250, 1968, Chem. Abstrs., 1968, Vol. 68, pp 59318.
7. Author's Cetificate 1685920 USSR. RZhKhim [Abstract Journal of Chemistry], 1991, Vol. 29 [in Russian].
8. Emanuel, N. M.; Denisov, E. T.; Majzus, Z. K. *Chain Reactions of Oxidation of Hydrocarbons in a Fluid Phase*. Nauka, Moscow, 1965 [in Russian].
9. Gribov, L. A. *Elements of the Quantum Theory of a Constitution and Properties of Molecules*. Moscow, 2010; 172 [in Russian].
10. Volodkin, A. A.; Paramonov, V. I.; Egidis, F. M.; Popov, L. K. *Khim. Prom-st* [Chem. Industry. 1988; Vol. 12, 7 [in Russian].
11. Stewart, J. J. P. *J. Mol. Mod.* **2007**, *13,* 1173.
12. Denisov, E. T. *Methods of Definition of Dissociation Energy O-H of Communication in Phenolums*, the Receiving Tank of Lectures on V111 the International Conference Bioantioxidant. Moscow, 2010; 50 [in Russian].
13. Denisov, E. T. *J. Phys. Chim.*, **1995**, *69*, 623 [in Russian].
14. Denisov, E. T.; Denisova, T. G. *Handbook of Antioxidants*; CRC Press: London-New York, 2000; 88.
15. Vasserman, A. M.; Buchachenko, A. L.; Nikiforov, G. A.; Ershov, V. V.; Neumann, M. B. *J. Phys. Chim.*, **1967**, *41,* 705 [in Russian].
16. Hursan, S. L. Receiving Tank of Lectures on V111 the International Conference *Bioantioxidant.* Moscow, 2010; 195 [in Russian].
17. Jp. Patent 104348; 1984, Hemispherical Irganox 1010. Chem. Abstrs., 1984; Vol. 101, 191363.
18. Tsepalov, V. F. *Research of Synthetic and Connatural Antioxidants In Vivo and In Vitro.* Nauka, Moscow, 1992, 16 [in Russian].
19. Dyubchenko, O. I.; Nikulina, V. V.; Terah, E. I.; Prosenko, A. E.; Grigoriev, I.A. Izv. The Russian Academy of Sciences, Ser. Khim., 2007; 1107 [in Russian].
20. Storozhok, N. M.; Perevozkina, M. G.;.Nikiforov, G. A. *Izv. Akad. Nauk, Ser. Khim.*, **2005**, 323 [in Russian].
21. Tsepalov, V. F.; Haritonova, A. A.; Gladyshev, G. P.; Emanuel, N. M. *Kinetics Catalysis*, **1977**, *18*, 1261 [in Russian].

CHAPTER 8

ON RECEPTION OF THE METHYL ESTER 3-(3′, 5′-DI-TRET.BUTYL-4′-HYDROXYPHENYL)-PROPIONIC ACID

G. E. ZAIKOV, A. A. VOLODKIN, S. M. LOMAKIN,
and S. KUBICA

CONTENTS

8.1 INTRODUCTION

Antioxidants are a component of the majority of known polymerous stuffs and consumer properties of products depend on their efficacy. Antioxidants of phenolic structure are additives to polymers because of high performance to protect polymers from thermal-oxidative degradation processes. From this class of stabilizers the most widespread are derivatives of a methyl ester 3-(3',5'-di-tert.buyil-4'-hylroxyphenyl)-propionic acid, productions which are based on ester exchanges polyatomic alcohols. To the most significant is pentaeritril-tetrakis-3-(3',5'-di-tert.buyil-4'-hylroxyphenyl)-propionate. From the beginning of works on synthesis of derivatives of a methyl ester 3-(3',5'-di-tert.buyil-4'-hylroxyphenyl)-propionic acid on the present are executed a series of basic researches on catalysis and technology. These workings out can appear perspective for improvement of technology on the basis of building of continuous process of reception of a methyl ester 3-(3',5'-di–tert.buyil-4'-hylroxyphenyl)-propionic acid. In the conditions of catalysis with use of potassium or sodium hydroxides a methyl ester 3-(3',5'-di-tert.buyil-4'-hylroxyphenyl)-propionic acid is formed with yields to 98%. That allows using a main product further in ester exchange by pentaerythritol. In this technology power-intensive stages of allocation by fractionation in vacuo and stages of use of solvents that can render positive influences on the cost price and an ecological situation by production of the specified antioxidant are excluded. In the present work the basic results of researches of reaction of alkylation are stated 2,6-di-tert.butylphenol by methyl acrylate on the basis of computer modeling and experimental data.

Methyl Ester *3-(3′,5′-Di-Tert.Buyil-4′-Hylroxyphenyl)*-Propionic Acid

Reception of the methyl ester 3-(3',5'-di-tert.buyil-4'-hylroxyphenyl)-propionic acid (ArOH) is based on alkylation reaction 2,6-di-tert.butylphenol (Ar'OH) by methyl acrylate [1]. For the first time in patents [2–4] reception ArOH at interaction Ar'OH with methyl acrylate in the presence of Bu^tOK and CH_3ONa is described, Use of solvents (tret-butanol,

2-propanol, TGF, DMSO, DMF) is declared in patents [5–7] and publications [8]. Dilating of possibilities of synthesis ArOH educed at the expense of augmentation of assortment of alkaline components that in certain degree defined novelty and patentability various before the described means of reception ArOH. Such approach first of all became possible in communication to specific properties of hydroxyl group of a molecula 2,6-di-tert. butylphenol (pK =14) [9] and configurations of frame of a molecula owing to influence steric factors [10]. In this connection in solution 2,6-di-tert. butylphenol does not react with hydroxides of alkali metals. However, at interaction Ar'OH with alcoholates, according to some explorers, it is formed 2,6-di-tert.butylphenolate alkali metal and this alkaline component participates in reaction of alkylation Ar'OH [8]. Reactivity of such system is defined by properties of ion pairs, the nature of cation, solvent and other factors influencing reaction rate, its selectivity and conversion Ar'OH. In practice of realization of a mean of reception of propionate ArOH oxy-hydroxides of potassium [2, 11] and sodium [11], hydride of lithium [8], amid sodium [12], sodium phenolate [13], borhydride sodium [14] have been used sodium methoxide in methanol [2]. At presence above the specified alkaline components the main product (ArOH) is formed with a yield of 75–90% during reaction in the range of 3–18 h. In the presence of lithium hydride at 120°C it is formed about 0,5% 3-(3',5'-di-tert.buyil-4'-hylroxyphenyl)-dimethylglutorate (Ar'OH). At interaction Ar'OH with methyl acrylate in the presence of lithium hydride at 160°C yield Ar'OH ~ 9% [15]. The following stage of development of the yielded direction has been bound to workings out of accelerator on the basis of interaction of potassium (KOH) or sodium (NaOH) hydroxides with Ar'OH at elevated temperature in the conditions of water branch removing with participation of solvent [16–19]. In these conditions there is a particulate conversion of hydroxides of alkali metals in the corresponding 2,6-di-tert.butylphenolate potassium (Ar'OK) or sodium (Ar'ONa). Kinetics and the mechanism of reaction of alkylation 2,6-di-tert.butylphenol by methyl acrylate at presence 2,6-di-tert.butylphenolate lithium investigated in solution DMSO [20]. On the basis of this data the assumption is come out that reaction proceeds on the bimolecular mechanism. In the conditions of "a pseudo-catalytic" mode accumulation ArOH speed of its formation linearly depends on concentration 2,6-di-tert.butylphenol lithium. Optimum parameters of alkylation are: time-7–8 h, an interrelation of reagents: Ar'OH, LiH, the methyl acrylate, equal 1: 1: 3, temperature of 90–95°C. In these conditions yield ArOH reaches 80–85%. In the course of reaction methyl acrylate oligomers are formed. The further improvement of process of reception ArOH it is bound to building of more perfect accelerator, keeping 2,6-di-tert.butylphenolate potassium (or sodium) and the conforming hydroxide of alkali metal [11, 21, 22]. The yielded kinetics of reaction and results of researches of influence of various factors on selectivity of process have given the grounds to assume the Schema 1.

Schema 1

$$\text{Ar'OH+KOH} \rightleftharpoons \text{Ar'OH} + H_2O \quad (1)$$

$$\text{Ar'OK+MA} \rightleftharpoons \text{Ar'OK.MA} \quad (2)$$

$$\text{Ar'OK.MA} + H_2O \overset{K_5}{\rightleftharpoons} \text{ArOH} + \text{KOH} \quad (3)$$

$$\text{Ar'OK.MA} + H_2O \longrightarrow \text{Ar'AlkOMe} + CH_3OH \quad (4)$$

The decision of system of the differential equations of the kinetic schema including the Eqs. (1)–(3), at initial conditions [Ar'OH] = $[\text{Ar'OH}]_o$, [MA] = $[\text{MA}]_o$, [Ar'OK] = $[\text{Ar'OK}]_o$, [KOH] = $[\text{KOH}]_o$, $[H_2O] = 0$, [Ar'OK.MA] = 0 $[\text{Ar'OH}]_t = 0$ and $[\text{KOH}] + [H_2O] = [\text{KOH}]_o$

Leads to expression:

$$W_o = k_5 \{[\text{Ar'OK}]_o + [\text{KOH}]_o\} \cdot [\text{KOH}]$$

That will be coordinated with experimental data. On Eq. (4) there is an irreciprocal expense of the alkaline agent.

Results of Alkylation 2,6-Di-Tert.Butylphenol by Methyl Acrylate In Concentrated Solutions

Reaction realization in concentrated solutions has the features bound to participation in the elementary certificate of contact ion pairs. Classical methods of the analysis of results of kinetics of investigated reaction in concentrated solutions, as a rule, are bound to certain conditions that are bound to specificity of the mechanism of investigated reactions. Alternative is the analysis of the kinetic results based on use of computer programs. The first work [23] these series is published in 1989 and devoted research of the mechanism of reaction of alkylation of Ar'OH by methyl acrylate in the presence of Ar'OK and KOH. The kinetic schema contained 28 unit steps describing the mechanism of formation of propionate ArOH and others, both mediate, and reaction finished products [11, 21, 22]. According to Figs. 1 and 2 [24, 25] in the kinetic schema used 30 component and 62 stages describing the mechanism of reaction. In comparison with [23, 26] it is included dimerization stages 2,6-di-tert.butylphenolate potassium and formation of cations of metals in dimers. In the presence of water or alcohol the processes bound to formation of hydroxide of potassium, products of its interaction with water (ion pairs) and alcoholates become more awake. Taking into account possible processes the yielded kinetic schema has allowed predicting results of reaction Ar'OH with methyl acrylate in a wide range of concentration of components of reaction, including accelerator. On optimization of constants of speeds satisfactory coincidence settlement and experimental data has been reached.

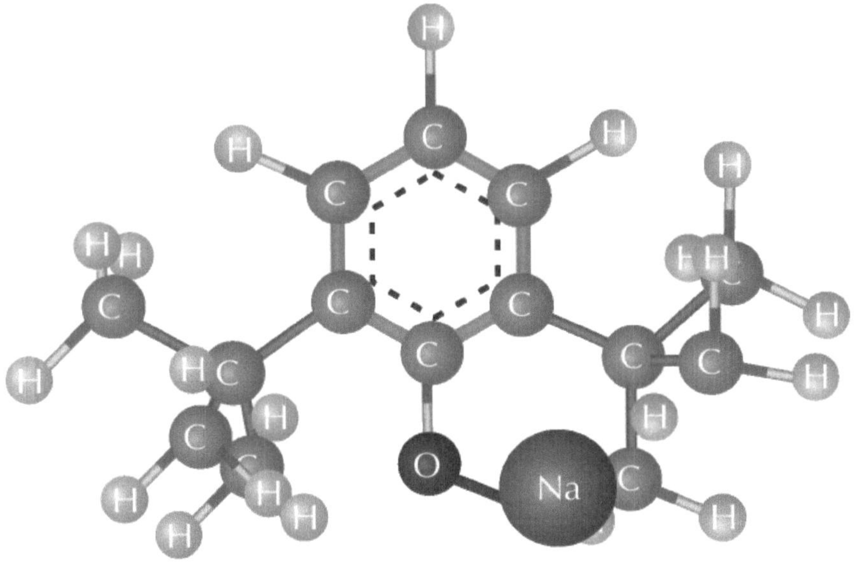

FIGURE 1 Monomer Ar'ONa.

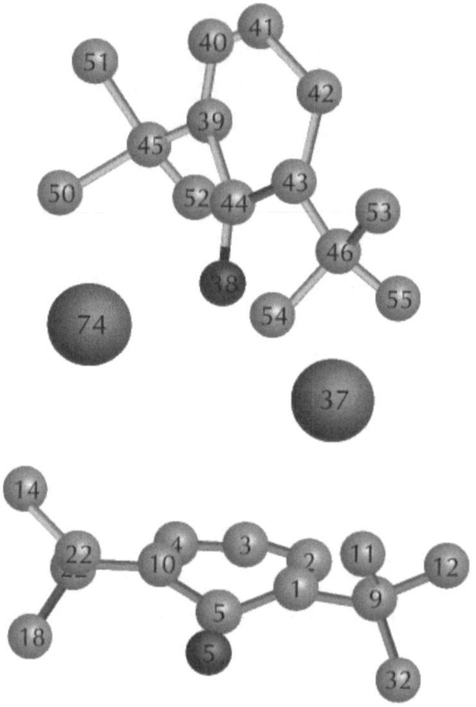

FIGURE 2 Dimer Ar'ONa.

Computer Modeling of a Constitution And Properties of 2,6-Di-Tert-Butylphenolates Alkali Metals

Computer modeling is based on calculation of frames by the quantum-chemical method, allowing to calculate geometrical and power parameters. From results of definition of dependence of energy of formation (H_f^o) from frame follows that the local minimum (H_f^o = –366.4 kJ·mol^{-1}) 2,6-di-tret-butylphenolate sodium corresponds to frame of monomer Ar'ONa (1*) local minimum H_f^o = –926,7 kJ·mol^{-1} corresponds to dimer Ar'ONa (2*).

Dimerization is accompanied by changes of charges, sizes of lengths of communications and angles between communications of atoms (Tables 1–3).

TABLE 1 Charges and lengths of communications of frames of monomer (1) and dimer (2).

Frame Na*/K**	Charge *q*				Length of communication *d* / Å		
	M (37)	M (74)	O (6)	C (3)	M (37) – (6)	WITH (5) – (6)	M (74)–O (6)
1*	+1,055		–0,719	–0,141	2, 122	1,279	
2*	+1,425	+0,933	–0,724	–0,170	1,259	2,378	3,495
3**	+1,221		–0,802	–0,180	3,481	1,249	
3a**	+1,028		–0,648	–0,844	2,978	1,223	
4**	+1,204	+1,204	–0,749	–0,173	3,694	1,263	2,633

TABLE 2 Torsionne angles in frames of monomer (1) and dimer (2).

	Angle ω/ hailstones			
	M (37)–O (6)–C (5)	M (37)–C (5)–C (1)	O (6)–C (5)–C (1)	C (3)–C (1)–C (9)
1*	110.9	119.6	119.6	18.2
2*	49.1	76.9	121.5	20.7
3 **	106.2	80.8	121.3	18.3
3a **	78.6	98.8	122.2	16.2
4 **	53.0	80.0	120.4	19.1

TABLE 3 Torsionne angles in dimers 2* and 4**.

	M (37)–O (38)–C (44)	M (74)–O (6)–C (5)	O (6)–M (74)–M (37)	O (38)–M (37)–M (74)
2*	127.2	108.5	66.7	78.2
4**	105.5	105.5	75.5	75.5

At dimerization Ar'ONa energy of Gibbs increases (ΔG^o = 57.3 kJ·mol^{-1}), entropy decreases (ΔS^o = –39.7 cal/K/mol). In balance: monomer (1) ⇔dimer (2) K_{298} ~ 0.6, where K_{298} – a thermodynamic equilibrium constant.

The augmentation of radius of cation in frame of monomer Ar'OK **(3)** leads to geometry change in comparison with Ar'ONa **(1)** owing to redistribution of electronic density between charges. Local minimum of potential energies (H_f^o) are equal – 361.6 and –393.1 kJ·mol^{-1} accordingly. In frame **3** length of communication (*d*) About (6) – with (5) = 1,223 Å, in frame **3à** *d* = 1,249 Å, in frame **1** *d* = 1,279 Å. The charge on atom with (3) in hexatomic cycles decreases: q = 0,1796 (**3**); q = –0,1411 (**1**). From comparison of geometry of frames of dimers **2** and **4** follows that the frame **4** is symmetric for example, angles To (37) – About (38) – with (44) and To (74) – About (6) – WITH (5) are accordingly equal 105.5°, whereas in frame **2** angle Na (74) – About (38) – WITH (44) – = 127.2° and angle Na (74) – O (6) – C (5) = 108.5°. At dimerization Ar'OK energy of Gibbs increases (ΔG^o = 51,9 kJ·mol^{-1}), entropy decreases (ΔS^o = –54.6 cal/K/mol). In a thermodynamic equilibrium: monomer (3) ⇔dimer (4), K_{298} ~ 0.6.

Dependence of geometry of frames on metal cation in moleculas 2.6-di-tert-butyl-phenolates also possibility formation of dimers does this bunch of bonds perspective for research of frames and complexes with participation of polar solvents. From calculations of complexes of monomer Ar'OK with one molecula DMSO follows that there can be some local minimum of potential energies from which to the steadiest formation there corresponds frame 5a (Fig. 3).

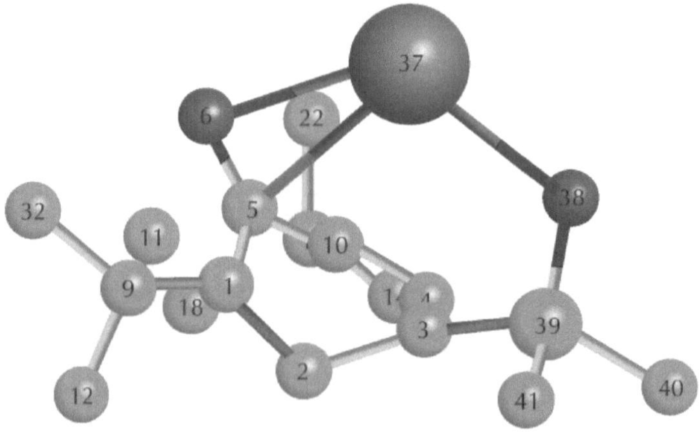

FIGURE 3 5a – the complex frame 2,6-di-tert.butylphenolate potassium with DMSO.

In this frame *d* About (6) – To (37) = 2,842, To (37) – with (3) = 1,536 Å. And *d* with (3) – S (18) = 1,536 Å. This data testifies that in a complex 5a communication with (3) – S (18) is strong enough. Geometry of complexes 2,6-di-tret.butyilphfenolate potassium with 2 moleculas DMCO (**5b**) and 3 moleculas DMCO (**5c**, Figs. 4 and 5) differ on angles and charges.

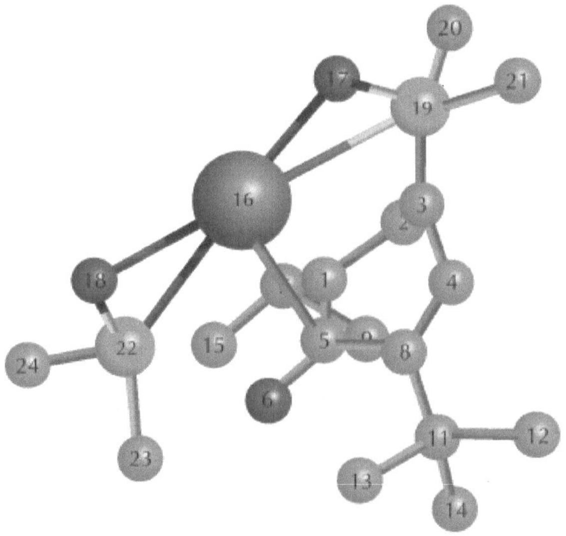

FIGURE 4 Structure a complex oxyhydroxide 5b Structure a complex oxyhydroxide with 2 DMCO oxyhydroxide potassium with 3 DMCO.

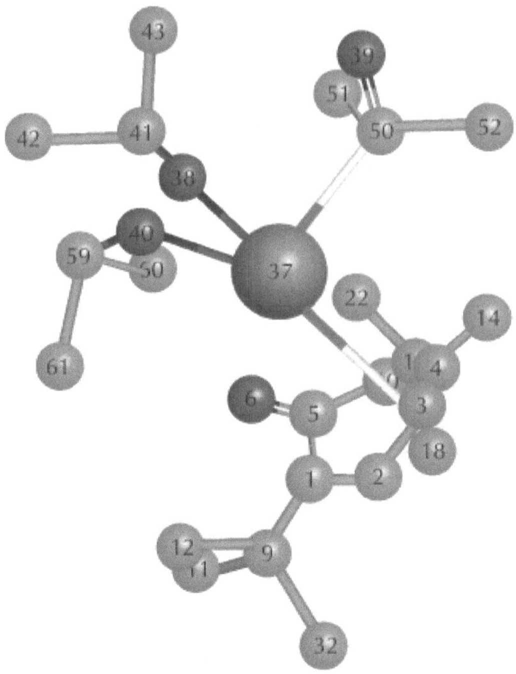

FIGURE 5 Structure of samples studied.

With augmentation of number of moleculas DMSO participating in salvation of cation reactivity should increase 2,6-di-tert.butylphenolate in combination reactions on carbon atom in pair position of an aromatic cycle.

Monomers 2,6-di-tert-butylphenolate it is formed at interaction 2,6-di-tert.butyphfenol with sodium hydroxide (or potassium) at temperature >180°C, and at refrigerating dimerization [27] that will be coordinated with the calculation data spontaneously proceeds. Dimers 2,6-di-tert.butylphenolate are steady in an individual kind, them use as accelerators of reactions of alkylation in polar solvents, in particular DMSO [6, 11]. The information on a constitution and properties of products of interaction of the specified phenolates with DMSO is actual enough. The frame of a complex of dimer 2,6-di-tert.butylphenolate sodium with DMSO is confirmed by a X-ray diffraction technique [28]. Results of calculation of geometry of frame of a complex 2,6-di-tert.butylphenolate sodium with two moleculas DMCO (**6**) are presented in Tables 4 and 5.

TABLE 4 Charges and lengths of communications of frames of complexes of dimers **(6)** and **(7)** with DMSO.

	Charge *q*				**Length of communication *d* / Å**		
	M (37)	**M (74)**	**O (6)**	**C (3)**	**M (37) – (6)**	**WITH (5) – (6)**	**M (74)–O (6)**
6*	+1,151	+0,732	–0,832	–0,155	3,124	1,264	5,627
7 **	+1,155	+1,204	–0,829	–0,166	3,755	1,266	5,941

TABLE 5 Torsionne angles in frames of complexes of dimers **(6)** and **(7)** with DMSO.

	Angle ω/ hailstones				
	M (37)–O (6)–C (5)	**O (38)–C44)–M (74)**	**O (6)–M (37)–O (38))**	**O (85)–M (74)–O (75)**	**M (74)–O (6)–C (5)**
6*	57.1	92.3	126.0	38.1	107.2
7**	52.3	93.1	145.6	51.3	92.2

Comparison of results of calculation and experiment on separate positions are close enough. In calculation angle Na (37) – O (85) – Na (74) =

96,1°, in experiment – 95.5°; in calculation *d* with (5) – About (6) = 1,264 Å. In experiment – *d* = 1,257 Å; In calculation Å *d* Na (37) – O (38) – = 2,342 Å, In experiment – *d* = 2,366 Å.

To geometry of complexes of dimers with DMSO (**6** and **7**) are close on communications: About (6) – with (5), About (38) – with (44); to angles: About (38) – with (44) – Na (74) = 92.3° and O (38) – C (44) – K (74) = 93.1° and to other positions (Figs. 6 and 7).

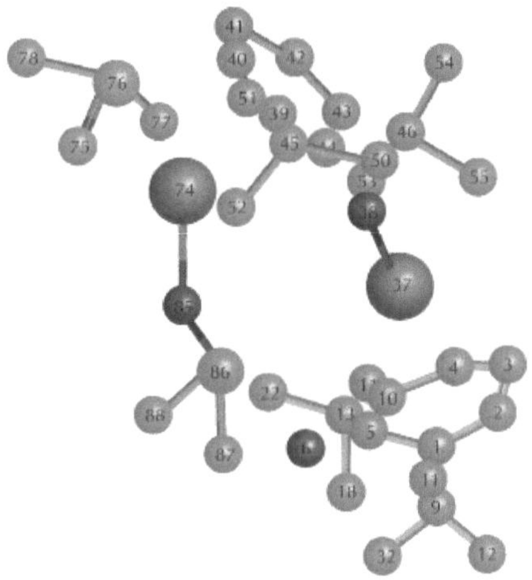

FIGURE 6 6 – a complex of dimer 2 with DMSO.

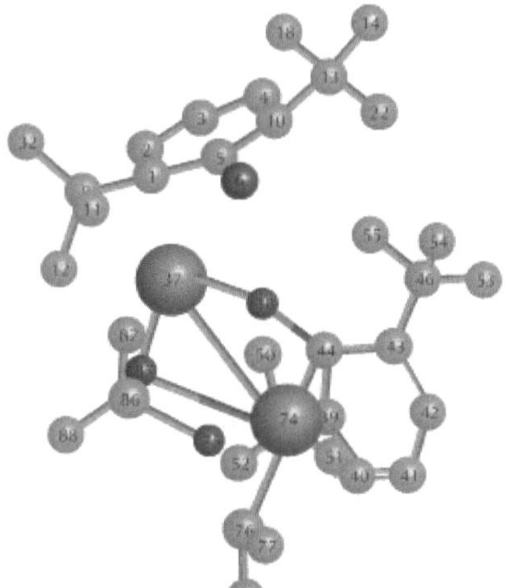

FIGURE 7 7 – a complex of dimer 4 with DMSO.

Apostatis in geometry with participation of atoms of metal, for example, on communications is revealed: Na (37) – O (6) = 3,124 Å and K (37) – O (6) = 3,755 Å; Na (74) – O (6) = 5,627 Å è K (74) – O (6) = 5,941 Å because of distinction in radius of atom of sodium.

Synthesis and Properties of Monomers and Dimers 2,6-Di-Tert. Butylphenolate Potassium and Sodium

The quantitative data characterizing change of composition of solution in the course of interaction 2,6-di-tert.butylphenol c hydroxides of alkali metals, have allowed analyzing possibilities of this reaction [29]. Equilibrium concentrations of components at 100°C in reaction Ar'OH c KOH are positioned within 25–30 min after mixing of reagents and solution thermostatic control. Reaction Ar'OH c NaOH proceeds with smaller speed and balance is positioned for 55–65 min. Values of equilibrium constants (K_{eff}) reactions Ar'OH c KOH and reactions Ar'OH c NaOH have been calculated. For the purpose of finding-out of a role of diffusion in static conditions the system is investigated: condensed – a gaseous (vaporous) phase. In 30 min of interaction Ar'OH c KOH in a fluid phase the above-stated value becomes equal To $_{эфф}$. And practically does not change in system. The similar result is received at research of reaction Ar'OH c NaOH. From this data follows that diffusion does not make essential impact on an interrelation of formed and initial components in the range of temperatures of 100–140°C. Hence, at interaction of Ar'OH with hydroxides of alkali metals at temperature 100–120°C in the conditions of vacuum blowing in time only the admixture phenolate and the alkali metal hydroxide which efficacy is considered earlier till 2 o'clock can be formed. However at temperature of 180–200°C speed of interaction of Ar'OH with KOH or NaOH has appeared sufficient for formation of the fixed water volume in the form of a vaporous phase, to allocation of water vapor from reactionary mass and to formation 2,6-di-tert.butylphenolate alkali metals with a quantitative yield [30, 31] At high temperature in the course of interaction of Ar'OH c hydroxides of alkali metals the monomeric form which in the course of temperature depressing is transformed to steadier dimer form is formed.

From the yielded NMR ^{1}H spectrums 2,6-di-tert.butylphenolare potassium and sodium follows that signals from meta- and para-protons are located in more stronger sex in comparison with meta- and the pair-protons 2,6-di-tert.butylphenol. The kind of electronic spectrums of dimers 2,6-di-tert.butylphenolate potassium and sodium in DMSO changes in time that specifies in interaction of the above-stated phenolates with polar solvent. The dimer spectrum 2,6-di-tert.butylphenoatel alkali metal, after bond dissolution in solvent, is presented by two strips with maximum in the field of 260–270 nanometers, 311 322 nanometers and a wide strip with a maximum of 761–764 nanometers which are transformed with a roping with maximum of 421–436 nanometers and 475–481 nanometers. The strip with a maximum of 475–481 nanometers is transformed to a strip with maximum 421–436 nanometers. Change of optical density of strips begins with the moment of preparation of solution and proceeds within 4–6 h. Dependence of value of optical density (D) absorption of a strip with λ_{max} =320–317 nanometer from time is linear and proceeds with a constant k_1 which value depends by nature solvent. Similar results are received at research of electronic spectrums 2,6-di-tert.butylphenolate sodium in solutions DMSO and DMF. The specified changes in electronic spectrums have been interpreted as consequence of interaction of dimers of phenolates with solvent which lead to formation of the most stable frames depending on pair: phenolate metals – solvent

Laws and the Mechanism of Reaction of Dimers 2,6-Di-Tert. Butylphenolate Alkali Metals with Methyl Acrylate in Solutions Polar Solvents.

In solution DMSO or DMF dimers 2,6-di-tert.butylphenolate potassium or sodium as reagents interact with methyl acrylate with formation of ArOH. In absence of solvent the basic direction is polymerization of methyl acrylate [23, 27]. The kinetics of reaction of dimer Ar'OK with methyl acrylate in solutions DMSO and DMF is investigated and on the basis of comparison of the experimental and settlement data of the kinetic schema taking into account interaction of dimer Ar'OK with solvent calculates effective constants (k_2) reactions of associates Ar'OK-solvent with methyl acrylate. In this case the kinetic schema 2 is presented by four reactions.

THE SCHEMA 2

The schema 2

$$(ArOK)_2 + \text{ solvent} \underset{k_{-1}}{\overset{k_1}{\rightleftharpoons}} 2\ ArOK.\text{- solvent} \qquad (5)$$

$$ArOK.\text{- solvent } + MA \underset{k_{-2}}{\overset{k_2}{\rightleftharpoons}} ArOK\text{- } MA + \text{solvent} \qquad (6)$$

$$ArOK\text{- } MA \xrightarrow{k_3} Ar'OK \qquad (7)$$

$$ArOK + MA \longrightarrow Ar'OH \qquad (8)$$

Mathematical modeling of the experimental kinetic data of accumulation (ArOK) taking into account value $k_1 = 7{,}5\cdot10^{-5}\ s^{-1}$, found of dependence of reduction of optical density of a strip with λ_{max}=320 nanometer of dimer Ar'OK in DMSO is made?.

At initial concentration: $[Ar'OK]_{2o} = 0.1$ mol·l^{-1}; $[MA]_o = 0.3$ mol·l^{-1}; $[DMSO]_o = 12$ mol·l^{-1} $k_1 = 7.5\cdot10^{-5}\ s^{-1}$; $k_2 = 2.3\cdot10^{-}$ l·mol^{-1}·s^{-1}; $k_{-2} = 4\cdot10^{-3}$ l·mol^{-1}·s^{-1}; $k_3 = 6\cdot10^{-3}$.

In solution DMF: $[Ar'OK]_{2o} = 0.1$ mol·l^{-1}; $[MA]_o = 0.3$ mol·l^{-1} $[DMF]_o = 12$ mol·l^{-1} $k = 3.8\cdot10^{-4}\ s^{-1}$; $k_2 = 6.3\cdot10^{-3}$ l·mol^{-1}·s^{-1}; $k_{-2} = 4\cdot10^{-3}$ l·mol^{-1}·s^{-1}; $k_3 = 2\cdot10^{-3}$.

Calculation of the kinetic schema in reaction of methyl acrylate with dimer Ar'ONa in DMSO leads to result: $k_1 = 9{,}1\cdot10^{-4}\ s^{-1}$; $k_2 = 9.7\cdot10^{-4}$; $k_{-2} = 7.4\cdot10^{-4}$ l·mol^{-1}·s^{-1}; $k_3 = 8.2\cdot10^{-3}$.

Reaction of dimer **2** or **4** with methyl acrylate in solution proceeds through stages of formation of complexes of monomers with DMSO. At the same time there is opened a question, concerning to a stage of addition of methyl acrylate to a complex and the mechanism of desolvation DMSO. The data of calculation of some the frames containing fragments DMSO and methyl acrylate gives the grounds to assume that in the conditions of surplus of solvent reaction proceeds through stages of addition of a molecula of methyl-acrylate to potassium cation, bound to three moleculas DMSO. According to calculation complexes: 2,6-di-tert.butylphenolate potassium – 3 moleculas DMSO-methyl akrylate, correspond to frames **8**a and **8b** (Figs. 8 and 9).

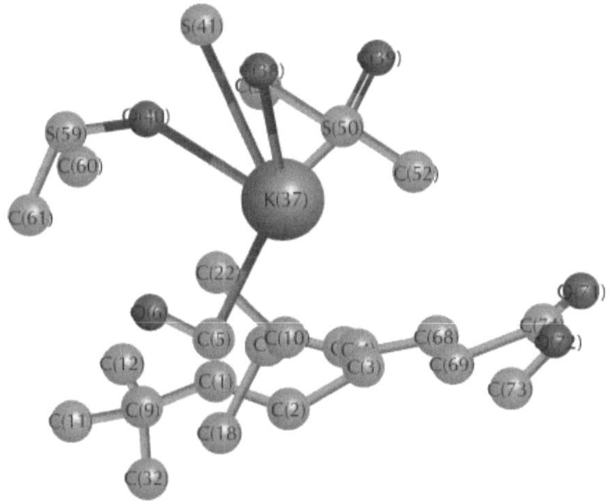

FIGURE 8 8a – frame of complex Ar'OK+3DMSO+MA.

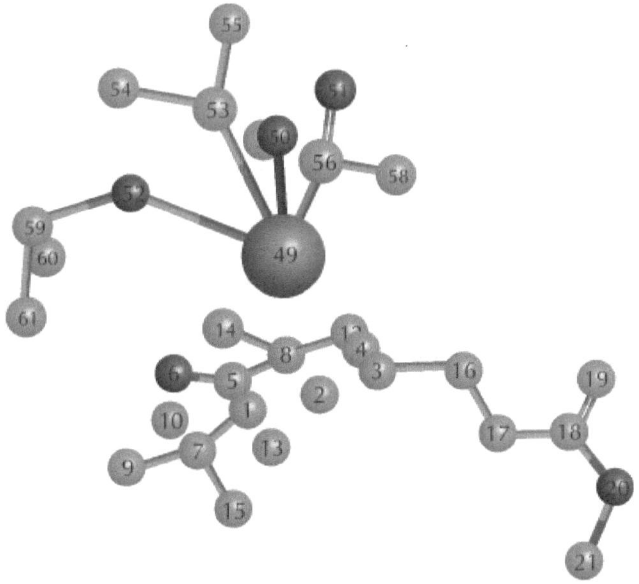

FIGURE 9 Frame 8b.

In complexes **8a** and **8b** cation is bound to three moleculas DMSO and bonding of a molecula of methyl acrylate with a hexatomic cycle; communication With (3) – With (68). The local minimum (H_{8a}^{o}) = –1296,1 kJ·mol^{-1} corresponds to frame **8a**, H_{8b}^{o} = –1392,4 kJ·mol^{-1} – frame **8b** which represents an ion pair: phenolate-ion of a methyl ester 3-(3',5'-*di-tert-butyl*-4'-hydroxyphenyl)-propionic acid –potassium cation. This data has allowed to specify the mechanism of reaction of dimer Ar'OK (4) with methyl acrylate taking into account a stage of formation of complexes **7** and **5c** with the subsequent additions of methyl acrylate to a complex **5c** (the Schema 3).

THE SCHEMA 3

The schema 3

$$4 + 2.\text{DMSO} \xrightleftharpoons{K=1.54} 7 \qquad (9)$$

$$7 + \text{DMSO} \xrightleftharpoons{K=3.1} 5c + 3 \qquad (10)$$

$$3 + \text{DMSO} \xrightleftharpoons{K=1.43} 5a \qquad (11)$$

Equilibrium constants of reaction of a complex formation with solvent participation are calculated from the settlement yielded enthalpies and entropies of frames of the bonds participating in reversible reactions (the equation of Boltsmana).

From the settlement data follows that in Eqs. (9)–(11) and, hence, this data confirms complex formation **5c** from dimer **4** in the conditions of salvation. Mathematical modeling of kinetics of reactions under the schema 3 on the basis of the decision of the differential equations for initial conditions: $[\mathbf{4}]_{o} = 0{,}1$; $[\text{MA}]_{o} = 0{,}3$; $[\text{ДMSO}]_{o} = 12$ mol·l^{-1}, yields result: $k_3 = 2{,}3 \cdot 10^{-3}$; $k_{-3} = 4 \cdot 10^{-3}$ l·mol^{-1}·s^{-1}. The presented data is based on complex formation **8b** (H_{8b}^{o} =–1392,4 kJ·mol^{-1}) in the course of interaction **5c** with methyl acrylate. However the frame **8a** with a local minimum of energy is possible: H_{8a}^{o} =–1296,1 kJ·mol^{-1}. The frame **8a** is formed at complex interaction **5c** with methyl acrylate and this formation can be considered as a transition state at formation **8b**. Process through a transition state proceeds in time and, hence, has the kinetic laws. In this connection the calculated constants k_3 and k_{-3} are effective. At frame isomerization **8a** in **8b** geometrical parameters variate, and process should be considered within the limits of the theory isomer–isomeasured of states [32]. In the case under consideration the model of connected isomers with incorporated level of energy, apparently, takes place. On catalytic activity dimer Ar'OK is similar to an interaction product 2,6-di-tert.butylphenol with Bu^tOK.

Monomer Ar'OK is most effective of earlier known accelerators of reaction of alkylation 2,6-di-tert.butylphenol methyl acrylate. At the maintenance of monomer Ar'OK in number of 1.5–3% a pier from Ar'OH and temperature of 110–115°C reaction concludes for 15–20 min with a yield of propionate ArOH to 98%. Reaction with 2,6-di-tert.butylphenolate potassium, received of ArOH and Bu^tOK (>5% the mole) in simulated condition proceeds for 2,5–3 h with yield ArOH no more than 85%. Catalytic properties of monomeric form ArONa are similar ArOK whereas reaction of Ar'OH c methyl acrylate in the presence of dimer Ar'ONa proceeds with more low speeds.

Influence of Polar Solvents and the Mechanism of Catalytic Alkylation 2,6-Di-Tert.Butylphenol With Methyl Acrylate in the Presence of Monomers 2,6-Di-Tert/Butylphenolate Potassium or Sodium.

Distinctions in properties of two forms Ar'OK are most distinctly displayed at their use as accelerators of alkylation of phenol Ar'OH. Monomer Ar'OK is formed at interaction KOH with Ar'OH at temperature of 180–190°C and a molar interrelation the KOH: Ar'OH = 0.03–1, and further, after refrigerating of reactionary mass to temperature of 143–120°C, to the received admixture added methyl acrylate. In these conditions reaction of catalytic alkylation proceeds for 15–20 min with formation ArOH with yields to 98%. Results of researches of kinetics of this reaction in the range of 105–125°C have shown that the kind of kinetic curves does not depend on temperature of reaction that can testify to "tunnel effect" in the mechanism of catalytic alkylation in the presence of monomer Ar'OK.

Results interaction Ar'OH with methyl acrylate in the presence of dimer form Ar'OK are similar to results of reaction Ar'OH c methyl acrylate at presence 2,6-di-tert.butylphenolate potassium received at interaction of Ar'OH with Bu^tOK. Polar solvents (DMSO, DMF) were used in reaction of alkylation of phenol Ar'OH in the presence of alkaline accelerator and thus yield ArOH 80–85%. Polar solvents lead to inhibition of reaction of catalytic alkylation Ar'OH by methyl acrylate, and negative influence of solvents is displayed in microquantities. In the presence of solvent speed of interaction Ar'OH with methyl acrylate in the conditions of catalytic alkylation decreased, and the inhibiting effect of additives of polar solvents increased among: MeCN > DMF > DMSO (Fig. 10).

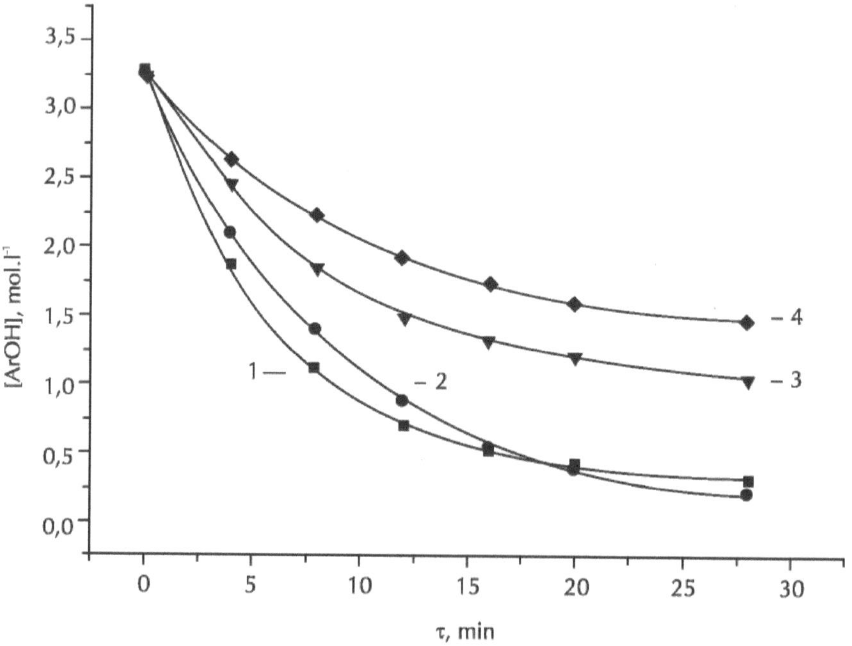

FIGURE 10 Kinetics of reaction Ar'OH c methyl acrylate (MA) in the presence of Ar'ONa and additives of solvents, 110°C. 1 solvent is absent; 2 – MeCN; 3 DMF 4 – DMSO. $[Ar'OH]^o$ = 3.26–3.28; $[MA]^o$ = 3.5–3.6; $[Ar'ONa]^o$ = 0.1 $mol \cdot l^{-1}$; a line – calculation; points – experiment. $[solvent]^o$ = 0 (1); $[MeCN]^o$ = 0.24 (2); $[DMF]^o$ = 0.21 (3); $[DMSO]^o$ = 0.23 (4); $mol \cdot l^{-1}$.

This result is bound to interaction of monomer Ar'OK with a polar component and this interaction is competing to a stage of formation of a transition state of phenolate Ar'OK, Ar'OH and methyl acrylate. The specified laws are characteristic for reactions with a polar component in the concentration comparable to concentration of phenolate Ar'OK. Under this schema formation of resultant of reaction ArOH proceeds through a stage of formation of a complex with participation of moleculas of monomer Ar'OK, Phenol Ar'OH, methyl acrylate and moving of cation of potassium to a template cage. However possibility of a yield of cation of potassium in volume with formation of kineticly independent particle is not excluded. In this case reaction of catalytic alkylation 2,6-di-tert.butylphenol methyl acrylate will proceed on the is ionic–chain mechanism.

The augmentation of concentration Ar'OK in reactionary mass above optimum (about 5 and more % a pier) leads to changes of character of interaction of the reagents bound to activation of side reactions and formation of oligomers of methyl acrylate. At interrelation Ar'OH: methyl acrylate > 3 and maintenance Ar'OK more than 10% an alkylation product yield – ArOH compounds a pier no more than 15% with simultaneous gain in yield of oligomers of methyl acrylate. Stimulating influence of polar solvents in the conditions of bimolecular addition is known [33]. However in this case in the conditions of catalytic alkylation polar solvents have negative influence that is displayed in decrease in speed of expense Ar'OH and reduction of a product yield of alkylation – ArOH. This fact contradicts the classical mechanism, however will be co–ordinated with the ionic-chain mechanism as, disassociating metal cation participates in reactivation of the catalyst Ar'OK and is that particle which provides reaction on the is ionic–chain mechanism.

Continuous Process of Reception Methyl Ester 3-(3′,5′-Di-Tret. Butyl-4′-Hydroxyphenyl)-Propionic Acid – The Following Stage of Technology

In respect of the technological decision of production of a methyl ester 3-(3',5'-di-tert.buyil-4'-hylroxyphenyl)-propionic acid (ArOH) it is represented to the most perspective the mean of catalytic alkylation 2,6-di-tert. butylphenol (Ar'OH) in the presence of monomer 2,6-di-tert.butylphenolate potassium (Ar'OK). Till now production of propionate ArOH based on technology of batch process and catalysis conditions on a basis compositions from 2,6-di-tert.butylphenolate potassium (sodium) and hydroxides of potassium (sodium) of uncertain composition. To predict in these conditions a net result it is represented problematic. In continuous process probably to carry out continuous control over reaction passing. Proceeding from the same initial components (2,6-di-tert.butylphenol, methyl acrylate and alkali metal (hydroxide) at the expense of technology improvement is obviously possible to improve industrial indexes. For comparison results of reception of propionate ArOH on known and yielded a pilot plant [34] developed at Institute of the biochemical physics of the Russian Academy of Sciences are presented, on continuous process of reception of propionate ArOH (Table 6). From the resulted data follows that one of limiting factors is the accelerator preparation time on reaction of Ar'OH with hydroxide of alkali metal in the conditions of branch of water in the course of reaction. In continuous process time compounds few min, and during reaction of alkylation during 30–40 mins reaction enters to 98% initial Ar'OH.

TABLE 6 The relative data of results of reception of a methyl ester 3-(3',5'-di-tret.butyl-4'-hydroxyphenyl)-propionic acid (ArOH).

Ar'OH/MA/ ArO'K (Na) mol.% Pier	Accelerator °C, Time, h		Alkylation °C, Time, h		Yield ArOH mol% Pier.	References
1: 1,27: 0,05	120–130	3–4,5	110–120	2–3	88	11
1: 1,1: 0,02	100	1	115	>2	93	16
1: 1: 0,06	130–135	3	90–120	1	96	35
1: 1,2: 0,1			160	5	77	8
1:1,15: 0,03			110	7–8	90	4
1:1,2: 0,02	180–190	0,08	100–120	0,5	98	34
1: 1,2: 0,03			50	5	87	2

The quantity of impurities in a kind dimethylglutorate (Ar''OH) does not exceed 0,2% that allows to use reactionary mass as a reagent in ester exchange for antioxidant – reception «phenosan 23». From results of Table 6 follows that in continuous process [34] the quantity KOH in 2.5 times less in comparison with batch process in the presence of accelerator on the basis of hydroxide of potassium [11] is used. Further, in the presence of the accelerator prepared as a result remove of water by vacuum blowing at 100–135°C and use in reaction of alkylation Ar'OH by methyl acrylate about 7% of impurities [11] are formed. In this connection in technology of reception ArOH used stages of branch and main product purification.

Use as initial formulation constituents of hydroxides of potassium and sodium improves quality indicators of technological process in a periodic duty. It is necessary to notice that on use compositions from potassium and sodium hydroxide the USSR [35], and with a priority earlier, than the patent "Siba-Gaigy" is received patents "Siba-Gajgi" [16] and Institute of chemical physics.

Installation for continuous process represents a design from the stainless steel consisting of the reactor in a kind filling station from three sections, bridged among themselves valves for moving of a reaction mixture from one section in another (Fig. 11).

FIGURE 11 Installation for realization of continuous process of reception of a methyl ester 3-(3',5'-di-tert.buyil-4'-hylroxyphenyl)-propionic acid.

Into the top section from container pomp introduce to 180–190°C fixed volume 2,6-di-tert.butylphenol, and from a batcher inject solution of hydroxide of alkali metal. In this section in the conditions of branch of water vapor by a inert gas current the admixture is formed 2,6-di-tert.butylphenol also monomer 2,6-di-tert.butylphenolate. At valve crack the admixture moves to the second section into which inject methyl acrylate. In the second section the reactionary mass which moves further at discovering of the second valve to the third section where alkylation reaction proceeds is formed. The main product of its own accord through a reducer in a continuous mode follows from a horizontal part of section.

The mechanism of regulation is placed by work of valves, pompes in blocks inert gas repeatedly moves in the isolated system. From water vapor inert gas separates in container with an absorbent and further the compressor moves in section.

The installation design allows to variate a temperature schedule in each of three sections, time of reaction and an interrelation of reagents on a continuous process course that has allowed to optimize process of reception of a main product.

KEYWORDS

- **Ion pairs**
- **Pentaerythritol**
- **Phenolate**
- **Pseudo-catalytic**

REFERENCES

1. Volod'kin, A. A.; Zaikov, G. E. *Ros.Khim. Zh.*, **2000**, *44*, 81–88 [*Mendeleev chem. J.*, **2000**, *44*, 81–88 (Engl. Transl.)].
2. Dexter, M.; Meier, E. *Alkylphenols*. U.S. Patent 3364250, 1968. Chem. Abstrs., 1968, Vol. 68, pp 59318.
3. Haeberli, J.; Park, K.; Vellturo, A. *Nurnberg G., 2,6-dialkilphenols.* Ger. Patent 2364126, 1972. Chem. Abstrs., 1974; Vol. 81, pp 120260.

4. Burton, L. *4-hydroxy-3,5-dialkylphenylpropionic Acid Ester Derivation.* U.S. Patent. 4659863, 1987; Chem. Abstrs, 1989; Vol. 107, pp 77439.
5. Nachitani, M. *2,6-dialkylphenols Ester Derivation.* Jap. Patent 12341, 1981. Chem. Abstrs., 1985, Vol. 95, pp 42686.
6. Yamada, H.; Nanide, Y. *Method Prepare of 4-hydroxy-3,5-dialkylphenylpropionic Acid Esters.* Jap. Patent 84785, 1973; Chem. Abstrs., 1974; Vol. 81, pp 50593.
7. Tahara, K.; Motoya, Sh. I. *Hindered Phenols.* Jap. Patent 7933, 1971. Chem. Abstrs, 1971; Vol. 75, pp 35453.
8. Titova, T. F.; Krysin, A. P.; Shakirov, V. I.; et al. *J. Org. Chim.*, **1984**, *20*, 331–335.
9. Coffild, T.; Fielbey, A.; Ecke, E.; Kolka, A. *J. Am. Chem. Soc.*, **1957**, *79*, 5019–5022.
10. Cohen, L.; Jones, W. *J. Am. Chem. Soc.*, **1960**, *82*, 1907.
11. Volod'kin, A. A.; Gorbunov, B. N.; Ershov, V. V. *Methyl Ester 3- (3,5-di-tert.butyl-4 hydroxyphenyl)-propionic Acid.* Author's Cetificate USSR 1001649, 1981. Bjull. Izobr. 1986, 23, 298.
12. Orban, I. *Antioxydant of Polypropylene.* Eur. Patent 206789, 1987. Chem. Abstrs., 1987, Vol. 107, pp 58657.
13. Swatari, K.; Oda, S.; Rikumaru, K. *New Method Prepare Hindered Phenols.* Jap. Patent 62731, 1973. Chem. Abstrs., 1973, Vol. 79, pp 146244.
14. Sagawa, S.; Fujiyoshi, K. *4-hydroxy-3,5-dialkylphenylpropionic Acid Esters.* Jap. Patent 76037, 1975. Chem. Abstrs., 1976, Vol. 84, pp 30688.
15. Titova, T. F.; Krysin, A. P.; Bulgakov, V. A.;. Mamatyuk, V. I. *Zh. Org. Khim.*, **1984**, *20*, 1899–1902. [*J. Org. Chem. USSR*, **1984**, 20 (Engl. Transl.)].
16. Broogli, F.; Kalin, G. *Method of Preparing 4-hydroxy-3,5-di-tert.butylphenylpropionic Acid Methyl Ester.* US. Patent 5177247, 1993. Chem. Abstrs., 1993, Vol. 116, pp 6249.
17. Nachitani, M.; Charikava, M. *Phenol-alkyl-ester Derivation Prepared.* Jap. Patent 161350, 1981. Chem. Abstrs., 1982, Vol. 96, pp 162344.
18. Eric, E.; Roman, K. *Method of Preparing 4-hydroxy-3,5-di-tert.butylphenylpropionic Acid Methyl Ester.* U.S. Patent 5264612, 1993. Chem. Abstrs., 1994, Vol. 120, pp 191356.
19. Erschov, V. V.; Gorbunov, B. N.; Volod'kin, A. A. *Method of Preparing 4-hydroxy-3,5-di-tert.butylphenylpropionic Acid Alkyl Esters.* UK Patent 2161112, 1983. Chem. Abstrs, 1984, Vol. 100, pp 183456.
20. Bulgakov, A. V.; Gorodetskaja, N. N.; Nikiforov, A. G.; et al. *Izv. Akad. Nauk SSSR, Ser. Khim.*, **1983**, 71 [*Bull. Acad. Sci. USSR, Div. Chem. Sci.*, **1983**, 32 (Engl. Transl.)].
21. Volod'kin, A. A.; Popov, L. K.; Egidis, F. M.; et al. *2,6-di-tert.butylphenol in Organic Synthesis of Stabilizers for Polymers*, NIITKHIM, Moscow, 1987, pp 1–43 [in Russian].
22. Volod'kin, A. A.; Paramonov, I. V.; Egidis, F. M.; et al. *Khim. Prom-st [Chem. Industry]*, **1988**, *12*, pp. 7–8 [in Russian].
23. Volod'kin, A. A.; Zajtsev, A. S.; Rubajlo, L. V.; Belyakov, V. A.; Zaikov, G. E. *Izv. Akad. Nauk SSSR, Ser. Khim.*, **1989**, 1829 [*Bull. Acad. Sci. USSR, Div. Chem. Sci.*, **1989**, *38*, 1677 (Engl. Transl.)].
24. Volod'kin, A. A.; Zaitsev, A. S.; Belyakov, V. A.; Zaikov, G. E. *Polimer Degradation and Stability*, **1989**, *26*, 89–100.
25. Volod'kin, A. A. *Izv. Akad. Nauk SSSR, Ser. Khim.*, **1991**, 989–996.

26. Volod'kin, A. A.; Zaikov, G. E.; Deev, A. S. *Polimer Degradation and Stability,* **1992**, *36*, 25–130.
27. Volod'kin, A.A.; Zaikov, E. G. *Izv. Izv. Akad. Nauk SSSR, Ser. Khim.*, **2006**, 2138–2143.
28. Matilainen, L.; Leskela, M.; Klinga, M. *J. Chem. Commun.*, **1995**, 421–423.
29. Volod'kin, A. A.; Zaikov, G. E. *Dokl. Russian Acad. Sci.*, **2007**, *414*(2), 1–3.
30. Volod'kin, A. A.; Zaikov, E. G. *Izv. Akad. Nauk, Ser. Khim.*, **2002**, **2031–2037** [*Russ. Chem. Bull., Int. Ed.*, **2002**, *51*, 2189].
31. Volod'kin, A. A.; Zaikov, E. G. *Izv. Russian Acad. Sci,, Ser. Khim.*, **2007**, 1971–1878.
32. Gribov, L. A. *Elements of the Quantum Theory of a Constitution and Properties of Molecules*. Moscow, Intelligence, 2010, p 284.
33. Bekker. *Introduction in the Electronic Theory of Organic Reactions*, "World", Moscow, 1977, pp 166–172.
34. Volod'kin, A. A.; Zaikov, E. G. *The Mean of Reception of Methyl Ester 3- (3,5-di-tert.butyl-4-hydroxyphenyl)-propionic Acid.* the Stalemate. The Russian Federation 22331522, 2004.
35. Volod'kin, A. A.; Deev, A. S.; Bulgakov, V. A. *The Mean of Reception of Methyl Ester 3- (3,5-di-tert.butyl-4-hydroxyphenyl)-propionic Acid.* Author's Cetificate.1685920 USSR. RZhKhim. [Abstr. J. Chem.], 1991, Vol. 29 [in Russian].

CHAPTER 9

ACTIVATING BY PARA-AMINOBENZOIC ACID OF SOWING PROPERTIES OF SEED OF WINTER GRAIN CROPS AND FORAGE CEREALS

S. A. BEKUSAROVA, N. A. BOME, L. I. WEISFELD, F. T. TZOMATOVA, and G. V. LUSCHENKO

CONTENTS

9.1 INTRODUCTION

In the process of selection for the speed-up study of material it is necessary to get the sufficient amount of seed in the earliest possible dates. For the speed-up estimation of plant-breeding material it is necessary to get the sufficient amount of seed in the earliest possible dates. To that end conducted sowing of inflorescences without threshing from the plants of cereal cultures selected for a selection. The method of sowing by ears was previously tested during 2 yr in Tyumen research station of N.I. Vavilov Research Institute of Plant Industry on more than 100 samples of winter wheat from world collections. The productivity of winter form of cereals depends on a set of biotic (pathogenic microorganisms) and abiotic (temperature, amount of precipitation etc.) factors. We obtained positive results for winter hardiness, resistance to snow mold, which main pathogen is *Microdochium nivale* (Fr.) Samuels and I.C. Hallett (= *Fusarium nivale* Ces. ex Berl. and Voglino).

Para-aminobenzoic acid (PABA) discovered by J.A. Rapoport as modifier of metabolic processes already in 1940s. He showed on the example of experimental object, *Drosophila*, that PABA evokes positive changes of non-hereditary character (i.e., it is not a mutagen) in the organism development [1]. Rapoport [2] proposed a scheme of relations between genotype, ferments and phenotype: "genes → their hetero catalysis (on the substrate of ribonucleic acid (RNA) molecule) → messenger RNA (mRNA) → mRNA catalysis (substrate of amino acids) → ferments → their catalysis (different substrates) → phenotype".

Works on the application of PABA in the agriculture continue and deepen. The experiments performed in the Republic of North Ossetia [4–9] showed, that the addition of PABA to the nutritive substances (potassium humate, irlit, leskenit, corn extract, juice of ambrosia, melted snow water and others) gives a positive effect. Treatment with PABA seeds of pea and honey plant sverbiga east (*Bunias orientalis*) before planting [4], extra nutrition of clover [5], seed of leguminoze grasses [6] stimulates the germination plants in a greater degree, then nutritive substances without PABA. A positive result was got at treatment of sprouts of potato by mixture of the melted water with juice of ambrosia, leskenit and PABA [7]. This method resulted in the increase of harvest of potato and decline of disease of Fusarium. Addition PABA in nutrient medium for treatment seed of triticale [8] results in the increase of maintenance of protein in green mass. Protracted treatment by PABA of handles of dogwood [9] increased engraftment of grafts. PABA was used for the receipt of potato without viruses [10].

In the joint research of scientists of the Institute of Biochemical Physics RAS and the North Caucasus Research Institute of mountain and foothill agriculture it was shown that the treatment by PABA of potatoes tubers with the subsequent enveloping in an ash [11] increase yield of potato. Treatment of seed and seedlings of vegetable cultures by solution of PABA in mixture with boric acid and permanganate potassium [12] improves resistance to diseases of young plantlets and increases harvest of carrot, beet, cucumbers, and tomatoes, seeds and seedlings of vegetables.

After treatment by PABA of binary mixture of seed of winter wheat and winter vetch [13] the productivity and quality of green feed increases at mowing of mixture in a period from the beginning of exit in a tube of wheat and beginning of budding of vetch to forming of grain of milky ripeness of wheat.

In the conditions of the northern forest-steppe of Tyumen region, where beside of fertile soils occur saline, jointed impact of salt solutions and PABA (0.01% solution) on the seeds of three barley varieties with low salinity resistance considerably increased the salinity resistance of germs independently on NaCl concentration [14]. The positive results were obtained under spraying of inflorescences of mother plants of barley by the solutions of PABA before realization of crossing. In a series of hybrid combinations the exceeding over control of length and width of flag leaf, number of leaves by plant and plant height was observed [15]. Spraying of inflorescences of four amaranth samples by PABA solutions increased seed productivity, the concentration of 0.02% showed to be the most effective [15].

In the present study we researched the effect of PABA on germination, germination energy and winter-hardiness of cereals, namely ears of winter cereals and panicles of millet species. In contrast with most of above-mentioned studies, where PABA was dissolved in the hot water following the method developed by Rapoport with collaborators [3], in the present study PABA was dissolved in an acetic acid. The combination of PABA and an acetic acid creates an acid media and allows preventing of a set of fungi illnesses simultaneously with the preservation of genotype of samples under multiplying.

Below we present the results of experiments performed on the experimental base of the North Caucasus Research Institute of mountain and foothill agriculture where perennial crops of legumes herbs were precursor.

9.2 MATERIAL AND METHODS

Cereals the most widely applied in the agricultural production and selection were applied as material. We tested the following cultures of yield of 2009: winter wheat varieties Ivina, Vassa, Bat'ko, Don 107, Kollega, Kalym and winter barley variety Bastion – the varieties designed for North-Caucasian region. We also tested the following varieties introduced in North Ossetia: selection samples of millet: Japanese millet (*Echinochloa frumentacea*), panic (*Setaria italica Panicumitalicum*) Italian millet (*Setaria italica*).

PABA represents a fine-grained powder; it easily and completely dissolves in 3% solution of an acetic acid without heating. We dissolved one tee spoon of dry PABA powder (10 g) in a small volume (20–25 ml) of 3% solution of an acetic acid. Then we dissolved the obtained mixture in 1 liter of tap water under room temperature. Thus we obtain the PABA concentration of 0.1%. In order to obtain the concentration of 0.2% we dissolve 20 g of dry powder.

We soaked 10 samples for the inflorescences without trashing of cereals with mature seeds, namely ears of cereals and panicles of millet species, in the solutions of PABA (0.1 or 0.2%) during 2–2.5; 3–4 and 5–6 h. Then the inflorescences were planted in the in the soil in the open field. We kept the distance of 20–25 cm between samples. Untreated inflorescences, soaked in water during 3–4 h, were applied as control (control I). When preparing control II we dissolved PABA in hot water and soaked inflorescences during 2–2.5 h.

In tables the variants of experiments were marked: in variant "0" is presented Control of I, where inflorescences were soaked in water (without treatment by PABA); in variant "1" is presented Control of II, where inflorescences were soaked in PABA dissolved in hot water; variants 2–5 are a soakage of inflorescences in PABA dissolved in an acetic acid of concentrations of 0.1–0.2% through 2–6 h.

Simultaneously we placed thrashed seeds for germination in a Petri dish with PABA dissolved in hot water (control I) or in an acetic acid (control II). For the control II the data averaged for PABA concentrations of 0.1% and 0.2% are presented because their difference was insignificant.

9.3 RESULTS

The PABA concentration of 0.1% or 0.2% in an acetic acid was sufficient for the penetration of substance in the embryo when treating cereal inflorescences. High concentrations (9–11%) inhibit a height and development of plants [10].

Germination of threshed seeds soaked in PABA dissolved in an acetic acid, appeared 3–4 days earlier than when soaked in water. On seedlings from seed germinated in an aqueous solution of PABA developed fungal microflora and they died. When dissolved in an acetic acid PABA seedlings persisted during long time, they continued to grow.

Germinating capacity of seeds in inflorescences under soaking in PABA, dissolved in an acetic acid, considerably exceeded germinating capacity in control variants (Table 1). The effectiveness of treatment by PABA solutions depended on its duration and concentration of solutions. The same trend concerning germination energy was observed for all five cereals under study.

The middle percent of wintering as compared to control of I (water without PABA) increases with the increase of concentration of PABA and durations of treatment. However at more long-continued soakage of plants – during 5–6 h (variant 5) increase in relation to both Controls was less, than at a soakage during 3–4 h (variant 4). At a soakage in water solution of PABA of 0.1% increase observed also, but on more low level, then at dissolution in an acetic acid.

The energy of germination, presented to the Table 2, exceeded control without treatment of PABA in all variants of dissolution of PABA dissolved in an acetic acid and in a variant with treatment of PABA, dissolved in hot water (control of II, variant 1), and was higher, than in control without addition of PABA (zero variant) at all five investigated cultures. Exceeding of indexes above control of I in control of II was below, than in variants 1–5.

Winter-hardness of winter cereals (Table 3) – wheat and barley in all variants with treatment of PABA dissolved in an acetic acid (variants 2–5) and in control of II (dissolution is in hot water) was higher then in control of 1 (without PABA). As well as other experiments indexes at dissolution of PABA in hot water in control II were below, than in other variants (3–5).

The dynamics of activating of phenotype under act of PABA is evidently shown on a histogram (Fig. 1), reflecting the middle indexes of levels of increases of energy of germination, germination and winter-hardness. The tendency of growth of middle indexes is evidently.

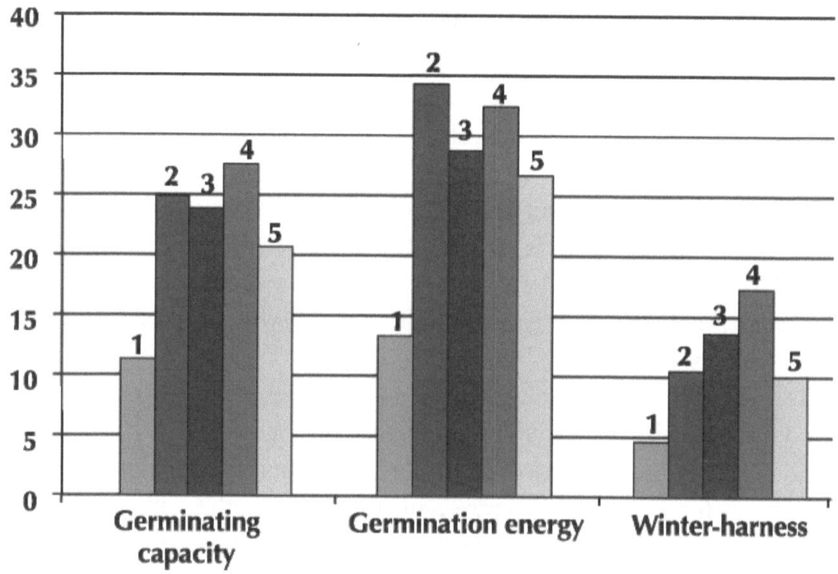

FIGURE 1 Comparison of averages (totally on all cultures, %) of germinated capacity, energies of germination of seed and winter-hardness of plants at the different methods of treatment of ears of winter crops and panicles of kinds millet by para-aminobenzoic acid in the variants of experiments from 1 to 5 (see Tables 1–3) by comparison to control of I (without treatment, in tables – a zero variant). Ordinate axis –%.

9.4 DISCUSSION

The higher indexes of germinating capacity, energy of germination at all cultures and winter-hardness cold of winter crops are marked – winter wheat and barley in variants (2–5 in diagram) after treatment of PABA, dissolved in an acetic acid, as compared to control variants – without treatment of PABA (variant 0 – control of I without PABA) and at dissolution of PABA in hot water (variant 1 – control of II) in analogical variants (see Tables 1–3) were got. All indexes at a soakage in PABA in an acetic acid most long time – 5–6 h (variant 5) as compared to analogical treatment during 3–4 h (variant 4) has a tendency to go down (see Tables 1–3). Maybe, in this case braking of development of plantules takes place [10].

TABLE 1 Germinating capacity of seeds of winter grain crops and millet species under different ways of treatments of cereal inflorescences by para-aminobenzoic acid.

Variant of experiment	Winter wheat		Winter barley		Japanese millet		Panic		Italian millet	
	%	+/%	%	+/%	%	+/%	%	+/%	%	+/%
0. Control I without treatment: soaking in water for 3–4 h	62	0	75	0	67	0	72	0	68	0
1. Control II: soaking in PABA (0.1–0.2%) dissolved in hot water for 2–2.5 h	74	12/16	84	9/11	76	9/12	82	10/1.2	72	4/6
2. Soaking in PABA (0.1%) dissolved in an ac. a. for 2–2.5 h	82	20/24	93	18/19	87	20/23	92	20/22	84	16/19
3. Soaking in PABA (0.2%) dissolved in an ac. a. for 2–2.5 h	86	24/28	95	20/21	92	25/27	92	20/22	87	19/24
4. Soaking in PABA (0.2%) dissolved in an ac. a. for 3–4 h	96	34/35	98	23/23	95	28/29	95	23/23	92	24/25
5. Soaking in PABA (0,2%) dissolved in an ac. a. for 5–6 h	82	20/24	86	11/13	90	23/26	90	18/20	86	18/21
LSD_{05}	4.5		2.4		2.8		2.1		2.2	4.5

Note. LSD – the least substantial difference; ac. a. – an acetic acid.

TABLE 2 Germination energy of seeds of winter grain crops and millet species under different ways of treatments of cereal inflorescences by para–aminobenzoic acid.

Variant of experiment	Winter wheat		Winter barley		Japanese millet		Panic		Italian millet	
	%	+/%	%	+/%	%	+/%	%	+/%	%	+/%
0. Control I without treatment: soaking in water for 3–4 h	54	0	68	0	53	0	60	0	54	0
1. Control II: soaking in PABA (0.1–0.2%), dissolved in hot water 2–2.5 h	65	11/17	73	5/7	62	9/15	73	12/16	63	8/13
2. Soaking in PABA (0.1%), dissolved in an ac. a. 2–2.5 h	72	55/76	81	13/16	75	22/29	81	22/31	73	19/26
3. Soaking in PABA (0.2%), dissolved in an ac. a. 2–2.5 h	80	27/34	84	16/19	79	26/33	87	26/30	76	22/29
4. Soaking in PABA (0.2%), dissolved in an ac. a. 3–4 h	85	32/38	89	21/24	83	30/36	89	29/33	79	25/32
5. Soaking in PABA (0.2%), dissolved in an ac. a. 5–6 h	78	25/32	80	11/14	78	26/33	84	24/29	75	21/28
LSD_{05}		3.4		2.1		2.6		2.9		2.5

TABLE 3 Winter–hardiness (% of plants overwintered) of winter cereals under different ways of treatments of ears by para-aminobenzoic acid.

Variant of experiment	Winter wheat		Winter barley	
	%	+/%	%	+/%
0. Control I without treatment: soaking in water for 3–4 h	78	0	84	0
1. Control II: soaking in PABA (0.1–0.2%), dissolved in hot water 2–2.5 h	82	4/4.8	88	4/4.8
2. Soaking in PABA (0.1%), dissolved in an ac. a. 2–2.5 h	86	8/9.3	95	11/22.6
3. Soaking in PABA (0.2%), dissolved in an ac. a. 2–2.5 h	90	12/13.3	97	13/13.4
4. Soaking in PABA (0.2%), dissolved in an ac. a. 3–4 h	98	20/20.4	98	14/14.5
5. Soaking in PABA (0.2%), dissolved in an ac. a. 5–6 h	88	10/11.4	92	8/8.7
LSD_{05}		3.1		0.8

This work executed in connection with the necessity of speed-up creation of new varieties steady to the ecologically diverse conditions and simultaneously responsive on the additional fertilizing. Work is important from the economic point of view. Sowing of inflorescences without threshing considerably simplifies work of breeder both in a scientific plan and in organizational: the methods of selection are simplified, reproduction of the best plants is accelerated for a selection and forming of new varieties, diminish expenses of labor. Usually at the unfavorable condones of growing, at introduction of plants, at the high doses of mutagens or during distant hybridization of cultural plants with wild sorts fall down number in inflorescences, germination of seeds, viability of plants. At sowing of inflorescences – ears or panicles possibility to distinguish the most perspective families on the number of germinating seed in them appears with mature grain. From every separate inflorescence it is possible to collect more seeds for the receiving of posterity, the terms of estimation of material and terms of reproduction of valuable selection samples grow short. A method especially touches the freshly material, when at preparation to sowing of winter crops in the year of harvest time not enough for threshing of ears and additional treatment of grain. Such quite often happens in those districts, where short vegetation period, for example, in the Tyumen area.

KEYWORDS

- **Cereals**
- **Fusarium**
- **Plantules**
- **Tyumen**

REFERENCES

1. Rapoport, I. A. *Phenogenetic Analysis of Dependent and Independent Differentiation.* Proceedings of the Institute of Cytology, Histology, Embryology of Academy of Sciences. **1948**, *2(1)*, 3–128.

2. Rapoport, I. A. *The Action of PABA in Connection with the Genetic Structure.* Chemical Mutagens and Para-aminobenzoic Acid to Increase the Yield of Crops. Moscow. Science, 1989. pp. 3–37.
3. Rapoport, I. A. *Chemical Mutagens and Para-aminobenzoic Acid in Enhancing the Productivity of Crops.* Moscow, Science, 1989, 253pp.
4. Bekuzarova, S. A.; Abiyeva, T. S.; Tedeeva, A. A. *The Method of Pre-treatment of Seeds.* Patent No. 2270548. Published on 2006.
5. Bekuzarova, S. A.; Farniyev, A. T.; Basiyeva, T. B.; Gaziyev, V. I.; Kaliceva, D. N. *The Method of Stimulation and Development of Clover Plants.* Patent No. 2416186. Published on 2011.
6. Bekuzarova, S. A.; Shtchedrina, D. I.; Farnoyev, A. T.; Pliyev, M. A. *Method of Additional Fertilizing of Leguminous Grasses.* Patent No. 2282342. Published on 2006.
7. Ikaev, B. V.; Marzoev, A. I.; Bekuzarova, S. A.; Basayev, I. B.; Bolieva, Z. A.; Kizinov, F. I. *The Method of Treatment of Pre-plant Shoots of Potato Tubers.* Patent No. 2385558. Published on 2010.
8. Bekuzarova, S. A.; Antonov, O. V.; Fedorov, A. K. *Method of Increase of Content of Protein in Green Mass of Winter Triticale.* Patent No. 2212777. Published on 2003.
9. Cabolov, P. H.; Bekuzarova, S. A.; Tigiyeva, I. F.; Tadtayeva, E. A.; Eiges, N. S. *The Method of Reproduction of Dogwood Drafts.* Patent No. 2294619. Published on 2007.
10. Shcherbinin, A. N.; Soldatova, T. B. *The Nutrient Medium for Micropropagation of Potato.* Patent No. 2228354. Published 2004.
11. Eiges, N. S.; Weisfeld, L. I.; Volchenko, G. A.; Bekuzarova, S. A. *Method of Pre-treatment of Tubers of Potatoes.* Patent No. 2202701. Published on 2007.
12. Eiges, N. S.; Weissfeld, L. I.; Volchenko, G. A.; Bekuzarova, S. A. *The Method of Pre-treatment of Seeds and Seedlings of Vegetable Crops.* Patent No. 2200392. Published 2003.
13. Eiges, N. S.; Weissfeld, L. I.; Volchenko, G. A.; Bekuzarova, S. A.; Pliyev, M. A.; Hadarceva, M. V. *The Method of Receiving of Feeds in Green Conveyer.* Patent No. 2330410. Published on 2008. Bull. No. 22.
14. Bome, N. A.; Govorukhin, A. A. *The Effectiveness of the Influence of Para-aminobenzoic Acid on the Ontogeny of Plants under Stress.* **Bulletin of** Tyumen State University. Tyumen: Tyumen State University Publishing House, 1998, Vol. 2. pp 176–182.
15. Bome, N. A.; Bome, A. J.; Belozerova, A. A. *Stability of Crop Plants to Adverse Environmental Factors*; Monograph. Tyumen State University Publishing House: Tyumen, 2007, 192 p.

CHAPTER 10

ON THE QUANTUM-CHEMICAL CALCULATION

V. A. BABKIN, D. S. ZAKHAROV, and G.E. ZAIKOV

CONTENTS

10.1 INTRODUCTION

The aim of this work is a study of electronic structure of molecules α-cyclopropyl-p-izopropylstyrene, α-cyclopropyl-p-izopropylstyrene, α-cyclopropyl-2,4-dimethylstyrene, α-cyclopropyl-p-ftorstyrene [1] and theoretical estimation its acid power by quantum-chemical method MNDO. The calculation was done with optimization of all parameters by standard gradient method built-in in PC GAMESS [2]. The calculation was executed in approach the insulated molecule in gas phase. Program MacMolPlt was used for visual presentation of the model of the molecule [3].

10.2 METHODICAL PART

Geometric and electronic structures, general and electronic energies of molecules α-cyclopropyl-p-izopropylstyrene, α-cyclopropyl-2,4-dimethylstyrene, α-cyclopropyl-p-ftorstyrene was received by method MNDO and are shown on Figs. 1–3 and in Tables 1–4, respectively. The universal factor of acidity was calculated by formula: pKa = 49,4–134,61*$q_{max}H^+$ [4], which used with success, for example, in Refs. [5–35] (where, $q_{max}H^+$ – a maximum positive charge on atom of the hydrogen (by Milliken [1]), R = 0.97, R – a coefficient of correlations, q_{max}^{H+} = +0.06, +0.06, +0.08, respectively. pKa = 30–33.

Quantum-chemical calculation of molecules α-cyclopropyl-p-izopropylstyrene, α-cyclopropyl-2,4-dimethylstyrene, α-cyclopropyl-p-ftorstyrene by method MNDO was executed for the first time. Optimized geometric and electronic structures of these compound was received. Acid power of molecules α-cyclopropyl-p-izopropylstyrene, α-cyclopropyl-2,4-dimethylstyrene, α–cyclopropyl-p-ftorstyrene was theoretically evaluated (pKa = 30–33). These compound pertain to class of very weak H-acids (pKa > 14).

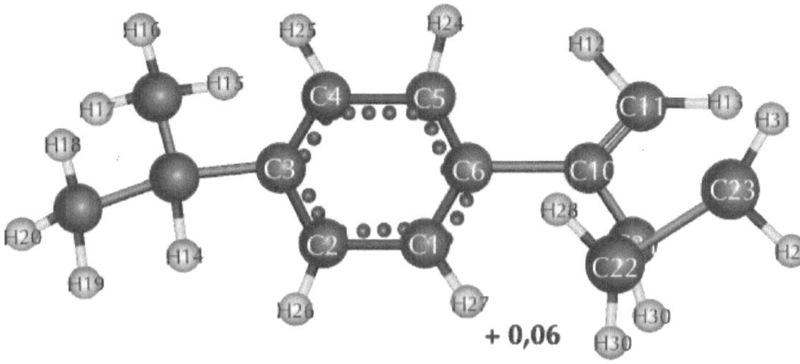

FIGURE 1 Geometric and electronic molecular structure of a-cyclopropyl-p-izopropylstyrene.
(E0 = –196,875 kDg/mol, Eel = –1,215,375 kDg/mol).

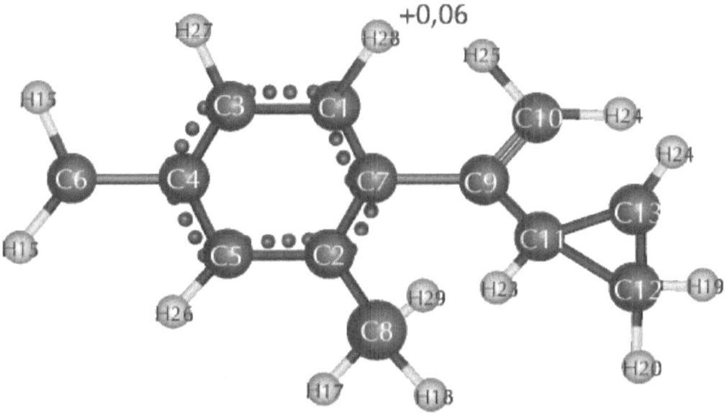

FIGURE 2 Geometric and electronic molecular structure of a-cyclopropyl-2,4-dimethylstyrene.
(E0 = –181,125 kDg/mol, Eel = –1,084,125 kDg/mol).

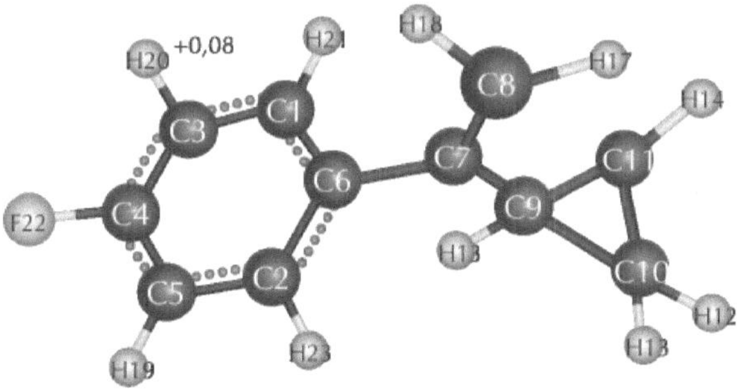

FIGURE 3 Geometric and electronic molecular structure of a-cyclopropyl-p-ftorstyrene.
(E0 = –196,875 kDg/mol, Eel = –1,155,000 kDg/mol).

TABLE 1 Optimized bond lengths, valence corners and charges on atoms of the molecule of a-cyclopropyl-p-izopropylstyrene.

Bond lengths	R,A	Valence corners	Grad	Atom	Charge (by Milliken)
C(1)–C(6)	1.40	C(6)–C(1)–C(2)	121	C(1)	–0.04
C(2)–C(1)	1.39	C(1)–C(2)–C(3)	121	C(2)	–0.05
C(3)–C(2)	1.40	C(2)–C(3)–C(4)	119	C(3)	–0.07
C(3)–C(7)	1.50	C(3)–C(4)–C(5)	121	C(4)	–0.05
C(4)–C(3)	1.40	C(4)–C(5)–C(6)	121	C(5)	–0.05
C(5)–C(4)	1.39	C(2)–C(3)–C(7)	120	C(6)	–0.03
C(6)–C(5)	1.40	C(3)–C(7)–C(8)	111	C(7)	–0.02
C(6)–C(10)	1.47	C(3)–C(7)–C(9)	111	C(8)	+0.04
C(7)–C(8)	1.52	C(8)–C(7)–C(9)	110	C(9)	+0.04
C(7)–C(9)	1.52	C(1)–C(6)–C(10)	121	C(10)	–0.06
C(10)–C(11)	1.34	C(6)–C(10)–C(11)	122	C(11)	–0.03
C(10)–C(21)	1.47	C(10)–C(11)–H(12)	122	H(12)	+0.04
H(12)–C(11)	1.10	C(10)–C(11)–H(13)	122	H(13)	+0.04
H(13)–C(11)	1.10	C(3)–C(7)–H(14)	108	H(14)	+0.01
H(14)–C(7)	1.13	C(7)–C(8)–H(15)	110	H(15)	0.00
H(15)–C(8)	1.12	C(7)–C(8)–H(16)	111	H(16)	0.00
H(16)–C(8)	1.12	C(7)–C(8)–H(17)	110	H(17)	–0.01
H(17)–C(8)	1.12	C(7)–C(9)–H(18)	111	H(18)	0.00
H(18)–C(9)	1.12	C(7)–C(9)–H(19)	110	H(19)	0.00
H(19)–C(9)	1.12	C(7)–C(9)–H(20)	110	H(20)	–0.01
H(20)–C(9)	1.12	C(6)–C(10)–C(21)	116	C(21)	–0.07
C(21)–C(22)	1.51	C(11)–C(10)–C(21)	123	C(22)	–0.05
C(22)–C(23)	1.50	C(22)–C(23)–C(21)	60	C(23)	–0.06
C(23)–C(21)	1.51	C(10)–C(21)–C(22)	121	H(24)	+0.06
H(24)–C(5)	1.10	C(21)–C(23)–C(22)	60	H(25)	+0.06
H(25)–C(4)	1.10	C(21)–C(22)–C(23)	60	H(26)	+0.06
H(26)–C(2)	1.10	C(22)–C(21)–C(23)	60	**H(27)**	**+0.06**
H(27)–C(1)	1.10	C(4)–C(5)–H(24)	120	H(28)	+0.04
H(28)–C(22)	1.10	C(3)–C(4)–H(25)	120	H(29)	+0.04
H(29)–C(23)	1.10	C(1)–C(2)–H(26)	120	H(30)	+0.04
H(30)–C(22)	1.10	C(2)–C(1)–H(27)	120	H(31)	+0.04
H(31)–C(23)	1.10	C(21)–C(22)–H(28)	119	H(32)	+0.05
H(32)–C(21)	1.11	C(21)–C(23)–H(29)	118		
		C(21)–C(22)–H(30)	119		
		C(21)–C(23)–H(31)	120		
		C(10)–C(21)–H(32)	111		

TABLE 2 Optimized bond lengths, valence corners and charges on atoms of the molecule of a-cyclopropyl-2,4--imethylstyrene.

Bond lengths	R,A	Valence corners	Grad	Atom	Charge (by Milliken)
C(1)–C(7)	1.42	C(1)–C(7)–C(2)	119	C(1)	–0.0508
C(2)–C(5)	1.42	C(7)–C(1)–C(3)	121	C(2)	–0.0835
C(3)–C(1)	1.40	C(1)–C(3)–C(4)	121	C(3)	–0.0412
C(4)–C(3)	1.41	C(2)–C(5)–C(4)	123	C(4)	–0.1009
C(5)–C(4)	1.41	C(3)–C(4)–C(5)	118	C(5)	–0.0280
C(6)–C(4)	1.51	C(3)–C(4)–C(6)	121	C(6)	0.0814
C(7)–C(2)	1.42	C(5)–C(2)–C(7)	119	C(7)	–0.0187
C(7)–C(9)	1.50	C(5)–C(2)–C(8)	119	C(8)	0.0807
C(10)–C(9)	1.35	C(1)–C(7)–C(9)	118	C(9)	–0.0545
C(11)–C(9)	1.50	C(7)–C(9)–C(10)	120	C(10)	–0.0416
C(12)–C(11)	1.54	C(7)–C(9)–C(11)	115	C(11)	–0.0617
C(13)–C(12)	1.52	C(9)–C(11)–C(12)	125	C(12)	–0.0561
C(13)–C(11)	1.54	C(9)–C(11)–C(13)	125	C(13)	–0.0568
H(14)–C(6)	1.11	C(4)–C(6)–H(14)	111	H(14)	–0.0028
H(15)–C(6)	1.11	C(4)–C(6)–H(15)	111	H(15)	–0.0027
H(16)–C(6)	1.11	C(4)–C(6)–H(16)	113	H(16)	–0.0050
H(17)–C(8)	1.11	C(2)–C(8)–H(17)	112	H(17)	–0.0072
H(18)–C(8)	1.11	C(2)–C(8)–H(18)	111	H(18)	–0.0002
H(19)–C(12)	1.10	C(11)–C(12)–H(19)	121	H(19)	0.0389
H(20)–C(12)	1.10	C(11)–C(12)–H(20)	118	H(20)	0.0368
H(21)–C(13)	1.10	C(11)–C(13)–H(21)	121	H(21)	0.0387
H(22)–C(13)	1.10	C(11)–C(13)–H(22)	118	H(22)	0.0370
H(23)–C(11)	1.10	C(9)–C(11)–H(23)	111	H(23)	0.0451
H(24)–C(10)	1.09	C(9)–C(10)–H(24)	124	H(24)	0.0394
H(25)–C(10)	1.09	C(9)–C(10)–H(25)	123	H(25)	0.0425
H(26)–C(5)	1.09	C(2)–C(5)–H(26)	119	H(26)	0.0550
H(27)–C(3)	1.09	C(1)–C(3)–H(27)	119	H(27)	0.0581
H(28)–C(1)	1.09	C(3)–C(1)–H(28)	119	**H(28)**	**0.0600**
H(29)–C(8)	1.11	C(2)–C(8)–H(29)	111	H(29)	–0.0019

TABLE 3 Optimized bond lengths, valence corners and charges on atoms of the molecule of a-cyclopropyl-p-ftorstyrene.

Bond lengths	R,A	Valence corners	Grad	Atom	Charge (by Milliken)
C(1)–C(3)	1.40	C(1)–C(6)–C(2)	119	C(1)	–0.02
C(2)–C(6)	1.41	C(4)–C(5)–C(2)	120	C(2)	–0.02
C(3)–C(4)	1.42	C(5)–C(4)–C(3)	120	C(3)	–0.09
C(4)–C(5)	1.42	C(1)–C(3)–C(4)	120	C(4)	+0.15
C(5)–C(2)	1.40	C(6)–C(2)–C(5)	121	C(5)	–0.09
C(6)–C(7)	1.49	C(3)–C(1)–C(6)	121	C(6)	–0.06
C(7)–C(9)	1.50	C(1)–C(6)–C(7)	121	C(7)	–0.06
C(8)–C(7)	1.35	C(6)–C(7)–C(8)	120	C(8)	–0.04
C(9)–C(10)	1.54	C(6)–C(7)–C(9)	115	C(9)	–0.07
C(10)–C(11)	1.52	C(7)–C(9)–C(10)	124	C(10)	–0.06
C(11)–C(9)	1.54	C(7)–C(9)–C(11)	125	C(11)	–0.06
H(12)–C(10)	1.10	C(9)–C(10)–H(12)	121	H(12)	+0.04
H(13)–C(10)	1.10	C(9)–C(10)–H(13)	118	H(13)	+0.04
H(14)–C(11)	1.10	C(9)–C(11)–H(14)	121	H(14)	+0.04
H(15)–C(11)	1.10	C(9)–C(11)–H(15)	118	H(15)	+0.04
H(16)–C(9)	1.10	C(7)–C(9)–H(16)	111	H(16)	+0.04
H(17)–C(8)	1.09	C(7)–C(8)–H(17)	124	H(17)	+0.04
H(18)–C(8)	1.09	C(7)–C(8)–H(18)	123	H(18)	+0.04
H(19)–C(5)	1.09	C(2)–C(5)–H(19)	120	H(19)	+0.08
H(20)–C(3)	1.09	C(1)–C(3)–H(20)	120	**H(20)**	**+0.08**
H(21)–C(1)	1.09	C(3)–C(1)–H(21)	119	H(21)	+0.07
F(22)–C(4)	1.33	C(3)–C(4)–F(22)	120	F(22)	–0.18
H(23)–C(2)	1.09	C(5)–C(2)–H(23)	119	H(23)	+0.07

TABLE 4 General and energies (E_0), maximum positive charge on atom of the hydrogen ($q_{max}H^{+}$), universal factor of acidity (pKa).

Molecules of aromatic olefins	E_0	$q_{max}H^+$	pKa
a-cyclopropyl-p-izopropylstyrene.	–196,875	+0.06	33
a-cyclopropyl-2,4-dimethylstyrene	–181,125	+0.06	33
a-cyclopropyl-p-ftorstyrene.	–196,875	+0.08	30

KEYWORDS

- **Electronic structures**
- **Geometric structures**
- **Quantum-chemical**
- **Weak H-acids**

REFERENCES

1. Kennedy, J. *Cation Polymerization of Olefins*. The World, Moscow, 1978, 430p.
2. Shmidt, M. W.; Baldrosge, K. K.; Elbert, J. A.; Gordon, M. S.; Enseh, J. H.; Koseki, S.; Matsvnaga, N.; Nguyen, K. A.; Su; S. J.; et al. *J. Comput. Chem.*, **1993**, *14*, 1347–1363.
3. Babkin, V. A.; Fedunov, R. G.; Minsker, K. S.; et al. *Oxidation Commun.*, **2002**, *1*(25), 21–47.
4. Bode, B. M.; Gordon, M. S. *J. Mol. Graphics Mod.*, **1998**, *16*, 133–138.
5. Babkin, V. A.; Dmitriev, V. Yu.; Zaikov, G. E. In *Quantum Chemical Calculation of Unique Molecular System*. Vol. I. VolSU, c. Volgograd, 2010, pp 93–95.
6. Babkin, V. A.; Dmitriev, V. Yu.; Zaikov, G. E. In *Quantum Chemical Calculation of Unique Molecular System*. Vol. I. VolSU, c. Volgograd, 2010, pp 95–97.
7. Babkin, V. A.; Dmitriev, V. Yu.; Zaikov, G. E. In *Quantum Chemical Calculation of Unique Molecular System*. Vol. I. VolSU, c. Volgograd, 2010, pp. 97–99.
8. Babkin, V. A.; Dmitriev, V. Yu.; Zaikov, G. E. In *Quantum Chemical Calculation of Unique Molecular System*. Vol. I. VolSU, c. Volgograd, 2010, pp 99–102.
9. Babkin, V. A.; Andreev, D. S. In *Quantum Chemical Calculation of Unique Molecular System*. Vol. I. VolSU, c. Volgograd, 2010, pp 176–177.
10. Babkin, V. A.; Andreev, D. S. In *Quantum Chemical Calculation of Unique Molecular System*. Vol. I. VolSU, c. Volgograd, 2010, pp 177–179.
11. Babkin, V. A.; Andreev, D. S. In *Quantum Chemical Calculation of Unique Molecular System*. Vol. I. VolSU, c. Volgograd, 2010, pp 179–180.
12. Babkin, V. A.; Andreev, D. S. In *Quantum Chemical Calculation of Unique Molecular System*. Vol. I. VolSU, c. Volgograd, 2010, pp 181–182.
13. Babkin, V. A.; Dmitriev, V. Yu.; Zaikov, G. E. In *Quantum Chemical Calculation of Unique Molecular System*. Vol. I. VolSU, c. Volgograd, 2010, pp 89–90.
14. Babkin, V. A.; Dmitriev, V. Yu.; Zaikov, G. E. In *Quantum Chemical Calculation of Unique Molecular System*. Vol. I. VolSU, c. Volgograd, 2010, pp 93–95.
15. Babkin, V. A.; Dmitriev, V. Yu.; Zaikov, G. E. In *Quantum Chemical Calculation of Unique Molecular System*. Vol. I. VolSU, c. Volgograd, 2010, pp 103–105.
16. Babkin, V. A.; Dmitriev, V. Yu.; Zaikov, G. E. In *Quantum Chemical Calculation of Unique Molecular System*. Vol. I. VolSU, c. Volgograd, 2010, pp 105–107.

17. Babkin, V. A.; Dmitriev, V. Yu.; Zaikov, G. E. In *Quantum Chemical Calculation of Unique Molecular System*. Vol. I. VolSU, c. Volgograd, 2010, pp 107–108.
18. Babkin, V. A.; Dmitriev, V. Yu.; Zaikov, G. E. In *Quantum Chemical Calculation of Unique Molecular System*. Vol. I. VolSU, c. Volgograd, 2010, pp 108–109.
19. Babkin, V. A.; Andreev, D. S. In *Quantum Chemical Calculation of Unique Molecular System*. Vol. I. VolSU, c. Volgograd, 2010, pp 235–236.
20. Babkin, V. A.; Andreev, D. S. In *Quantum Chemical Calculation of Unique Molecular System*. Vol. I. VolSU, c. Volgograd, 2010, pp 236–238.
21. Babkin, V. A.; Andreev, D. S. In *Quantum Chemical Calculation of Unique Molecular System*. Vol. I. VolSU, c. Volgograd, 2010, pp 238–239.
22. Babkin, V. A.; Andreev, D. S. In *Quantum Chemical Calculation of Unique Molecular System*. Vol. I. VolSU, c. Volgograd, 2010, pp 240–241.
23. Babkin, V. A.; Andreev, D. S. In *Quantum Chemical Calculation of Unique Molecular System*. Vol. I. VolSU, c. Volgograd, 2010, pp 241–243.
24. Babkin, V. A.; Andreev, D. S. In *Quantum Chemical Calculation of Unique Molecular System*. Vol. I. VolSU, c. Volgograd, 2010, pp 243–245.
25. Babkin, V. A.; Andreev, D. S. In *Quantum Chemical Calculation of Unique Molecular System*. Vol. I. VolSU, c. Volgograd, 2010, pp 245–246.
26. Babkin, V. A.; Andreev, D. S. In *Quantum Chemical Calculation of Unique Molecular System*. Vol. I. VolSU, c. Volgograd, 2010, pp 247–248.
27. Babkin, V. A.; Andreev, D. S. In *Quantum Chemical Calculation of Unique Molecular System*. Vol. I. VolSU, c. Volgograd, 2010, pp 249–250.
28. Babkin, V. A.; Andreev, D. S. In *Quantum Chemical Calculation of Unique Molecular System*. Vol. I. VolSU, c. Volgograd, 2010, pp 251–252.
29. Babkin, V. A.; Andreev, D. S. In *Quantum Chemical Calculation of Unique Molecular System*. Vol. I. VolSU, c. Volgograd, 2010, pp 252–254.
30. Babkin, V. A.; Andreev, D. S. In *Quantum Chemical Calculation of Unique Molecular System*. Vol. I. VolSU, c. Volgograd, 2010, pp 254–256.
31. Babkin, V. A.; Andreev, D. S. In *Quantum Chemical Calculation of Unique Molecular System*. Vol. I. VolSU, c. Volgograd, 2010, pp 256–258.
32. Babkin, V. A.; Andreev, D. S. In *Quantum Chemical Calculation of Unique Molecular System*. Vol. I. VolSU, c. Volgograd, 2010, pp 260–262.
33. Babkin, V. A.; Andreev, D. S. In *Quantum Chemical Calculation of Unique Molecular System*. Vol. I. VolSU, c. Volgograd, 2010, pp 262–264.
34. Babkin, V. A.; Andreev, D. S. In *Quantum Chemical Calculation of Unique Molecular System*. Vol. I. VolSU, c. Volgograd, 2010, pp 264–265.
35. Babkin, V. A.; Andreev, D. S. In *Quantum Chemical Calculation of Unique Molecular System*. Vol. I. VolSU, c. Volgograd, 2010, pp 266–267.

CHAPTER 11

ON THE QUANTUM-CHEMICAL MODELING

A. A. TUROVSKY, A. R. KYTSYA, L. I. BAZYLYAK, and G. E. ZAIKOV

CONTENTS

11.1 INTRODUCTION

Especially great successes under investigation of the driving forces of the enzymatic catalysis were achieved in a case of the Chemotrypsin. Chemotrypsin – this is endopeptidase, which cleaves the peptide bonds into peptides. The most important information about the structure of the Chemotrypsin's molecule has been obtained with the use of the roentgen investigations [1–4]. It was found, that the all charged groups into the molecule of the ferment are directed sideways to the aqueous solution (with the except for the three, which have the special functions into a mechanism action of the active center). The successes in kinetic investigations in the most cases were caused by the works of M. Bergmann, D. Frugonn and H. Neyrag, who determined that the Chemotrypsin can hydrolyze also the simple low–molecular products (amides, esters).

The force of the Chemotrypsin catalytic action under the esters hydrolysis approximately in 10^6 times exceeds the catalytic action both of OH^-, and H_3O^+.

Hydrolysis of the substrate (amides, esters) on the active center of the Chemotrypsin proceeds in some stages. The first stage of the enzymatic process includes the sorption (so–called formation of the Michaels's complex ES). The next stages include the chemical transformation of the sorbed molecule with the formation of the intermediate compound of the acyl-ferment EA in accordance with the following kinetic scheme:

$$E + S \underset{k_{-1}}{\overset{k_1}{\Leftrightarrow}} ES \underset{\downarrow P_1}{\overset{k_2}{\rightarrow}} EA \underset{\uparrow H_2O}{\overset{k_3}{\rightarrow}} E + P_2 \qquad (1)$$

$$K_S = \frac{k_{-1}}{k_1}$$

P_1 and P_2 are products of the hydrolysis.

The equilibrium position is determined only by the non-valence interactions with the protein of side chemically inert fragments of the substrate's molecule. The intermediate product represents by itself the acyl-ferment, which is unstable compound (lifetime ~ 0.01 s. [5]).

Finally, the enzymatic hydrolysis can be presented in accordance with the following scheme:

$$\underset{(E)}{EH} + \underset{(S)}{RC(O)OCH_3} \overset{K_S}{\Leftrightarrow} \underset{(E \cdot S)}{HE \cdot RC(O)OCH_3} \quad (2)$$

$$E \cdot S \overset{k_2}{\rightarrow} \underset{(EA)}{E - C(O)R} + \underset{(P_1)}{CH_3OH} \quad (3)$$

$$EA \underset{H_2O}{\overset{k_3}{\rightarrow}} E + RCOOH \quad (4)$$

However, the stages (3) and (4) are not elementary [6–9] and include the quick (and equilibrium) formation of the intermediate position, which corresponds to a new conformational state of the ferment.

Under hydrolysis of the molecule of a substrate, which is sorbing on an active center, the OH-group of the Serine stands out as the attacking nucleophile [6, 10–12]. It is assumed that the high activity of the Serine related with its surroundings into the active center. Along with the Serine, the imidazole group of His also takes part in its activity [6, 10, 11, 13]. At this, the Nitrogen atom of the Histidine forms the hydrogen bond with Oxygen of the Serine hydroxyl. Accordingly to Blow [14] the second hydrogen bond exists between the atoms N and Histidine and carbonyl group of the remains of Asp, which is located into the depths of the ferment globule. The system of the hydrogen bonds leads to increase of the negative charge on the OH group of the Serin that promotes to its nucleophilicity.

On the acidulating stage the nucleophilic attack of the carboxyl Carbon of the substrate by generalized nucleophile of active center: Ser, His, Asp proceeds. As a result of the active center acidulating, the turn of the Ser remains around the bonds C_α–C_β which is accompanying by the displacement of the Oxygen atom on ~ 2,5 A, takes place. At this, imidazole group of the His displaces sideways to the solvent [3]. As a result, the imidazole group of the His is included into a free ferment (and into the Michaels's complex) in hydrogen bond. Ser in acyl-ferment contributes one's own atom N for the formation of the hydrogen bond with water. As a result, the activated molecule of water has the ability to effectively attack the carbonyl Carbon of the substrate on the deacidulating stage. At this, it is formed the product of the hydrolysis and it is regenerated a free ferment. The above-said chemical mechanism of the hydrolytic action of Chemotrypsin is described in such a way in references.

We were interested into the analysis of such approach to the enzymatic catalysis with the use of the quantum-chemical calculations.

11.2 EXPERIMENTAL TECHNIQUES

The aim of the presented work was the modeling of the kinetics and chemical mechanism of the enzymatic catalysis process for reaction of hydrolysis of methyl acetate by Chemotrypsin with the use of the quantum-chemical method.

Firstly, it was interesting to consider the stage-by-stage modeling of the process in accordance with the well-known schemes accepted in references and, as far as possible, to append them.

Secondly, it was interesting to estimate the kinetic parameters of the enzymatic catalysis reaction.

Thirdly, it was interesting to compare the kinetics of processes of enzymatic and homogene-ous catalysis of methyl acetate.

For quantum-chemical calculations the semiempirical method PM6 was used.

Modeling Objects:

- methyl acetate;
- generally accepted active form of Chemotrypsin containing of three fragments (Fig. 1). On Fig. 1 presented also the geometry and electronic characteristics of Chemotrypsin.

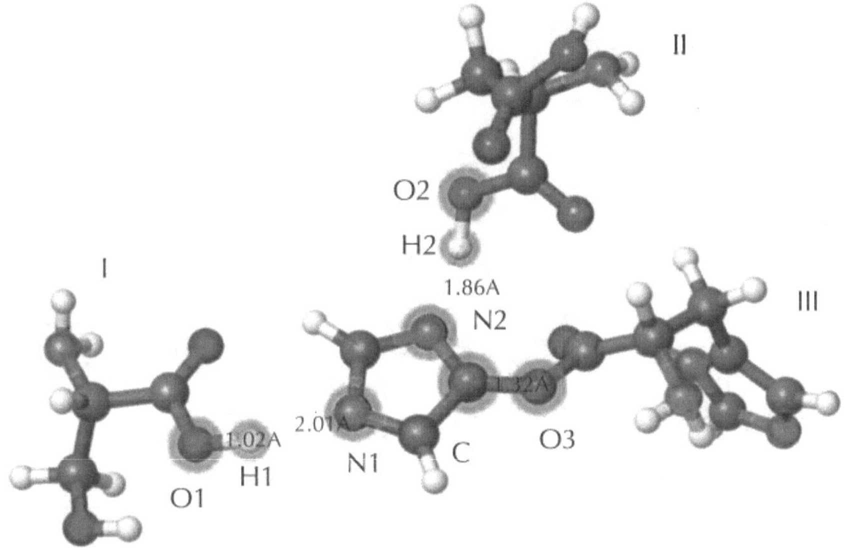

Charge on	H1	O1	N1	H2	O2	N2	C	O3
	+0.385	−0.546	−0.363	+0.403	−0.515	−0.474	+0.338	−0.429
Distance	O1 – H1	N1 – H1	O2 – H2	N2 – H2	C – O3			
A	1.02	2.01	1.05	1.86	1.32			

FIGURE 1 Geometry and electronic characteristics of the Chemotrypsin molecule: I – Ser; II – Asp; III – His. Ionization potential 9.605 eV; electron affinity: −2.601. Angle O2–N2–O3 104.8^0 eV.

11.3 RESULTS AND DISCUSSION

It can be seen from the optimal geometry of the active fragment of Chemotrypsin (Fig. 1), that the hydrogen bonds correspondingly to the distances exist between the atoms H1 and N1, and also between atoms H2 and N2. Probably, the second bond is stronger (the distance is shorter). Let's note, that the electron affinity in the Chemotrypsin fragment is enough high. At the fitting of the methyl acetate molecule to the Chemotrypsin along the coordinate C1–O1 (Fig. 2a) at the distance 1.60 A a change of the complex's geometry takes place, and also the elongation of hydrogen bond O1...H...N1 takes place, that is caused by the conformational transformations of the Chemotrypsin molecule. Such conformational transformation proceeds with enough great energy charge (see peak № 1 on Fig. 3).

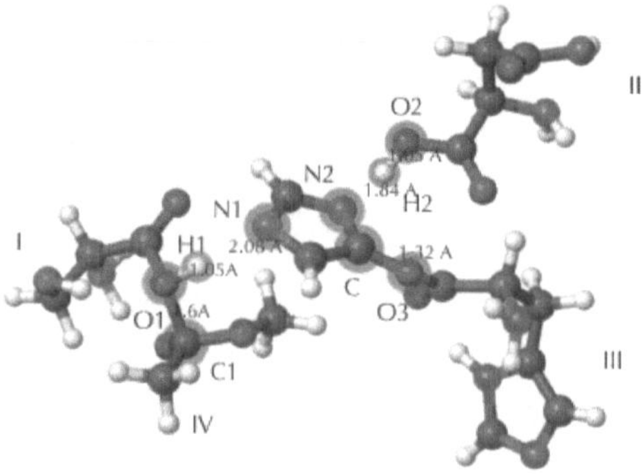

(a)

Distance C1 – O1: 1.60 A. Angle O2 – N2 – O3: 104.7^0

Charge on	H1	O1	N1	C1	H2	O2	N2	C	O3
	+0.399	–0.531	–0.358	+0.758	+0.401	–0.519	–0.476	+0.339	–0.432
Distance	O1 – H1	N1 – H1	O2 – H2	N2 – H2	C – O3				
A	1.05	2.08	1.05	1.84	1.32				

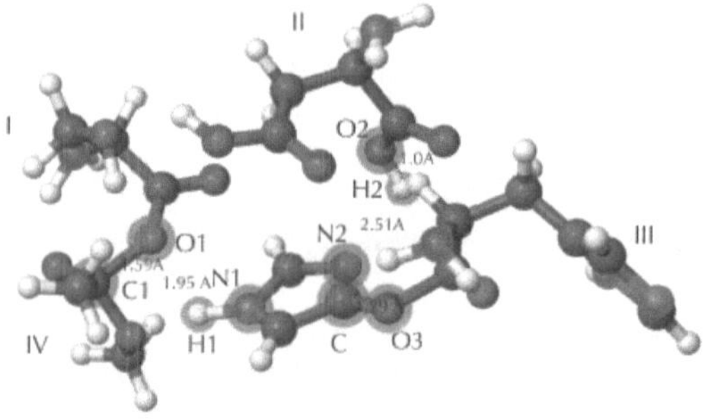

(b)

Distance C1 – O1: 1.59 A. Angle O2 – N2 – O3: 122.9^0

Charge on	H1	O1	N1	C1	H2	O2	N2	C	O3
	+0.405	–0.706	–0.212	+0.811	+0.360	–0.500	–0.457	+0.423	–0.420
Distance	O1 – H1	N1 – H1	O2 – H2	N2 – H2	C – O3				
A	1.95	1.13	1.00	2.51	1.29				

FIGURE 2 Change of the geometry of complex in point 1 (correspondingly to Fig. 3). ^{2} – Ser, II – Asp, III – His, IV – methyl acetate.

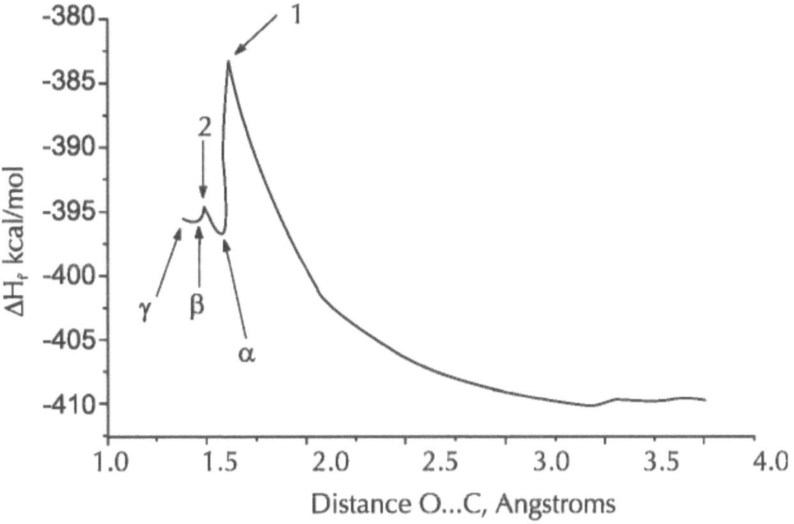

FIGURE 3 Formation of complex [Kh–CH_3CO]. **1** – complex coordination [Kh–CH_3COOCH_3] (1.60 A), transfer of Hydrogen atom from Serin to Nitrogen atom; **2** – detach of methanol (1.48 A).

It can be seen from Fig. 2 (a), that at the distance C1–O1 1.60 A the charges on H1 and O1 are some increased; the charge on N1 is decreased. The change of angle O2–N2–O3 is insignificant. The interatomic distance O1–H1 is also enhanced. Insignificant change of a distance C1–O1 leads to considerable changes into geometry and values of charges on atoms of the Chemotrypsin fragment. At the distance C1–O1 1.59 A as a result of new conformational changes a new complex (α) is formed (Fig. 2 (b) and Fig. 3 (α)). From Fig. 2 (b) it can be seen a great increase of a charge on the reactive center O1. It is also increased a charge on C1 of methyl acetate. A charge on N1 is considerably decreased as a result of a great decrease of a distance N1–H1. The charges on N2 and H2 are correspondingly decreased at the expense of the hydrogen bond distance increasing as a result of the change of the interatomic distances and angles of the surroundings atoms. It is observed an essential change of the dihedral angle O2–N2–O3 that leads to more stable conformation of Chemotrypsin (Fig. 3 (α)).

At the fitting of atom C1 to O1 the atom O_4 (Fig. 4 (a); Fig. 3, peak № 2) is neared to the atom H1, and at the distance 1.48 A the transition of atom H1 to O4 and detach of methanol molecule from the complex take place. Such process is illustrated by a peak № 2 on Fig. 3.

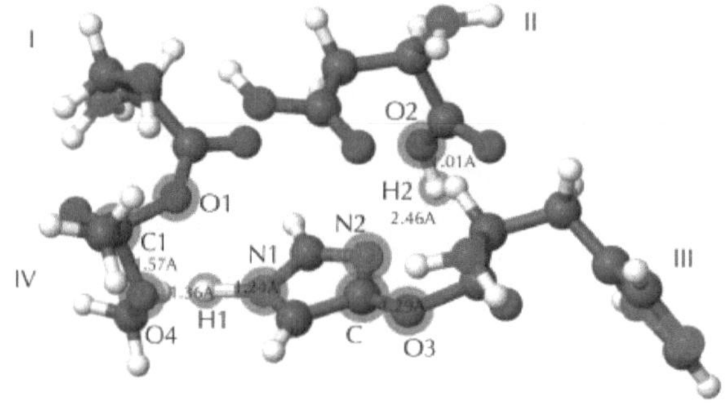

(a)

Distance C1 – O1: 1.48 A. Angle O2 – N2 – O3: 123.3^{0}

Charge on	H1	O1	N1	C1	O4	H2	O2	C	O3
	+0.420	–0.648	–0.280	+0.801	–0.550	+0.363	–0.502	+0.399	–0.425
Distance	O4 – H1	C1 – O4	N2 – H1	O2 – H2	N2 – H2	C – O3			
A	1.36	1.57	1.24	1.01	2.46	1.29			

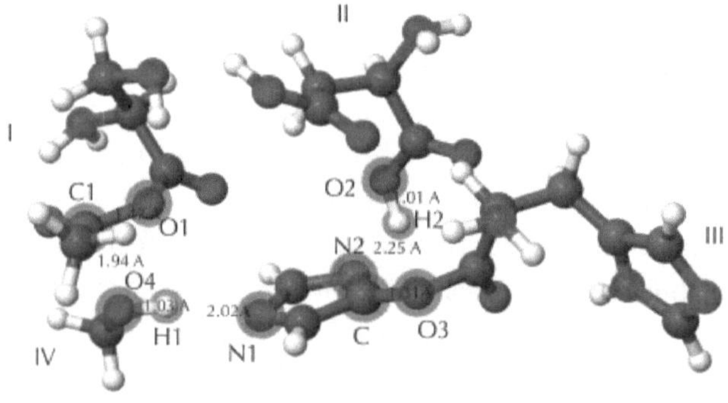

(b)

(b)

Distance C1 – O1: 1.47 A. Angle O2 – N2 – O3: 117.8^{0}

Charge on	H1	O1	N1	C1	O4	H2	O2	C	O3	N2
	+0.376	–0.623	–0.374	+0.794	–0.523	+0.378	–0.513	+0.349	–0.434	–0.481
Distance	O1 – H1	C1 – O4	N1 – H1	O2 – H2	N2 – H2	C – O3				
A	1.03	1.94	2.02	1.01	2.25	1.31				

FIGURE 4 The change of the geometry of complex in point 2. (a) ^{2} – Ser, II – Asp, III – His, IV – methyl acetate; (b) ^{2} – Ser + CH_3CO, II – Asp, III – His, IV – methanol.

For complex (Fig. 4 (**a**)), which represents by itself the activated complex for the formation of the intermediate unstable compound, the charges on atoms H1 and N1 are increased the most essentially. The charges on C1 and O1 are decreased. As a result of the hydrogen atom to O4 of methyl acetate addition (Fig. 4 (**a**)) the distances for N1–H1 are increased, and for N2–H2 are decreased correspondingly.

In complex (β) (Fig. 3) at distance C1–O1 = 1.47 A the distances O4–H1, N2–H2 are considerably decreased and the interatomic distances C1–O4, N1–H1, C–O3 are increased as a result of the essential changes of the general geometry of complex. At this, the charges on N4, N2 and H2 atoms are increased and on H1, O1, C1 and O4 atoms are decreased. The angle O2–N2–O3 is essentially changed.

The complex (β) (Fig. 3) transfers into the intermediate compound via the conformation (γ) (Fig. 3). The formation of the intermediate compound [Kh–CH_3CO] (Fig. 5) leads to decrease of the charges on atoms C1, O1, N1, N2, O3 and to increase of the charges on atoms N2, O2, H2 and O1.

Interatomic distance C1–O1 for intermediate compound consists of 1.39 A. The distance N2…H2 is considerably increased. The angle O2 – N2 – O3 is greatly changed. The electron affinity is decreased in comparison with the starting active center of the Chemotrypsin.

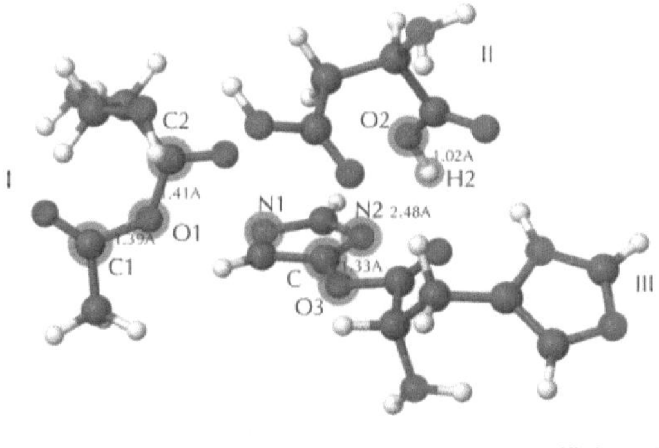

Angle O2 – N2 – O3: 83.2°. Ionization potential: 9.050 eV. Electron affinity: –2.120 eV.

Charge on	C2	O1	N1	C1	H2	O2	C	O3	N2
	+0.595	–0.559	–0.324	+0.717	+0.388	–0.520	+0.302	–0.412	–0.419
Distance	O1 – C1	C2 – O1	O2 – H2	N2 – H2	C – O3				
A	1.39	1.41	1.02	2.48	1.33				

FIGURE 5 Geometry and electronic characteristics of complex [Kh–CH_3CO]. ² – Ser + CH_3CO, II – Asp, III – His.

Let's consider the thermodynamics of the enzymatic catalysis by Chemotrypsin for the reaction of C–O bond breakdown in methyl acetate. Below there is a scheme of a process, thermodynamics of which calculated without taking into account the interactions of complexes with the methanol (1) and acetic acid (2):

$$Kh + CH_3COOCH_3 \rightarrow [Kh{-}CH_3CO] + CH_3OH \qquad (1)$$

(ΔH_p = + 12,9 kilocalorie/mol; ΔS_p = + 6,0 calorie/mol·K; ΔF_p = + 14,7 kilocalorie/mol)

$$[Kh{-}CH_3CO] + H_2O \rightarrow Kh + CH_3COOH \qquad (2)$$

(ΔH_p = – 10,6 kilocalorie/mol; ΔS_p = + 10,1 calorie/mol·K; ΔF_p = –13,6 kilocalorie/mol)

Here: Kh – Chemotrypsin; Kh–CH_3CO – Acylchemotrypsin

$\Sigma\Delta F$ = + 1,1 kilocalorie/mol; K = exp(–ΔF/RT) = 0,2.

Thermodynamic characteristics of the starting substances, of the reaction products and of the intermediate compound (Acylchemotrypsin) are represented in Table 1. As we can see from the calculations, the Acylchemotrypsin formation process is endothermic. Such intermediate compound thermodynamically is not stable and is easy transformed into the final products of the deacidulating reaction.

TABLE 1 Thermodynamic characteristics of the reagents and of the products of the methyl acetate hydrolysis reaction.

№	Formula / substance	ΔH, kilocalorie/mol	ΔS, calorie/mol·K	ΔF, kilocalorie/mol
1	Chemotrypsin	−306.6	238.9	−377.8
2	$CH_3C(O)OCH_3$	−97.5	75.8	−120.1
3	H_2O	−54.3	45.0	−67.7
4	CH_3OH	−48.3	55.9	−65.0
5	CH_3COOH	−101.2	69.0	−121.8
6	[Kh–CH_3CO]	−342.9	252.8	−418.2

Let's consider the kinetics of the acidulating and deacidulating reactions of the fermentative part of Chemotrypsin. Kinetic data for the acylferment formation reaction and for the stages of its deacidulating are represented in Table 2.

TABLE 2 Kinetic parameters of the enzymatic catalysis of the acylferment formation.

Stage	Point of calculation	ΔH, kilocalorie/mol	ΔS, calorie/mol·K	ΔH≠, kilocalorie/mol	ΔS≠, calorie/mol·K
1	Starting substances	−410.1	275.0	15.2	24.5
	Activated complex	−394.9	250.0		
	Products of the reaction	398.9	262.8		
2	Starting substances	406.4	212.8	11.2	37.6
	Activated complex	395.2	250.4		
	Products of the reaction	433.2	242.3		

On Fig. 3 presented the potential curve of the reaction of acylferment formation. It can be seen from the presented curve that as a result of the interaction of Chemotrypsin and methyl acetate it is formed a series of intermediate complexes, which are caused by the conformational changes of the ferment under the action of the methyl acetate. Energy of the process of the substrate adsorption on the ferment is sufficient for the formation of Michael's complex (conformation on Fig. 3 (1)). Complex (1), which is characterized by enough high free energy, is easy transformed into the complex (α). At this, the transition from the complex (1) into the complex (α) is the exothermal process with the heat effect ~ 10 kilocalorie/mol. The complex (α) is easy transformed into the activated complex (2) (see Fig. 3, peak № 2) with the activation energy ~ 2 kilocalorie/mol. Activated complex (2) transfers into the conformation (β), and (β) is transformed into the conformation (γ), which with a little barrier is transformed into the intermediate compound – acylferment. Activation enthalpies of the acylferment formation are some overestimate as a result of the inaccuracy of the semiempirical calculation methods. However, kinetic characteristics of the process may be trusted qualitatively (Table 2). Free activation energy of the acidulating reaction consists of 7.8 kilocalorie/mol, and of the deacidulating reaction this value consists of $\Delta F^{\neq} = 4{,}0$ kilocalorie/mol, in other words the deacidulating reaction rate is considerably great. The acidulating reaction is the limiting stage of the methyl acetate hydrolysis process.

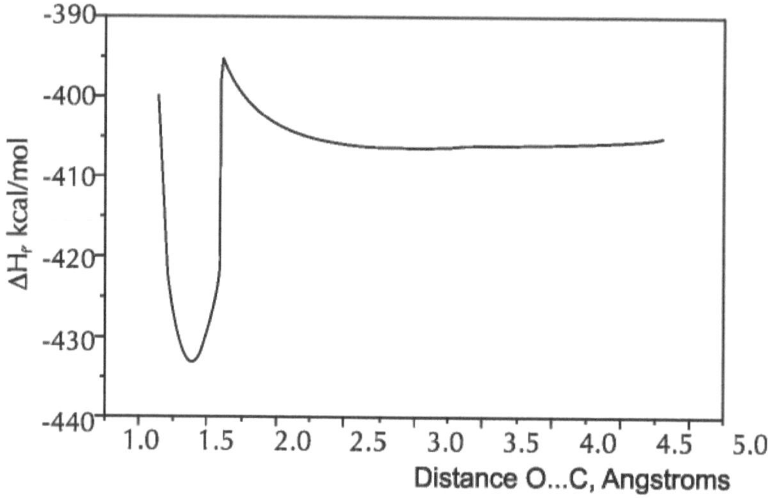

FIGURE 6 Potential curve of the deacidulating for Chemotrypsin complex.

It can be seen from the Fig. 6, that the activated complex of the intermediate product deacidulating reaction is formed at the interatomic distance C–O 1,75 A. The deacidulating process is exothermal and proceeds with a heat effect 35 kilocalorie/mol. Probably, the deacidulating process is one–stage, in other words it can be considered as elementary one. A great difference in activation entropies for acidulating and deacidulating reactions is explained by more distended structure of the activated complex for the deacidulating reaction.

Generally, it is necessary to note, that a great contribution into the values of the rate constants in Chemotrypsin acidulating and deacidulating reactions is characterized by the entropic factors; this is connected with a great contribution of the activation oscillating constituents of entropy, which caused by the low–frequency vibrations of the bonds, in $\Delta S^{\neq}$.

It was interesting for comparison to study the modeling of the reaction of the methyl acetate acidic homogeneous catalysis.

The objects of investigation:

– methyl acetate;

– Serin.

It was assumed, that the process of homogeneous catalysis proceeds also via the stage of acidulating and deacidulating. Calculations of the thermodynamic parameters of the reagents, of the intermediate products and of the final products are represented in Table 3.

TABLE 3 Thermodynamic characteristics of the reagents.

№	Formula / substance	ΔH, kilocalorie/mol	ΔS, calorie/mol·K	ΔF, kilocalorie/mol
1	Ser–H	−138.7	87.8	−164.9
2	$CH_3C(O)OCH_3$	-97.5	75.8	−120.1
3	H_2O	−54.3	45.0	−67.7
4	CH_3OH	−48.3	55.9	−65.0
5	CH_3COOH	−101.2	69.0	−121.8
6	CH_3CO–O–Ser	−175.6	108.9	−208.0

Scheme of the process:

CH_3COOCH_3 + Ser–H –> CH_3CO–O–Ser + CH_3OH –(+ H2O)–> CH_3COOH + Ser–H

$$CH_3COOCH_3 + \text{Ser–H} \to CH_3CO\text{–O–Ser} + CH_3OH \quad (1)$$

(ΔH_p = + 12.3 kilocalorie/mol, ΔS_p = + 1.2 calorie/mol·K, ΔF_p = + 11.9 kilocalorie/mol)

$$CH_3CO\text{–O–Ser} + H_2O \to CH_3COOH + \text{Ser–H} \quad (2)$$

(ΔH_p = – 10.0 kilocalorie/mol, ΔS_p = + 2.9 calorie/mol·K, ΔF_p = – 10.8 kilocalorie/mol)

ΣΔF = + 1.1 kilocalorie/moe, $K = \exp\left(-\frac{\Delta F}{RT}\right) = 0{,}16.$

It can be seen from the calculations, that the stage of the intermediate product formation is endothermic and its formation is not thermodynamically stable. Probably that the time of the intermediate product existing is very little. The hydrolysis reaction of the intermediate compound is thermodynamically efficient. Generally, free energy of the hydrolysis process consists of +1.1 kilocalorie/mol. Approximately within the limits of experimental error the methyl acetate hydrolysis reaction is thermodynamically allowed.

Let's consider the kinetics of the catalysis process. On Fig. 7 there are interatomic distances and electronic characteristics of the Serin.

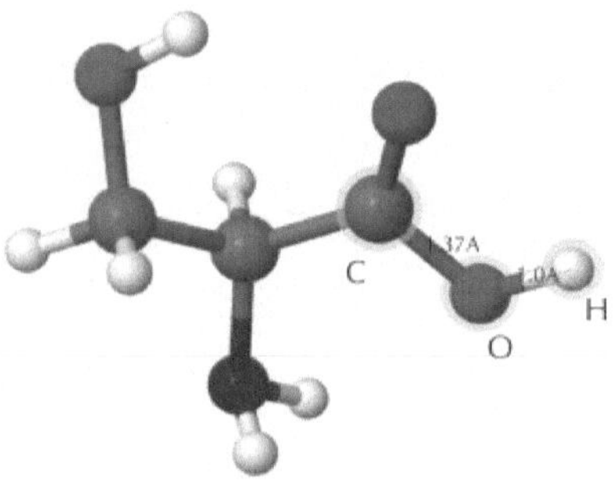

Angle C – O – H: 113.0⁰. Ionization potential: 10.232 eV. Electron affinity: –0.066 eV

Charge on	H	O	C	Distance	O–H	C–O
	+0.348	–0.536	+0.561	A	1.00	1.37

FIGURE 7 Geometry and electronic characteristics of the Serin.

Ionization potential is some higher, than in a case of the Chemotrypsin. Electron affinity is sufficient low.

In activated complex of the acylserin (see Fig. 8) it is observed a great change of the charges on the reactive center C–O, in comparison with the starting system.

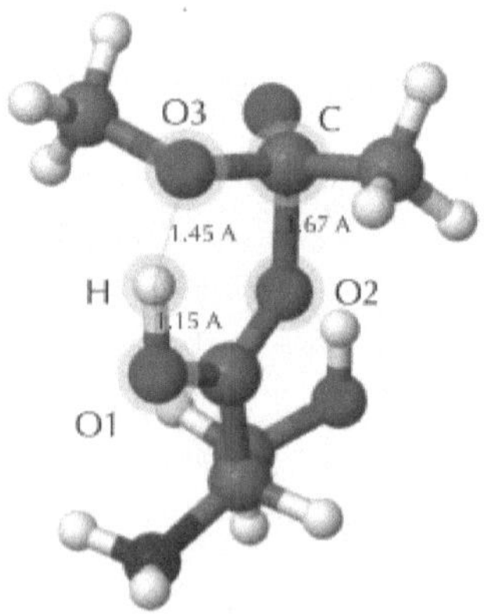

Charge on	H	O1	O2	C	O3	Distance	C–O2	O1–H	H–O3
	+0.418	–0.537	–0.570	+0.771	–0.585	A	1.67	1.15	1.45

FIGURE 8 Geometry and the values of charges for activated complex of the acylserin formation.

The final intermediate product – acylserin – is formed at the distance C–O ~ 1.5 A (Fig. 9).

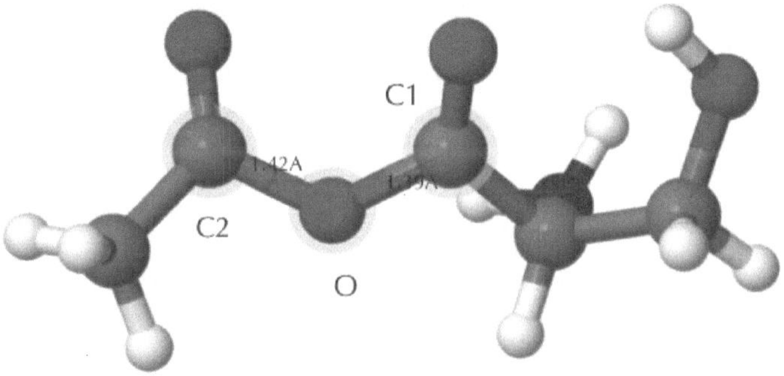

Angle C – O – C: 126.1^{0}. Ionization potential: 10.012 eV. Electron affinity: –-0.401 eV

Charge on	C1	O	C2	Distance	C1–O1	O–C2
	+0.584	–0.570	+0.703	A	1.390	1.42

FIGURE 9 Geometry and electronic characteristics of the complex CH_3CO–O–Ser (acidulating reaction).

The process of the intermediate product formation is endothermic (Fig. 10) with the reaction heat ~ 11 kilocalorie/mol.

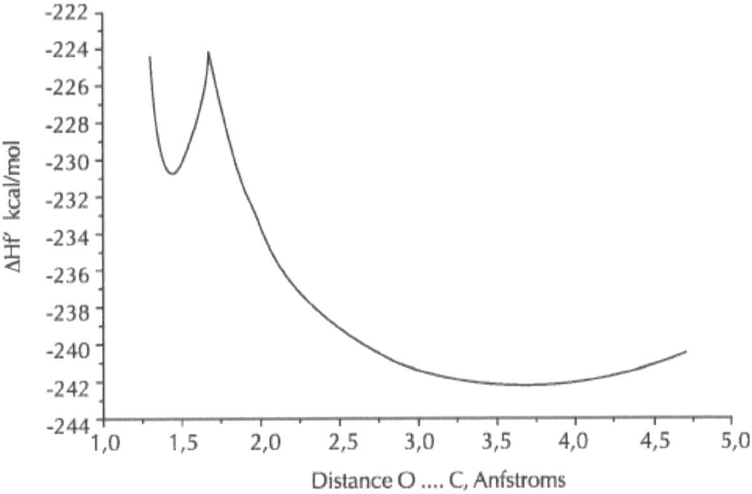

FIGURE 10 Potential curve of the intermediate product acylserin CH_3CO–O–Ser formation

Transformation of the activated complex into the intermediate product proceeds exothermally with heat ~ 6 kilocalorie/mol.

Activated complex of the hydrolysis reaction (Fig. 11) of the intermediate product is formed at the internuclear distance C–O ~ 1.6 A. Reaction of the hydrolysis is exothermal and proceeds with heat ~ 6.5 kilocalorie/mol.

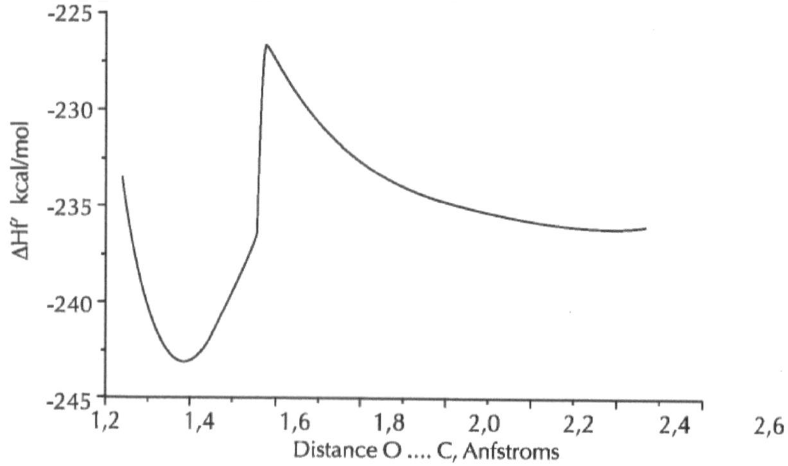

FIGURE 11 Potential curve of the deacidulating CH_3CO–O–Ser reaction

Kinetic parameters of the reaction of intermediate product formation (acidulating) and final products (hydrolysis) are represented in Table 4. The potential curves of the stages are represented on Figs. 10 and 11, respectively.

TABLE 4 Kinetic parameters of the acidulating and deacidulating reactions of Serin.

Stage	Point of calculation	ΔH, kilocalorie/mol	ΔS, calorie/mol·K	E_A, kilocalorie/mol	$\Delta S^{\#}$, calorie/mol·K
	Starting substances	–242.3	137.1		
1	Activated complex	–224.4	106.1	17.9	31.0
	Products of reaction	–230.7	136.6		
	Starting substances	–236.0	122.7		
2	Activated complex	–226.3	114.4	9.7	8.3
	Products of reaction	–243.1	103.7		

Heat of the activation of acidulating reaction is considerably greater, than the $\Delta H^{\neq}$ of the deacidulating reaction. However, the activation entropy of the first reaction is considerably greater.

Free activation energy of the acidulating reaction consists of − 0.7 kilocalorie/mol. Free activation energy of the deacidulating reaction at T = 298 K consists of + 4.7 kilocalorie/mol. In other words, the second stage is the limiting one, whereas for the enzymatic reaction the limiting stage is the acidulating stage.

More electron affinity of the active fragment of Chemotrypsin (−2.601 eV) explains the acceleration of the enzymatic catalysis in comparison with the homogeneous catalysis by Serin, electron affinity of which is sufficiently low (−0.0066 eV).

If to assume, that the reactive centers C1−O1 for Chemotrypsin and Serin are characterized by approximately the same resonance integrals and by the overlapping integrals, then the orbital energy of these both reactions will be determined accordingly to formula **[15]** $E_{orbit} \approx \frac{const}{I-E}$, where I is the ionization potential (donor), and E in an electron affinity (acceptor). For the methyl acetate I ≈ 11.66 eV; E for the Chemotrypsin consists of − 2.66 eV, whereas for Serin this value is equal to − 0.0066 eV. Correspondingly, the orbital energies of interaction for the reaction of intermadiare acylcompounds formation will be consist $\frac{const}{14,32}$ and $\frac{const}{11,67}$, respectively; in other words, the orbital energy of the acylchemotrypsin formation will be in 1.22 times lesser, in comparison with the intermediate energy of the acylcompound formation in a case of the homogeneous catalysis by Serin.

It is necessary to note, that the electron structure of the reagents has an essential influence on the hydrolysis reaction of C−O bond in methyl acetate both via the enzymatic and via the homogeneous (acidic) catalysis. For example, the values of charges on the active atoms of the Oxygen of hydroxyl groups of the serinic fragment of Chemotrypsin and free Serin itself are respectively equal to −0.546 i −0.536 that in some way leads to higher nucleophilicity of the Chemotrypsin.

During the enzymatic catalysis the electronic characteristics of the reagents are greatly changed, that leads to the essential changes of the nucleophilicity of the reactive center of Chemotrypsin and of the reactive center of the substrate (methyl acetate). If to assume the respective charges as the characteristics of such reactive centers, then we will obtain the following picture. The values of the atoms charges along the reaction way $\overset{+}{C}1-\overset{-}{O}1$ (equilibrium state) is changed from + 0.666, – 0.559 to + 0.801, – 0.648 into the activated complex.

For the reaction of homogeneous catalysis of methyl acetate by Serin into the equilibrium state the charges on the reactive centers C and O are equal to + 0.666 and – 0.536, respectively, and for the activated complex + 0.771 and – 0.537.

It is following from the all above-said, that the nucleophilicity of the reactive center (atom of Oxygen of Serin) of Chemotrypsin is some more than the same characteristic for Serin. Generally, it can be concluded, that the stabilization of the activated complex at the expense of the charges of reactive center C–O will be higher in comparison with the homogeneous catalysis by Serin molecule. Accordingly to this also heat of the activated complex formation will less in a case of the enzymatic catalysis that leads to the less activation heat of the acylferment formation.

Let's note, that the activated complex of the acylferment in a case of the enzymatic catalysis is stabilized also at the expense of the formation of hydrogen bond of His and atom of oxygen of methyl acetate (finally, atom of hydrogen transfers on the oxygen with the formation of methanol).

Let's consider rather more detailed the reaction of Chemotrypsin deacidulating. The atom of Carbon C1 (Fig. 12) is attacked by the atom of Oxygen O3 of water molecule with the simultaneous attack of Hydrogen of water the Nitrogen atom of the imidazole group of the His. Geometry and the charges distribution into the activated complex of acylchemotrypsin are represented on Fig. 12.

From the comparison of both reactions of acylchemotrypsin and acylserin hydrolysis it can be concluded that most stabilization of the activated complex is observed in a case of the acylchemotrypsin deacidulating stage. Due to such higher stabilization of the activated complex of the deacidulating reaction it is observed some less activation energy for the enzymatic reaction.

The values of charges on the reactive center for activated complex of the acylchemotrypsin C1–O3 are respectively equal to + 794 i – 0,554, whereas for the activated complex of the acylserin the values of such charges are equal to C2 (+0,757) – O2 (–0.549) (see Fig. 13).

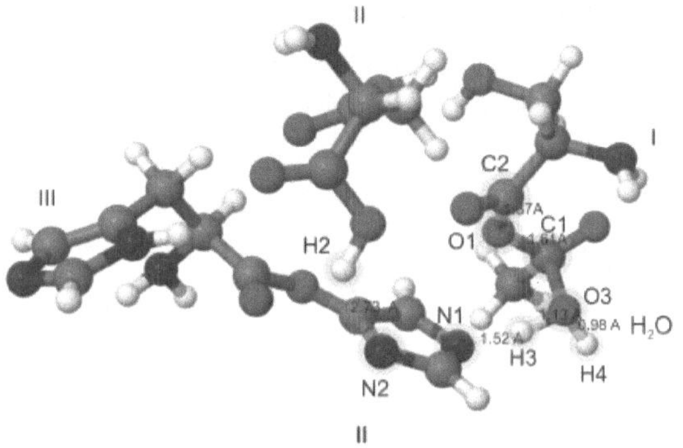

Charge on	C2	O1	N1	C1	H2	H3	H4	O3	N2
	+0.633	–0.651	–0.420	+0.794	+0.374	+0.412	+0.336	–0.554	–0.401
Distance	O1 – C1	C2 – O1	O3 – C1	O3 – H3	O3 – H4	N1 – H3	N2 – H2		
A	1.49	1.37	1.61	1.13	0.98	1.52	2.73		

FIGURE 12 Geometry of activated complex and the charges of some atoms [Kh–CH_3CO]–H_2O. ^{2} – Ser + CH_3CO, II – Asp, III – His.

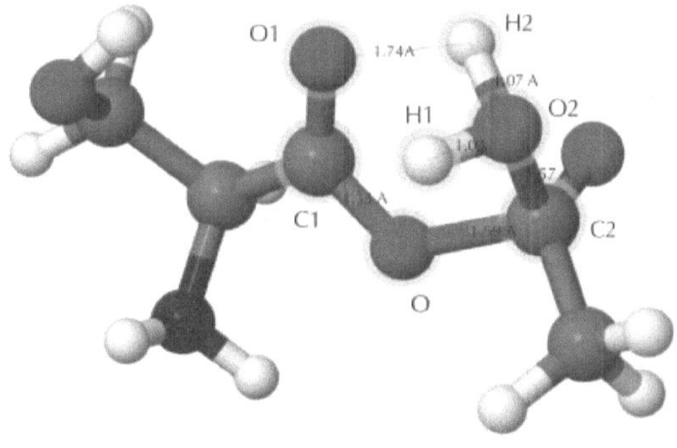

Charge on	C1	O	C2	O1	O2	H1	H2
	+0.634	–0.636	+0.757	–0.645	–0.459	+0.346	+0.404
Distance	C1 – O	O – C2	O2 – C2	O2 – H1	O2 – H2	O1 – H2	
A	1.33	1.58	1.57	1.00	1.07	1.74	

H_2O	H_3O^+
Charge on O: –0.6188	Charge on O: –0.122
Charge on H: +0.3094	Charge on H: +0.374
Distance O – H: 0.95 A	Distance O – H: 1.04 A
Ionization potential: 11.906 eV	Ionization potential: 22.869 eV
Electron affinity: 4.068 eV	Electron affinity: –6.346 eV

FIGURE 13 Geometry of the activated complex [CH_3CO–O–Ser] – H_2O

In a case of the acylcompounds hydrolysis a molecule of water plays the role of the nucleophile. Electronic characteristics of water's molecule are represented in legends to the Fig. 13. Taking into account, that the ionization potential of the acylchemotrypsin is equal to 9.05 eV, and for the acylserin ~ 10 eV (the difference is 23 kilocalorie/mol), it can be concluded that the deacidulating reaction will be proceed with the less activation energy for the acylchemotrypsin. It possible that the deacidulating reactions will took place with the participation of the hydronium ion, which is characterized with a high electron affinity.

Also it was considered the reaction of the methyl acetate hydrolysis in the presence of the sulphuric acid. It was shown, that the first acidulating stage proceeds with $\Delta H_{\text{реакц.}} = +1.4$ kilocalorie/mol, $\Delta S_{\text{реакц.}} = -2.9$ calorie/mol·K, $\Delta F_{\text{реакц.}} = +2.3$ kilocalorie/mol. The second deacidulating stage is characterized by the following thermodynamic parameters: $\Delta H_{\text{реакц.}} = +0.9$ kilocalorie/mol, $\Delta S_{\text{реакц.}} = 7.0$ calorie/mol·K, $\Delta F_{\text{реакц.}} = -1.2$ kilocalorie/mol. Total reaction: $\Sigma\Delta F_{\text{реакц.}} = +1.1$ kilocalorie/mol; $K = \exp\left(-\frac{\Delta F}{RT}\right) = 0{,}16$.

Activation parameters of the acidulating reaction are: $\Delta H^{\neq} = 34.2$ kilocalorie/mol, $\Delta S^{\neq} = 10.1$ calorie/mol·K, $\Delta F^{\neq} = \sim 31$ kilocalorie/mol.

Activation parameters of the deacidulating reaction are: $\Delta H^{\neq} = 10.3$ kilocalorie/mol, $\Delta S^{\neq} = 8.3$ calorie/mol·K, $\Delta F^{\neq} = 7.8$ kilocalorie/mol.

At the comparison of the methyl acetate hydrolysis kinetics in the presence of the sulphuric acid, it is necessary to note, that the deacidulating process proceeds with the same kinetic parameters as in a case of Serin. However, the activation energy for H_2SO_4 is practically in two times higher, and the activation entropy is too lesser.

It is seen from the value $\Delta F^{\neq}$ that the limiting stage in the methyl acetate hydrolysis with the sulphuric acid reaction is the stage of the acylproduct formation whereas in a case of the Serin it is observed the opposite situation. Probably, it is not be unambiguously stated, that for the all reactions of the ethers hydrolysis by different substrates the reaction of the intermediate product (namely, acylferment) formation is the limiting stage.

Quantum–chemical calculations show, that, really, into the activated state of the ferment there are hydrogen bonds, at the expense of which the nucleophilicity of the reactive atom of Oxygen of Serin is increased. These bonds substantially determine the chemical mechanism of the acylferment formation and have an influence on the values of their kinetic parameters for the enzymatic catalysis and on the geometry of conformers on the reaction way.

KEYWORDS

- **Acylchemotrypsin**
- **Asp**
- **Chemotrypsin**
- **His**
- **Histidine**
- **Serine**

REFERENCES

1. Blow, D. M.; Steitz, T. A. *Ann. Rev. Biochem.*, **1970**, *39*, 63.
2. Matthews, B. W.; et al. *Nature*, **1967**, *214*, 652.
3. Henderson, R. J. *Mol. Biol.*, **1970**, *54*, 341.
4. Birktoft, J. J.; et al. *Phil. Trans. Roy. Soc. (London)*, **1970**, *B 257*, 67.
5. Miller, Ch. G.; Bender, M. L. *J. Amer. Chem. Soc.*, **1968**, *90*, 6850.
6. Bernhard, S. *Struct. Funct. Enzymes*, **1971**, *45*, 438 (in Russian).
7. Hess, G. P.; et al. *Phil. Trans. Roy. Soc. (London)*, **1970**, *B 257*, 89.
8. Bernhard, S. A.; Gutfreund, H. *Trans. Phil. Trans. Roy. Soc. (London)*, **1970**, *B 257*, 105.
9. Himoe, A.; Brandt, K.; Hess, G. P. *J. Mol. Biol.*, **1971**, *55*, 215.
10. Mosolov, V. V. *Proteolytic Enzymes* Nauka, **1971** (in Russian).
11. Cunningham, L. *Comprehensive Biochem.*, **1965**, *16*, 85.
12. Bender, M. L.; Kezdy, F. J. *J. Am. Chem. Soc.*, **1964**, *86*, 3704.
13. Berezin, I. V.; Martynek, K. In Colln.: *Structure and Functions of Enzymes.* Moscow State University Edition, **1972** (in Russian).
14. Blow ,D. M.; Birktoft, J. J.; Hartley, B. S. *Nature*, **1969**, *221*, 337.
15. *Reactive Ability And The Reaction Ways* Klopmann G., Ed., M.: «Myr», 1977, 383p. (in Russian).
16. Berezin, I. V.; Martynek, K. *The Principles of the Physical Chemistry of the Enzymatic Catalysis.* M.: «Vysshaya Shkola», 1978, 350p. (in Russian).
17. Stiepukhovich, A. D.; Ulitsky, V. A. *Kinetics and Thermodynamics of Cracking Radical Reactions* M.: «Khimiya», 1975; 255 p. (in Russian).

CHAPTER 12

SOME ASPECTS OF BIO-DECOMPOSED POLYMERS AND AGRICULTURE'S WASTE

I. A. KIRSH, D. A. POMOGOVA, and D. A. SOGRINA

CONTENTS

12.1 INTRODUCTION

There is a problem of polymeric waste's recycling. One of the most promising directions in the field of packing waste's recycling is creation of biodecomposed polymeric materials. Nowadays synthesis of biopolymers is expensive, and these materials have the limited usage. Therefore, the greatest interest is gathered by the filled biodecomposed compositions. Such compositions are referred to partially decomposed or punched materials. Getting to the environment, these materials are exposed to impacts of external factors and bacteria, the filler completely assimilates and polymer is destroyed. That leads to reduction of dumps by means of desomposition polymer's time reduction.

Much waste of agriculture is being accumulated now. Approximately 70% of wastes are processed into forages and fertilizers, and 30% are utilized by way of dumping and thermal methods, which have negative effect on our environment. That's why it is expedient to use agroindustrial complex waste as a filler for creation of biodecomposed polymeric materials, and as a polymeric matrix we should use waste of packing branch.

12.2 EXPERIMENTAL PART

Waste of polyethylene film and agricultural waste have been chosen for getting biodegradated compositions: cacao bean husk, beet bin pulp, buckwheat, rice, millet, sunflower shuck, potato and corn mar. The maximum size of the filler's particles made less than 150 microns.

There were complexities while processing on standard laboratorial extrusion-type equipment for all compositions, for example there was bad distribution of a filler: its agglomeration led to extruding press's productivity decrease and to formation of defects in material.

The basic criterion is durability for creation of secondary qualitative raw polymeric materials and products made from it. It is characterized by breakdown voltage's magnitude at monoaxial stretching (σp). By data researches that were carried out this magnitude should be not less than 4 MPa.

As a result of researches in physicomechanical characteristics compositions that have fillers like cacao bean husk, beet bin pulp and rice schuck have been selected.

In the process of compositions' research it was noticed that agroindustrial complex waste which is "incompatible" with polymeric waste, reduce durability that is connected with specialty of fillers' distribution from agroindustrial complex waste in the supramolecular structure of polymer which leads to formation of non-uniform material's structure.

For clearing this lack while processing mixed compositions it is expedient to use special additives which can lead not only to improvement of compositions' workability, but also to updating of secondary raw materials' properties.

In this case additive's selection has been determined not only by its impact on physicomechanical properties of polymeric compositions, but also by its ability to biodegradation. It is known that colloidal clay with an advanced surface is a natural filler, a good adsorbent and a dispergator from treatise [1]. Colloidal clay belongs to alumosilicic hydrates of layered structure [2].

12.3 RESULTS AND DISCUSSION

Investigations were conducted on compositions on basis difference concentration colloidal clay. As a result of the researches that were carried out optimum concentration colloidal clay has been established. It has made 2%. It can be explained that colloidal clay is created by uniform distibution of a filler in concentration of 2% in the conditions of processing highly filled (30–40% of agroindustrial complex waste) polymeric compositions, that the conducted researches have proved by a method of optical microscopy (Fig. 1a and 1b). In the Fig. 1(a) we can see more uniform distribution of a filler in a polyethylene composition with colloidal clay in comparison with the structure of Fig. 1(b). In case of concentration's increase in polymeric compositions the effect of forming of colloidal clay's own structures in polymer-filled compositions is observed. Herewith, there are complexities of filler's distribution in polymer.

There is a composition on the basis of polyethylene waste containing 30% of agroindustrial complex waste (rice shuck) in the Fig. 1(a); and 2% of colloidal clay in the Fig. 1 (b).

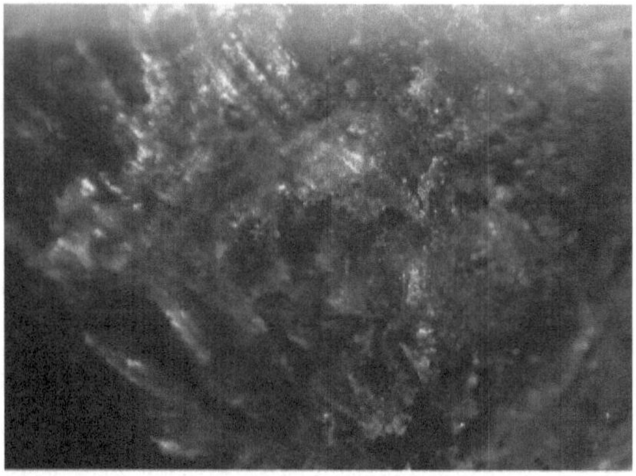

(a)

(b)

FIGURE 1 Optical microphotos of the filled polymeric compositions' samples filled with agroindustrial complex waste (zooming in 250 times).

Consequently, infusion of colloidal clay has led not only to uniform distribution of agroindustrial complex waste, but also has allowed to increase breakdown voltage and percent elongartion with rupture in 2–3 times (Table 1).

TABLE 1 Physicomechanical properties of polymer compositions on the basis of agroindustrial complex waste and polyethylene.

Filler's name*	Colloidal clay in polymer compositions	Breakdown voltage σ_p, MPa	Percent elongation with rupture, ε_p,%
Beet bin pulp	–	1.30±0.08	9.80±0.10
	+	4.75±0.08	11.25±0.09
Rice shuck	–	2.00±0.07	8.70±0.09
	+	6.00±0.15	16.50±0.11
Cacao bean husk	–	1.60±0.08	11.00±0.12
	+	6.00±0.09	16.00±0.09

* – The amount of filler in polymer compositions is 30%.

To estimate dynamics of the filled polymeric compositions' biodegradation composting was used. Samples were placed in special mallets with biohumus at temperature 23 ± 2°C and humidity 70 ± 10%. Degree of polymeric compositions' biodegradation was estimated by change of physicomechanical properties such as breakdown voltage and percent elongation with rupture. Calculation of biodegradation degree of composition was made according to the following formula:

$$\Delta = \frac{a_1 - a_0}{a_0} \cdot 100, (\%)$$

where, a_1 is a parameter's meaning before composting;

a_0 is parameter's meaning after composting.

In the Table 2 results of polymer compositions' biodegradation research on the basis of polyethylene's containing waste, agroindustrial complex waste and colloidal clay as a filler for 12 months are presented.

TABLE 2 Changes in physicomechanical properties after composting.

Filler's name*	Changes in physicomechanical properties	
	Change of breakdown voltage, $\Delta\sigma_p$,%	Change of percent elongation with rupture, $\Delta\varepsilon_p$,%
Beet bin pulp	60 ± 5%	68 ± 0.1%
Rice shuck	65 ± 5%	72 ± 0.1%
Cacao bean husk	65 ± 5%	73 ± 0.1%

* – The amount of filler in polymer compositions is 30%.

It is obvious after composting breakdown voltage for all compositions has decreased on the average in 2.5 times and percent elongation with rupture has lowered in 3 times. That testifies about processing of polymer compositions' biodegradation with agroindustrial complex waste. The visual estimation has allowed to establish filler's destruction on all surface of samples which had friable structure after composting. They were fragile and some samples can be broken into small fractions in case of withdrawal from biohumus.

Thereby, received polymeric compositions on the basis of agroindustrial complex waste are partially biodecomposed polymeric compositions. Agroindustrial complex waste, namely cacao bean husk, rice shuck and beet bin pulp assimilates in the environment and the polymeric matrix is destroyed. Polymer's decomposition is reduced by that.

Besides, imposition of colloidal clay in polymeric compositions has multiplied physicomechanical properties of material. Using mathematical modeling underneath breakdown voltage and term of biodecomposition on amount of filler it has been established that filler's concentration (agroindustrial complex waste) in polymeric composition can be increased up to 40%.

It was necessary to eliminate presence of an unpleasant smell for making products from biodecomposed polymeric compositions on the basis of agroindustrial complex and polyethylene waste.

Charcoal was used as a sorbent. Samples of following structure have been received on extrusion-type equipment: 58% of polyethylene, 40% of filler (cacaon bean husk, rice shuck or beet bin pulp), 2% of sorbent and 59% of polyethylene, 40% of filler, 1% of sorbent. As introduction of 1% charcoal reduced a smell, but didn't eliminate it completely, compositions with containing 2% of coal that did not have any smell at all.

Then tests of received compositions for durability were conducted at monoaxial stretching (Table 3).

TABLE 3 Compositions for durability.

Filler's name*	Breakdown voltage σ_p, MPa	Percent elongation with rupture, ε_p,%
Beet bin pulp	6.25±0.41	4.50±0.35
Rice shuck	5.00±0.36	6.80±0.78
Cacao bean husk	6.25±0.43	6.80±0.59

* – The amount of filler in polymer compositions is 40%.

It has been established that breakdown voltage (σp) for all samples made more than 5 MPa that meets the requirements while manufacturing products of technical purpose, for example, trays of small extract for storage of hardware or for thin slab.

Further researches of received polymeric compositions biodegradation for 6 months were conducted. Biodegradation of compositions was held by a method of composting (Table 4).

TABLE 4 Changes in physicomechanical properties after composting.

Filler's name*	Changes in physicomechanical properties	
	Change of breakdown voltage, $\Delta\sigma_p$,%	Change of percent elongation with rupture, $\Delta\varepsilon_p$,%
Beet bin pulp	66 ± 5%	78 ± 0.1%
Rice shuck	72 ± 5%	74 ± 0.1%
Cacao bean husk	67 ± 5%	83 ± 0.1%

* – The amount of filler in polymer compositions is 40%.

In all cases the breakdown voltage magnitude has decreased on the average in 3 times, the magnitude of percent elongation has decreased on the average in 4.5–5 times.

Also the method of optical microscopy was conducted the research of polymer's structure. Microphotos of samples before and after composting (Fig. 2 (a and b)) were received.

Structure of biodecomposed polymeric composition contains 56% of polyethylene, 40% of filler (rice shuck), 2% of colloidal clay and 2% of charcoal.

From the data we can see that initial structure of polymeric compositions' samples containing agroindustrial complex waste, is more homogeneous, and emptiness formed after composting (dark areas in photos) tell about biodecomposition of filler in the conditions of constant temperature and humidity.

(a)

(b)

FIGURE 2 Microphotos of biodecomposed material's structure before (a) and after (b) composting (zooming in 250 times).

On the basis researches held it has been established that biodecomposed polymeric compositions can be received on the basis of polyethylene's and agricultural waste. Biodecomposed polymeric compositions possess the physicomechanical properties, satisfying conditions of making secondary qualitative raw polymeric materials and products from it.

KEYWORDS

- **Cacao bean husk**
- **Beet bin pulp**
- **Rice schuck**
- **Microphotos**

REFERENCES

1. Anan'ev, V.; Kirsh, I. In *Recycle Polymer Materials (in Rus.),* Moscow State University of Food Industry Publishing House: Moscow, 2006; 250 pp.
2. Andrianova, G. In *Technology Of Polymer Composition (in Rus.)*, Kolos Publishing House: Moscow, 2008; 270 pp.

CHAPTER 13

SOME ASPECTS OF SECONDARY POLYMERIC MATERIALS ON THE BASIS OF POLYPROPYLENE AND POLYETHYLENETEREPHTHALAT

I. A. KIRSH, D. A. POMOGOVA, and D. A. SOGRINA

CONTENTS

13.1 INTRODUCTION

A lot of attention is paid to recycling's problem or to secondary processing of polymer materials nowadays. On the most expensive recycling's procedures is sorting of waste. Identification of polymer's waste is really difficult in some cases. If we are speaking about multilayer materials, their identification is impossible. Waste is formed on the stage of materials' processing and after packing's usage. It is utilized by storing in the dumps and polygons or it can be burned. This has a negative effect on the environment. That's why there are a lot of researches, which study simultaneous processing of thermodynamically incompatible polymers and getting of new-composed polymeric materials.

There is a row of difficulties during processing of such polymers as polypropylene and polyethylene terephthalat and other polymeric compositions. They are connected with their thermodynamic incompatibility. It is known that ultrasonic oscillations can lead to reduction of molecular mass that can bring together polymers' solubility parameters during their processing. However, such researches of polymeric melts had been held small [1]. That's why the purpose of our investigation is to study influence of ultrasonic oscillations on polymers' properties for designing of joint polymeric waste's processing of different chemical nature.

13.2 EXPERIMENTAL PART

Polypropylene "Kaplen" and polyethylene terephthalat TU6-05-1984–85 were chosen as objects of research. They were exposed to multiple processing with ultrasonic vibro attachment and without it. Compositions of polypropylene and polyethylene terephthalat were made after processing's cycle of each polymer. Received samples were analyzed on physico-mechanical and rheological properties. We investigated molecular structure of the samples.

13.3 RESULTS AND DISCUSSION

Investigations were conducted on a joint recycling polypropylene (PP) and polyethyleneterephthalate (PET). Results of tests on physical and mechanical properties are shown in Table 1.

TABLE 1 Physical and mechanical properties of samples.

The number of cycles of treatment	The damaging stress, M Pa		The elongation at break, %	
	Without ultrasonic vibrations	With ultrasonic vibrations	Without ultrasonic vibrations	With ultrasonic vibrations
Polypropylene (PP)				
1	37±3	33±2	13±1	11.44±0.7
2	36±10	47±10	14±2	11.11±1.2
3	33±5	63±7	12±3	11.52±0.5
4	36±7	60±5	12±3	10.85±1.2
Polyethyleneterephthalate (PET)				
1	49±3	48±3	302±5	370±2
2	73±7	84±2	263±4	230±4
3	54±4	66±4	28±5	237±2
4	28±3	53±2	19±2	145±3

Breakdown voltage of polypropylene processed with ultrasound is gradually increasing in two times from the first to the second cycles and it does no change on the fourth cycle. Breakdown voltage of polypropylene processed without ultrasonic vibro attachment does not change practically. Percent elongation with rupture of polypropylene received with ultrasound and without it does not change practically form the first to the fourth processing's cycles.

We can see from the results that breakdown voltage of polyethylene terphthalat received with ultrasonic vibro attachment and without it increases in the second cycle. After this I decreases on the third and on the fourth processing's cycles. We should note that the meanings of polyethylene terephthalat's physico-mechanical properties processed with ultrasonic influence is higher than for polyethylene terephthalat processed without it. Percent elongation with rupture of polyethylene terephthalat processed with ultrasonic vibro attachment is gradually decreasing from the second processing's cycle. Percent elongation with rupture of polyethylene terephthalat received without ultrasonic oscillations does not change practically on the first and on the second cycles, then it decreases abruptly.

The raiting of rheological properties was melt's flaw index. In the Figs. 1 and 2 there is a dependence of received samples' melt's flaw index on the amount of processing's cycles.

Conducted researches showed that melt's flaw index of polypropylene (Fig. 1) received with ultrasonic oscillations does not change practically from cycle to processing's cycle. Whereas melt's flaw index at samples received without ultrasonic oscillations increases in two times on the second processing's cycle.

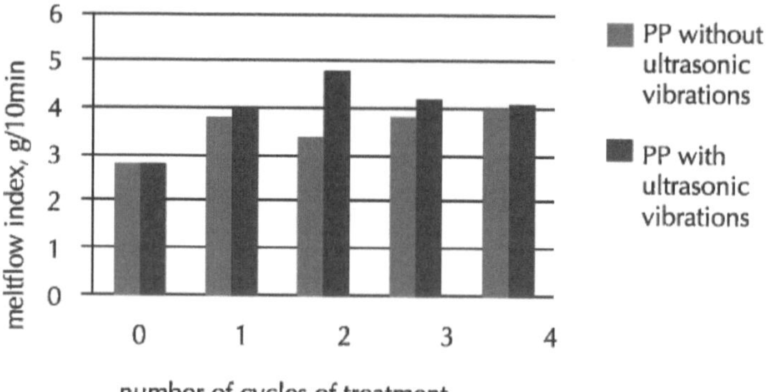

FIGURE 1 Dependence of polypropylene's melt's flaw index on the amount of processing cycles.

Conducted researches of melt's flaw index of polyethyleneterephthalat processed with ultrasonic vibro attachment and without it (Fig. 2) showed that melt's flaw index increases from cycle to cycle. This indicates destructive processes in polyethyleneterephthalat, polymer changed its color from white to putty with the growth of processing's multiplicity.

Tests to determine density showed that the density of polypropylene obtained with ultrasound and without it, virtually unchanged throughout the treatment cycles. The density of polyethylene in both cases is greatly reduced after the first cycle of treatment, and the next three cycles vary slightly.

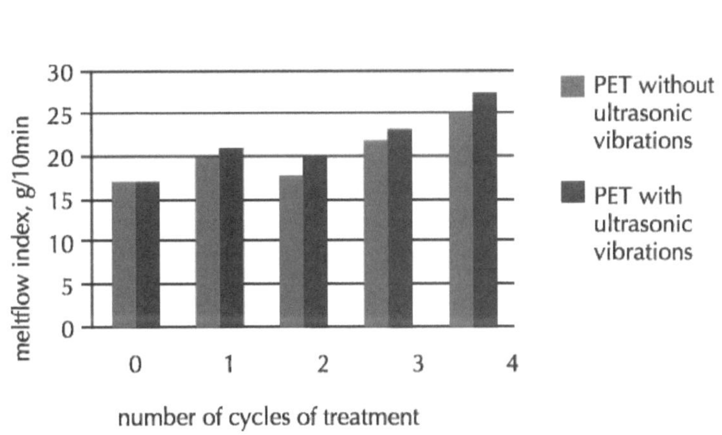

FIGURE 2 Dependence of polyethylene terephthlat's melt's flaw index on the amount of processing cycles.

Figure 3 shows the relative values of the absorption bands of polyethyleneterephthalate groups on the number of cycles of treatment.

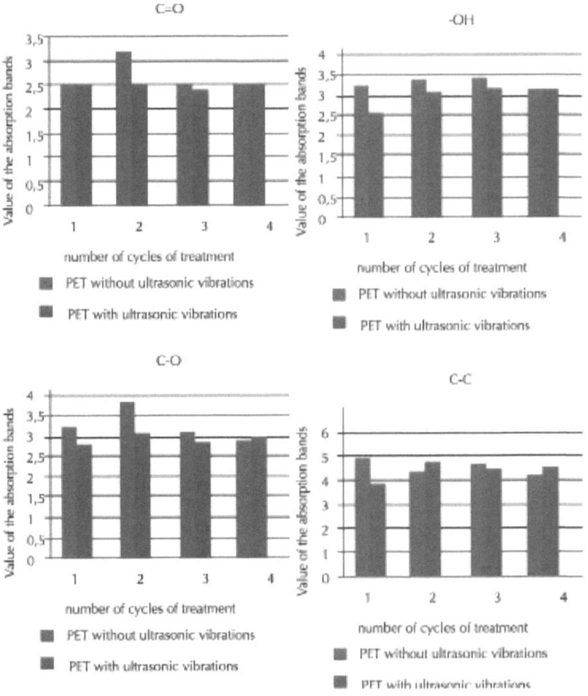

FIGURE 3 Depending on the relative values of the absorption bands of polyethyleneterephthalate groups on the number of cycles of treatment.

After analyzing the data obtained, it should be noted that from cycle to cycle processing the content of oxygen-containing groups increases. When comparing the values of the oxygen-containing groups of samples obtained with and without ultrasound may be noted that the samples obtained by ultrasonic treatment contain less oxygen-containing groups. Comparing the values of group C–C samples of the first cycle and subsequent cycles of treatment, reduction of these groups can be traced in the case of processing of samples without ultrasound and an increase in group C–C in samples treated with ultrasound in the melt. Thus, it was found that from cycle to cycle processing is an intensify process of degradation in polymer, and the ultrasonic treatment leads to partial recovery of the polymer macromolecules and slow degradation in recycling.

Changes in the structure of polypropylene during multiple processing studied by thermo mechanical curves (TC). Thermomechanical curves of PP 1 and 4 cycles of treatment are presented in Figs. 4, and 5.

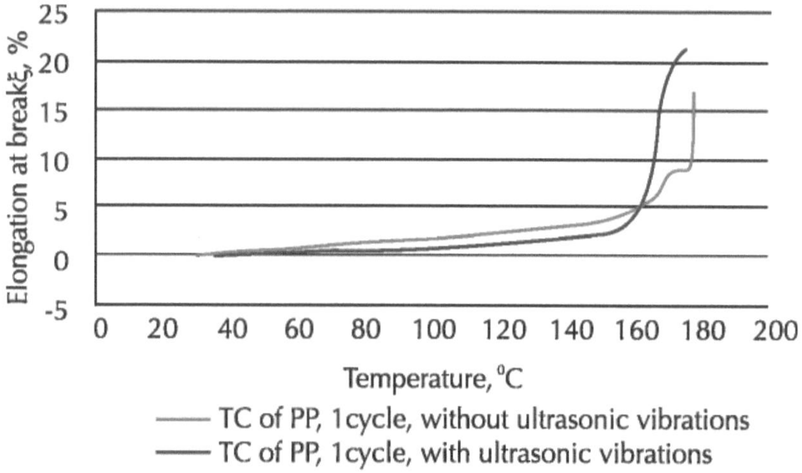

FIGURE 4 The thermomechanical curves of the first cycle of recycling polypropylene.

From the analysis of these curves shows that with increasing multiplicity of processing an ultrasound helps reduce the flow temperature of polypropylene. Furthermore, it should be noted that the ultrasonic treatment leads to recovery of macromolecules of the polymer, as without ultrasound polypropylene at 180°C (fourth cycle) is irreversibly destroyed due to an increase in the number of defects in the sample when it is repeated processing, and polypropylene, treated with ultrasound is not destroyed by this temperature is due to the restoration of macromolecules under the influence of ultrasonic vibrations.

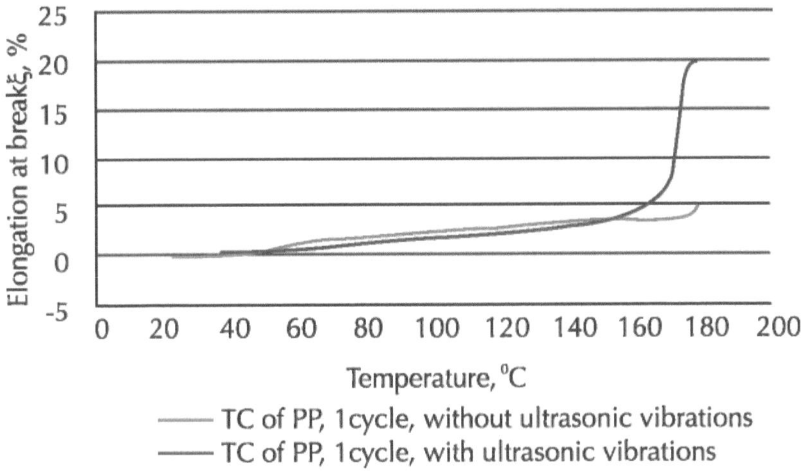

FIGURE 5 The thermomechanical curves of the fourth cycle of recycling polypropylene.

Since a mixture of polypropylene and polyethyleneterephthalat were obtained at the treatment temperature of last, then further studies were carried out flow curves of polypropylene treated with or without ultrasound at 265°C. Figures 6, and 7 show the flow curves of polypropylene at 230°C and 265°C.

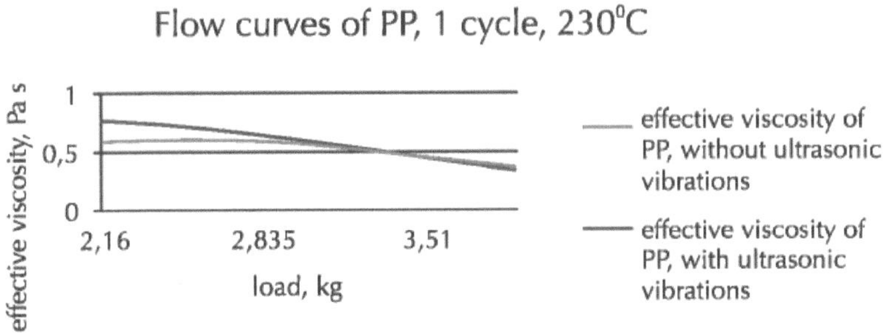

FIGURE 6 The curves of flow of the first cycle of processing of polypropylene at 230°C

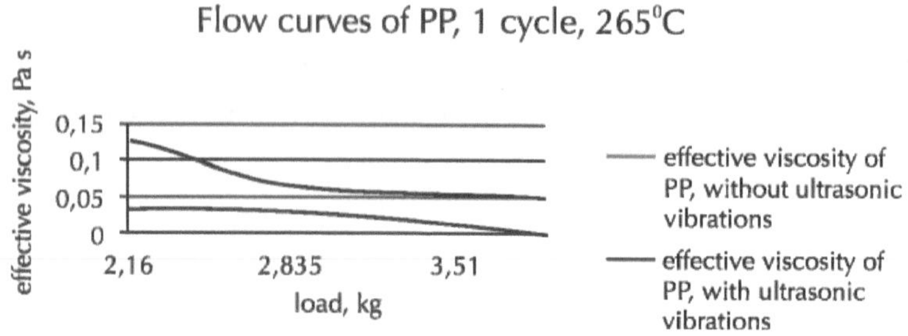

FIGURE 7 The curves of flow of the first cycle of processing of polypropylene at 265oC

At the temperature 230°C the flow curves of polypropylene treated with ultrasound and without it are in the area of the confidence interval. At a temperature of 265°C values of the effective viscosity of the samples differ significantly. Since polypropylene, treated without ultrasound, has very low values of this characteristics and ultrasonic treatment increases the effective viscosity of the polypropylene. When comparing the MFI of polypropylene obtained with the use of ultrasonic vibrations generator and PET found that MFI of these samples are comparable, that might improve their compatibility with the joint processing.

On the next stage of research it was necessary to receive and investigate the properties of composed mixes that were made by blending of original components and increasing of polypropylene's content in polyethyleneterephthalat. Received samples were researched on physico-mechanical properties like original components. In the Figs. 8 and 9 there are dependences of breakdown voltage (Fig. 8) and percent elongation with rupture (Fig. 9) on polypropylene's content in polyethyleneterephthalat.

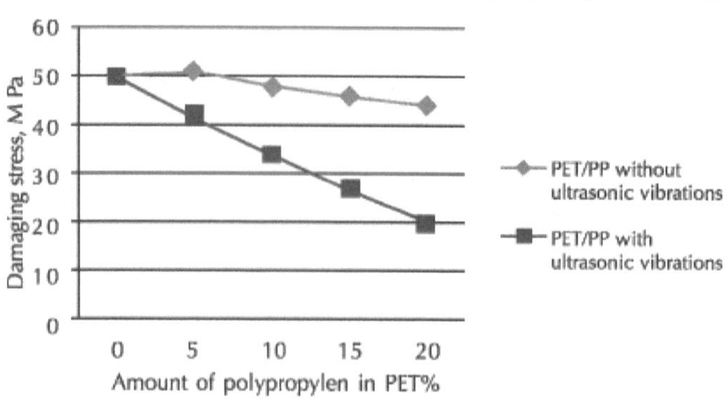

FIGURE 8 Dependence of polymeric composition's breakdown voltage on the amount of processing's cycles.

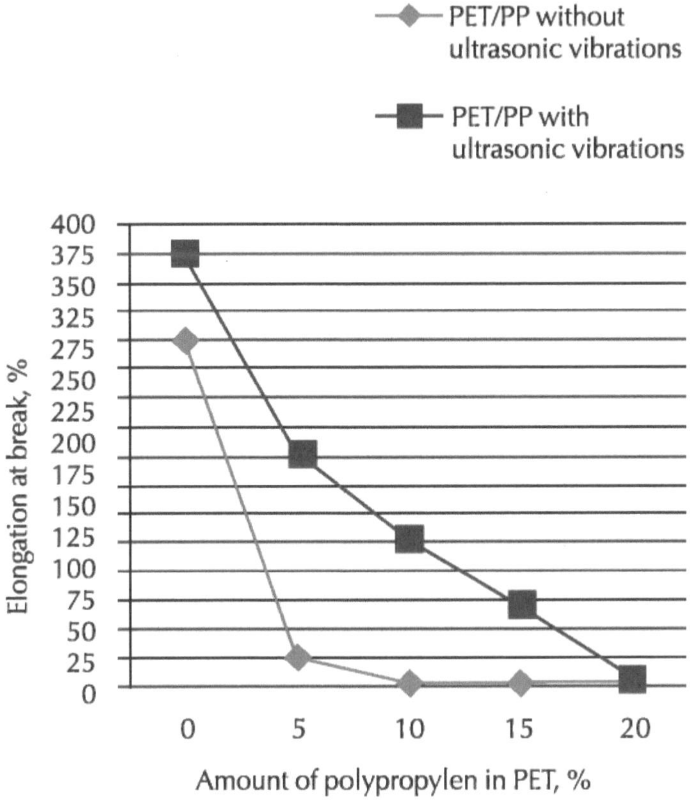

FIGURE 9 Dependence of polymeric composition's percent elongation with rupture on the amount of processing's cycles.

We can see that even a small content of polypropylene in polyethyleneterephthalat can lead to reduction of polymeric compositions' deformatively strengthening indexes. It is allowed to note that percent elongation with rupture decreases to 8% with containing of polypropylene in polyethyleneterephthalat makes 10%. It is insufficient characteristic for getting secondary raw polymeric materials. During the influence of ultrasonic oscillations composition's percent elongation with rupture is considerably higher than without its impact. So, it is allowed to make a conclusion about broadening of technological compatibility's interval of such polymers like polypropylene and polyethyleneterephthalat.

KEYWORDS

- **Kaplen**
- **Polyethylene terephthalate**
- **Polypropylene**
- **Thermomechanical curves**

REFERENCE

1. Anan'ev, V.; Gubanova, M.; Kirsh, I. Influence ultrasonic on polyetelene's properties. *J. Polym. Plastic. (in Rus.),* **2008**, *6*, 126–128.

CHAPTER 14

HIGH-ENERGY BONDS FORMATION IN ATP BASED ON EXPERIMENTAL AND QUNTUM-MECHANICAL DATA

G. A. KORABLEV, N. V. KHOKHRIAKOV, G. E. ZAIKOV, and YU. G. VASILIEV

CONTENTS

14.1 INTRODUCTION

During the interaction of oppositely charged heterogeneous systems the certain compensation of volume energy of interacting structures takes place which leads to the decrease in the resulting energy (e.g., during the hybridization of atomic orbitals). But this is not the direct algebraic deduction of the corresponding energies. The comparison of multiple regularities of physical, chemical and biological processes allow assuming that in such and similar cases the principle of adding the reciprocals of volume energies or kinetic parameters of interacting structures are executed.

Lagrangian equation for the relative movement of the system of two interacting material points with the masses m_1 и m_2 in coordinate x is as follows:

$$M_{red}x'' = -\frac{\partial U}{\partial x}, \quad \text{where} \quad \frac{1}{m_{red}} = \frac{1}{m_1} + \frac{1}{m_2} \qquad (1)\,(1a)$$

where U – mutual potential energy of material points; m_{red} – reduced mass. Herein $x'' = a$ (system acceleration).

For the elementary areas of interactions Δx we can accept: $\frac{\partial U}{\partial x} \approx \frac{\Delta U}{\Delta x}$.

$$\text{Then: } m_{red}a\Delta x = -\Delta U; \quad \frac{1}{1/(a\Delta x)} \cdot \frac{1}{(1/m_1 + 1/m_2)} \approx -\Delta U$$

$$\text{or: } \frac{1}{1/(m_1 a\Delta x) + 1/(m_2 a\Delta x)} \approx -\Delta U$$

Since the product $m_i a\Delta x$ by its physical sense equals the potential energy of each material point ($-\Delta U_i$),

then:

$$\frac{1}{\Delta U} \approx \frac{1}{\Delta U_1} + \frac{1}{\Delta U_2} \qquad (2)$$

Thus the resulting energy characteristic of the system of two interacting material points is found by the principle of adding the reciprocals of initial energies of interacting subsystems.

"The electron with the mass m moving about the proton with the mass M is equivalent to the particle with the mass: $m_{red} = \frac{mM}{m+M}$" [1].

Therefore, modifying the Eq. (2) we can assume that the energy of atom valence orbitals (responsible for interatomic interactions) can be calculated [2] by the principle of adding the reciprocals of some initial energy components based on the equations:

$$\frac{1}{q^2/r_i}+\frac{1}{W_i n_i}=\frac{1}{P_E} \text{ or } \frac{1}{P_0}=\frac{1}{q^2}+\frac{1}{(Wrn)_i}; \; P_E=P_0/r_i \qquad (3),(4),(5)$$

where, W_i – orbital energy of electrons [3]; r_i – orbital radius of i orbital [4]; $q=Z^*/n^*$ – by [5, 6], n_i – number of electrons of the given orbital, Z^* and n^* – nucleus effective charge and effective main quantum number, r – bond dimensional characteristics.

The value P_O is called a spatial-energy parameter (SEP), and the value P_E – effective P-parameter (effective SEP). Effective SEP has a physical sense of some averaged energy of valence orbitals in the atom and is measured in energy units, e.g., in electron-volts (eV).

The values of P_0-parameter are tabulated constants for electrons of the given atom orbital.

For SEP dimensionality:

$$[P_0]=[q^2]=[E]\cdot[r]=[h]\cdot[\upsilon]=\frac{kgm^3}{s^2}=Jm$$

where [E], [h] and [υ] – dimensionalities of energy, Planck's constant and velocity.

The introduction of P-parameter should be considered as further development of quasi-classical concepts with quantum-mechanical data on atom structure to obtain the criteria of phase-formation energy conditions. For the systems of similarly charged (e.g., orbitals in the given atom) homogeneous systems the principle of algebraic addition of such parameters is preserved:

$$\sum P_E=\sum(P_0/r_i); \qquad (6),$$

$$\sum P_E=\frac{\sum P_0}{r} \qquad (7)$$

or:
$$\sum P_0 = P_0' + P_0'' + P_0''' + \ldots ; \tag{8}$$

$$r \sum P_E = \sum P_0 , \tag{9}$$

Here P-parameters are summed up by all atom valence orbitals.

To calculate the values of P_E-parameter at the given distance from the nucleus either the atomic radius (R) or ionic radius (r_I) can be used instead of r depending on the bond type.

Let us briefly explain the reliability of such an approach. As the calculations demonstrated the values of P_E-parameters equal numerically (in the range of 2%) the total energy of valence electrons (U) by the atom statistic model. Using the known correlation between the electron density (β) and intra-atomic potential by the atom statistic model [7], we can obtain the direct dependence of P_E-parameter on the electron density at the distance r_i from the nucleus. The rationality of such technique was proved by the calculation of electron density using wave functions by Clementi [8] and comparing it with the value of electron density calculated through the value of P_E-parameter.

The modules of maximum values of the radial part of Ψ-function were correlated with the values of P_0-parameter and the linear dependence between these values was found. Using some properties of wave function as applicable to P-parameter, the wave equation of P-parameter with the formal analogy with the equation of Ψ-function was obtained [9].

14.2 WAVE PROPERTIES OF P-PARAMETERS AND PRINCIPLES OF THEIR ADDITION

Since P-parameter has wave properties (similar to Ψ-function), the regularities of the interference of the corresponding waves should be mainly fulfilled at structural interactions.

The interference minimum, weakening of oscillations (in antiphase) occurs if the difference of wave move (Δ) equals the odd number of semi-waves:

$$\Delta = (2n+1)\frac{\lambda}{2} = \lambda\left(n+\frac{1}{2}\right), \text{ where n} = 0, 1, 2, 3, \ldots \quad (10)$$

As applicable to P-parameters this rule means that the interaction minimum occurs if P-parameters of interacting structures are also "in antiphase" – either oppositely charged or heterogeneous atoms (e.g., during the formation of valence-active radicals CH, CH_2, CH_3, NO_2 ..., etc) are interacting.

In this case P-parameters are summed by the principle of adding the reciprocals of P-parameters – Eqs. (3) and (4).

The difference of wave move (Δ) for P-parameters can be evaluated via their relative value $\left(\gamma = \frac{P_2}{P_1}\right)$ of relative difference of P-parameters (coefficient α), which at the interaction minimum produce an odd number:

$$\gamma = \frac{P_2}{P_1} = \left(n + \frac{1}{2}\right) = \frac{3}{2}; \frac{5}{2} \ldots \text{ When } n = 0 \text{ (main state) } \frac{P_2}{P_1} = \frac{1}{2} \quad (11)$$

It should be pointed out that for stationary levels of one-dimensional harmonic oscillator the energy of these levels $\varepsilon = h\nu(n + \frac{1}{2})$, therefore, in quantum oscillator, in contrast to the classical one, the least possible energy value does not equal zero.

In this model the interaction minimum does not provide zero energy corresponding to the principle of adding reciprocals of P-parameters – Eqs. (3) and (4).

The interference maximum, strengthening of oscillations (in phase) occurs if the difference of wave move equals the even number of semiwaves:

$$\Delta = 2n\frac{\lambda}{2} = \lambda n \text{ or } \Delta = \lambda(n+1).$$

As applicable to P-parameters the maximum interaction intensification in the phase corresponds to the interactions of similarly charged systems or systems homogeneous by their properties and functions (for example, between the fragments or blocks of complex inorganic structures, such as CH_2 and NNO_2 in octogene).

And then:
$$\gamma = \frac{P_2}{P_1} = (n+1) \tag{12}$$

By the analogy, for "degenerated" systems (with similar values of functions) of two-dimensional harmonic oscillator the energy of stationary states:

$$\varepsilon = h\nu(n+1)$$

By this model the interaction maximum corresponds to the principle of algebraic addition of P-parameters – Eqs. (6)–(8). When n=0 (main state) we have $P_2=P_1$, or: the interaction maximum of structures occurs if their P-parameters are equal. This concept was used [2] as the main condition for isomorphic replacements and formation of stable systems.

14.3 EQUILIBRIUM-EXCHANGE SPATIAL-ENERGY INTERACTIONS

During the formation of solid solutions and in other structural equilibrium-exchange interactions the unified electron density should be established in the contact spots between atoms-components. This process is accompanied by the re-distribution of electron density between valence areas of both particles and transition of a part of electrons from some external spheres into the neighboring ones.

It is obvious that with the proximity of electron densities in free atoms-components the transition processes between the boundary atoms of particles will be minimal thus contributing to the formation of a new structure. Thus the task of evaluating the degree of such structural interactions in many cases comes down to comparative assessment of electron density of valence electrons in free atoms (on the averaged orbitals) participating in the process.

Therefore the maximum total solubility evaluated via the structural interaction coefficient α is defined by the condition of minimal value of coefficient α, which represents the relative difference of effective energies of external orbitals of interacting subsystems:

$$\alpha = \frac{P_O'/r_i' - P_O''/r_i''}{(P_O'/r_i' + P_O''/r'')/2}100\%, \tag{13}$$

or
$$\alpha = \frac{P_S' - P_S''}{P_S' + P''} 200\%, \tag{14}$$

where P_S – structural parameter is found by the equation:

$$\frac{1}{P_S} = \frac{1}{N_1 P_E'} + \frac{1}{N_2 P_E''} + \ldots, \tag{15}$$

where N_1 and N_2 – number of homogeneous atoms in subsystems.

The nomogram of the dependence of structural interaction degree (ρ) upon the coefficient α, the same for the wide range of structures, was prepared by the data obtained (the figure is not available).

Isomorphism as a phenomenon is usually considered as applicable to crystalline structures. But obviously the similar processes can also take place between molecular compounds where the bond energies can be assessed via the relative difference of electron densities of valence orbitals of interacting atoms. Therefore, the molecular electronegativity is rather easily calculated via the values of corresponding P-parameters.

In complex organic structures the main role in intermolecular and intramolecular interactions can be played by separate "blocks" or fragments considered as "active" areas of the structures. Therefore, it is necessary to identify these fragments and evaluate their spatial-energy parameters. Based on wave properties of P-parameter, the total P-parameter of each element should be found following the principle of adding the reciprocals of initial P-parameters of all the atoms. The resulting P-parameter of the fragment block or all the structure is calculated following the rule of algebraic addition of P-parameters of their constituent fragments.

Apparently, spatial-energy exchange interactions (SEI) based on leveling the electron densities of valence orbitals of atoms-components have in nature the same universal value as purely electrostatic coulomb interactions and complement each other. Isomorphism known from the time of E. Mitscherlich (1820) and D.I. Mendeleev (1856) is only a special demonstration of this general natural phenomenon.

The quantitative side of evaluating the isomorphic replacements both in complex and simple systems rationally fits into P-parameter methodology. More complicated is the problem of evaluating the degree of structural SEI for molecular structures, including organic ones. Such structures and their fragments are often not completely isomorphic to each other. Nevertheless, SEI is going on between them and its degree can be evaluated either semi-quantitatively numerically or qualitatively. By the degree of isomorphic similarity all the systems can be divided into three types:

I Systems mainly isomorphic to each other – systems with approximately the same number of heterogeneous atoms and cumulatively similar geometric shapes of interacting orbitals.

II Systems with organic isomorphic similarity – systems which:

1) either differ by the number of heterogeneous atoms but have cumulatively similar geometric shapes of interacting orbitals;
2) or have certain differences in the geometric shape of orbitals but have the same number of interacting heterogeneous atoms.

III Systems without isomorphic similarity – systems considerably different both by the number of heterogeneous atoms and geometric shape of their orbitals.

Taking into account the experimental data, all SEI types can be approximately classified as follows:

Systems I

1. $\alpha < (0–6)\%$; $\rho = 100\%$. Complete isomorphism, there is complete isomorphic replacement of atoms-components;
2. $6\% < \alpha < (25–30)\%$; $\rho = 98 - (0–3)\%$.
   There is wide or unlimited isomorphism.
3. $\alpha > (25–30)\%$; no SEI.

Systems II

1. $\alpha < (0–6)\%$;
   a) There is reconstruction of chemical bonds that can be accompanied by the formation of a new compound;
   b) Cleavage of chemical bonds can be accompanied by a fragment separation from the initial structure but without adjoinings and replacements.
2. $6\% < \alpha < (25–30)\%$; the limited internal reconstruction of chemical bonds is possible but without the formation of a new compound and replacements.
3. $\alpha > (20–30)\%$; no SEI.

Systems III

1. $\alpha < (0–6)\%$;
   a) The limited change in the type of chemical bonds in the given fragment is possible, there is an internal re-grouping of atoms without the cleavage from the main molecule part and replacements;
   b) The change in some dimensional characteristics of the bond is possible;
2. $6\% < \alpha < (25–30)\%$;
   A very limited internal re-grouping of atoms is possible;
   $\alpha > (25–30)\%$; no SEI.

When considering the above systems, it should be pointed out that they can be found in all cellular and tissue structures in some form but are not isolated and are found in spatial-time combinations.

The values of α and ρ calculated in such a way refer to a definite interaction type whose nomogram can be specified by fixed points of reference systems. If we take into account the universality of spatial-energy interactions in nature, such evaluation can have the significant meaning for the analysis of structural shifts in complex bio-physical and chemical processes of biological systems.

Fermentative systems contribute a lot to the correlation of structural interaction degree. In this model the ferment role comes to the fact that active parts of its structure (fragments, atoms, ions) have such a value of P_E-parameter, which equals the P_E-parameter of the reaction final product. That is the ferment is structurally "tuned" via SEI to obtain the reaction final product, but it will not enter it due to the imperfect isomorphism of its structure (in accordance with III).

The important characteristics of atom-structural interactions (mutual solubility of components, chemical bond energy, energy of free radicals, etc.) for many systems were evaluated following this technique [10, 11].

14.4 CALCULATION OF INITIAL DATA AND BOND ENERGIES

Based on the Eqs. (3–5) with the initial data calculated by quantum-mechanical methods [3–6] we calculate the values of P_0-parameters for the majority of elements being tabulated, constant values for each atom valence orbital. Mainly covalent radii – by the main type of the chemical bond of interaction considered were used as a dimensional characteristic for calculating P_E-parameter (Table 1). The value of Bohr radius and the value of atomic ("metal") radius were also used for hydrogen atom.

TABLE 1 P-parameters of atoms calculated via the bond energy of electrons.

Atom	Valence electrons	W (eV)	r_i (Å)	q^2_0 (eVÅ)	P_0 (eVÅ)	R (Å)	P_0/R (eV)
						0.5292	9.0644
H	$1S^1$	13.595	0.5295	14.394	4.7985	0.28	17.137
						R^-_I=1.36	3.525
	$2P^1$	11.792	0.596	35.395	5.8680	0.77	7.6208
						0.67	8.7582
	$2P^2$	11.792	0.596	35.395	10.061	0.77	13.066
C						0.67	15.016
	$2S^2$				14.524	0.77	18.862
	$2S^2+2P^2$				24.585	0.77	31.929
					24.585	0.67	36.694
	$2P^1$	15.445	0.4875	52.912	6.5916	0.70	9.4166
	$2P^2$				11.723	0.70	16.747
	$2P^3$				15.830	0.70	22.614
N						0.55	28.782
	$2S^2$	25.724	0.521	53.283	17.833	0.70	25.476
	$2S^2+2P^3$				33.663	0.70	48.09
	$2P^1$	17.195	0.4135	71.383	6.4663	0.66	9.7979
	$2P^1$					R_I=1.36	4.755
	$2P^1$					R_I=1.40	4.6188
	$2P^2$	17.195	0.4135	71.383	11.858	0.66	17.967
						0.59	20.048
O						R_I=1.36	8.7191
						R_I=1.40	8.470
	$2P^4$	17.195	0.4135	71.383	20.338	0.66	30.815
						0.59	34.471
	$2S^2$	33.859	0.450	72.620	21.466	0.66	32.524
	$2S^2+2P^4$				41.804	0.66	63.339
						0.59	70.854

TABLE 1 *(Continued)*

Atom	Valence electrons	W (eV)	r_i (Å)	q^2_0 (eVÅ)	P_0 (eVÅ)	R (Å)	P_0/R (eV)
	$4S^1$	5.3212	1.690	17.406	5.929	1.97	3.0096
Ca	$4S^2$				8.8456	1.97	4.4902
	$4S^2$					R^{2+}=1.00	8.8456
	$4S^2$					R^{2+}=1.26	7.0203
	$3P^1$	10.659	0.9175	38.199	7.7864	1.10	7.0785
	$3P^1$					R^{3-}=1.86	P_3=4.1862
P	$3P^3$	10.659	0.9175	38.199	16.594	1.10	15.085
	$3P^3$					R^{3-}=1.86	8.9215
	$3S^2+3P^3$				35.644	1.10	32.403
	$3S^1$	6.8859	1.279	17.501	5.8568	1.60	3.6618
Mg	$3S^2$				8.7787	1.60	5.4867
						R^{2+}=1.02	8.6066
	$4S^1$	6.7451	1.278	25.118	6.4180	1.30	4.9369
Mn	$4S^1+3d^1$				12.924	1.30	9.9414
	$4S^2+3d^2$				22.774	1.30	17.518
		4.9552	1.713	10.058	4.6034	1.89	2.4357
Na	$3S^1$					R^{1+}_I=1.18	3.901
						R^{1+}_I=0.98	4.6973
K	$4S^1$	4.0130	2.612	10.993	4.8490	2.36	2.0547
						R^{1+}_I=1.45	3.344

In some cases the bond repetition factor for carbon and oxygen atoms was taken into consideration [10]. For a number of elements the values of P_E-parameters were calculated using the ionic radii whose values are indicated in column 7. All the values of atomic, covalent and ionic radii were mainly taken by Belov-Bokiy, and crystalline ionic radii – by Batsanov [12].

The results of calculating structural P_S-parameters of free radicals by the Eq. (15) are given in Table 2. The calculations are done for the radicals contained in protein and amino acid molecules (CH, CH_2, CH_3, NH_2 etc.), as well as for some free radicals formed in the process of radiolysis and dissociation of water molecules.

TABLE 2 Structural P_S–parameters calculated via the bond energy of electrons.

Radicals, molecule fragments	$P_i'(eV)$	$P_i''(eV)$	$P_S(eV)$	Orbitals
OH	9.7979 17.967	9.0644 17.138	4.7080 8.7712	O $(2P^1)$ O $(2P^2)$
H_2O	2·9.0644	17.967	9.0227	O $(2P^2)$
CH_2	17.160 31.929	2·9.0644 2·17.138	8.8156 16.528	C $(2S^1 2P^3_r)$ C $(2S^2 2P^2)$
CH_3	15.016 40.975	3·9.0644 3·9.0644	9.6740 16.345	C $(2P^2)$ C $(2S^2 2P^2)$
CH	31.929	12.792	9.1330	C $(2S^2 2P^2)$
NH	16.747 19.538	17.138 17.132	8.4687 9.1281	N$(2P^2)$ N$(2P^2)$
NH_2	19.538 28.782	2·9.0644 2·17.132	9.4036 18.450	N$(2P^2)$ N$(2P^3)$
CO–OH	8.4405	8.7710	4.3013	C$(2P^2)$
C=O	15.016	20.048	8.4405	C$(2P^2)$
C=O	31.929	34.471	16.576	O$(2P^4)$
CO=O	36.694	34.471	17.775	O$(2P^4)$
C–CH_3	17.435	19.694	9.2479	–
C–NH_2	17.435	18.450	8.8844	–
CO–OH	12.315	8.7712	5.1226	C$(2S^2 2P^2)$
$(HP)O_3$	23.122	23.716	11.708	O$(2P^2)$ P$(3S^2 3P^3)$
$(H_3P)O_4$	17.185	17.244	8.6072	O$(2P^1)$ P$(3P^1)$

TABLE 2 *(Continued)*

Radicals, molecule fragments	$P_i'(eV)$	$P_i''(eV)$	$P_S(eV)$	**Orbitals**
$(H_3P)O_4$	31.847	31.612	15.865	$O(2P^2)$ $P(3S^2 3P^3)$
H_2O	2·4.3623	8.7191	4.3609	$O(2P^2)$ r =1.36 Å
H_2O	2·4.3623	4.2350	2.8511	$O(2P^2)$ r=1.40
$C–H_2O$	2.959	2.8511	1.4520	–
$(C–H_2O)_3$ Lactic acid	–	–	1.4520·3= 4.3563	–
$(C–H_2O)_6$ Glucose	–	–	1.4520·6= 8.7121	–

The technique previously tested [10] on 68 binary and more complex compounds was applied to calculate the energy of coupled bond of molecules by the equations:

$$\frac{1}{\mathring{A}}=\frac{1}{P_S}=\frac{1}{\left(P_E\frac{n}{K}\right)_1}+\frac{1}{\left(P_E\frac{n}{K}\right)_2}; \tag{16}$$

$$P_E\frac{n}{K}=P \tag{17}$$

where n – bond average repetition factor, K – hybridization coefficient which usually equals the number of registered atom valence electrons.

Here the P-parameter of energy characteristic of the given component structural interaction in the process of binary bond formation.

"Non-valence, non-chemical weak forces act … inside biological molecules and between them apart from strong interactions" [13]. At the same time, the orientation, induction and dispersion interactions are used to be called Van der Waals. For three main biological atoms (nitrogen, phosphorus and oxygen) Van der Waals radii numerically equal approximately the corresponding ionic radii (Table 3).

TABLE 3 Ionic and Van der Waals radii (Å).

Atom	Ionic radii			Van der Waals radii		
	Orbital	R_I	P_E/K (eV)	R_B	Orbital	P_E/K (eV)
H	$1S^1$	$R^-=1.36$	3.525	1.10	$1S^1$	4.3623
		r =0.5292	9.0644	1.32		3.6352
N				1.50	$2P^1$	4.3944/1
	$2P^3$	$R^{3-}=1.48$	10.696/3=3.5653	1.50	$2P^3$	10.553/3=3.5178
				1.50	$2S^22P^3$	22.442/5=4.4884
P				1.9	$3P^1$	4.0981/1
	$3P^3$	$R^{3-}= 1.86$	8.9215/3=2.9738	1.9	$3P^3$	8.7337/3=2.9112
				1.9	$3S^23P^3$	18.760/5=3.752
O		$R^{2-}=1.40$	8.470/2=4.2350	1.40	$2P^1$	4.6188/1
	$2P^2$			1.50	$2P^1$	4.3109/1
		$R^{2-}=1.36$	8.7191/2=4.3596	1.40	$2P^2$	8.470/2=4.2350
				1.50	$2P^2$	7.9053/2=3.9527
C				1.7	$2P^1$	3.4518/1
	$2S^22P^2$	d*/2=3.2/2=1.6	15.365/4=3.841	1.7	$2P^2$	5.9182/2=2.9591
				1.7	$2S^22P^2$	14.462/4=3.6154

d* – contact distance between C–C atoms in polypeptide chains [13].

It is known that one of the reasons of relative instability of phosphorus anhydrite bonds in ATP is the strong repulsion of negatively charged oxygen atoms. Therefore it is advisable to use the values of P-parameters calculated via Van der Waals radii as the energy characteristic of weak structural interactions of biomolecules (Table 3).

Bond energies for P and O atoms were calculated taking into account Van der Waals distances for atomic orbitals: $3P^1$ (phosphorus)-$2P^1$ (oxygen) and for $3P^3$ (phosphorus)-$2P^2$ (oxygen). The values of E obtained slightly exceeded the experimental, reference ones (Table 4). But for the actual energy physiological processes, e.g., during photosynthesis, the efficiency is below the theoretical one, being about 83%, in some cases [14,15].

TABLE 4 Bond energy (eV).

Atoms, structures, orbitals	Bond	Component 1		Component 2		Component 3		Calculation	E	E	Remarks
		P_E (eV)	n/K	P_E (eV)	n/K	P_E (eV)	n/K	E	[13,14,15]	[16] [17]	
1	2	3	4	5	6	7	8	9	10	11	12
								6.277			
P···O				70.854	1.5/6				6.1385		
	cov.	32.403	1.5/5			6.14		6.024			PO free molecule
$3S^23P^3$–$2S^22P^4$				63.339	1.5/6				6.14		
								<6.15>			
H_2O	cov.	2×9.0624	1/1	17.967	1/6			2.570	2.476		Decay
$1S^1$–$2P^2$	cov.	2×9.0624	1/1	20.048	2/2			9.520		10.04	of one molecule
H_3PO_4	cov.	3×9.0624	1/1	32.405	1/5	4×17.967	1/2	4.8779	4.708		
C···O ($2P^1$–$1S^1$)	cov.	7.6208	1.125/2	9.7979	1/1			4.2867			
C–N $2P^1$–$2P^1$	cov.	7.6208	1/4	9.4166	1/5			0.9471			
C···N $2P^1$–$2P^1$	cov.	7.6208	1.125/4	9.4166	1.1667/5			1.0898	0.870		
K–C–N $4S^2$–$2P^1$–$2P^1$	cov.	2.0547	1/1	7.6208	1/4	9.4166	1/5	0.648			
(C–H_2O)–(C–H_2O)	VdW	1.4520	1/1	1.4520	1/1			0.726			
C···O	cov.	31.929	1.125/4	20.048	1/2			4.7367			
$2S^22P^2$–$2P^2$		31.929	1/4	20.042	1/2			4.4437			

TABLE 4 Bond energy (eV).

Atoms, structures, orbitals	Bond	Component 1		Component 2		Component 3		Calculation	E	E	Remarks
		P_E (eV)	n/K	P_E (eV)	n/K	P_E (eV)	n/K	E	[13,14,15]	[16] [17]	
1	2	3	4	5	6	7	8	9	10	11	12
N···H	cov.	9.4166	1.1667/1	9.0644	1/1			4.9654			
$2P^1$–$1S^1$		9.4166	1/1	9.0644	1/1			4.6186			
C–H $2P^1$–$1S^1$	cov.	13.066	1/2	9.0644	1/1			3.797	3.772		
C–H $2P^2$–$1S^1$		13.066	1/2	17.137	1/1			4.7295			
N–H_2 $2P^3$–$1S^1$	cov.	22.614	1/3	2×9.0644	1/1			5.3238			
–H···O		3.525	3.525/17.037	4.6188	1/6			0.3730	0.3742		Hydrogen bond
P=O $3P^3$–$2P^2$	cov.	15.085	2/3	20.042	2/2			6.6970	6.504	6.1385	Free molecule
P–O $3P^3$–$2P^2$	VdW	8.7337	1/5	8.470	1/6			0.781	0.670		ΔG ATP
P–O $3P^1$–$2P^1$	cov.	7.0785	1/1	9.7979	1/1			4.1096	4.2059	4.2931	
P–O $3P^1$–$2P^1$	VdW	4.0981	1/5	4.6188	1/6			0.3970	0.34–0.35		Phospholyration

Perhaps the electrostatic component of resulting interactions at anion-anionic distances is considered in such a way. Actually the calculated value of 0.83E practically corresponds to the experimental values of bond energy during the phospholyration and free energy of ATP in chloroplasts.

Table 4 contains the calculations of bond energy following the same technique but for stronger interactions at covalent distances of atoms for the free molecule P···O (sesquialteral bond) and for the molecule P=O (double bond).

The sesquialteral bond was evaluated by introducing the coefficient $n = 1.5$ with the average value of oxygen P_E-parameter for single and double bonds.

The average breaking energy of the corresponding chemical bonds in ATP molecule obtained in the frameworks of semi-empirical method PM3 with the help of software GAMESS [16] are given in column 11 of Table 4 for comparison. The calculation technique is detailed in [17].

The calculated values of bond energies in the system K–C–N being close to the values of high-energy bond P~O in ATP demonstrate that such structure can prevent the ATP synthesis.

When evaluating the possibility of hydrogen bond formation, we take into account such value of n/K in which K =1, and the value n =3.525/17.037 characterizes the change in the bond repetition factor when transiting from the covalent bond to the ionic one.

14.5 FORMATION OF STABLE BIOSTRUCTURES

At equilibrium-exchange spatial-energy interactions similar to isomorphism the electrically neutral components do not repulse but approach each other and form a new composition whose α in the Eqs. (13) and (14).

This is the first stage of stable system formation by the given interaction type which is carried out under the condition of approximate equality of component P-parameters: $P_1 \approx P_2$.

Hydrogen atom, element No 1 with the orbital $1S^1$ determines the main criteria of possible structural interactions. Four main values of its P-parameters can be taken from Tables 1 and 3:

(1) for strong interactions: P'_E = 9.0644 eV with the orbital radius 0.5292 Å and P'''_E = 17.137 eV with the covalent radius 0.28 Å.
(2) for weaker interactions: P'_E = 4.3623 eV and P_E = 3.6352 eV with Van der Waals radii 1.10 Å and 1.32 Å. The values of P-parameters $P' : P'' : P'''$ relates as 1:2:4. In accordance with the concepts in Section 2, such values of interaction P-parameters define the normative functional states of biosystems, and the intermediary can produce pathologic formations by their values.

The series with approximately similar values of P-parameters of atoms or radicals can be extracted from the large pool of possible combinations of structural interactions (Table 5). The deviations from the initial, primary values of P-parameters of hydrogen atom are in the range ± 7%.

The values of P-parameters of atoms and radicals given in the Table define their approximate equality in the directions of interatomic bonds in polypeptide, polymeric and other multi-atom biological systems.

In ATP molecule these are phosphorus, oxygen and carbon atoms, polypeptide chains – CO, NH and CH radicals. In Table 5 you can also see the additional calculation of their bond energy taking into account the sesquialteral bond repetition factor in radicals C···O и N···H.

On the example of phosphorus acids it can be demonstrated that this approach is not in contradiction with the method of valence bonds, which explains the formation peculiarities of ordinary chemical compounds. It is demonstrated in Table 6 that this electrostatic equilibrium between the oppositely charged components of these acids can correspond to the structural interaction for H_3PO_4 $3P^1$ orbitals of phosphorus and $2P^1$ of oxygen, and for HPO_3 $3S^23P^3$ orbitals of phosphorus and $2P^2$ of oxygen. Here it is stated that P-parameters for phosphorus and hydrogen subsystems are added algebraically. It is also known that the ionized phosphate groups are transferred in the process of ATP formation that is apparently defined for phosphorus atoms by the transition from valence-active $3P^1$ orbitals to $3S^23P^3$ ones, i.e., 4 additional electrons will become valence-active. According to the experimental data the synthesis of one ATP molecule is connected with the transition of four protons and when the fourth proton is being transited the energy accumulated by the ferment reaches its threshold [18, 19]. It can be assumed that such proton transitions in ferments initiate similar changes in valence-active states in the system

TABLE 5 Bio-structural spatial-energy parameters (eV).

Series No	H	C	N	O	P	CH	CO	NH	Glucose	Lactic acid	OH	Remarks
I	9.0644	8.7582	9.4166	9.7979	8.7337	9.1330	8.4405	8.4687	8.7121			Strong inter-action
	$(1S^1)$	$(2P^1)$	$(2P^1)$	$(2P^1)$	$(3P^3)$	$(2S^22P^2–1S^1)$	$(2P^2–2P^2)$	$(2P^2–1S^1)$	$2P^2–$		8.7710	
		9.780						9.1281	$–(1S^1–2P^2)$			
		$(2P^1)$						$(2P^2–1S^1)$				
II	17.132	17.435	16.747	17.967	18.760	C and H blocks	16.576	N and H				Strong inter-action
	$(1S^1)$	$(2S^12P^1)$	$(2P^2)$	$(2P^2)$	$(3S^23P^3)$		$(2S^22P^2–2P^4)$	blocks				
III	(4.3623)	3.8696	4.3944	4.3109	4.0981	4.7295	4.4437	4.6186		4.3563	4.7084	Weak
	$(1S^1)$	$(2P^2)$	$(2P^1)$	$(2P^1)$	$(3P^1)$		4.7367	4.9654		$2P^2–$		interaction
				4.6188						$(1S^1–2P^2)$		
				$(2P^1)$								
IV	3.6352	3.4518	3.5178	4.2350	4.0981	4.7295	4.4437	4.6186				Effective
	$(1S^1)$	$(2P^1)$	$(2P^3)$	$(2P^2)$	$(3P^1)$		4.7367	4.9654				bond energy
		3.6154		3.6318	3.752							
		$(2S^22P^2)$		$(2P^4)$	$(3S^23P^3)$							

TABLE 6 Structural interactions in phosphorus acids.

Molecule	Component 1			Component 2			$\alpha=(\Delta P/<P>)*100\%$
	Atom	Orbitals	$P=P_1+P_2$ (eV)	Atom	Orbitals	P(eV)	
$(H_3P)O_4$	H_3P	$1S^1-3P^1$	4.3623*3+ 4.0981=17.185	O_4	$2P^1$	4.3109*4=17.244	0.34
		$1S^1-(3S^23P^3)$	4.3623*3+18.760=31.847	at r=1.50Å	$2P^2$	7.9053*4=31.612	0.74
$(HP)O_3$	HP	$1S^1-(3S^23P^3)$	4.3623+18.760=23.122	O_3 at r=1.50Å	$2P^2$	7.9053*3=23.716	2.54

P–O. In the process of oxidating phospholyration the transporting ATP-synthase uses the energy of gradient potential due to $2H^+$-protons, which in the given model for such a process, corresponds to the initiation of valence-active transitions of phosphorus atoms from $3P^1$ to $3P^3$-state.

In accordance with the Eq. (17) we can assume that in stable molecular structures the condition of the equality of corresponding effective interaction energies of the components by the couple bond line is fulfilled by the following equations:

$$\left(P_E\frac{n}{K}\right)_1 \approx \left(P_E\frac{n}{K}\right)_2 \rightarrow P_1 \approx P_2 \tag{18}$$

And for heterogeneous atoms (when $n_1 = n_2$):

$$\left(P_E/K\right)_1 \approx \left(P_E/K\right)_2 \tag{18a}$$

In phosphate groups of ATP molecule the bond main line comprises phosphorus and oxygen molecules. The effective energies of these atoms by the bond line calculated by the Eq. (18) are given in Tables 4 and 5, from which it is seen that the best equality of P_1 and P_2 parameters is fulfilled for the interactions $P(3P^3)$ – 8.7337 eV and $O(2P^2)$ – 8.470 eV that is defined by the transition from the covalent bond to Van der Waals ones in these structures.

The resulting bond energy of the system P–O for such valence orbitals and the weakest interactions (maximum values of coefficient K) is 0.781 eV (Table 4). Similar calculations for the interactions $P(2P^1)$ – 4.0981 eV and $O(2P^1)$ – 4.6188 eV produce the resulting bond energy 0.397 eV.

The difference in these values of bond energies is defined by different functional states of phosphorous acids HPO_3 and H_3PO_4 in glycolysis processes and equals 0.384 eV that is close to the phospholyration value (0.34–0.35 eV) obtained experimentally.

Such ATP synthesis is carried out in anaerobic conditions and is based on the transfer of phosphate residues onto ATP via the metabolite. For example: ATP formation from creatine phosphate is accompanied by the transition of its NH group at ADP to NH_2 group of creatine at ATP.

From Table 4 it is seen that the change in the bond energy of these two main radicals of metabolite is 5.3238 – 4.9654 = 0.3584 eV – taking the sesquialteral bond N···H into account (as in polypeptides) and 5.3238 – 4.6186 = 0.7052 eV – for the single bond N–H. This is one of the intermediary results of the high–energy bond transformation process in ATP through the metabolite. From Tables 4 and 6 we can conclude that the phosphorous acid H_3PO_4 can have two stationary valence-active states during the interactions in the system P–O for the orbitals with the values of P-parameters of weak and strong interactions, respectively. This defines the possibility for the glycolysis process to flow in two stages. At the first stage, the glucose and H_3PO_4 molecules approach each other due to similar values of their P-parameters of strong interactions (Table 2). At the second stage, H_3PO_4 P-parameter in weak interactions 4.8779 eV (Table 4) in the presence of ferments provokes the bond (H_2O–C)–(C–H_2O) breakage in the glucose molecule with the formation of two molecules of lactic acid whose P-parameters are equal by 4.3563 eV. The energy of this bond breakage process equalled to 0.726 eV (Table 4) is realized as the energy of high-energy bond it ATP.

According to the reference data about 40% of the glycolysis total energy, i.e. about 0.83 eV, remains in ATP.

By the hydrolysis reaction in ATP in the presence of ferments ($HPO_3 + H_2O \rightarrow H_3PO_4 + E$) for structural P_S-parameters (Table 2) E = 11.708 + 4.3609 – 15.865 = 0.276 eV.

It is known that the change in the free energy (ΔG) of hydrolysis of phosphorous anhydrite bond of ATP at pH = 7 under standard conditions is 0.311 – 0.363 eV. But in the cell the ΔG value can be much higher as the ATP and ADP concentration in it is lower than under standard conditions. Besides, the ΔG value is influenced by the concentration of magnesium ions, which is the acting co-ferment in the complex with ATP. Actually Mg^{2+} ion has the P_E-parameter equalled to 8.6066 eV (Table 1) which is very similar to the corresponding values of P-parameters of phosphorous and oxygen atoms.

The quantitative evaluation of this factor requires additional calculations.

KEYWORDS

- **Active**
- **Blocks**
- **Degenerated**
- **In antiphase**
- **Lagrangian equation**
- **Metal**
- **Tuned**

REFERENCES

1. Eyring, H.; Walter, J.; Kimball, G. E. *Quantum Chem..*, I. L., M., **1948**, 528p.
2. Korablev, G. A. *Spatial-Energy Principles of Complex Structures Formation Brill.* Academic Publishers and VSP: Netherlands, 2005, 426pp. (Monograph).
3. Fischer, C. F. *Atomic Data*, **1972**, *4*, 301–399.
4. Waber, J. T.; Cromer, D. T. *J. Chem. Phys.*, **1965**, *42*(12), 4116–4123.
5. Clementi, E.; Raimondi, D. L. *J. Chem. Phys.*, **1963**, *38*(11), 2686–2689.
6. Clementi, E.; Raimondi, D. L. *J. Chem. Phys.*, **1967**, *47*(14), 1300–1307.
7. Gombash, P. *Atom Statistical Model and its Application* M.: I. L., 1951, 398p.
8. Clementi, E. *Develop.* **1965**, *9*(2), 76.
9. Korablev, G. A.; Zaikov, G. E. *Spatial-Energy Parameter as a Materialised Analog of Wafe Function Progress on Chemistry and Biochemistry;* Nova Science Publishers, Inc.: New York, 2009, pp 355–376.
10. Korablev, G. A.; Zaikov, G. E. *J. Appl. Polym. Sci.*, **2006**, *101*(3), 2101–2107.
11. Korablev, G. A.; Zaikov, G. E. *Mech. Composite Mater. Struct.*, **2009**, *15*(1), 106–118.
12. Batsanov, S. S. *Structural chemistry. Facts and Dependencies*. M.: MSU, 2009.
13. Volkenshtein. *Biophysics*. M.: Nauka, 1988, 598p.
14. Govindzhi. M., Ed.: Mir, *Photosynthesis* **1987**, *1*, 728; **1987**, *2*, 460.
15. Clayton, R. *Photosynthesis. Physical Mechanisms and Chemical Models*. M.: Mir, 1984, 350p.
16. Schmidt, M. W.; Baldridge, K. K.; Boatz, J. A.; et al. *J. Comput. Chem.*, **1993**, *14*, 1347–1363.
17. Khokhriakov, N. V.; Kodolov, V. I. *Chem. Phys. Mesoscopy*, **2009**, *11*(3), 388–402.
18. Feniouk, B. A. *Biol. Nat. Sci.*, **1998**, *6*, 108.
19. Feniouk, B. A.; Junge, W.; Mulkidjanian, A. *Tracking of Proton Flow across the Active ATP-Synthase of Rhodobacter Capsulatus in Response to a Series of Light Flashes.* EBEC Reports, **1998**, *10*, 112.

CHAPTER 15

PRACTICAL HINTS ON PRODUCTION OF CARBON NANOTUBES AND POLYMER NANOCOMPOSITES (PART I)

Z. M. ZHIRIKOVA, V. Z. ALOEV, G. V. KOZLOV, and G. E. ZAIKOV

CONTENTS

15.1 INTRODUCTION

At present it is considered that carbon nanotubes (CNT) are one of the most perspective nanofillers for polymer nanocomposites [1]. The high anisotropy degree (their length to diameter large ratio) and low transverse stiffness are CNT_s specific features. These factors define CNT_s ring-like structures formation at manufacture and their introduction in polymer matrix. Such structures radius depends to a considerable extent on CNT_s length and diameter. Thus, the strong dependence of nanofiller structure on its geometry is CNT_s application specific feature. Therefore the present work purpose is to study the dependence of nanocomposites butadiene-styrene rubber/carbon nanorubes (BSR/CNT) properties on nanofiller structure, received by CVD method with two catalysts usage.

15.2 EXPERIMENTAL

The nanocomposites BSR/CNT with CNT content of 0.3 mass% have been used as the study object. CNT have been received in the Institute of Applied Mechanics of Russian Academy of Sciences by the vapors catalytic chemical deposition method (CVD), based on carbon – containing gas thermochemical deposition on non-metallic catalyst surface. Two catalysts – Fe/Al_2O_3 (CNT–Fe) and Co/Al_2O_3 (CNT–Co) – have been used for the studied CNT. The received nanotubes have diameter of 20 nm and length of order of 2 mcm.

The nanofiller structure was studied on force-atomic microscope Nano-DST (Pacific Nanotechnology, USA) by a semi-contact method in the force modulation regime. The received CNT size and polydispersity analysis was made with the aid of the analytical disk centrifuge (CPS Instrument, Inc., USA), allowing to determine with high precision the size and distribution by sizes in range from 2 nm up to 5 mcm. The nanocomposites BSR/CNT elasticity modulus was determined by nanoindentation method on apparatus Nano-Test 600 (Great Britain).

15.3 RESULTS AND DISCUSSION

In Fig. 1, the electron microphotographs of CNT coils are adduced, which demonstrate ring-like structures formation for this nanofiller. In Fig. 2, the indicated structures distribution by sizes was shown, from which it follows, that for CNT-Fe narrow enough monodisperse distribution with maximum at 280 nm is observed and for CNT-Co – polydisperse distribution with maximums at ~ 50 and 210 nm.

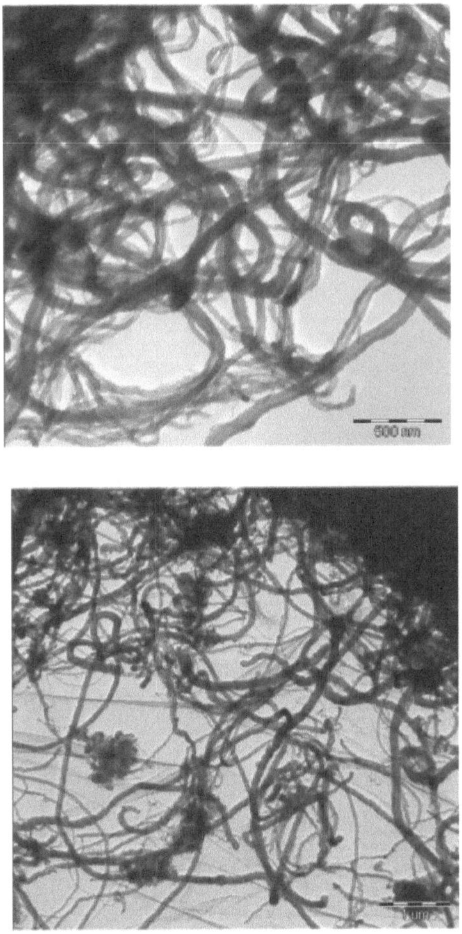

FIGURE 1 Electron micrographs of CNT structure, received on transmission electron microscope.

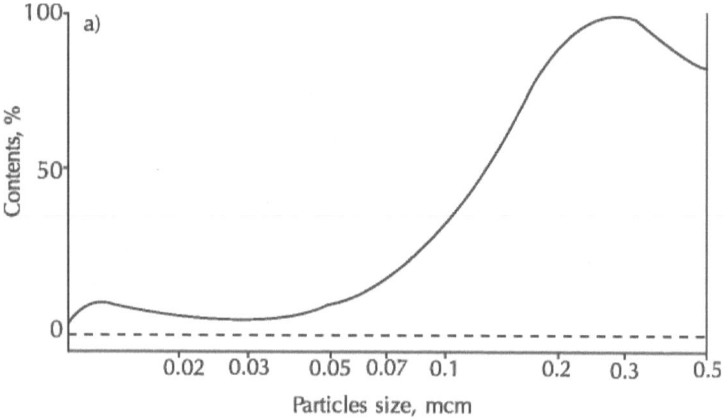

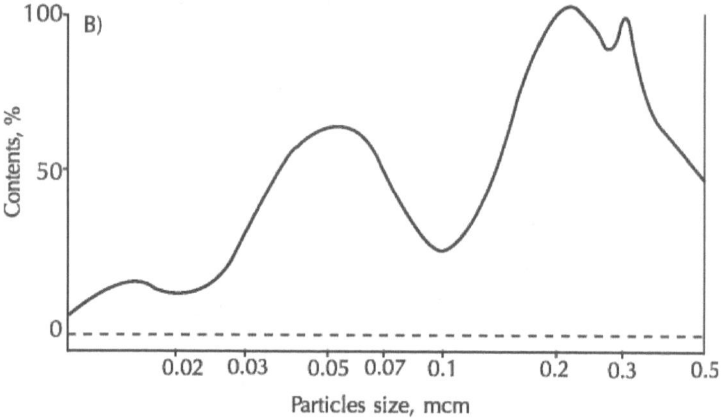

FIGURE 2 The particles distridution by sizes for CNT–Fe (1) and CNT–Co (2).

Further let us carry out the analytical estimation of CNT formed ring-like structures radius R_{CNT}. The first method uses the following formula, obtained within the frameworks of percolation theory [2]:

$$\varphi_n = \frac{\pi L_{CNT} r^2_{CNT}}{(2R_{CNT})^3}, \tag{1}$$

where φ_n is CNT volume content, L_{CNT} and r_{CNT} are CNT length and radius, respectively.

The value φ_n was determined according to the well-known equation [3]:

$$\varphi_n = \frac{W_n}{\rho_n}, \tag{2}$$

where W_n and ρ_n are mass content and density of nanofiller, respectively.

In its turn, the value ρ_n was calculated as follows [4]:

$$\rho_n = 0.188 D_{CNT}^{1/3}, \tag{3}$$

where D_{CNT} is CNT diameter.

The second method is based on the following empirical formula application [5]:

$$R_{CNT} = \left(\frac{D_{CNT}}{D_{CNT}^{st}}\right)^2 \left(0.64 + 4.5\times10^{-3}\varphi_n^{-1}\right),\ \text{mcm}, \tag{4}$$

where D_{CNT}^{st} is a standard nanotube diameter, accepted in paper [5] equal to CNT of the mark "Taunite" diameter (45 nm).

The values R_{CNT}, calculated according to the Eqs. (1) and (4), are adduced in Table 1, from which their good correspondence (the discrepancy is equal to ~ 15%) follows. Besides, they correspond well enough to Fig. 2 data.

In Table 1 the values of elasticity modulus E_n for the studied nanocomposites and E_m for the initial BSR are also adduced. As one can see, if for the nanocomposite BSR/CNT–Fe the very high (with accounting of the condition W_n=0.3 mass%) reinforcement degree E_n/E_m=1.485 was obtained, then for the nanocomposite BSR/CNT–Co reinforcement is practically absent (with accounting for experiment error): $E_n \approx E_m$. Let us consider the reasons of such essential distinction.

TABLE 1 The structural and mechanical characteristics of nanocomposites BSR/CNT.

Catalyst	E_n, MPa	E_n/E_m	E_n/E_m, the equation (5)	R_{CNT}, nm, the equation (1)	R_{CNT}, nm, the equation (4)	b, the equation (6)
Fe/Al_2O_3	4.9	1.485	1.488	236	278	8.42
Co/Al_2O_3	3.1	~ 1.0	1.002	236	278	0.27

The value E_m for BSR is equal to 3.3 MPa.

As it is known [4], the reinforcement degree for nanocomposites polymer/CNT can be calculated as follows:

$$\frac{E_n}{E_m} = 1 + 11\left(c\varphi_n b\right)^{1.7} \tag{5}$$

where c is proportionality coefficient between nanofiller φ_n and interfacial regions φ_{if} relative fractions, b is the parameter, characterizing interfacial adhesion polymer matrix-nanofiller level.

The parameter b in the nanocomposites polymer/CNT case depends on nanofiller geometry as follows [5]:

$$b = 80\left(\frac{R_{CNT}^2}{L_{CNT} D_{CNT}^2}\right) \tag{6}$$

Calculated according to the Eq. (6) values b (for BSR/CNT-Fe and BSR/CNT-Co R_{CNT} magnitudes were accepted equal to 280 and 50 nm, respectively) are adduced in Table 1. As one can see, R_{CNT} decreasing for the second from the indicated nanocomposites results to b reduction in more than 30 times.

The coefficient c value in the Eq. (5) can be calculated as follows [4]. First the interfacial layer thickness l_{if} is determined according to the Eq. [6]:

$$l_{if} = a\left(\frac{r_{CNT}}{a}\right)^{2\left(d-d_{surf}\right)/d} \tag{7}$$

where a is lower linear scale of polymer matrix fractal behavior, accepted equal to statistical segment length l_{st}, d is dimension of Euclidean space, in which fractal is considered (it is obvious that in our case d=3), d_{surf} is CNT surface dimension, which for the studied CNT was determined experimentally and equal to 2.89.

The indicated dimension d_{surf} has a very large absolute magnitude ($2 \le d_{surf} < 3$ [4]) that supposes corresponding roughness of CNT surface, which BSR macromolecule, simulated by rigid statistical segments sequence [7], cannot be "reproduced." Therefore, in practice the effective value d_{surf} (d_{surf}^{ef} for $d_{surf} > 2.5$) is used, which is equal to [7]:

$$d_{surf}^{ef} = 5 - d_{surf} \tag{8}$$

And at last, the statistical segment length l_{st} is estimated according to the equation [4]:

$$l_{st} = l_0 C_\infty \tag{9}$$

where l_0 is the length of the main chain skeletal bond, C_∞ is characteristic ratio. For BSR l_0=0.154 nm, C_∞=12.8 [8].

Further, simulating an interfacial layer as cylindrical one with external radius $r_{\mathrm{CNT}}+l_{if}$ and internal radius r_{CNT}, let us obtain from geometrical considerations the formula for φ_{if} calculation [6]:

$$\varphi_{if} = \varphi_n \left[\left(\frac{r_{\mathrm{CNT}} + l_{if}}{r_{\mathrm{CNT}}} \right)^3 - 1 \right], \tag{10}$$

according to which the value c is equal to 3.47.

The reinforcement degree E_n/E_m calculation results according to the Eq. (5) are adduced in Table 1. As one can see, these results are very close to the indicated parameter experimental estimations. From the Eq. (5) it follows unequivocally, that the values E_n/E_m distinction for nanocomposites BSR/CNT-Fe and BSR/CNT-Co is defined by the interfacial adhesion level difference only, characterized by the parameter b, since the values c and φ_n are the same for the indicated nanocomposites. In its turn, from the Eq. (6) it follows so unequivocally, that the parameters b distinction for the indicated nanocomposites is defined by R_{CNT} difference only, since the values L_{CNT} and D_{CNT} for them are the same. Thus, the fulfilled analysis supposes CNT geometry crucial role in nanocomposites polymer/CNT mechanical properties determination.

Let us note, that the usage of the average value R_{CNT} for nanocomposites BSR/CNT–Co according to Fig. 2 data in the Eq. (6), which is equal to 130 nm, will not change the conclusions made above. In this case E_n/E_m=1.036, that is again close to the obtained experimentally practical reinforcement absence for the indicated nanocomposite.

KEYWORDS

- **Euclidean space**
- **Nanocomposite**
- **Percolation theory**
- **Taunite**

REFERENCES

1. Yanovskii, Yu. G. *Nanomechanics and Strength of Composite Materials*. Publishers of IPRIM RAN: Moscow, 2008, 179p.
2. Bridge, B. *J. Mater. Sci. Lett.*, **1989**, *8*(2), 102–103.
3. Sheng, N.; Boyce, M. C.; Parks, D. M.; Rutledge, G. C.; Abes, J. I. *Cohen Polymer, R. E.*, **2004**, *45*(2), 487–506.
4. Mikitaev, A. K.; Kozlov, G. V.; Zaikov, G. E. *Polymer Nanocomposites: Variety of Structural Forms and Applications*. Nova Science Publishers, Inc.: New York, 2008, 319p.
5. Zhirikova, Z. M.; Kozlov, G. V.; Aloev, V. Z. *Materials of VII International Sci.-Practice Conference "New Polymer Composite Materials*". Nal'chik, KBSU, 2011, pp 158–164.
6. Kozlov, G. V.; Burya, A. I.; Lipatov, Yu. S. *Mekhanika Kompozitnykh Materialov*, **2006**, *42*(6), 797–802.
7. Van Damme, H.; Levitz, P.; Bergaya, F.; Alcover, J. F.; Gatineau, L.; Fripiat, J. J. *J. Chem. Phys.*, **1986**, 85(1), 616–625.
8. Yanovskii, Yu. G.; Kozlov, G. V.; Karnet, Yu. N. *Mekhanika Kompozitsionnykh Materialov i Konstruktsii*, **2011**, 17(2), 203–208.

CHAPTER 16

PRACTICAL HINTS ON PRODUCTION OF CARBON NANOTUBES AND POLYMER NANOCOMPOSITES (PART II)

Z. M. ZHIRIKOVA, G. V. KOZLOV, and V. Z. ALOEV

CONTENTS

16.1 INTRODUCTION

Carbon nanotubes (nanofibers) have two specific features. Firstly, they are normally supplied as little bundles of entangled nanotubes, dispersion of which is difficult enough [1]. Secondly, carbon nanotubes possess very high longitudinal stiffness (high elasticity modulus) while their flexural rigidity is very low for geometric reasons [2]. This property defines distortion (departure from rectilinearity) of the initial nanotubes geometry, namely, their rolling up in ring-like structures [2]. However, at small concentrations, as those used in the present work, the formation of aggregates is scarcely probable [2]. The purpose of the present paper is the development of a quantitative model including both the longitudinal stiffness and the flexibility as parameters for predicting the effectiveness of nanotubes reinforcement.

16.2 EXPERIMENTAL

Polypropylene (PP) "Kaplen" commercially available, having average-weight molecular weight of ~ $(2–3)\times10^5$ and polydispersity index of 4.5, was used as a matrix polymer. Two types of carbon nanotubes were used as nanofiller. Nanotubes of mark "Taunite" (CNT) have an external diameter of 20–70 nm, an internal diameter 5–10 nm and length of 2 mcm and more. Besides, multiwalled nanofibers (CNF), having layers number of 20–30, a diameter of 20–30 nm and length of 2 microns and more, have been used. The mass contents of carbon nanotubes of both types was changed in the range of 0.15–3.0 mass %.

16.3 RESULTS AND DISCUSSION

As it is noted above, carbon nanotubes rectilinearity distorsion or their bending (rolling up) is due to their very high flexibility, and can be one of the main model parameters. The distortion degree can be estimated with the help of forming ring-like structures radius R_{CNT} [3]:

$$R_{\mathrm{CNT}} = \left(\frac{\pi L_{\mathrm{CNT}} D_{\mathrm{CNT}}^2}{32\varphi_n} \right)^{1/3}, \tag{1}$$

where L_{CNT} and D_{CNT} are length and diameter of carbon nanotubes. For CNT and CNF the value L_{CNT} was fixed to 2 micrometers.

The scaling model, already presented in paper [4], can be used for quantitative treatment of carbon nanotubes rolling up processes in order to evaluate its influence on nanocomposites stiffness.

The gist of this model consists of the introduction of reduction factor α^{n-1}, connecting filler mass concentrations W_n in two equivalent composites A and B:

$$W_n^{\mathrm{B}} = \alpha^{n-3} W_n^{\mathrm{A}}, \tag{2}$$

where W_n^{A} and W_n^{B} is the mass content in composites A and B, respectively, α is the aspect ratio of aggregates or particles, n is the parameter characterizing the shape of the filler particles, assumed equal to 1 for short fibers, 2, for disk-like (flaky) particles and 3, for spherical particles.

Within the frameworks of the model, the relationship between the filler particles size D_p and composite stiffness E_c [4] is defined as follows:

$$E_c\left(\theta, W_n, D_p\right) = E_c\left(\theta, \alpha^{n-3}, W_n \alpha D_p\right), \tag{3}$$

where θ is the parameter reflecting the particles size distribution.

The model [4] assumes that the elasticity modulus depends on the reduced filler contents, corrected on the basis of particle shape (n) and size (D_p).

Carbon nanotubes behavior as nanofiller in polymer matrix, require a reduction factor α calculated as follows:

$$\alpha = \frac{R_{CNT}^{\max}}{R_{CNT}}, \tag{4}$$

where $R_{\mathrm{CNT}}^{\max}$ and R_{CNT} are maximum and current radii of ring-like structures, which form carbon nanotubes.

Calculation according to the Eq. (1) has shown that the value $R_{\mathrm{CNT}}^{\max} = 1.33$ microns for CNT and $R_{\mathrm{CNT}}^{\max} = 0.71$ microns for CNF. In CNT nanotubes, varying the mass content from 0.25 up to 3.0 mass % results in a R_{CNT} reduction from 1.33 up to 0.58 microns, while for CNF nanotubes with the mass content varying from 0.15 up to 3.0% R_{CNT} reduces from 0.71 down to 0.26 mcm. The exponent *n* in the Eqs. (2) and (3) was fixed equal to 1. Hence, the final formula for reduction factor has the following form:

$$\alpha^{n-3} = \left(\frac{R_{CNT}^{\max}}{R_{CNT}} \right)^{-2} \tag{5}$$

In Fig. 1, the reinforcement effectiveness, E_n/E_m, is plotted as function on the combined parameter $\alpha^{n-3}W_n$ for nanocomposites PP/CNT and PP/CNF. As one can see, the reinforcement effectiveness is described by a straight line, that, by best fitting the data, can be expressed analytically as follows:

$$\frac{E_n}{E_m} = 1 + 0.65\alpha^{-2}W_n \tag{6}$$

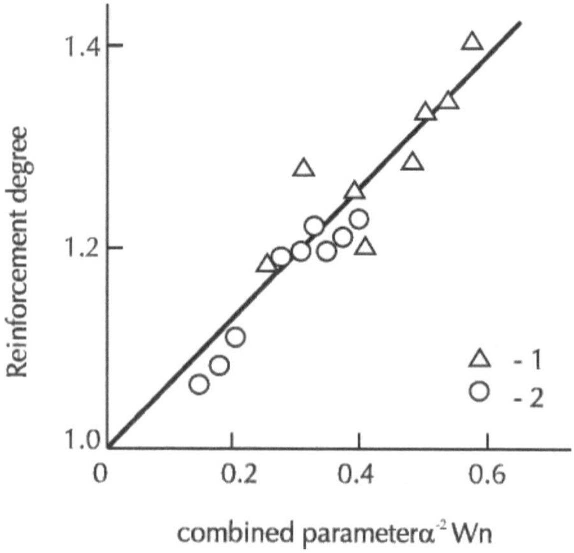

FIGURE 1 The dependence of reinforcement degree En/Em on combined parameter α–2Wn for nanocomposites PP/CNT (1) and PP/CNF (2).

KEYWORDS

- **Kaplen**
- **Multiwalled nanofibers**
- **Ring-like structure**
- **Taunite**

REFERENCES

1. Rakov, E. G. *Uspekhi Khimii*, **2007**, *76*(1), 3–19.
2. Sow, C.-H.; Lim, K.-Y.; Cheong, F.-C.; Saurakhiya, M. *Curr. Res. Nanotechnol.*, **2007**, *1*(2), 125–154.
3. Yang, Y.; D'Amore, A.; Di, Y.; Nicolais, L.; Li, B. *J. Appl. Polym. Sci.*, **1996**, *59*(7), 1159–1166.
4. Bridge, B. *J. Mater. Sci. Lett.*, **1989**, *8*(2), 102–103.
5. Koiwai, A.; Kawasumi, M.; Hyodo, S.; Motohiro, T.; Noda, S.; Kamigaito, O. *Materials of International Symposium "Benibana,"* Yamagata, Japan, 1990; pp. 105–110.
6. Netti, P.; D'Amore, A.; Ronca, D.; Ambrosio, L.; Nicolais, L. *J. Mater. Sci.: Mater. Med.*, **1996**, *7*(9), 525–530.
7. Caprino, G.; D'Amore, A.; Facciolo, F. *J. Composite Mater.*, **1998**, *32*(12), 1203–1220.

CHAPTER 17

A NOTE ON IRREVERSIBLE AGGREGATION MODEL FOR NANOFILLERS

G. V. KOZLOV, YU. G YANOVSKY, and G. E. ZAIKOV

CONTENTS

17.1 INTRODUCTION

The aggregation of the initial nanofiller disperse particles is more or less large particles aggregates always occurs in the course of technological process of making particulate-filled polymer composites in general [1] and elastomer composites in particular [2]. The aggregation process acts on composites (nanostructured composites) macroscopic properties [1, 3]. For composites a process of nanofiller aggregation gains a special significance, since its intensity can be the one that nanofiller particles aggregates size exceeds 100 nm – the value, which assumes (although and conditionally enough [4]) as an upper nanoscale limit for nanoparticle. In other words, the aggregation process can result to the situation, when primordially supposed nanostructured composite ceases to be the one. Therefore, at present several methods exist, which allowed to suppress nanoparticles aggregation process [2, 5]. Proceeding from this, in the present paper theoretical treatment of disperses nanofiller aggregation process in butadiene-styrene rubber matrix within the frameworks of irreversible aggregation models was carried out.

17.2 EXPERIMENTAL

The objects of the analysis were elastomer composites based on butadiene-styrene rubber (BSR). BSR of industrial production, mark SKS-30 ARK was used, which contains 7.0–12.3% cis- and 71.8–72.0% trans links with density of 920–930 kg/m^3. The rubber is completely amorphous one. Mineral shungite (Zazhoginskii deposit of III-th variety) makes up ~30% globular amorphous metastable carbon and ~60% high-disperse silicates (see Table 1). Its structure is fullerene-like one. Nano- and microdimensional disperse particles of schungite were obtained at the Institute of Applied Mechanics of the Russian Academy of Sciences from the industrially mined mineral by unique technology (Russian Federation patent No.2442657 of 20.02.2012) on a planetary ball mill Retsch PM 100 (Germany).

The process of the preparation of samples for testings' consisted of a few stages. A mixture of a polymeric matrix and fillers in equal volume ratios and of other ingredients was prepared in mixer. Thereafter for the mixtures obtained the optimum time of vulcanization was determined and the very process of vulcanization was carried out in special moulds. As a result, plates from the elastomer material of size 15×15 cm^2 and thickness of 2 mm were obtained. Then from these plates the specimens of size 1×1 cm^2 were cut and, to decrease roughness of the specimen surface, micro-cuts with the aid of a microtome were made. The resulting specimens were investigated by the AFM. For macrotestings standard specimens were cut from the 15×15 cm^2 2-mm thick plates investigated in accordance with the State Standards 270–275.

TABLE 1 The chemical composition of mineral shungite.

SiO_2	TiO_2	Al_2O_3	FeO	MgO	CaO	Na_2O	K_2O	S	C	H_2O
57.0	0.2	4.0	2.5	1.2	0.3	0.2	1.5	1.2	29.0	4.2

The analysis of the received in milling process shungite particles were monitored with the aid of analytical disk centrifuge (CPS Instruments, Inc., USA), allowing to determine with high precision the size and distribution by sizes within the range from 2 nm up to 50 mcm.

Nanostructure was studied on scanning probe microscopes Nano-DST (Pacific Nanotechnology, USA) and Easy Scan DFM (Nanosurf, Switzerland) by semi-contact method in the force modulation regime. Scanning probe microscopy results were processed with the aid of specialized software package SPIP (Scanning Probe Image Processor, Denmark). SPIP is a powerful program package for processing of images, obtained on scanning probe microscopy (SPM), atomic forced microscopy (AFM), scanning tunneling microscopy (STM), scanning electron microscopes, transmission electron microscopes, interferometers, confocal microscopes, profilometers, optical microscopes and so on. The given package possesses the whole functions number, which are necessary at images precise analysis, in the number of which the following are included:

(1) the possibility of three-dimensional reflected objects obtaining, distortions automatized leveling, including Z-error mistakes removal for examination of separate elements and so on;

(2) quantitative analysis of particles or grains, more than 40 parameters can be calculated for each found particle or pore: area, perimeter, average diameter, the ratio of linear sizes of grain width to its height distance between grains, coordinates of grain center of mass can be presented in a diagram form or in a histogram form.

17.3 RESULTS AND DISCUSSION

For theoretical treatment of the processes of nanofiller particles aggregates growth and final sizes the traditional irreversible aggregation models are inapplicable, since, it is obvious, that in composites aggregates a large number of simultaneous growth sites takes place. Therefore, the model of multiple growth sites, offered in paper [6], was used for a description of nanofiller aggregation.

In Fig. 1, the images of the studied composites, obtained in the force modulation AFM regime, and corresponding to them the distributions of a nanoparticles aggregates fractal dimension d_f are adduced (for detail see also Ref. [7]). As it follows from the values d_f (d_f=2.40–2.48), nanofiller particles aggregates in the composites under consideration are formed by a mechanism of a particle-cluster (P-Cl), i.e., they are Witten-Sander clusters [8]. The variant «a», of above model [6] was chosen for the modeling. According to this variant the mobile particles are added to the lattice, consisting of a large number of "seeds" with density of c_0 at the modeling beginning [6]. Such model generates structures, which have fractal geometry on short scales of length with value $d \approx 2.5$ (see Fig. 1) and homogeneous structure on large scales of length. A relatively high particles concentration c is required in the model for formation of uninterrupted network [6].

In case of "seeds" high concentration c_0 for the variant «a» of the model the following relationship was obtained [6]:

$$R_{max}^{d_f} = N = c / c_0, \tag{1}$$

where R_{max} is a maximal radius of nanoparticles cluster (aggregate), N is a number of nanoparticles per one aggregate, c is a concentration of nanoparticles, c_0 is "seeds" number, which is equal to nanoparticles clusters (aggregates) number.

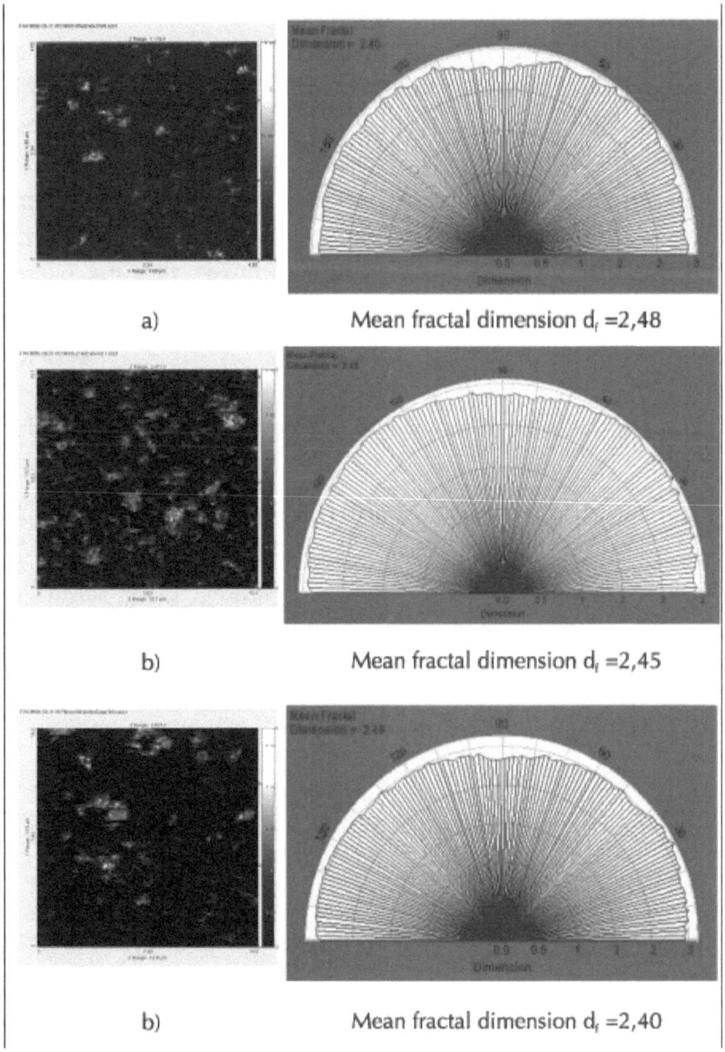

FIGURE 1 The images, obtained in the force modulation regime, for composites, filled with technical carbon (a), nanoshungite (b), microshungite (c) and corresponding to them fractal dimensions d_f.

The value N can be estimated according to the following equation [9]:

$$2R_{\max} = \left(\frac{S_n N}{\pi q} \right)^{1/2}, \tag{2}$$

where S_n is cross-sectional area of nanoparticles, from which aggregate consists, q is packing coefficient, equal to 0.74.

The experimentally obtained value for nanoparticles and theirs aggregates was taken from [7]. A diameter $2R_{agr}$ for nanoparticles aggregate was accepted as $2R_{max}$ (Table 2). The value S_n was also calculated according to the experimental values of nanoparticles radius r_n (Table 2). In Table 2 the values N for the studied nanofillers, obtained according to the above indicated method, were adduced. It is significant, that the value N is a maximum one for nanoshungite despite of the larger values r_n in comparison with technical carbon.

TABLE 2 The parameters of irreversible aggregation model of nanofiller particles aggregates growth.

Filler	Experimental radius of nanofiller aggregate R_{agr}, nm	Radius of nanofiller particle r_n, nm	Number of particles in one aggregate N	Radius of nanofiller aggregate R^T_{max}, the equation (1), nm	Radius of nanofiller aggregate R^T_{agr}, the equation (3), nm	Radius of nanofiller aggregate R_c, the equation (8), nm
Technical carbon	34.6	10	35.4	34.7	34.7	33.9
Nanoshungite	83.6	20	51.8	45.0	90.0	71.0
Microshungite	117.1	100	4.1	15.8	158.0	255.0

Further, the Eq. (1) allows to estimate the maximal radius R^T_{max} of nanoparticles aggregate within the frameworks of the aggregation model [6]. These values R^T_{max} are adduced in Table 2, from which their reduction in a sequence of technical carbon-nanoshungite-microshungite, that fully contradicts to the experimental data, i.e., to R_{agr} change (Table 2). However, we must not neglect the fact, that the Eq. (1) was obtained within the frameworks of computer simulation, where the initial aggregating particles sizes are the same in all cases [6]. For real composites the values r_n can be distinguished essentially (Table 2). It is expected, that the value R_{agr} or R^T_{max} will be the higher, than the larger radius of nanoparticles, forming aggregate. Then theoretical value of a radius R_{agr}^{T} of nanofiller particles cluster (aggregate) can be determined as follows:

$$R_{agr}^{T} = k_n r_n N^{1/d_f}, \tag{3}$$

where k_n is proportionality coefficient, in the present work accepted empirically equal to 0.9.

The comparison of experimental R_{agr} and calculated according to the Eq. (3) R_{agr}^{T} values of the studied nanofillers particles aggregates radius shows their good correspondence (the average discrepancy of R_{agr} and R_{agr}^{T} makes up 11.4%). Therefore, the theoretical model [6] gives a good correspondence to the experiment only in case of consideration of aggregating particles real characteristics and, in the first place, their size.

Let us consider two more important aspects of nanofiller particles aggregation within the frameworks of the model [6]. Some features of the indicated process are defined by nanoparticles diffusion at composites processing. Specifically, length scale, connected with diffusible nanoparticle, is correlation length ξ of diffusion. By definition, the growth phenomena in sites, remote more than ξ, are statistically independent. Such definition allows to connect the value ξ with the mean distance between nanofiller particles aggregates L_n. The value ξ can be calculated according to the equation [6] as

$$\xi^2 \approx \tilde{n}^{-1} R_{agr}^{d_f-d+2}, \tag{4}$$

where c is a concentration of nanoparticles, d is dimension of Euclidean space, in which a fractal is considered (it is obvious, that in our case d=3). The value c should be accepted equal to nanofiller volume contents φ_n, which is calculated [10] as follows:

$$\varphi_n = \frac{W_n}{\rho_n}. \tag{5}$$

Here W_n is nanofiller mass contents, ρ_n is its density, determined according to the equation [3]:

$$\rho_n = 0.188(2r_n)^{1/3}. \tag{6}$$

The values r_n and R_{agr} were determined experimentally (see graph of Fig. 2 [7]). In Fig. 3 the relation between L_n and ξ is adduced, which, as it is expected, proves to be linear and passing through coordinates origin. This means, that the distance between nanofiller particles aggregates is limited by mean displacement of statistical walks, by which nanoparticles are simulated. The relationship between L_n and ξ can be expressed analytically as follows:

$$L_n = 9.6\xi \text{, nm.} \tag{7}$$

The second important aspect of the model [6], in reference to simulation of nanofiller particles aggregation, is a finite nonzero initial particles concentration c or φ_n effect, which takes place in any real systems. This effect is realized at the condition $\xi \approx R_{agr}$, that occurs at the critical value $R_{agr}(R_c)$, determined according to the relationship [6]:

$$c \sim R_c^{d_f - d} \tag{8}$$

The relationship (8) right side represents cluster (particles aggregate) mean density. This equation establishes, that fractal growth continues only, until cluster density reduces up to medium density, in which it grows. The calculated according to the relationship (8) values R_c for the considered nanoparticles are adduced in Table 2, from which it follows, that they give reasonable correspondence with the experimental values R_{agr} (the average discrepancy of R_c and R_{agr} makes up 24%).

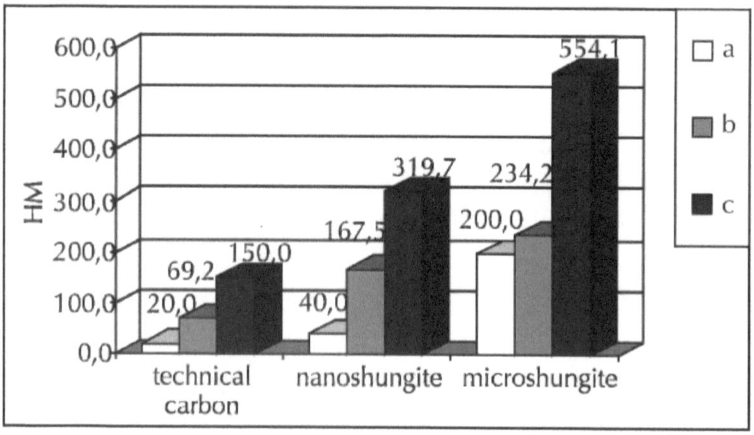

FIGURE 2 The initial particles diameter (a), their aggregates size in composite (b) and distance between nanoparticles aggregates (c) for composites, filled with technical carbon, nano– and microshungite.

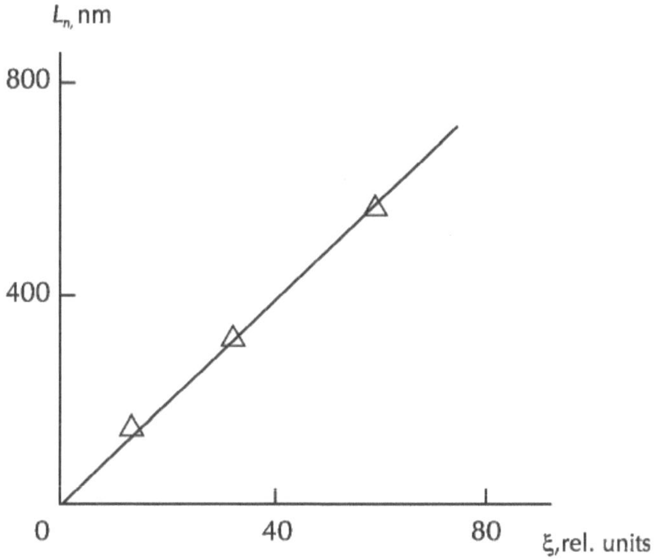

FIGURE 3 The relation between diffusion correlation length ξ and distance between nanoparticles aggregates Ln for studied composites.

Since, the treatment [6] was obtained within the frameworks of a more general model of diffusion-limited aggregation, then its correspondence to the experimental data indicated unequivocally, that aggregation processes in these systems were controlled by diffusion. Therefore let us consider briefly nanofiller particles diffusion. Statistical walkers diffusion constant ζ can be determined with the aid of the relationship [6]:

$$\xi \approx (\zeta t)^{1/2}, \tag{9}$$

where t is walk duration.

The Eq. (9) supposes (at t=const) ζ increase in a number technical carbon-nanoshungite-microshungite as 196-1069-3434 relative units, i.e. diffusion intensification at diffusible particles size growth. At the same time diffusivity D for these particles can be described by the well-known Einstein's relationship [11]:

$$D = \frac{kT}{6\pi\eta_{\bar{n}} r_n \alpha} \tag{10}$$

where k is Boltzmann constant, T is temperature, η_c is a composite medium viscosity, α is numerical coefficient, which further is accepted equal to 1.

In its turn, the value η can be estimated according to the equation [12]:

$$\frac{\eta_{\tilde{n}}}{\eta_0} = 1 + \frac{2.5\varphi_n}{1-\varphi_n}, \tag{11}$$

where η_0 and η_c are initial polymer and its mixture with nanofiller viscosities, accordingly, φ_n is nanofiller volume contents.

The calculation according to the Eqs. (10) and (11) shows, that within the indicated above nanofillers number the value D changes as 1.32–1.14–0.44 relative units, i.e., reduces in three times, that was expected. This apparent contradiction is due to the choice of the condition t=const (where t is composite production duration) in the Eq. (9). In real conditions the value t is restricted by nanoparticle contact with growing aggregate and then instead of t the value t/c_0 should be used, where c_0 is seeds concentration, determined according to the Eq. (1). In this case the value ζ for the indicated nanofillers changes as 0.288–0.118–0.086, i.e., it reduces in 3.3 times, that corresponds fully to the calculation according to the Einstein's relationship [Eq. (10)]. This means, that nanoparticles diffusion in polymer matrix obeys classical laws of Newtonian rheology [11].

KEYWORDS

- **Einstein's relationship**
- **Newtonian rheology**
- **Seeds**
- **Shungite particles**
- **Z-error**

REFERENCES

1. Kozlov, G. V.; Yu.Yanovskii, G.; Zaikov, G. E. *Structure and Properties of Particulate-Filled Polymer Composites: The Fractal Analysis;* Nova Science Publishers, Inc.: New York, 2010; 282p.

2. Edwards, D. C. *J. Mater. Sci.*, **1990**, *25*(12), 4175.
3. Mikitaev, A. K.; Kozlov, G. V.; Zaikov, G. E. *Polymer Nanocomposites: the Variety of Structural Forms and Applications*. Nova Science Publishers, Inc.: New York, 2008; 319p.
4. . Buchachenko, A. L. *Uspekhi Khimii*, **2003**, *72*(5), 419.
5. Kozlov, G. V.; Yanovskii, Yu. G.; Burya, A. I.; Aphashagova, Z. Kh. *Mekhanika Kompozitsionnykh Materialov i Konstruktsii*, **2007**, *13*(4), 479.
6. Witten, T. A.; Meakin, P. *Phys. Rev. B*, **1983**, *28*(10), 5632.
7. Yanovsky, Yu. G.; Valiev, Kh. Kh.; Kornev, Yu. V.; Karnet, Yu. N.; Boiko, O. V.; Kosichkina, K. P.; Yumashev, O. B. *Nanomech. Sci. Technol.: Int. J.*, **2010**, *1*(3), 187–211.
8. Witten, T. A.; Sander, L. M. *Phys. Rev. B*, **1983**, *27*(9), 5686.
9. Bobryshev, A. N.; Kozomazov, V. N.; Babin, L. O.; Solomatov, V. I. *Synergetics of Composite Materials*. Lipetsk, NPO ORIUS, 1994; 154p.
10. Sheng, N.; Boyce, M. C.; Parks, D. M.; Rutledge, G. C.; Abes, J. I.; Cohen, R. E. *Polymer*, **2004**, *45*(2), 487.
11. Happel, J.; Brenner, G. *The Hydrodynamics at Small Reynolds Numbers*. Moscow, Mir, 1976; 418p.
12. Mills, N. J. *J. Appl. Polym. Sci.*, **1971**, *15*(11), 2791.

CHAPTER 18

A REINFORCEMENT MECHANISM FOR NANOCOMPOSITES

G. V. KOZLOV, Z. KH. APHASHAGOVA, A. KH. MALAMATOV, and G. E. ZAIKOV

CONTENTS

18.1 INTRODUCTION

Very often a filler (nanofiller) is introduced in polymers with the purpose of the stiffness increase of the latter. This effect is called polymer composites (nanocomposites) reinforcement and it is characterized by reinforcement degree E_c/E_m (E_n/E_m), where E_c, E_n and E_m are elasticity moduli of composite, nanocomposite and matrix polymer, accordingly. The indicated effect significance results to a large number of quantitative models development, describing reinforcement degree: micromechanical [1], percolation [2] and fractal [3] ones. The principal distinction of the indicated models is the circumstance, that the first ones take into consideration filler (nanofiller) elasticity modulus and the last two don't. The percolation [2] and fractal [3] models of reinforcement assume, that a filler (nanofiller) role comes to modification and fixation of matrix polymer structure. Such approach is obvious enough, if to take into consideration the difference of elasticity moduli of the filler (nanofiller) and matrix polymer. So, for the considered in the present paper nanocomposites low-density polyethylene/calcium carbonate the matrix polymer elasticity modulus makes up 85 MPa [4] and nanofiller – of order of tens GPa [1], i.e., the difference makes up more than two orders. It is obvious, that at such conditions calcium carbonate strain is equal practically to zero and nanocomposite behavior in mechanical tests is defined by polymer matrix behavior.

Lately it has been offered to consider polymers amorphous state structure as natural nanocomposite [5]. Within the frameworks of cluster model of polymers amorphous state structure it is supposed, that the indicated structure consists of local order domains (clusters), immersed in loosely-packed matrix, in which the entire polymer free volume is concentrated [6, 7]. In its turn, clusters consist of several collinear densely-packed statistical segments of different macromolecules, i.e. they are amorphous analog of crystallites with the stretched chains. It has been shown [8], that cluster are nanoworld objects (true nanoparticles – nanoclusters) and in case of polymers representation as natural nanocomposites they play a nanofiller role and loosely-packed matrix – a nanocomposite matrix role. It is significant that the nanoclusters dimensional effect is identical to the indicated effect for particulate nanofiller in polymer nanocomposites – sizes decrease of both nanoclusters [9] and disperse nanoparticles [10] results to sharp enhancement of nanocomposites reinforcement degree (elasticity modulus). In connection with the indicated observations the question arises: how disperse nanofiller introduction in polymer matrix influences on nanoclusters size and how the variation of the latter influences on nanocomposite elasticity modulus value. The purpose of the present paper is these two problems solution on the example of particulate-filled nanocomposite low density polyethylene/calcium carbonate [4].

18.2 EXPERIMENTAL

Low density polyethylene (LDPE) of mark 10 803–020 was used as matrix polymer and nanodimensional calcium carbonate ($CaCO_3$) in the form of compound mark Nano–Cal NC–KO117 (China) with particles size of 80 nm and mass contents of 1–50 mass% was used as nanofiller.

Nanocomposites LDPE/$CaCO_3$ were prepared by components mixing in melt on twin–screw extruder Thermo Haake, model Reomex RTW 25/42, production of German Federal Republic. Mixing was performed at temperature 448–463 K and screw speed of 15–25 rpm during 5 min. Testing samples were obtained by casting under pressure method on casting machine Test Samples Molding Apparate RR/TS of firm Ray–Ran (Taiwan) at temperature 473 K and pressure 8 MPa.

Uniaxial tension mechanical tests have been performed on the samples in the shape of two–sided spade with sizes according to GOST-112 62–80. The tests have been conducted on universal testing apparatus Gotech Testing Machine CT-TCS 2000, production of German Federal Republic, at temperature 293 K and strain rate ~ 2×10^{-2} s^{-1}.

18.3 RESULTS AND DISCUSSION

For the solution of the first from the indicated problems the statistical segments number per one nanocluster n_{cl} and its variation at $CaCO_3$ contents change should be estimated. The parameter n_{cl} calculation consistency includes the following stages. At first the nanocomposite structure fractal dimension d_f is calculated according to the equation [11]:

$$d_f = (d-1)(1+\nu), \tag{1}$$

where d is dimension of Euclidean space, in which a fractal is considered (it is obvious, that in our case d=3), ν is Poisson's ratio, which is estimated according to mechanical tests results with the help of the following relationship [12]:

$$\frac{\sigma_Y}{E_n} = \frac{1-2\nu}{6(1+\nu)}, \tag{2}$$

where σ_Y and E_n are yield stress and elasticity modulus of nanocomposite, accordingly.

Then nanoclusters relative fraction φ_{cl} can be calculated by the following equation using [7]:

$$d_f = 3 - 6\left(\frac{\varphi_{cl}}{C_\infty S}\right)^{1/2}, \tag{3}$$

where C_∞ is characteristic ratio, which is a polymer chain statistical flexibility indicator [13], S is macromolecule cross-sectional area.

The value C_∞ is the function of d_f according to the relationship [7]:

$$C_\infty = \frac{2d_f}{d(d-1)(d-d_f)} + \frac{4}{3} \tag{4}$$

The value S for LDPE is accepted equal to 14.9 Å^2 [14]. Macromolecular entanglements cluster network density ν_{cl} can be estimated as follows [7]:

$$\nu_{cl} = \frac{\varphi_{cl}}{C_\infty l_0 S}, \tag{5}$$

where l_0 is the main chain skeletal bond length, which for polyethylene is equal to 0.154 nm [15].

Then the molecular weight of the chain part between nanoclusters M_{cl} was determined according to the equation [7]:

$$M_{cl} = \frac{\rho_p N_A}{\nu_{cl}}, \tag{6}$$

where ρ_p is polymer density, which for the studied LDPE is equal to 930 kg/m^3 [16], N_A is Avogadro number.

And at last, the value n_{cl} is determined according to the formula [7]:

$$n_{cl} = \frac{2M_e}{M_{cl}}, \tag{7}$$

where M_e is molecular weight of the chain part between entanglements traditional nodes (macromolecular "binary hooking"), which is equal to 1390 g/mol for LDPE [17].

In Fig. 1 the dependence of nanocomposite elasticity modulus E_n on value n_{cl} is adduced, from which E_n enhancement at n_{cl} decreasing follows. Such nanocomposites LDPE/$CaCO_3$ behavior is identical completely to natural nanocomposites behavior [9].

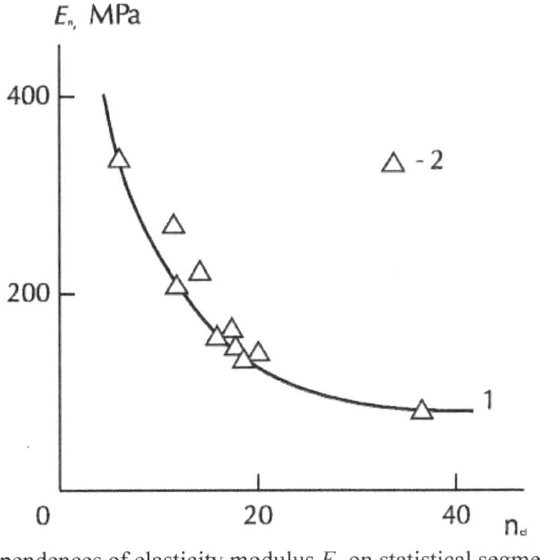

FIGURE 1 The dependences of elasticity modulus E_n on statistical segments number per one nanocluster n_{cl} for nanocomposites LDPE/$CaCO_3$. 1 – calculation according to the Eq. (8); 2 – the experimental data.

In paper [18] the theoretical dependence of E_n as a function of cluster model parameters for natural nanocomposites was obtained:

$$E_n = c\left(\frac{\varphi_{cl} \nu_{cl}}{n_{cl}}\right), \tag{8}$$

where c is constant, accepted equal to 1.15×10^{-26} m^3 for the studied LDPE.

In Fig. 1 the theoretical dependence $E_n(n_{cl})$, calculated according to the equation (8), for the studied nanocomposites is also adduced, which showed a good enough correspondence to the experiment (the average discrepancy of theory and experiment makes up ~ 15%, that is comparable with mechanical tests experimental error). Hence, at $CaCO_3$ mass contents W_n increasing within the range of 0–50 mass% n_{cl} value reduces from 36.0 up to 5.2, that is accompanied by nanocomposites LDPE/$CaCO_3$ elasticity modulus growth from 85 up to 340 MPa.

Let us consider the physical grounds of parameter n_{cl} reduction at nanofiller contents in nanocomposites LDPE/$CaCO_3$ growth. As it is known [19], the densely-packed regions maximum fraction φ_{dens} for polymeric materials is given by the following percolation relationship:

$$\varphi_{dens} = \left(\frac{T_m - T}{T_m}\right)^{\beta_T}, \tag{9}$$

where T_m and T are melting and testing temperatures, accordingly (for LDPE T_m=398 K [16]), β_T is a thermal cluster order index, which is equal to 0.55 for polymers [20].

It is quite obvious, that matrix polymer space filling by nanofiller decreases a polymer matrix fraction in nanocomposites structure, particularly at large (up to 50 mass%) the latter contents. Therefore the reduced value φ_{dens} (φ_{dens}^{red}) should be used, which is determined as follows [8]:

$$\varphi_{dens}^{red} = \frac{\varphi_{dens}}{1-\varphi_n}, \tag{10}$$

where φ_n is a nanofiller volume fraction, which can be calculated according to the equation [8]:

$$\varphi_n = \frac{W_n}{\rho_n}, \tag{11}$$

where W_n is nanofiller mass contents, ρ_n is its density, estimated according to the equation [8]:

$$\rho_n = 188 D_p^{1/3}, \text{ kg/m}^3, \tag{12}$$

where D_p is a diameter of particulate nanofiller particles, which is given in nm.

For nanocomposites LDPE/$CaCO_3$ as structure densely–packed regions the sum should be accepted [8]:

$$\varphi_{dens} = (1-K)\varphi_{cl} + \varphi_{if}, \tag{13}$$

where K is crystallinity degree, φ_{cl} and φ_{if} are nanoclusters and interfacial regions relative fractions, accordingly.

The value φ_{if} can be estimated with the help of the following percolation relationship [8]:

$$\frac{E_n}{E_m} = 1 + 11(\varphi_n + \varphi_{if})^{1.7} \tag{14}$$

In Fig. 2 the dependence $\varphi_{dens}^{red}(W_n)$ for nanocomposites LDPE/$CaCO_3$ is adduced. As one can see, the values φ_{dens}^{red}, determined according to the Eqs. (9), (10) and (13), agree well with each other. The Eq. (10) shows the cause of n_{cl} reduction at W_n (or φ_n) growth, i.e. polymer matrix fraction decrease, accompanied by φ_{cl} reduction, results to ν_{cl} decreasing according to the Eq. (5), M_{cl} increasing according to the Eq. (6) and n_{cl} reduction according to the Eq. (7).

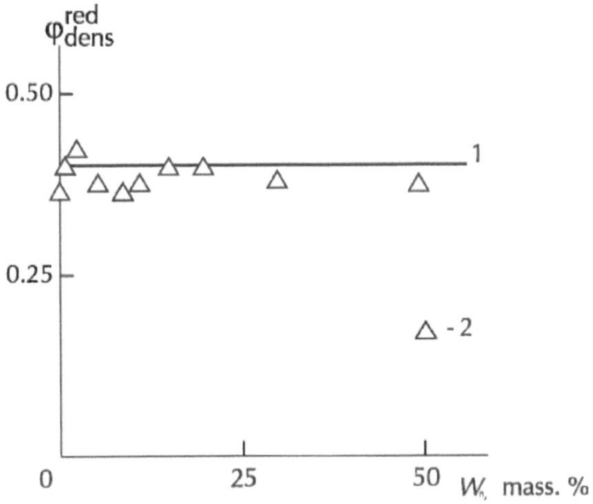

FIGURE 2 The dependences of densely–packed regions reduced relative fraction φ_{dens}^{red} on nanofiller mass contents W_n for nanocomposites LDPE/$CaCO_3$. Calculation: 1 – according to the Eqs. (9) and (10); 2 – according to the Eq. (13).

Let us note, that constant c in the Eq. (8) is the function of matrix polymer characteristics that was expected. In Fig. 3 the dependence of constant c on matrix polymer elasticity modulus E_m for two LDPE, polypropylene (PP) and polycarbonate (PC) is adduced. As one can see, the value c grows at E_m increasing and is described by the following empirical equation:

$$c = 9.71 \times 10^{-26} E_m, \quad (15)$$

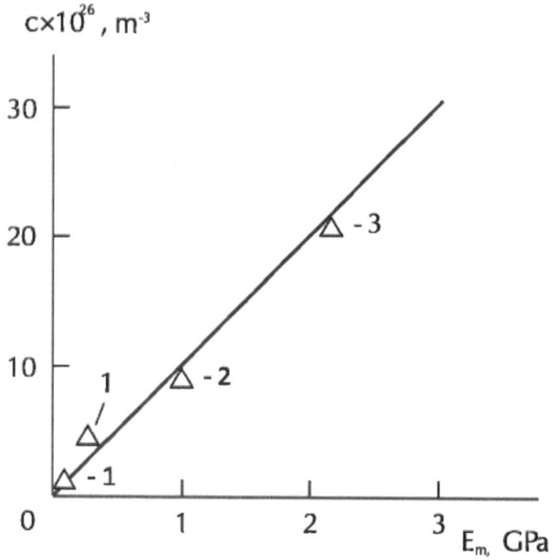

FIGURE 3 The dependence of constant c in the equation (8) on matrix polymer elasticity modulus E_m for LDPE (1), PP (2) and PC (3).

where E_m is given in GPa.

KEYWORDS

- **Binary hooking**
- **Euclidean space**
- **Nanocomposite matrix**
- **Poisson's ratio**

REFERENCES

1. Ahmed, S.; Jones, F. R. *J. Mater. Sci.*, **1990**, *25*(12), 4933–4942.
2. Bobryshev, A. N.; Kozomazov, V. N.; Babin, L. O.; Solomatov, V. I. *Synergetics of Composite Materials*. Lipetsk, NPO ORIUS, 1994; 154p.
3. Kozlov, G. V.; Yanovskii, Yu. G.; Zaikov, G. E. *Structure and Properties of Particulate-Filled Polymer Composites: The Fractal Analysis*. Nova Science Publishers, Inc.: New York, 2010; 282p.
4. Sultonov, N. Zh.; Mikitaev, A. K. *Materials VI International Science-Practice Conference "New Polymer Composite Materials."* Nal'chik, KBSU, 2010; pp. 392–398.
5. Kozlov, G. V.; Mikitaev, A. K. *Polymers as Natural Nanocomposites: Unrealized Potential*. Lambert Academic Publishing: Saarbrücken, 2010; 323p.
6. Kozlov, G. V.; Novikov, V. U. *Uspekhi Fizicheskikh Nauk*, **2011**, *171*(7), 717–764.
7. Kozlov, G. V.; Zaikov, G. E. *Structure of the Polymer Amorphous State*. Brill Academic Publishers: Utrecht, Boston, 2004; 465p.
8. Mikitaev, A. K.; Kozlov, G. V.; Zaikov, G. E. *Polymer Nanocomposites: Variety of Structural Forms and Applications*. Nova Science Publishers, Inc.: New York, 2008; 319p.
9. Magomedov, G. M.; Kozlov, G. V.; Zaikov, G. E. *Structure and Properties of Cross-Linked Polymers*. A Smithers Group Company: Shawbury, 2011; 492p.
10. Edwards, D. C. *J. Mater. Sci.*, **1990**, *25*(12), 4175–4185.
11. Balankin, A. S. *Synergetics of Deformable Body*. Publishers Ministry of Defence SSSR: Moscow, 1991; 404p.
12. Kozlov, G. V.; Sanditov, D. S. *Anharmonic Effects and Physical-Mechanical Properties of Polymers*. Nauka: Novosibirsk, 1994; 261p.
13. Budtov, V. P. *Physical Chemistry of Polymer Solutions*. Khimiya: Sankt-Peterburg, 1992; 384p.
14. Aharoni, S. M. *Macromolecules*, **1985**, *18*(12), 2624–2630.
15. Aharoni, S. M. *Macromolecules*, **1983**, *16*(9), 1722–1728.
16. Kalinchev, E. L.; Sakovtseva, M. B. *Properties and Processing of Thermoplastics*. Khimiya: Leningrad, 1983; 288p.
17. Wu, S. J. *Polymer Sci.: Part B: Polymer Phys.*, **1989**, *27*(4), 723–741.
18. Kozlov, G. V. *Recent Patents Chem. Eng.*, **2011**, *4*(1), 53–77.
19. Family, F. *J. Stat. Phys.*, **1984**, *36*(5/6), 881–896.
20. Kozlov, G. V.; Gazaev, M. A.; Novikov, V. U.; Mikitaev, A. K. *Pis'ma v ZhTF*, **1996**, *22*(16), 31–38.

CHAPTER 19

PRACTICAL HINTS ON PRODUCTION OF CARBON NANOTUBES AND POLYMER NANOCOMPOSITES (PART III)

G. V. KOZLOV, Z. M. ZHIRIKOVA, V. Z. ALOEV, and G. E. ZAIKOV

CONTENTS

19.1 INTRODUCTION

At it is known [1], a carbon nanotubes (CNT) being in their production process form aggregates, consisting of tangled separate nanotubes. For this effect weaking the number of methods is used: CNT functionalization [2], processing by ultrasound [3] and so on. Besides, it is well-known [2, 4], that possessing high anisotropy degree and small transversal stiffness CNT form ring-like structures. It is naturally to expect, that the indicated effects will influence on CNT structure in polymer nanocomposites and these nanomaterials properties. The present communication purpose is the study of the processing by ultrasound influence of nanocomposites epoxy polymer/carbon nanotubes and CNT ring-like structure formation.

19.2 EXPERIMENTAL

The data of paper [3] for nanocomposites epoxy polymer/carbon nanotubes with nanofiller supersmall contents (≤0.1 mass%) have been used. The epoxy diane resin ED-20 (ED) and diphenylolpropane diglycidyl ether (DDE) have been used as matrix polymer. Carbon nanotubes with diameter of ~50 nm and length of ~2 mcm and contents of 0.0009–0.10 mass% were dispersed by ultrasonic (US) vibrations with the frequency of 22 Mc/s. The details of nanocomposites ED/CNT and DDE/CNT and their testing methods are adduced in paper [3].

19.3 RESULTS AND DISCUSSION

As it has been shown in paper [5], CNT ring-like structures with radius R_n formation in polymer nanocomposite influences on these nanomaterials properties, particularly, on interfacial adhesion level, characterized by the parameter b_α. The inrercommunication of b_α and R_n is given by the following relationship [5]:

$$b_\alpha = 4.8\left(R_n^2 - 0.28\right), \qquad (1)$$

where R_n is given in mcm.

In its turn, the parameter b_α is determined with the help of the percolation relationship [6]:

$$\frac{E_n}{E_m} = 1 + 11\left(cb_\alpha \varphi_n\right), \tag{2}$$

where E_n and E_m are elasticity moduli of nanocomposite and matrix polymer, accordingly, and the ration E_n/E_m represents a reinforcement degree, c is proportionality coefficient between interfacial regions relative fraction φ_{if} and nanofiller volume content φ_n.

For nanocomposites polymer/CNT c=2.41 [6] and the value φ_n is determined according to the well-known formula [7]:

$$\varphi_n = \frac{W_n}{\rho_n}, \tag{3}$$

where W_n is nanofiller mass contents, ρ_n is its density, determined as follows [6]:

$$\rho_n = 188\left(D_n\right)^{1/3}, \text{ kg/m}^3, \tag{4}$$

where D_n is CNT diameter, which is given in nm.

Since the value R_n, determined according to the Eq. (1), was obtained according to the nanocomposites ED/CNT and DDE/CNT samples tests results, then it reflects CNT geometry, formed under US action (R_n^{US}). In its turn, the value R_n of CNT ring-like structures, which does not take into account US-processing, can be estimated with the help of the percolation relationship [8]:

$$\varphi_n = \frac{\pi L_n r_n^2}{\left(2R_n\right)^3}, \tag{5}$$

where L_n and r_n are CNT length and radius, accordingly.

In Fig. 1 the dependences of R_n^{US} and R_n on φ_n for nanocomposites ED/CNT and DDE/CNT are adduced. As it was to be expected, US-processing application results to R_n essential growth and this effect is expressed particularly strongly in case of CNT very small concentrations, namely, for $\varphi_n \leq 10^{-4}$. The estimations according to the Eq. (5) showed, that US-processing application was equivalent to φ_n reduction in 540 times at the smallest values φ_n from the using ones and in 115 times – at the greatest ones.

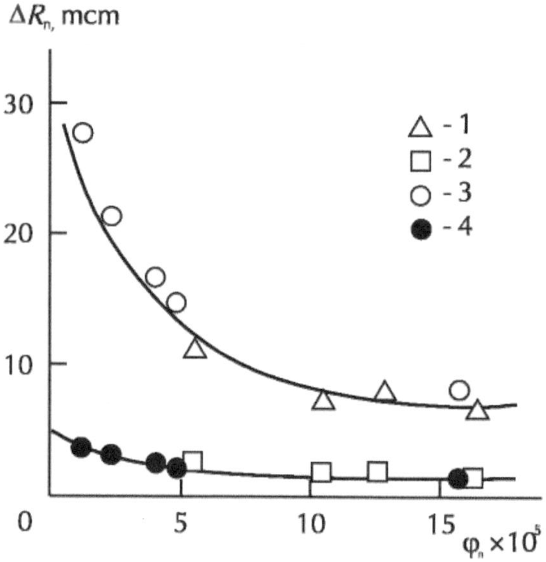

FIGURE 1 The dependences of CNT ring-like structures radius R_n at US-processing using (1, 3) and without it (2, 4) on nanofiller volume contents φ_n for nanocomposites ED/CNT (1, 2) and DDE/CNT (3, 4).

Since the US-processing efficiency is reduced at CNT contents growth, then in Fig. 2 the dependence of values R_n with the application of US and without application US ($\Delta R = R_n^{US} - R_n$) difference, characterizing the indicated efficiency, on the value $\varphi_n^{1/3}$ is adduced. Such form of the dependence $\Delta R_n(\varphi_n)$ was chosen with its linearization purpose. As it follows from the data of Fig. 2, the value ΔR_n reduces at φ_n growth that can be expressed analytically by the following equation:

$$\Delta R_n = 36 - 600\varphi_n^{1/3}, \text{ mcm.} \quad (6)$$

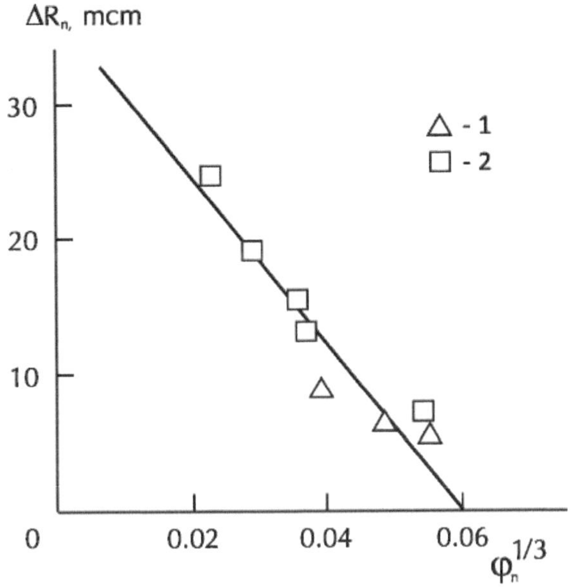

FIGURE 2 The dependence of CNT ring-like structures radius difference ΔR_n at US-processing using and without it on parameter $\varphi_n^{1/3}$ for nanocomposites ED/CNT (1) and DDE/CNT (2).

According to the Eq. (6) the value ΔR_n=0, i.e., when US-processing cesases to influence on CNT geometry, is equal to ~ 21.6×10^5. Let us consider physical aspect of this effect. The value of percolation threshold φ_c in case of CNT continuous network formation can be determined according to the equation [8]:

$$\varphi_{\tilde{n}} = \frac{\pi}{12}\left(\frac{D_n}{2R_n}\right) \tag{7}$$

The calculation according to the Eq. (7) shows, that the value φ_c varies within the limits of (29–94)×10^5 at R_n change within the range of 7–28 mcm. This value φ_c is approximately on an order of magnitude smaller than the similar parameter, determined by the authors [3] with the help of other methods.

It is assumed [3], that CNT aggregation process in ropes (bundles) begins at φ_n>φ_c. The stated above results allow assuming that US-processing does not influenced on CNT aggregation process (CNT bundles formation), but influences strongly on nanotubes geometry, characterized by ring-like structures radius R_n, at φ_n, which is smaller than percolation threshold φ_c.

In Fig. 3 the dependence of elasticity modulus E_n on CNT volume contents φ_n for the considered nanocomposites is adduced. As one can see, the systematic E_n change at φ_n variation more than one order of magnitude is not observed. This observation is explained by b_α reduction in 16 times at φ_n growth within the range of $(1.2–16)\times10^{-5}$ according to the Eq. (2) and corresponding R_n decrease in 4 times according to the Eq. (1). In other words, CNT contents increase is compensated by their geometry change, which is expressed by CNT ring-like structures radius R_n reduction.

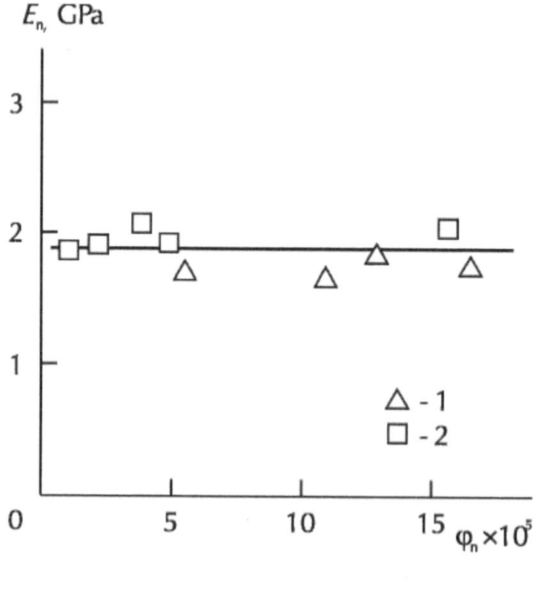

FIGURE 3 The dependence of elasticity modulus E_n on nanofiller volume contents φ_n for nanocomposites ED/CNT (1) and DDE/CNT (2).

KEYWORDS

- **Geometry**
- **Matrix polymer**
- **Ring-like structure**
- **US-processing**

REFERENCES

1. Eletsky, A. V. *Uspekhi Fizicheskikh Nauk*, **2007**, *177*(3), 223–274.
2. Sow, C.-H.; Lim, K.-Y.; Cheong, F.-C.; Saurakhiya, N. *Curr. Res. Nanotechnol.*, **2007**, *1*(2), 125–154.
3. Komarov, B. A.; Dzhavadyan, E. A.; Irzhak, V. I.; Ryabenko, A. G.; Lesnichaya, V. A.; Zvereva, G. I.; Krestinin, A. V. *Vysokomolek. Soed. A*, **2011**, *53*(6), 897–905.
4. Zhirikova, Z. M.; Aloev, V. Z.; Kozlov, G. V.; Zaikov, G. E. In *The Problems of Nanochemistry for the Creation of New Materials*. Lipanov, A. M.; Kodolov, V. I.; Kubica, S.; Zaikov, G. E., Eds.; Torun, REKPOL, 2012; pp. 37–43.
5. Kozlov, G. V.; Yanovsky, Ty. G.; Zhirikova, Z. M.; Aloev, V. Z.; Karnet, Yu. N. *Mekhanika Kompozitsionnykh Materialov i Konstruktsii*, **2012**, *18*(1), 131–152.
6. Mikitaev, A. K.; Kozlov, G. V.; Zaikov, G. E. *Polymer Nanocomposites: Variety of Structural Forms and Applications*. Science Publishers, Inc.: New York, 2008; 319p.
7. Sheng, N.; Boyce, M. C.; Parks, D. M.; Rutledge, G. C.; Abes, J. I. Cohen R. E. *Polymer*, **2004**, *45*(2), 487–506.
8. Bridge, B. *J. Mater. Sci. Lett.*, **1989**, *8*(2), 102–103.

CHAPTER 20

ON THE SYNERGETICS LAW AND NANOCOMPOSITES

G. M. MAGOMEDOV, KH. SH. YAKH'YAEVA, G. V. KOZLOV, and G. E. ZAIKOV

CONTENTS

20.1 INTRODUCTION

One of the main trends of nanoworld objects properties study is their features registration for on the basis of synergetics principles [1, 2]. At present it is known [3], that nanoparticles structure is defined by a chemical interactions between atoms nature forming them. The fundamental properties of nanoparticles, forming in strongly nonequilibrium conditions is their ability to:

(1) structures self-organization by adaptation to external influence;
(2) an optimal structure self-choice in bifurcation points, corresponding to preceding structure stability threshold and new stable formation;
(3) a self-operating synthesis (self-assembly) of stable nanoparticles, which is ensured by information exchange about system structural state in the previous bifurcation point at a stable structure self-choice in the following beyond it bifurcation point [3].

These theoretical postulates were confirmed experimentally. In particular, it has been shown [3–5] that nanoparticles sizes are not arbitrary ones, but change discretely and obey to synergetics laws. This postulate is important from the practical point of view, since nanoparticles size is the information parameter, defining surface energy critical level [3].

Let us consider these general definitions in respect to polymer particulate-filled nanocomposites [1], for which there are certain distinctions with the considered above criterions. As it is well-known [1], nanofiller aggregation processes in either form are inherent in all types of polymer nanocomposites and influence essentially on their properties. In this case, although nanofiller initial particles have size (diameter) less than 100 nm, but these nanoparticles aggregates can exceed essentially the indicated above boundary value for nanoworld objects [6]. Secondly, nanofiller particles aggregates are formed at the expense of physical interactions, but not chemical ones. Therefore the present work purpose is the study of the synergetics laws applicability for nanofiller aggregation processes and interfacial phenomena description in particulate-filled polymer nanocomposites on the example of nanocomposite polypropylene/calcium carbonate [7].

20.2 EXPERIMENTAL

Polypropylene (PP) of industrial production mark Kaplen 01 030 with weight – average molecular weight M_w of (2–3) $\times 10^5$ and polydispersity index of 4.5 was used as matrix polymer and nanodimensional calcium carbonate ($CaCO_3$) in the form of compound mark Nano-Cal R-1014 (China) with particles size of 80 nm and mass contents of 1–10 mass% was used as nanofiller.

Nanocomposites PP/$CaCO_3$ were prepared by components mixing in melt on twin-screw extruder Thermo Haake, model Reomex RTW 25/42, production of German Federal Republic. Mixing was performed at temperature 463–503 K and screw speed of 50 rpm during 5 min. Testing samples were obtained by casting under pressure method on casting machine Test Samples Molding Apparate RR/TS of firm Ray-Ran (Taiwan) at temperature 483 K and pressure 43 MPa.

The rester-type electron microscopy (REM) method was used for nanocomposites PP/$CaCO_3$ structure study. Study objects were prepared in liquid nitrogen with the purpose of microscopic sections obtaining. The scanning electron microscope with autoemissive cathode of high resolution JSM–7500F of firm JEOL (Japan) was used for microscopic sections surface images obtaining. Images were obtained in the mode of low-energetic secondary electrons, since this mode ensures the highest resolution.

Uniaxial tension mechanical tests have been performed on the samples in the shape of two-sided spade with sizes according to GOST 112 62–80. The tests have been conducted on universal testing apparatus Gotech Testing Machine CT-TCS 2000, production of German Federal Republic, at temperature 293 K and strain rate ~ 2×10^{-3} s^{-1}.

20.3 RESULTS AND DISCUSSION

A particulate nanofiller particles aggregate size (diameter) D_{agr} estimation can be performed according to the following formula [6]:

$$k(r)\lambda = \left[\left(\frac{25.1\pi D_{agr}^{1/3}}{W_n}\right)^{1/3} - 2\right]\frac{D_{agr}}{2}, \tag{1}$$

where $k(r)$ is an aggregation parameter, λ is distance between nanofiller particles, W_n is nanofiller mass contents in mass%.

In its turn, the value $k(r)\lambda$ is determined within the frameworks of strength dispersion theory with the help of the relationship [1]:

$$\tau_n = \tau_m + \frac{Gb_B}{k(r)\lambda}, \tag{2}$$

where τ_n and τ_m are yield stress in compression testing of nanocomposite and matrix polymer, accordingly, G is shear modulus, b_B is Burders vector.

The included in the equation (2) parameters are determined as follows. The general relationship between normal stress σ and shear stress τ has the following look [8]:

$$\tau = \frac{\sigma}{\sqrt{3}}. \tag{3}$$

Young's modulus E and shear modulus G are connected between themselves by the simple relationship [9]:

$$G = \frac{E}{d_f}, \tag{4}$$

where d_f is nanocomposite structure fractal dimension, which is determined according to the equation [9]:

$$d_f = (d-1)(1+\nu), \tag{5}$$

where d is the dimension of Euclidean space, in which fractal is considered (it is obvious, that in our case d=3), ν is Poisson's ratio, which is estimated by the mechanical testing results with the help of the relationship [10]:

$$\frac{\sigma_Y}{E_n} = \frac{1-2\nu}{6(1+\nu)}, \tag{6}$$

where σ_Y and E_n are yield stress and elasticity modulus of nanocomposite, respectively.

Burgers vector value b_B for polymeric materials is determined according to the equation [11]:

$$b_B = \left(\frac{60.5}{C_\infty} \right)^{1/2}, \text{Å}, \tag{7}$$

where C_∞ is characteristic ratio, connected with d_f by the equation [11]:

$$C_\infty = \frac{2d_f}{d(d-1)(d-d_f)} + \frac{4}{3}. \tag{8}$$

The calculation according to the equations (1)–(8) showed $CaCO_3$ nanoparticles aggregates mean diameter growth from 85 up to 190 nm within the range of W_n=1–7 mass% for the considered nanocomposites PP/$CaCO_3$. These calculations can be confirmed experimentally by the electron microscopy methods. In Fig. 1 the nanocomposites PP/$CaCO_3$ with nanofiller contents W_n=1 and 4 mass% sections electron micrographs are adduced. As one can see, if at W_n=1 mass% nanofiller particles are not aggregated practically, that is, their diameter is close to $CaCO_3$ initial nanoparticles diameter (~80 nm), then at W_n=4 mass% the initial nanoparticles aggregation is observed even visually and these particles aggregates sizes are varied within the limits of 80–360 nm. The adduced above estimations correspond to the results of calculation according to the Eq. (1) at the indicated $CaCO_3$ contents: 85 and 142 nm, accordingly. Hence, the considered above technique gives reliable enough estimations of nanofiller particles aggregates diameter.

It has been shown earlier on the example of different physical-chemical processes, that the self-similarity function has an iteration type function look, connecting structural bifurcation points by the relationship [12]:

$$A_m = \frac{\lambda_n}{\lambda_{n+1}} = \Delta_i^{1/m}, \tag{9}$$

where A_m is the measure of aggregate structure adaptability to external influence, λ_n and λ_{n+1} are preceding and subsequent critical values of operating parameter at the transition from preceding to subsequent bifurcation point, Δ_i is the structure stability measure, remaining constant at its re-organization up to symmetry violation, m is an exponent of feedback type; the value m=1 corresponds to linear feedback, at which transitions on other spatial levels are realized by multiplicative structure reproduction mechanism and at $m \geq 2$ (nonlinear feedback) – replicative (with structure improvement) one.

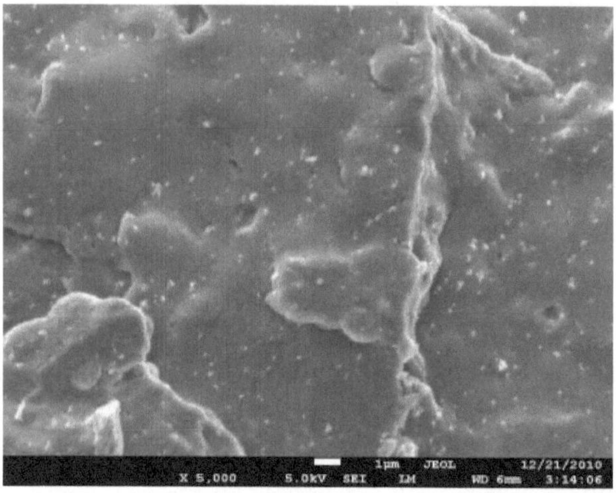

a)

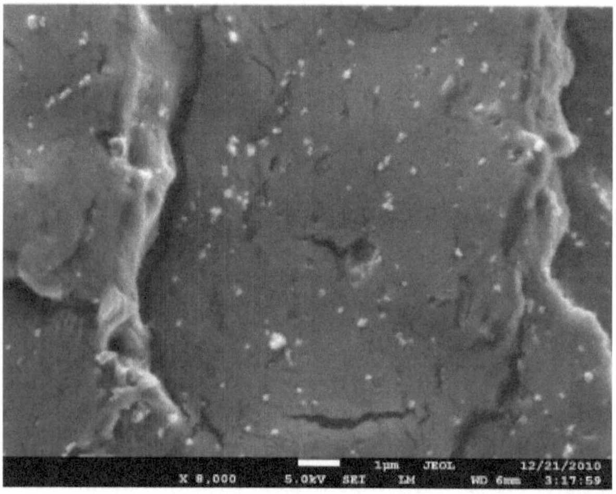

b)

FIGURE 1 Electron micrographs of sections of nanocomposites $PP/CaCO_3$ with nanofiller mass contents W_n=1 (a) and 4 (b) mass%.

Selecting as the operating parameter critical value nanoparticles aggregates diameter D_{agr} [3] at successive W_n change, the dependence of adaptability measure A_m on W_n can be plotted, which is shown in Fig. 2. As one can see, for the considered nanocomposites the condition is fulfilled

$$A_m = \frac{D_{agri}}{D_{agri+1}} = \text{const} = 0.899 \tag{10}$$

with precision of 2%. This means, that aggregation processes in the considered nanocomposites obey to the synergetics laws, although their aggregates size exceeds the boundary value of 100 nm for nanoworld [3]. Let us note the important aspect of the dependence $A_m(W_n)$, adduced in Fig. 2. The condition A_m=const is kept irrespective of gradation, with which W_n changes – 0.5 or 1.0 mass%.

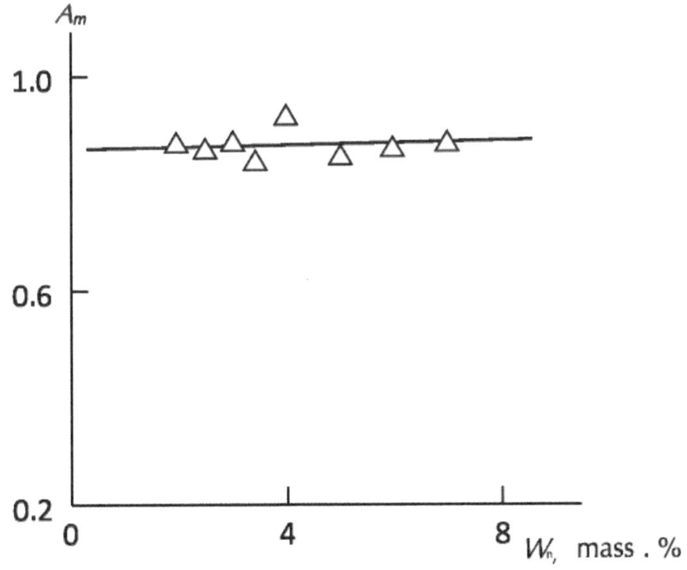

FIGURE 2 The dependence of adaptability measure A_m on nanofiller mass contents W_n for nanocomposites PP/$CaCO_3$.

As it is known [13], an interfacial layer in polymer nanocomposites can be considered as a result of two fractal objects (polymer matrix and nanofiller particles surface) interaction, for which there is the only linear scale l, defining these objects interpenetration distance [14]. Since the filler elasticity modulus is, as a rule, considerably higher than the corresponding parameter for polymer matrix, then the indicated interaction comes to filler indentation in polymer matrix and then $l=l_{if}$, where l_{if} is interfacial layer thickness [1, 13]. In this case it can be written [14]:

$$l_{if} = a\left(\frac{D_{agr}}{2a}\right)^{2(d-d_{surf})/d}, \tag{11}$$

where a is a lower linear scale of fractal behavior, which for polymeric materials is accepted equal to statistical segment length l_{st} [11], d_{surf} is nanofiller particles (aggregates of particles) surface fractal dimension.

l_{st} is determined according to the equation [15]:

$$l_{st} = l_0 C_\infty , \tag{12}$$

where l_0 is the main chain skeletal bond length, equal to 0.154 nm for PP [16].

The dimension d_{surf} is calculated in the following succession. First the nanofiller particles aggregate density ρ_n is estimated according to the formula [1]:

$$\rho_n = 188\left(D_{agr}\right)^{1/3}, \text{ kg/m}^3, \tag{13}$$

where D_{agr} is given in mcm and then the indicated aggregate specific surface S_u is determined [17]:

$$S_u = \frac{6}{\rho_n D_{agr}}. \tag{14}$$

And at last, the value d_{surf} calculation can be fulfilled with the help of the equation [1]:

$$S_u = 410\left(\frac{D_{agr}}{2}\right)^{d_{surf}-d}, \tag{15}$$

where S_u is given in m^2/g, D_{agr} – in nm.

The calculation according to the offered technique has shown l_{if} increase from 1.78 up to 5.23 nm at W_n enhancement within the range of 1–7 mass%. The estimations according to the Eq. (9), where as λ_n and λ_{n+1} the values $l_{if\,n}$ and $l_{if\,n+1}$ were accepted, showed that the following condition was fulfilled:

$$A_m = \frac{l_{if\,n}}{l_{if\,n+1}} = 0.880 \tag{16}$$

with the precision of 7%. Hence, an interfacial layers formation in polymer nanocomposites, characterizing interfacial phenomena in these nanomaterials, obeys to the synergetics laws with the same adaptability measure, as nanofiller particles aggregation. Nevertheless, it should be noted, that this analogy is not complete for the considered nanocomposites within the range of W_n=1–7 mass% the value D_{agr} increases in 2.24 times, whereas the value l_{if} does almost in three times.

Let us consider further the exponent m in the Eq. (9) choice, characterizing feedback type in aggregation process. As it has been noted above, this exponent is equal to 2 in case of aggregates structure improvement, which can be characterized by their fractal dimension d_f^{agr}. This dimension can be calculated with the help of the equation [18]:

$$\rho_n = \rho_{dens}\left(\frac{D_{agr}}{2a}\right)^{d_f^{agr}-d}, \tag{17}$$

where ρ_{dens} is massive material density, which is equal to 2000 kg/m^3 for $CaCO_3$, a is a lower linear scale of fractal behavior, accepted equal to 10 nm [19].

In Fig. 3 the dependence of d_f^{agr} on $CaCO_3$ mass contents W_n for the considered nanocomposites is adduced. As one can see, within the studied range of W_n the essential d_f^{agr} growth (from 2.34 up to 2.73 at general d_f^{agr} variation within the limit of 2.0 to 2.95 [9]) is observed, that can be classified as nanofiller particles aggregates structure improvement, as a minimum, by two reasons: their disaggregation level reduction and critical structural defect sizes decreasing [20]. Therefore, proceeding from the said above, it should be accepted that m=2, which according to the Eq. (9) gives Δ_i=0.808. Let us note, that this Δ_i value defines very stable nanostructures. So, for self-operating nano-solid solutions synthesis the values Δ_i=0.255–0.465 at m=2 were obtained and in addition it has been shown that an optimal technological regime indicator is Δ_i=0.465 attainment at nonlinear feedback (m=2) realization [4].

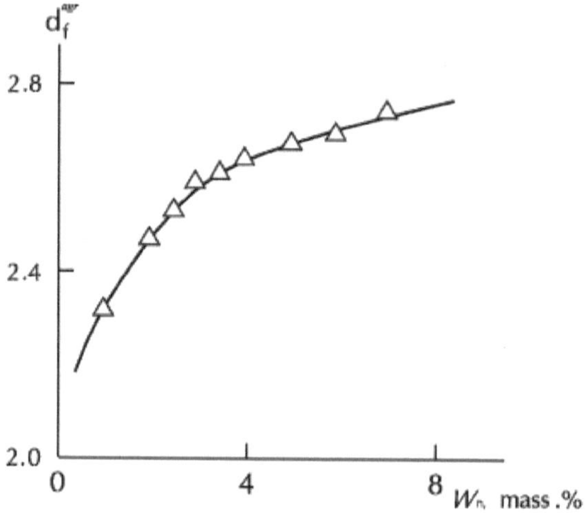

FIGURE 3 The dependence of nanofiller particles aggregates structure fractal dimension d_f^{agr} on its mass contents W_n for nanocomposites PP/$CaCO_3$.

Let us consider in conclusion the possibility of $CaCO_3$ nanoparticles aggregation process prediction within the frameworks of synergetic treatment. In Fig. 4 the comparison of $CaCO_3$ aggregates diameter values, calculated according to the Eqs. (1) D_{agr} and (9) D_{agr}^{syn} at A_m=const=0.899. As one can see, this comparison demonstrates very good conformity of nanofiller particles aggregates diameter, calculated by both indicated methods (the average discrepancy of D_{agr} and D_{agr}^{syn} makes up 2%).

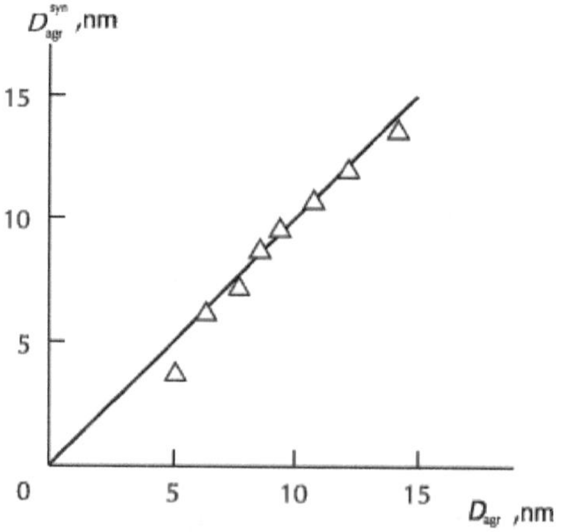

FIGURE 4 The comparison of nanofiller particles aggregates diameter, calculated according to the Eq. (1) D_{agr} and (9) D_{agr}^{syn}, for nanocomposites PP/$CaCO_3$.

KEYWORDS

- **Autoemissive cathode**
- **Burders vector**
- **Polypropylene**
- **Synergetic treatment**

REFERENCES

1. Mikitaev, A. K.; Kozlov, G. V.; Zaikov, G. E. *Polymer Nanocomposites: Variety of Structural Forms and Applications*. Nova Science Publishers, Inc.: New York, 2008; 319p.
2. Kozlov, G. V.; Mikitaev, A. K. *Polymers as Natural Nanocomposites: Unrealized Potential*. Lambert Academic Publishing: Saarbrücken, 2010; 323p.
3. Folmanis, G. E. *Proceedings of Intern. Interdisciplinary Symposium "Fractals and Applied Synergetics."* Publishers MSOU: Moscow, 2003; pp. 303–308.
4. Korzhikov, A. V.; Ivanova, V. S. *Proceedings of Intern. Interdisciplinary Symposium "Fractals and Applied Synergetics."* Publishers MSOU: Moscow, 2003; pp. 278–280.
5. Folmanis, G. E. *Proceedings of Intern. Interdisciplinary Symposium "Fractals and Applied Synergetics."* Publishers MSOU: Moscow, 2003; pp. 284–286.
6. Kozlov, G. V.; Sultonov, N. Zh.; Shoranova, L. O.; Mikitaev, A. K. *Naukoemkie Tekhnologii,* **2011**, *12*(6), 32–36.
7. Kozlov, G. V.; Mikitaev, A. K. *Nanotekhnologii. Nauka i Proizvodstvo,* **2011**, *4*, 57–63.
8. Honeycombe, R. W. K. *The Plastic Deformation of Metals*. Edward Arnold Publishers, Ltd.: Cambridge, 1968; 403p.
9. Balankin, A. S. *Synergetics of Deformable Body*. Publishers Ministry of Defence SSSR: Moscow, 1991; 404p.
10. Kozlov, G. V.; Sanditov, D. S. *Anharmonic Effects and Physical-Mechanical Properties of Polymers*. Nauka: Novosibirsk, 1994; 261p.
11. Kozlov, G. V.; Zaikov, G. E. *Structure of the Polymer Amorphous State*. Brill Academic Publishers: Utrecht, Boston, 2004; 465p.
12. Ivanova, V. S. *Strength and Fracture of Metal Materials*. Nauka: Moscow, 1992, 160p.
13. Kozlov, G. V.; Burya, A. I.; Lipatov, Yu. S. *Mekhanika Kompozitnykh Materialov*, **2006**, *42*(6), 797–802.
14. Hentschel, H. G. E.; Deutch, J. M. *Phys. Rev. A*, **1984**, *29*(12), 1609–1611.
15. Wu, S. J. *Polymer Sci.: Part B: Polymer Phys.*, **1989**, *27*(4), 723–741.
16. Aharoni, S. M. *Macromolecules*, **1983**, *16*(9), 1722–1728.
17. Bobryshev, A. N.; Kozomazov, V. N.; Babin, L. O.; Solomatov, V. I. *Synergetics of Composite Materials*. NPO ORIUS: Lipetsk, 1994; 154p.

18. Brady, L. M.; Ball, R. C. Nature, **1984**, *309*(5965), 225–229.
19. Avnir, D.; Farin, D.; Pfeifer P. Nature, **1983**, *308*(5959), 261–263.
20. Kozlov, G. V.; Yanovskii, Yu. G.; Zaikov, G. E. *Structure and Properties of Particulate-Filled Polymer Composites: The Fractal Analysis*; Nova Science Publishers, Inc.: New York, 2010; 282p.

CHAPTER 21

ON THE EFFECT OF SUBSTITUTES ON PHENOL TOXICITY

I. S. BELOSTOTSKAJA, E. B. BURLAKOVA, V. M. MISIN, G. A. NIKIFOROV, N. G. KHRAPOVA, and V. N. SHTOL'KO

CONTENTS

21.1 INTRODUCTION

Notwithstanding general point of view, peroxide oxidation of lipids is a physiological process of normal metabolizing tissues and follows the same laws as lipids oxidation does *in vitro* in liquid phase. Fine regulation of lipid oxidation in cell takes place both by chemical ways and with the help of enzymes. The rate of lipid oxidation is supported by a few systems, each of which is working within strick limits on relating stages of the process, having its own limiting reactions [1].

The lipid oxidation rate depends mainly on concentration changes of the substances, breaking the oxidation chains. Only natural antioxidants (basically phenols) have the ability to react with peroxide radicals. There is no interaction with pyroxile radicals of lipids in any other system of defense. None of the enzymes with antioxidant action break oxidation chain. Only natural antioxidants (AO) are inhibitors of radical processes. They can destroy redundant pyroxile radicals of lipids. Lipid oxidation rate is significantly affected by AO more than by other systems. It defines the unique role of natural AO in regulation the intensity of lipid oxidation [2].

The low level quantity of lipids peroxide in normal tissues is explained by well-balanced processes of peroxide formation and expenses. Breaking this balance results in changing the antioxidant status of the organism and may be the cause of some pathologies [2, 3]. In some studies it was established that AO therapy was necessary for treating many diseases, and free radical processes do play a certain role.

For inhibition reactions of chain oxidation *in vivo* as substances, phenol derivatives are widely used .They possess physiological activity in mammals. There appeared Biologically Active Supplements and medicines of the antioxidant effect, which are of great help to treat many diseases, for example: dibunoli, mexidol, emoxipin, trimexidini, probucoli [4–10]. In Russia the substances group of the antioxidant effect was chosen from Biologically Active Supplements [11].

For many substances toxicity was carefully studied; clinical, preclinical tests and experiments were carried out. However, in literature there is little information on toxicity of numerous synthesized phenols. They are joined into different groups and have various phenols, having substituents in position 2,4,6 to the group OH. For example, in Ref. [12], pulmonary toxicity of 24 phenols was studied on mice, which have various lipophil alkyl substituents in positions 2,4,6 and injected I.P in mammals.

In this study systematic competitive investigations were carried out for estimation toxicity of 27 phenol compounds, having in p-position substituents with different heteroatoms. They have the same formula:

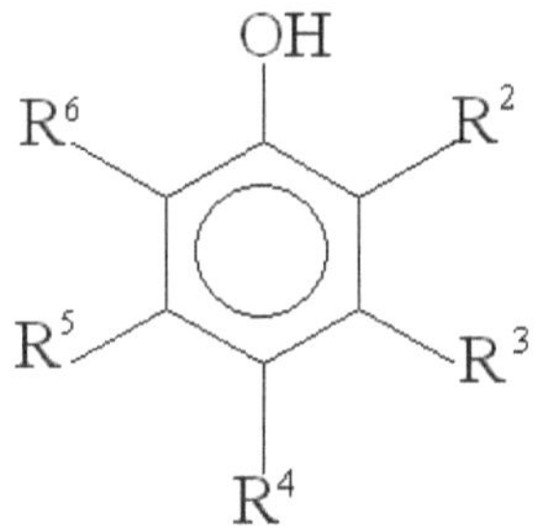

R^2, R^4, R^6 – various substituents; R^3=R^5=H

21.2 EXPERIMENTAL PART

Synthesis of studied phenols was described in [4, 13, 14]. Toxicity of all compounds was estimated by a value LD_{50}, measured in mg/kg weight of animal. For some compounds, values as much as possible transferable dose (MTD) and LD_{100} were additionally determined. Toxicity was determined on mice (males) of the line *Bulb* (mass 18–22 gr) with a single I.P. These mammals were housed in standard conditions of vivarium. Water-soluble preptions were injected as solutions of distilled water; liposoluble – 10% in solution Twin-80, used as solubilisator. Each doze off this preption was tested on not less than 4 mammals. For calculation, Beren's method was used (the method "frequency accumulation"), because it was simple and rather reliable [15].

21.3 RESULTS AND DISCUSSIONS

All investigated substances are resulted in Table 1. For better comparison of the obtained results on toxicity, investigated compounds were grouped separately in Tables 2–9. It was necessary to estimate toxicity of the substituted phenols depending on:

TABLE 1 Toxicity of the phenols containing various substituents. $R^3 = R^5 = H$.

№ patt.	Substituents			MTD, mg/kg	LD_{50}, mg/kg	LD_{100}, mg/kg
	R^2	R^4	R^6			
1	$C(CH_3)_3$	CH_3	$C(CH_3)_3$	250	375	650
2	$C(CH_3)_3$	NH_2	$C(CH_3)_3$	40	80	100
3	$C(CH_3)_3$	CH_2NH_2	$C(CH_3)_3$	30	70	130
4	$C(CH_3)_3$	$(CH_2)_2NH_2$	$C(CH_3)_3$	60	75	80
5	$C(CH_3)_3$	$(CH_2)_3NH_2$	$C(CH_3)_3$	50	60	–
6	$C(CH_3)_3$	$(CH_2)_4NH_2$	$C(CH_3)_3$	–	50	–
7	$C(CH_3)_3$	$CH(CH_3)NH_2$	$C(CH_3)_3$	80	105	130
8	$CH(CH_3)_2$	$CH(CH_3)NH_2$	$C(CH_3)_3$	30	80	100
9	$CH(CH_3)_2$	$CH(CH_3)NH_2$	$CH(CH_3)_2$	27	35	45
10	CH_3	$CH(CH_3)NH_2$	$C(CH_3)_3$	–	50	100
11	$C(CH_3)_3$	$CH(C_2H_5)NH_2$	$C(CH_3)_3$	40	70	110
12	H	$(CH_2)_3NH_2$	$C(CH_3)_3$	100	165	–
13	$C(CH_3)_3$	$CH_2NHCOCH_3$	$C(CH_3)_3$	350	425	–
14	$C(CH_3)_3$	$(CH_2)_2NHCOCH_3$	$C(CH_3)_3$	–	175	–
15	$C(CH_3)_3$	$(CH_2)_3NHCOCH_3$	$C(CH_3)_3$	75	125	–
16	$C(CH_3)_3$	CN	$C(CH_3)_3$	300	450	525
17	$C(CH_3)_3$	CH_2CN	$C(CH_3)_3$	50	95	180
18	$C(CH_3)_3$	$(CH_2)_2CN$	$C(CH_3)_3$	200	360	475
19	H	$(CH_2)_2CN$	$C(CH_3)_3$	250	285	350
20	H	$(CH_2)_2CN$	H	200	302	400
21	CH_3	CH_2CN	$C(CH_3)_3$	50	152	250
22	CH_3	CN	$C(CH_3)_3$	50	150	–
23	H	CN	$C(CH_3)_3$	250	185	–
24	H	CN	H	200	300	–
25	$C(CH_3)_3$	OH	$C(CH_3)_3$	–	390	–
26	$C(CH_3)_3$	$(CH_2)_2OH$	$C(CH_3)_3$	–	300	–
27	$C(CH_3)_3$	$(CH_2)_3OH$	$C(CH_3)_3$	–	225	–

(1) the type of p-substituents under identical 2,6-di-tert-butyl substituents ;
(2) spacer length $-(CH_2)_n-$ between phenol ring and p-substituents under identical 2,6-di-tert-butyl substituents;
(3) the type of both o-substituents under identical p-substituents.

First, phenol toxicity was compared for various p-substituents and identical 2,6-di-tert-butyl substituents, possessing maximum screening effect (Table 2). Toxicity estimated by the value LD_{50} was found to decrease in a row (in brackets values LD_{50} were given) of 2,6-di-tert-butylphenols having next n-substituents.

TABLE 2 Dependence of toxicity of the phenols having identical assistants $R^2=R^6 = C(CH_3)_3$, from type of p– substituents; $R^3=R^5=H$.

Substituents			MTD, mg/kg	LD_{50}, mg/kg	LD_{100}, mg/kg
R^2	R^4	R^6			
$C(CH_3)_3$	NH_2	$C(CH_3)_3$	40	80	100
$C(CH_3)_3$	CH_2NH_2	$C(CH_3)_3$	30	70	130
$C(CH_3)_3$	CH_3	$C(CH_3)_3$	250	375	650
$CH(CH_3)_2$	OH	$CH(CH_3)_2$	–	390	–
$C(CH_3)_3$	$C(CH_3)_3$	$C(CH_3)_3$	250	400	–
$C(CH_3)_3$	CN	$C(CH_3)_3$	300	450	525

$$-NH_2\ (80) \approx -CH_2NH_2\ (70) > -CH_3\ (375) > -OH\ (390) \approx -C(CH_3)_3\ (400) > -CN\ (450)$$

Toxicity decreasing of some phenol compounds from this row was approved with increasing the value of another parameter toxicity-MTD (in brackets the value MTD is given).

$$-NH_2\ (40) \approx -CH_2NH_2\ (30) >> -CH_3\ (250) = -C(CH_3)_3\ (250) > -CN\ (300\text{–}350)$$

The same decrease in toxicity was being observed for parameter LD_{100} in a row.

$$-NH_2\ (110) > -CH_2NH_2\ (130) > > -CN\ (525) > -CH_3\ (650)$$

For phenols, having identical 2,6-di-tert-butyl substituents, dependence of their toxicity on the distance of the functional p-substituents from benzene ring was investigated (Table 3). The tendency was found to increase phenol toxicity connected with the spacer length $-(CH_2)_n-$ for all investigated types of substituents ($-NH_2$, $-NHCOCH_3$, $-OH$, $-CN$) both electronodonors and electronoacceptors (Table 3).

TABLE 3 Dependence of toxicity of phenols on remoteness of p-substituents on a benzene ring. $R^4 = (CH_2)_nX$; $R^2=R^6= C(CH_3)_3$; $R^3=R^5=H$.

No	X=CN			$X=NH_2$			$X=NHCOCH_3$			X=OH
	MTD, mg/kg	LD_{50}, mg/kg	LD_{100}, mg/kg	MTD, mg/kg	LD_{50}, mg/kg	LD_{100}, mg/kg	MTD, mg/kg	LD_{50}, mg/kg	LD_{100}, mg/kg	LD_{100}, mg/kg
0	300	450	525	40	80	100	–	–	–	390
1	50	95	180	30	70	130	350	425	500	–
2	200	360	475	60	75	80	–	175	–	300
3	–	–	–	50	60	–	75	125	200	225
4	–	–	–	–	50	–	–	–	–	–

However, for 2,6-di-tert-butylphenol with p-substituent – CN, separated only by one group $-CH_2-$ (n=1) from benzene ring, a sharp toxicity increase, as compared with phenol toxicity, having n=0, was being observed. Further, for substances with the increasing bridges $-(CH)_n-$ toxicity sharply diminishes. But it remained greater than the phenol toxicity did, having p-substituent – just near benzole ring (n=0). It is interesting that the observed extreme dependence was repeating for all values, characterizing the toxicity of these compounds (MTD, LD_{50}, LD_{100}).

The observed occurrence of extreme phenol toxicity, having methylene (or methyl) group in p-position, could possibly be explained by greater ability of these phenols in comparison with other phenols. Thus there can be reactions of appearing of phenol radicals with the following reaction of demerization or reaction of disproportionation with formation cyclohexadienone-2,5 [4, 16, 17].

For 2,6-di-tert-butylphenol with p-substituent $-NH_2$ the same tendency of toxicity increase, according to increase of spacer length $-(CH_2)_n-$ has been observed (Table 3). Toxicity increase was observed for this phenol (n=1) with the help of values MTD and LD_{50}. It was interesting that sharper toxicity increase has been observed for similar phenol, but with CN substituent.

Identical dependence of sufficient toxicity increase with the spacer length $-(CH_2)_n-$ was observed for 2,6-di-tert-butylphenol having p-substituents $-(CH_2)_nNHCOCH_3$ (Table 3).

Symmetric 2,6-di-tert-butylphenol (as an example), containing group $-CH_2NH_2$ in p-position, toxicity dependence on the type of α-substituents in p-methylene group was investigated (Table 4). It was found phenol toxicity to decrease after introduction CH_3 group in α-position of the substitute CH_2NH_2. The introduction of C_2H_5 group returns the substituted phenol toxicity to its initial meaning.

TABLE 4 Toxicity of the phenols containing various R^4 substituents. $R^3 = R^5 = H$.

№ patt.	Substituents			MTD, mg/kg	LD_{50}, mg/kg	LD_{100}, mg/kg
	R^2	R^4	R^6			
1	$C(CH_3)_3$	CH_2NH_2	$C(CH_3)_3$	30	70	130
2	$C(CH_3)_3$	$CH(CH_3)NH_2$	$C(CH_3)_3$	80	105	130
3	$C(CH_3)_3$	$CH(C_2H_5)NH_2$	$C(CH_3)_3$	40	70	110

It was noted that the introduction of acetyl group to N-position of symmetrical 4-aminomethyl-2,6-di-tert-butylphenol has decreased the substance toxicity according to the value LD_{50} in 6 times (Table 5).

TABLE 5 Toxicity of the phenols containing various R^4 substituents. $R^3 = R^5 = H$.

№ patt.	Substituents			MTD, mg/kg	LD_{50}, mg/kg	LD_{100}, mg/kg
	R^2	R^4	R^6			
1	$C(CH_3)_3$	CH_2NH_2	$C(CH_3)_3$	30	70	130
2	$C(CH_3)_3$	$CH_2NHCOCH_3$	$C(CH_3)_3$	350	425	500

It was interesting to study the dependence of phenol toxicity on the type of substituents in o-positions.

For phenols (Table 6), having identical p-substituent $–CH(CH_3)NH_2$, it was investigated phenol toxicity influenced by the substituents of the types R_2 and R_6 (tert-butyl, iso-propyl, methyl). Over all the values, describing toxicity (MTD, LD_{50}, LD_{100}), phenol toxicity increases with the quantity decrease of tert-butyl groups, which are in both o-positions of these phenols in a row.

tert-butyl + tert-butyl < tert-butyl + iso-propyl < tert-butyl + methyl < iso-propyl + iso-propyl

Such toxicity increase can be explained both by decrease of steric hindrance, created to electronodonor group –OH by the volume tert-butyl substituents, and by hyper conjugation effect of o-substituents in OH-group [18].

TABLE 6 Toxicity of the phenols containing various R^2, R^6 substituents. $R^3 = R^5 = H$.

№ patt.	Substituents			MTD, mg/kg	LD_{50}, mg/kg	LD_{100}, mg/kg
	R^2	R^4	R^6			
1	$C(CH_3)_3$	$CH(CH_3)NH_2$	$C(CH_3)_3$	80	105	130
2	$CH(CH_3)_2$	$CH(CH_3)NH_2$	$C(CH_3)_3$	30	80	100
3	CH_3	$CH(CH_3)NH_2$	$C(CH_3)_3$	–	50	100
4	$CH(CH_3)_2$	$CH(CH_3)NH_2$	$CH(CH_3)_2$	25	35	45

For n-CN of the substituted phenols (Table 7) after decreasing the number of tert-butyl substituents in o-position it was found the sharp toxicity increase with the following decrease of it. Nevertheless the toxicity of CN-substituted 2,6-di-tert-butylphenol is less than toxicity of others CN-substituted phenol derivatives.

TABLE 7 Toxicity of the phenols containing various R^2, R^6 substituents. $R^3 = R^5 = H$.

№ patt.	Substituents			MTD, mg/kg	LD_{50}, mg/kg	LD_{100}, mg/kg
	R^2	R^4	R^6			
1	$C(CH_3)_3$	CN	$C(CH_3)_3$	300	450	525
2	CH_3	CN	$C(CH_3)_3$	50	150	–
3	H	CN	$C(CH_3)_3$	250	185	–
	H	CN	H	200	300	–

The analogous dependence has been observed under investigation according to values MTD, LD_{50} and LD_{100} of three phenols, having group-CH_2CH_2CN in p-position. They differ in quantity of tert-butyl substituents in o-position (Table 8). Toxicity increasing as found when the quantity of tert-butyl substituents decrease. However, extreme dependence of mono-tert-butyl-substituted phenol toxicity has been observed.

TABLE 8 Toxicity of the phenols containing various R^2, R^6 substituents. $R^3 = R^5 = H$.

№ patt.	Substituents			MTD, mg/kg	LD_{50}, mg/kg	LD_{100}, mg/kg
	R^2	R^4	R^6			
1	$C(CH_3)_3$	$(CH_2)_2CN$	$C(CH_3)_3$	200	360	475
2	H	$(CH_2)_2CN$	$C(CH_3)_3$	250	285	350
3	H	$(CH_2)_2CN$	H	200	300	400

So, the least toxicity of sterically hindred 2,6-di-tert-butilphenols has been observed for a few phenol groups with different p-substitutes. It can be explained by their worst bioaccessibility [4].

However, only for 2,6-di-tert-butyl-4-cyanophenol and 2-tert-butyl-6-methyl-4-cyanophenol (Table 9) reverse dependence was being observed: the change of one tert-butyl group to methyl group resulted in essential toxicity decrease. This was proved by the growth of both measured values LD_{50} and LD_{100}.

TABLE 9 Toxicity of the phenols containing various R^2, R^6 substituents. $R^3 = R^5 = H$.

№ patt.	Substituents			MTD, mg/kg	LD_{50}, mg/kg	LD_{100}, mg/kg
	R^2	R^4	R^6			
1	$C(CH_3)_3$	CH_2CN	$C(CH_3)_3$	50	95	180
2	CH_3	CH_2CN	$C(CH_3)_3$	50	150	250

The observed property dependence on the structure for various o-substituted phenols with polar substitutes in p-position does not correlate with the same dependence observed for those with nonpolar alkyl substitutes

[12]. Perhaps after introducing the substituents of various types to phenols, the competition of a few tendencies is being observed, each of them influencing on selected toxicity of the investigated substances [19].

First, the change of lipophility takes place, which results in the change of phenol transport through lipid membrane layers to the following receptors. For example, in Ref. [20] the influence of AO on the activity of lipodependent proteinkinase was noted not only through lipid membrane changes, but through direct interaction with enzyme. As the substituted phenols are accumulated in lipids, so lipophility dropping of substituted phenol should promote it to supply those fields of lipid membranes, which are enriched with oxidized lipids, and the products of their metabolism [19].

Second, the structure change results in the reaction ability change in substituted phenol groups, responsible for metabolizing with these phenols. OH-groups, particularly, are responsible for conjugation reactions glucuronic acid and sulfates_ [21, 22]. In turn heteratom groups in p-position take place participate in the processes on the first stage of metabolism with the following stage of metabolism by conjugation with possible metabolites – 2,5-cyclohexadienones formation [4, 12].

Third, the structure change of substituted phenols results in constant of ionization change of OH-group. It is sufficiently influences upon selective toxicity, connected with substrate ionization [19].

So, the above stated factors make difficult to find the true phenol toxicity dependence on structure. Besides it is difficult to compare these results with those obtained for phenols with lipophilic alkyl substituents [12]. Indeed, from these results in [12], it follows that 2,6-di-tert-butyl-4-methylphenol is one of the strongest toxicants. Similar principal result difference may be explained that strong phenol toxicity has been studied on mice in this study, and in [12] specific mice organs (lungs) have been investigated there; its affects on lungs.

For full explanation obtained, it is desirable to investigate specific affect of these phenols on various targets in some organs with help of pharmakokinetics. Nevertheless, the obtained results in this study may be of help in pharmacological investigations and planning studies connected with synthesis of new, nontoxical, biologically active phenol compounds.

KEYWORDS

- **Bulb**
- **Frequency accumulation**
- **Pharmacokinetics**
- **Proteinkinase**

REFERENCES

1. Burlakova, E. B.; Khrapova, N. G. *Russian Chem. Rev.,* **1985**, *54*(9), 1540–1558.
2. Khrapova, N. G. In *Chemical and Biological Kinetics. New Horizons. V. 2. Biological Kinetics*, Burlakova, E. B., Eds.; Khimiya: Moscow, 2005; pp. 46–60.
3. Burlakova, E. B.; Alesenko, A. V.; Molochkina, E. M.; et al. *Bioantioxidant in Radiation Sickness and Malignant Disease*, Burlakova, E. B., Eds.; Moscow: Nauka, 1975.
4. Ershov, V. V.; Nikiforov, A. G.; Volodkin, A. A. *Sterically Hindred Phenols;* Khimiya, Moscow, 1972.
5. Zarudij, F. S.; Gilmutdinov, Z. G.; Zarudy, R. F.; et al. *Khimiko-Farmacevtichesky Zh.*, **2001**, *35*(3), pp. 42–48.
6. Zorkina, A. V.; Kostin, J. V.; Inchina, V. I.; et al. *Khimiko-Farmacevtichesky Zh.*, **1998**, *32*(5), pp. 3–6.
7. Kotljarov, A. A.; Smirnov, L. D.; Smirnova, L. E.; et al. *Exp. Clin. Pharmacol.: Two-Month Scientific-Theoretical Magazine,* **2002**, *65*(5), 31–34.
8. Zenkov, N. K.; Kandalintseva, N. V.; Lankin, V. Z.; et al. *Phenolic Bioantioxidants*, SO RAMN, Novosibirsk, 2003.
9. Burlakova, E. B. In *Chemical and Biological Kinetics. New Horizons. V. 2. Biological Kinetics,* Vol. 2, Burlakova, E. B., Eds.; Khimiya: Moscow, 2005, pp. 10–45.
10. Kravchuk, E. A.; Keselyova, T. N.; Ostrovsky, M. A.; et al. *Refrac. Surg. Ophthalmol.*, **2008**, *8*(1), 36–41.
11. *The Russian Federal Register of Biologically Active Additives to Food*, 2nd ed., processed and added. 2001.
12. Mizutani, T.; Ishida, I.; Yamamoto, K.; Tajima, K. *Toxicol. Appl. Pharmacol.*, **1982**, 62, 273–281.
13. Ershov, V. V.; Belostotskaja, I. S. *Izv. Akad. Nauk SSSR, Ser. Khim.* **1965**, *7*, 1301–1303.
14. Belostotskaja, I. S.; Volodkin, A. A. *Izv. Akad. Nauk SSSR, Ser. Khim.* **1966**, *10*, 1833–1835.
15. Belen'ky, M. L. *Ed. Acad. Sci. Latv. SSR*, **1959**.
16. Roginsky, V. A. *Phenolic Antioxidants: Reactionary Ability and Efficiency;* Nauka: Moscow, 1988.
17. Takahashi, O.; Hiraga, K. *Fd. Cosmet. Toxicol.*, **1997**, *17*, 451–454.

18. Temnikova, T. I. *Course of Theoretical Bases of Organic Chemistry;* Khimiya: Leningrad, 1968.
19. Albert, A. *Selective Toxicity*, 7th ed. Chapman and Hall: London, 1985.
20. Hohlov, A. P. *Bull. Exp. Biol. Med.*, **1988**, *10*, 440–444.
21. Kabiev, K, O.; Balmuhanov, S. B. *Natural Phenols – A Perspective Class of Antineoplastic and Radio Protective Connections;* Meditsina: Moscow, 1975.
22. Vergejchik, T. H. *Toxicological Chemistry*; MEDpress-inform: Moscow, 2009.

CHAPTER 22

ON NEW APPLICATION OF NANOFILLERS

A. A. OLKHOV, A. L. IORDANSKII, R. YU. KOSENKO,
YU. S. SIMONOVA, and G. E. ZAIKOV

CONTENTS

22.1 INTRODUCTION

In last decade has occurred significant number of the works devoted to the quantitative description of processes of liberation of low molecular weight medicinal substances from polymeric matrixes. One of the major problems of modern medicine is creation of new methods of the treatment based on purposeful local introduction of medical products in a certain place with set speed. The basic requirement shown to the carrier of active material, its destruction and a gradual conclusion from an organism is [1, 2].

Medicinal forms (m.f.) with controllable liberation (medicinal forms with operated liberation, medicinal forms with programmed liberation) – group of medicinal forms with the modified liberation, characterized by elongation of time of receipt m.f. in a biophase and its liberation, corresponding to real requirement of an organism.

The decision of a problem of controllable liberation m.f. will allow designing a polymeric matrix of various degree of complexity, setting thereby programmed speed of allocation of a medicine in surrounding biological environment. Regulation of transport processes in polymers taking into account their morphological features is one of actual problems physical chemistry polymers. Research of interaction of polymeric materials with water important for many reasons, but the main things there are two: this interaction, which plays the important role in the processes providing ability to live of the person. It influences operational properties of polymeric materials.

The mechanism of address delivery m.f. includes diffusive process of its liberation of a polymeric system, therefore, research study and structures of a composite with transport characteristics m.f. are a necessary condition of creation new components means. [3]. It is known that essential impact makes on diffusive properties of polymeric films morphology and crystallinity of the components forming a composite. Besides, at composite formation the nature of the filler changing not only structure of the most polymeric matrix is very important.

The purpose of the present work was studying of influence finely divided schungite on structure, mechanical characteristics and kinetics of liberation of a medicinal drug – furacilinum from films polyhydroxybutyrate.

22.2 RESEARCH METHODS

As a polymeric matrix in work used polyhydroxybutyrate (PHB) – biodecomposed polymer. Thanking these properties PHB it is applied as packing, to the biomedical appointment, self-resolving fibers and films, etc.

Basic properties PHB: melting point of 173–180 wasps, temperature of the beginning of thermal degradation of 150 wasps, degree of crystallinity 65–80%, molecular weight 10^4–10^6 g/mol, ultimate tensile strength 40 MPa, an elastic modulus 3.5 GPa, tensile elongation 6–8%.

The basic lack of products from PHB is low specific elongation at the expense of high crystallinity and formation of large sphemlitic aggregates.

Introduction in PHB, for example, finely divided mineral particles of filler can reduce fragility at the expense of formation of fine-crystalline polymer structure.

As a medical product in work used antibacterial means – furacilinum (m.f.). Filler – fine-grained schungite technical specifications 2169–001–5773937–natural formation, on 30% consisting of carbon and 70% of silicates, the Zagozhinsky deposit [4].

It is necessary to note high probability of presence in schungit carbon of appreciable quantities фуллеренов, their chemical derivative and molecular complexes which including can play a role of structural plastifiers.

In work investigated a film drug of matrix type which can be used наружно at direct superposition on a wound or internally, for example, at intravitreal in a fabric [5].

Film made as follows:

Powder PHB filled in with chloroform (500–600 mg PHB on 15–20 ml $CHCl_3$) and agitated on a magnetic stirrer before formation of homogeneous weight (~10 mins). Then a solution lead up to boiling and at a working agitator brought furacilinum, and behind it – schungite, agitated even 10 min). A hot solution filtrated through two beds of kapron and poured out in Petri dish D = 9 sm which seated in a furnace at temperature 25°C, densely covered with the second Petri dish and dried up to constant weight.

For all samples defined the maintenance furacilinum, liberation furacilinum in water, density of samples, physical-mechanical characteristics on corresponding state standard specification, technical specifications and to laboratory techniques [6].

22.3 RESULTS AND DISCUSSION

In Fig. 1, typical dependence of size of liberation (M_{∞}/M_t) furacilinum from time is shown at the various maintenance schungite. From it follows that all values of desorption monotonously increase in due course releases.

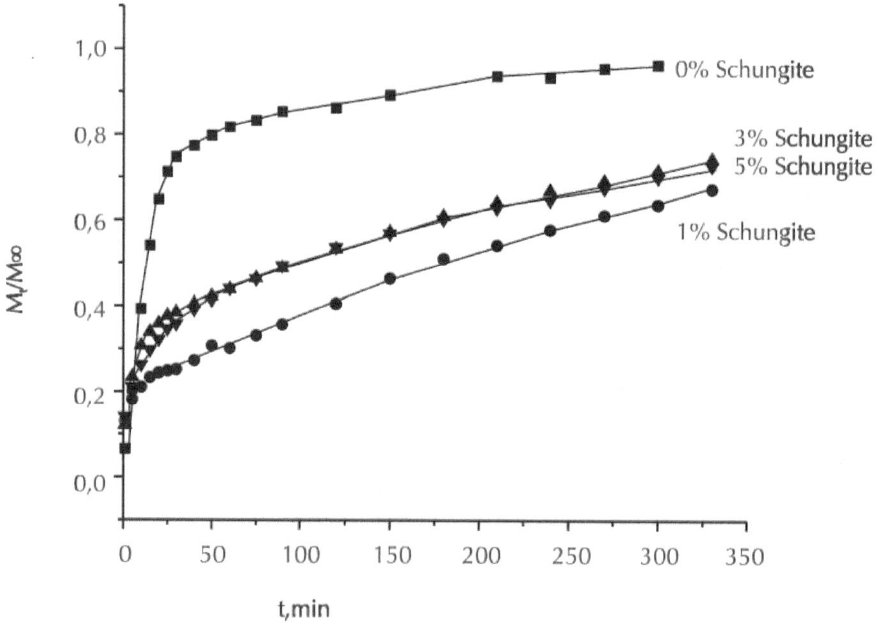

FIGURE 1 Kinetic curve releases of a medical product from films PHB at concentration furacilinum 3%

Apparently from drawing, schungite and, hence, appreciable impact on diffusion rate f.m. makes on its speed of liberation. To maintenance growth schungite there is a significant falling of diffusivities that is reduction of speed of allocation of medicinal substance. This effect can be connected with immobilization (interaction) of molecule m.f. with a surface schungite (with oxygen groups Si-O_2). It is follow-up possible to expect, that impenetrable particles schungite represent a barrier to diffusion M.F., and, hence, speed of transport in a composite drops.

On Fig. 2a, dependence of density of films PHB on the maintenance schungite in absence furacilinum for single-stage and two-phasic films is represented.

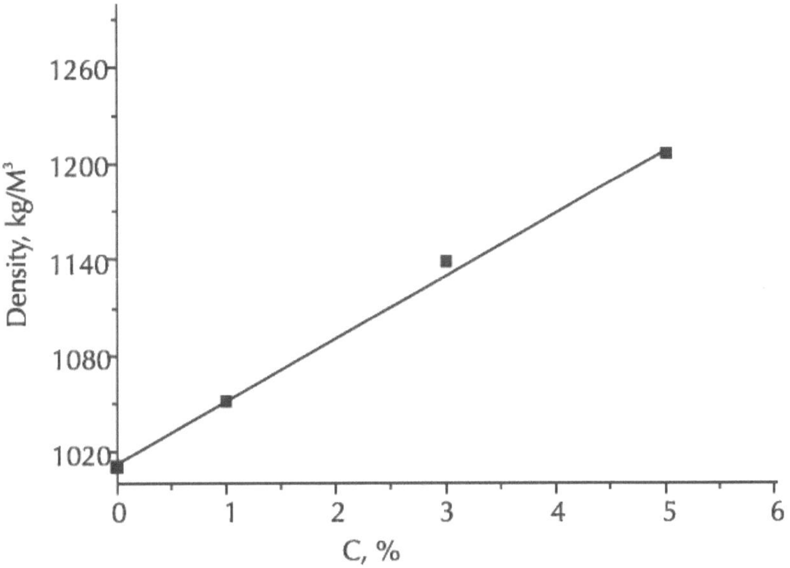

FIGURE 2A Dependence of density of films PHB on the maintenance schungite without furacilinum.

From drawing it is visible, that to maintenance growth schungite there is directly proportional increase in density of films. Increase of density of composite films is caused by increase in their concentration of more dense filler.

On Fig. 2b, dependence of density of films PHB on the maintenance schungite Is presented at 5% furacilinum. From drawing it is visible, that with maintenance increase schungite, the density decreases. It is possible to explain density decrease by a synergism resulting interference schungite and furacilinum. Yielded synergism it is possible to explain formation of large associates furatsilin-shungit. Associates can arise at the expense of adsorption of molecules furacilinum on particles schungite, having an active surface, in solvent at formation of composite films. Associates can interfere with crystallization PHB and by that to increase quantity of the loosened amorphous phase of polymer.

Under condition of a share constancy schungite in composites with maintenance increase furacilinum the density varies like the curve on Fig. 2b.

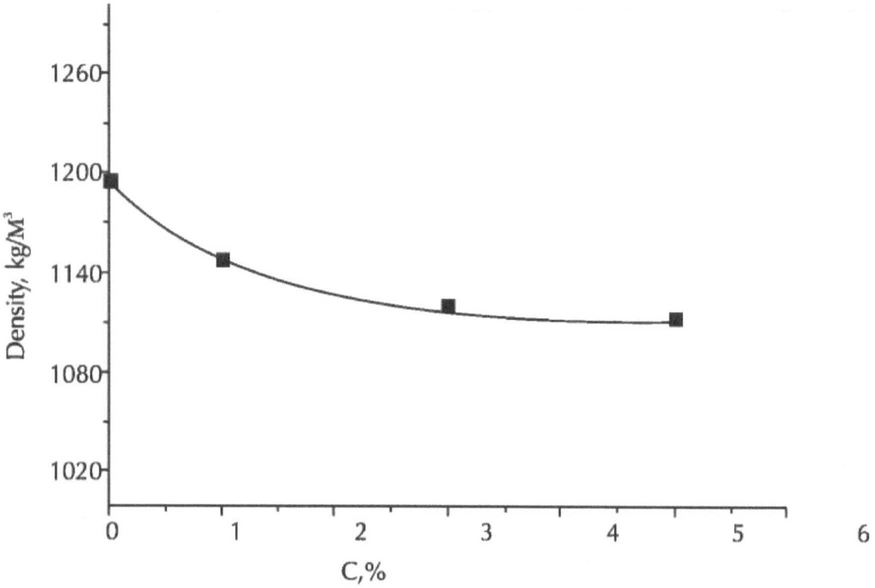

FIGURE 2B Dependence of density of films PHB on the maintenance schungite at 5% furacilinuma.

So at introduction 0.5% furacilinum are more narrow there is a reduction of density of a composition by 1%, and then at the further increase the maintenance furacilinum to 5% the density of films drops on 3–4%. Density reduction is caused, possibly, polymeric structures PGB and formation of associates furatsilin-shungit which, increasing quantity of an amorphous phase, create additional free volume in polymer – a matrix.

The establishment of influence of the maintenance schungite and furacilinum on mechanical characteristics of composite films was one of the purposes supplied in work. Dependence of ultimate tensile strength of films PHB on maintenance m.f. is presented in Fig. 3a.

As follows from Fig. 3a, hardness increases from the minimum value 1.71 MPa in absence schungite to 2.5 MPa, i.e. increases in 1.5 times. The hardness increase, possibly, is caused by formation of hydrogen bridges between furan groups LV and It - groups PHB. In more details the mechanism of formation of associates will be stated further.

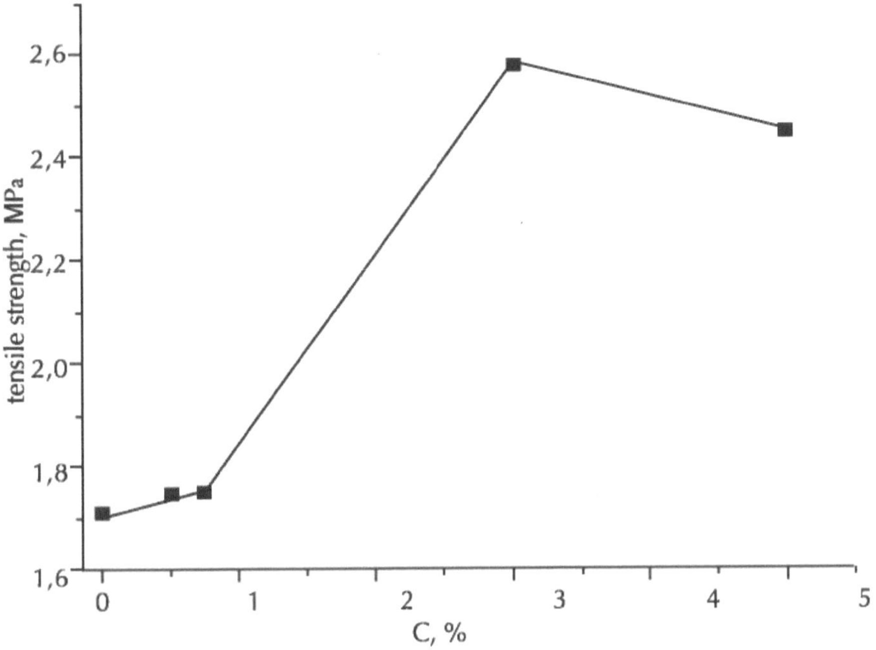

FIGURE 3A Dependence of tensile strength of films PHB on the maintenance furacilinum without schungite.

On Fig. 3b, dependence of hardness of films PHB on the maintenance schungite is presented at 5% furacilinum in films.

From Fig. 3b, also the increase in hardness with maintenance increase schungite more, than in 2 times is visible, that also speaks formation of fine-crystalline structure of films a preparation stage. As it was already marked above, additional hardening also follows the account of formation of associates PHB-furacilinum-schungite.

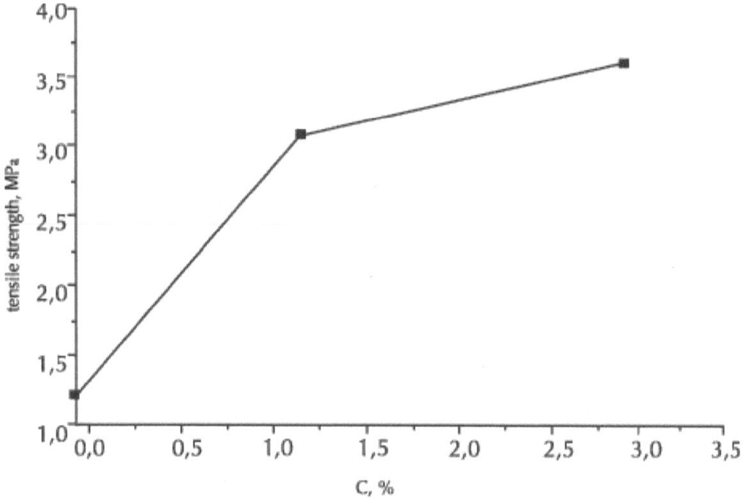

FIGURE 3B Dependence of hardness of films PHB on the maintenance schungite at 5% furacilinum.

Schungite presence in a composition result in to hardening of films PHB (data Fig. 4).

On Fig. 4 dependence of hardness of films PHB on the maintenance schungite is presented.

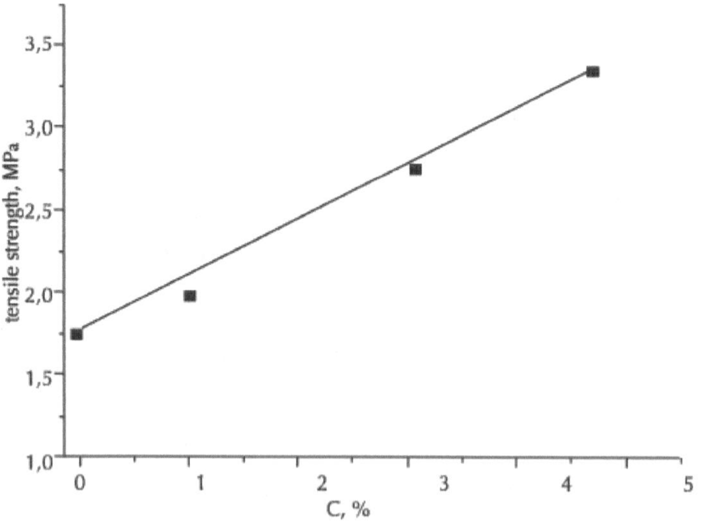

FIGURE 4 Dependence of hardness of films PHB on the maintenance schungite without furacilinum.

The increase in hardness of films schungite occurs to maintenance increase in 2.5 times. Apparently, at a stage of preparation of films schungite forms more perfect heterogeneous fine-crystalline structure. Schungite, being the nucleation center, creates set of nuclei, thereby, doing structure of films more ordered, that promotes hardness increase [7].

Considering collaterally Figs. 3 and 4 it is possible to assume following structural changes of a composite at influence on PHB schungite and m.f.:

Apparently, in composites PHB – schungite-furacilinum difficult associates, that is, formation of hydrogen bridges between oxygen-containing groups furacilinum and similar groups PHB are formed. As a result formed communications act in a role of the crosslinking localized in amorphous regions of a polymeric matrix. Presence of such crosslinking should result in to increase in hardness of composite films. Under our assumptions, at first molecules furacilinum form hydrogen bridges with trailer It-groups, and then are adsorbed by other end on a surface schungite (Fig. 5).

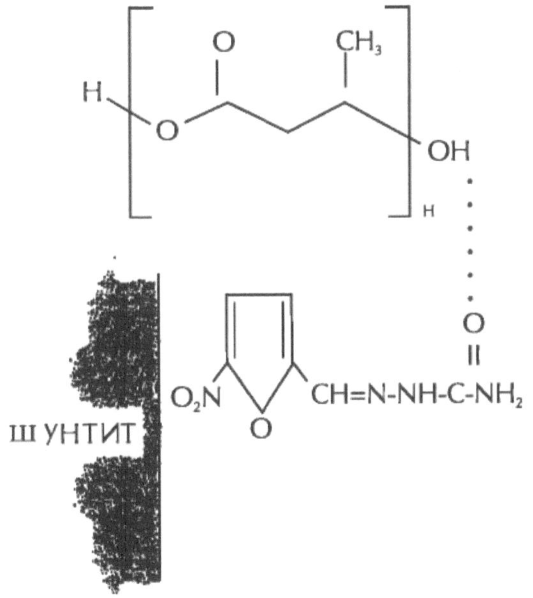

FIGURE 5 Schema of formation of associate PHB-furacilinum-schungite (Шунгит).

Growth of number of a crosslinking does not occur in direct ratio since there is a restriction by quantity of endgroups PHB. The yielded assumption is confirmed by means of IR-spectroscopy methods, DSC and EPR.

KEYWORDS

- **Furacilinum-schungite**
- **Medicinal forms**
- **Petri dish**
- **Zagozhinsky deposit**

REFERENCES

1. Plate, N. A.; Vasiliev, A. E. *Physiologically Active Polymers. M: Chemistry*, 1986.
2. Feldshtein, M, M.; Plate, N. N. In *Nuclear, Biological, and Chemical Risks*, Sohn, T.; Voicu, V. A., Eds.; Kluver Academic Publ.: Dordrecht, Boston, London; Vol. 25, 1999; pp. 441–458.
3. Zaikov, G. E.; L.Iordanskii, A.; Markin, V. S. *Diffusion of Electrolytes in Polymers. Ser. New Concepts in Polymer Science;* VSP Science Press: Utrecht, Tokyo Japan, 1988; 321p; Rogers, K. In *Problems of Physics and Chemistry of Solidity of Organic Matters* (The Lane with English) M: The World, 1968; pp. 229–328.
4. A site *Chemist.* http://www.xumuk.ru/farmacevt/**1420**.html – Furacilinum.
5. Koros, W. J.; Ed. *Barrier Polymers and Structures*. ACS Symposium, series 423. American Chemical Society: Washington, D. C., 1990.
6. Simonova, Ju. S. Moscow State Academy of Fine Chemical Technology: Moscow, 2008; Reg. No. 60, 82p.
7. Kuleznev, N. V.; Shershenev, V. A. *Chemistry and Physics of Polymers: Studies*. For Chemical-tenol. Vuzov. M.: Higher School, 1988; 312 p.

CHAPTER 23

UPDATES ON MODIFICATION OF ETHYLENE COPOLYMERS

S. N. RUSANOVA, O. V. STOYANOV, S. JU. SOFINA, and G. E. ZAIKOV

CONTENTS

23.1 INTRODUCTION

It is possible to extend the sphere of application of ethylene copolymers by the modification of the original polymer or by the development of composite materials on their basis. Traditionally the compositions formation is effective in the presence of additives, interacting with the polymer during the processing. It results in the regulation of the material properties. One of the methods of chemical modification is the introduction of organosilicon compounds into polyolefins.

Earlier [1, 2], we studied the modification of ethylene copolymers with vinylacetate by ethylsilicates containing various amounts of ester groups. Similar studies were conducted by V. Bounor-Legaré et al. [3], so it was interesting to study changes in the structure and properties of ethylene copolymers with vinylacetate containing other functional groups such as anhydride, to identify general trends and specific differences during the modification of ethylene copolymers, containing different reactive segments, by limiting alkoxysilane.

23.2 EXPERIMENTAL

23.2.1 MATERIALS

Copolymers of ethylene with vinylacetate (EVA) grades Evatane 2020 and Evatane 2805 (Arkema) and sevilen 11306-075 brand of JSC "Sevilen" (TU 6-05-1636-97); copolymer of ethylene with vinylacetate and maleic anhydride (EVAMA) grades Orevac 9305 and Orevac 9707 (Arkema) were used as the objects of the study. The main characteristics of the polymers are listed in the Table 1.

TABLE 1 Main characteristics of the polymers.

Characteristics	Sevilen 11306–075	Evatane 2020	Evatane 2805	Orevac 9305	Orevac 9307
Symbol	EVA 14	EVA 20	EVA 27	EVAMA 26	EVAMA 13
The content of vinyl acetate,%	14	19–21	27–29	26–30	12–14

TABLE 1 *(Continued)*

Characteristics	Sevilen 11306–075	Evatane 2020	Evatane 2805	Orevac 9305	Orevac 9307
Symbol	**EVA 14**	**EVA 20**	**EVA 27**	**EVAMA 26**	**EVAMA 13**
Melt Flow Rate, g/10 min, 125°C	0.85	2.23	0.53	14.06	1.10
Density, kg/m^3	0.935	0.936	0.945	0.963	0.924
Melting temperature (max), °C	97	82	72	67	90
Tensile strength, MPa	18.15	14.28	17.40	6.72	18.88
Breaking elongation, %	650	660	830	800	670

Ethylsilicate ETS-32 (ETS) (TU 6-02-895-78) being a mixture of tetraethoxysilane with geksaethoxydisiloxane with a small admixture of ethanol and oktaethoxytrisiloxane was used as a modifier. The silicon content in terms of silicon dioxide is 30–34%, tetraethoxysilane – 50–65%. Its density is 1.062 kg/m^3 and viscosity is 1.6 cP. Ethylsilicate is manufactured by Production JSC "Khimprom" of Novocheboksarsk.

23.2.2 SAMPLES OBTAINING

Reactive blending of polymers with ethylsilicate was carried out on the laboratory rolls at a rotational rolls speed of 12.5 m/min and at a friction of 1:1.2 during 10 min in the range of 100–120°C. The content of the modifier was varied in the range of 0–10 mass%. The samples for investigations were prepared by the direct pressing in the restrictive frameworks. Pressing regime is under the temperature of 160°C and the unit pressure of 15 MPa; the time of preheating, injection boost time and the time of cooling is 1 min. for each 1 mm of the sample thickness. After the pressing all the compositions were subjected to aging at a room temperature for 24 h.

Since the rolling of two-component systems as the formation of products of chemical interaction between the components and the simple mechanical mixtures is possible. That is why a purification of the modified polymers was carried out by fivefold reprecipitation under cold conditions by the ethanol from the solution in CCl_4. The samples for IR-spectroscopy absorption were prepared by watering from solution in carbon tetrachloride on a substrate of KBr. The film samples with the thickness of 0.07–0.12 mm for IR-spectroscopy ATR (Attenuated Total Reflection) were prepared by the direct pressing without restrictive frameworks on fluoroplastic plates.

23.2.3 RESEARCH METHOD

IR spectra were registered by the infrared Fourier spectrometer "Spectrum BXII" of Perkin Company by absorption spectroscopy in the range of 450–4000 cm^{-1} and with the method of ATR on the ZnSe crystal in the range of 650–4400 cm^{-1} with the subsequent transformation by Kubelka-Munk. All spectra were normalized according to an internal standard and the intensity of the band of 720 cm^{-1} related to the deformation vibrations of CH_2 groups of the main chain not involved in the chemical reaction was assumed as the internal standard [1, 4]. The original spectra in the coordinates of the optical density – wave number were processed using the software package ACD/SpecManager (ACD/UV–IR Manager and UV-IR Processor. Version 6.0 for Microsoft Windows) to separate the individual components of the spectrum in the areas corresponding to strongly overlapping absorption bands. The contours shape during the spectra simulation is a mixed Gaussian–Lorentzian. After converting the spectra to eliminate the influence of the penetration depth of radiation and automatically determine the main peaks position a preliminary decomposition of the spectrum on these bands was conducted. The first derivative of the experimental contour and the deviation of the calculated spectrum from the experimental spectrum were analyzed. The most probable position of the characteristic peaks not recorded in the preliminary decomposition was determined by the deviation of the resulting deviation from the zero level and the position of the peaks on the graph of the derivative, taking into account the literature data. The addition of a decomposition component was carried out step by step with the expansion of the spectrum conduction, taking into account the added peak and the deviations of the calculated spectrum analysis.

The Melt Flow Rate (MFR) was determined by capillary viscometer IIRT-5M according to GOST 11645-73 under the temperature of 125°C and the load of 2.16 kg.

The intrinsic viscosity was determined by the standard method [5] in carbon tetrachloride solution at 25°C.

23.3 RESULTS AND DISCUSSION

The change of the macromolecules chemical structure as a result of polymer-analogous transformations can be analyzed by various spectroscopic methods. Qualitative differences between the IR spectra of modified and unmodified samples confirm this fact (Figs. 1, and 2). On spectra of polymers modified with ethylsilicates the bands related to silicone fragments in the areas of 1020–1090 cm^{-1}, 780–830 cm^{-1}, specific for the stretching vibrations of Si–O, Si–O–Si, Si–O–C appears, as well as the band of 971 cm^{-1} appears, characterizing Si–O and Si–O–Si bonds in the cross-linked siloxane fragments. The difference in the nature of optical density changes is possibly due to the different structures formed during the reaction of organosilicon fragments.

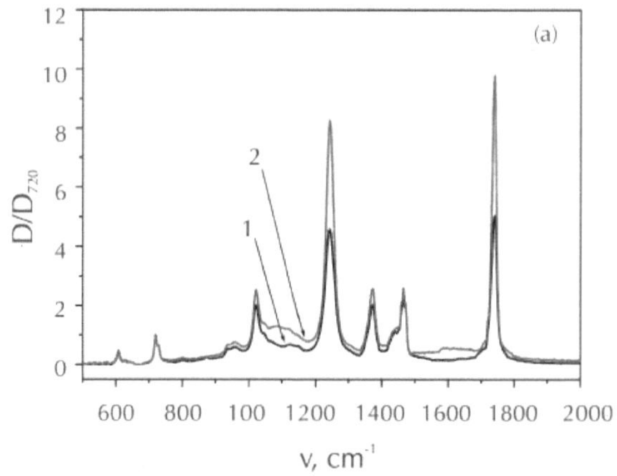

FIGURE 1 *(Continued)*

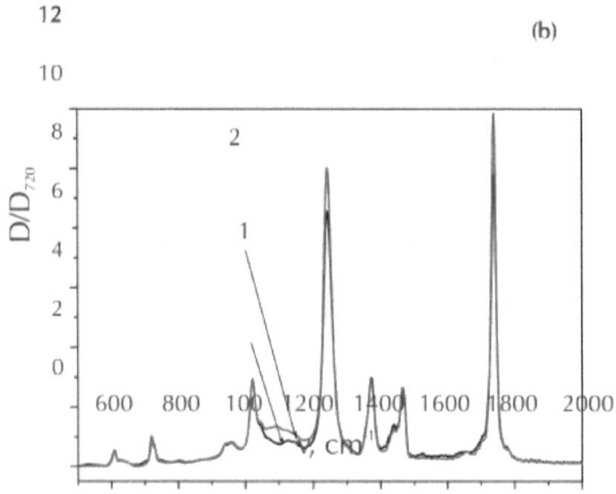

FIGURE 1 IR spectra absorption of EVA 27 (a) and EVAMA 26 (b) initial (1) and modified by ethylsilicates (2).

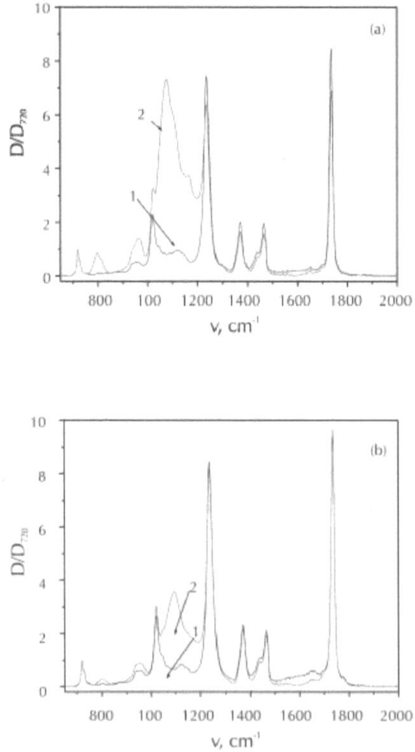

FIGURE 2 IR-spectra ATR of EVA 27 (a) and EVAMA 26 (b) initial (1) and modified by ethylsilicates (2).

It was established earlier [1, 2] that the introduction of tetraethoxysilane in the sevilen leads to a splitting in the IR spectra of characteristic band of 1240 cm^{-1} corresponding to the stretching vibrations of C–O bond in ester groups caused by the substitution of acetyl fragment of the copolymer for the remainder of organosilicon modifier. A similar splitting was observed for ethylene copolymers with vinylacetate and maleic anhydride (Fig. 3). The vibrations of C–O bonds are in the strong interaction with the vibrations of C–C bonds due to the small differences in power coefficients and the prox-

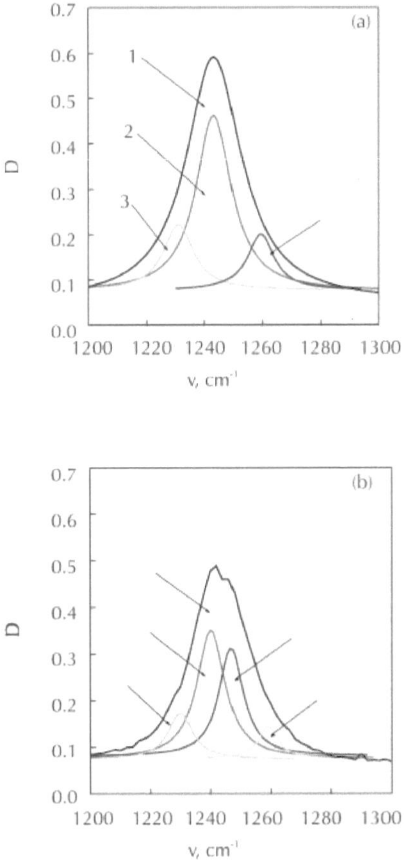

FIGURE 3 Computer decomposition of the infrared spectra in the region of stretching vibrations of C–O bond in ester groups of EVAMA initial (a) and modified by ethylsilicates (b): Observed absorption band of 1240 cm^{-1} (1); individual components of the spectrum: 1240 cm^{-1} (2), 1230 cm^{-1} (3), 1259 cm^{-1} (4), 1246 cm^{-1} (5).

imity of the atoms masses forming the bond. Therefore, the contour of the C–O bands is characterized by the presence of satellites due to rotational isomerism with respect to σ-bonds [6, 7]. The characteristic band shift of the stretching vibrations of C–O bond in the direction of higher frequencies during the substitution of the acetate fragment by the silicone is due to the tension connection by the steric interactions of volume replacement groups [8]. A decrease in the intensity of the characteristic band of 1462 cm^{-1} corresponding to the deformation vibrations of methyl groups in vinylacetate has been observed for EVAMA as well as in the modification of EVA (Fig. 4), which confirms the participation of both copolymers EVA and EVAMA in the transesterification reaction of acetyl fragment.

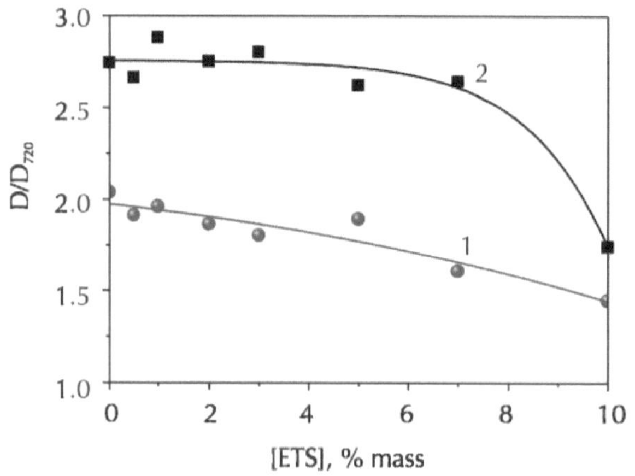

FIGURE 4 Dependence of the relative optical density of the band of 1462 cm^{-1} in EVA 27 (1) and EVAMA 26 (2) from the content of ETS (IR-spectroscopy absorption).

The absence of characteristic bands doublets of 1790 cm^{-1} and 1850 cm^{-1} related to the stretching vibrations of C=O groups of maleic anhydride, both in the modified and unmodified terpolymer is of great interest, though in IR – spectra of films obtained from EVAMA granules, it is presented. We can assume that it is due to the intense thermomechanical effects in the oxygen environment during the rolling, which results in the anhydride cycle opening with the formation of carboxylic acid, reacting further with alkoxy groups of the modifier.

In the spectra of organosilicon compounds the band of 1070 cm^{-1} corresponding to the stretching vibrations of Si–O–Si groups is indicated vividly.

This band presents, respectively, in the spectra of the modified copolymers. The introduction of the modifier into the polymer, naturally, leads to an increase of the optical density of this band corresponding to the total content of Si–O–Si links, which is in the direct ratio to the amount of the injected additive (Fig. 5). However, it was found that the surface layers of the polymer are enriched by siloxane phase, as evidenced by the differences in the growth of the characteristic band of 1070 cm^{-1} in the spectra obtained by ATR (surface layer) and by the absorption IR spectroscopy (integral). It was established earlier [9] that EVA modified by ethylsilicate is a two-component heterophase system. Since polysiloxanes and polyolefins have different segmental mobility and large differences in free surface energy, it was suggested that the migration of grafted siloxane fragments into the subsurface layers of material is possible during the formation of the samples. The saturation of the polymer surface layer during the silanol modification may be of great interest in the development of adhesive materials.

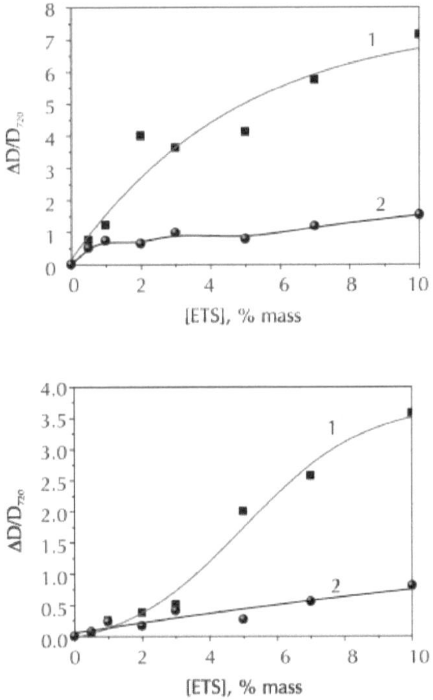

FIGURE 5 Dependence of the relative optical density of the band of 1070 cm^{-1} in the ATR spectra (1) and IR spectra absorption (2) in EVA 27 (a) and EVAMA 26 (b) from the content of ETS.

KEYWORDS

- **Ester groups**
- **Methyl groups**
- **Infrared Fourier spectrometer**
- **Carbon tetrachloride**

REFERENCES

1. Stoyanov, O. V.; Rusanova, S. N.; Petukhova, O. G.; *Zhurn. Prikl. Khimii,* **2000**, *71*(4), 998–1014.
2. Stoyanov, O. V.; Rusanova, S. N.; Petukhova, O. G.; Remisov, A. B. *Zhurn. Prikl. Khimii,* **2001**, *74*(7), 1174–1177.
3. Bounor-Legaré, V.; Ferreira, I.; Verbois, A.; Cassagnau, Ph.; Michel, A. *Polymer,* **2002**, *43*, 6085–6092.
4. Bogatyrev, V. L.; Maksakova, G. A.; Villevald, G. V.; Logvinenko, V. A. *Izv. Sib. otd. AN SSSR Ser. khim. Nauk,* **1986**, *2*(1), 104–105.
5. Kurenkov, V. F.; Budarina, L. A.; Zaikin, A. E. *Praktikum po fizike i khimii polimerov Khimiya-Koloss*, Moscow, 2008.
6. Barnsa, A.; Tomas U. Orvill (Ed) *Kolebatelnaya spektroskopiya. Sovremennye vozzreniya. Tendencii razvitiya*. Mir, Moscow, 1981.
7. Tomas Dzh. Orvill (Ed) *Vnutrennee vrashenie molecul*. Mir, Moscow, 1987.
8. Belopolskaya, T. V. *Vysokomolek. soed.,* **1972**, *14*, 640–645.
9. Chalykh, A. E.; Gerasimov, V. K.; Rusanova, S. N.; Stoyanov, O. V.; Petukhova, O. G.; Kulagina, G. S.; Pisarev, S. A. *Vysokomolek. soed. A*. **2006**, *48*(10), 1801–1810.

CHAPTER 24

CHEMICAL INTERACTION OF ORGANOSILICON COMPOUNDS AND ETHYLENE COPOLYMERS

S. N. RUSANOVA, O. V. STOYANOV, A. B. REMIZOV, S. JU. SOFINA, V. K. GERASIMOV, A. E. CHALYKH, and G. E. ZAIKOV

CONTENTS

24.1 INTRODUCTION

It is possible to extend the sphere of application of ethylene copolymers by the modification of the original polymer or by the development of composite materials on their basis. Traditionally the compositions formation is effective in the presence of additives, interacting with the polymer during the processing. It results in the regulation of the material properties. One of the methods of chemical modification is the introduction of organosilicon compounds into polyolefins. Earlier [1, 2], we studied the modification of ethylene copolymers with vinylacetate by ethyl silicates containing various amounts of ester groups. Similar studies were conducted by V. Bounor-Legaré et al. [3], so it was interesting to study changes in the structure of ethylene copolymers with other comonomers of ethereal nature, to identify general trends and specific differences during the modification of ethylene copolymers, containing different reactive segments, by limiting alkoxysilane.

24.2 EXPERIMENTAL

24.2.1 MATERIALS

Copolymer of ethylene and butyl acrylate (EBA) grade Lotryl 35 VA 320 (Arkema) and copolymer of ethylene and ethyl acrylate (EEA) CAS № 9010–86–0 (Sigma Aldrich) were used as the objects of the study. The main characteristics of the polymers are listed in the Table 1.

TABLE 1 Characteristics of the investigated polymers.

Name index	EBA	EEA
Inherent viscosity, dl/g	0.64	0.77
Average molecular weight, M_η	25,000	34,000
Melting point, °C	65	116
Density, g/cm^3	0.93	0.93
Melt Flow Rate, g/10min	263.1	19.3

TABLE 1 *(Continued)*

Name index	EBA	EEA
The content of butyl acrylate / ethyl acrylate, mol.%	10.54	5.79
Tensile strength, MPa	1.5	6.6
Elongation at break, %	200	530

Ethyl silicate ETS-32 (ETS) (TU 6-02-895-78) was used as a modifier. The silicon content in terms of silicon dioxide is 30–34%, tetraethoxysilane – 50–65%. Its density is 1.062 kg/m^3 and viscosity is 1.6 cP. Ethyl silicate is manufactured by Production JSC "Khimprom" of Novocheboksarsk.

24.2.2 SAMPLES OBTAINING

Reactive blending of polymers with ethyl silicate was carried out on the laboratory rolls at a rotational rolls speed of 12.5 m/min and at a friction of 1:1.2 during 10 min in the range of 100–120°C. The content of the modifier was varied in the range of 0–10 mass.%. The samples for investigations were prepared by the direct pressing in the restrictive frameworks. Pressing regime is under the temperature of 160°C and the unit pressure of 15 MPa; the time of preheating, injection boost time and the time of cooling is 1 min. for each 1 mm of the sample thickness. After the pressing all the compositions were subjected to aging at a room temperature for 24 h.

Since the rolling of two-component systems as the formation of products of chemical interaction between the components and the simple mechanical mixtures is possible. That is why a purification of the modified polymers was carried out by fivefold reprecipitation under cold conditions by the ethanol from the solution in CCl_4. In this case the ethyl silicate and its oligomers in the residual solution were not detected by infrared spectroscopy. Spectra of purified and unpurified polymers are the same.

24.2.3 RESEARCH METHOD

IR spectra in the region of 650–4400 cm^{-1} were registered by the infrared Fourier spectrometer "Spectrum BXII" of Perkin Elmer Company using the method of ATR (Attenuated Total Reflection) on the ZnSe crystal. This method was chosen because of the high intensity of characteristic peaks in the IR spectra absorption that does not allow carrying out an analysis in the areas of strongly overlapping bands. In the ATR spectra the intensity of the absorption bands depends on the depth of penetration of radiation into the sample and on the quality of the sample clamping to the crystal ZnSe. To eliminate the influence of these factors when comparing the spectra of different samples after the Kubelka-Munk conversion, the spectra were normalized according to an internal standard and the intensity of the band of 720 cm^{-1} related to the deformation vibrations of CH_2 groups of the main chain not involved in the chemical reaction was assumed as the internal standard [1, 4]. The original spectra in the coordinates of the optical density – wave number were processed using the software package ACD/SpecManager (ACD/UV-IR Manager and UV-IR Processor. Version 6.0 for Microsoft Windows) to separate the individual components of the spectrum in the areas corresponding to strongly overlapping absorption bands. The contours shape during the spectra simulation is a mixed Gaussian–Lorentzian. After converting the spectra to eliminate the influence of the penetration depth of radiation and automatically determine the main peaks position a preliminary decomposition of the spectrum on these bands was conducted. The first derivative of the experimental contour and the deviation of the calculated spectrum from the experimental spectrum were analyzed. The most probable position of the characteristic peaks not recorded in the preliminary decomposition was determined by the deviation of the resulting deviation from the zero level and the position of the peaks on the graph of the derivative, taking into account the literature data. The addition of a decomposition component was carried out step by step with the expansion of the spectrum conduction, taking into account the added peak and the deviations of the calculated spectrum analysis.

24.3 RESULTS AND DISCUSSION

The change of the macromolecules chemical structure as a result of polymer–analogous transformations can be analyzed by various spectroscopic methods. On the course of the reaction between the polymers and the organosilicon modifier was judged by the change in the intensity of the characteristic bands of stretching vibrations of C=O groups in the area of 1740 cm^{-1}, as well as by the change in the intensity of the band of 1070 cm^{-1} related to the stretching vibrations of Si–O–Si and Si–O–C bonds. The proportion of ester fragments, reacting by polymer–analogous transformations, were estimated by changes of the reduced relative optical densities of the bands in the region of stretching vibrations of C=O groups [5]. Identification of organosilicon groups was performed by the bands of stretching vibrations of Si–O, Si–O–Si, Si–O–C bonds in the area of 1100–1190, 1020–1090 and 780–840 cm^{-1} [6, 7]. Dependence of the absorption band of carbonyl groups position from the intramolecular effects and intermolecular interactions, isolation of these bands and their high intensity determine the importance of this absorption area in the spectrochemical studies [8]. Frequency of the absorption band of stretching vibrations of C=O bond is determined by the structure of the nearest groups surrounding the carboxy group, so during the study of this characteristic band position you can get the data about the environment of the carbonyl group [6].

The introduction of a modifier in the studied copolymers of ethylene with acrylates has no effect on the integrated optical density of the characteristic band of 1736 cm^{-1} (Fig. 1). In computer decomposition of the initial copolymers spectrum revealed that, in addition to the band of 1733 cm^{-1} relating to the vibrations of esters carboxy group there are the bands of 1738 cm^{-1} and 1721 cm^{-1} identified as belonging to the vibrations of the carboxy group of a saturated carboxylic acid [9] and a ketone with an open chain [5]. We assume that these groups are formed during the oxidation of acrylate and ethylene unit, respectively, as a result of shear and temperature effects during rolling (Figs. 2a, and 3a). During the chemical reaction of ETS and copolymer both alkoxy group of unoxidized acrylate unit and hydroxyl group of oxidized acrylate unit are replaced by triethoxysisilil. These functional groups are the closest "neighbors" of carboxy group. As a result of changes in the nearest environment of the carboxy group during the introduction of a reactive organosilicon modifier in the spectra of copolymers of ethylene with ethyl acrylate and butyl acrylate a striping of characteristic peaks takes place and new peaks of 1744 cm^{-1} and 1728 cm^{-1} of stretching vibrations of C=O groups associated with siloxane fragments (Figs. 2b, and 3b) appear. Observed in this case the redistribution of optical densities of characteristic bands at a constant integrated absorption (Figs. 1, 4, and 5) and the absorption of carboxy groups of oxidized ethylene units, suggests that the chemical interaction of copolymer macromolecule with ETS occurs on C–O link of acrylate group. Since the modification of ethylene copolymers with acrylates was carried out in the melt a significant impact on the reaction should provide steric factors obstructing the interaction of ester units of copolymers with the functional groups of ETS. As shown in Ref. [5], by the change of the optical densities, defined to the molar content of reactive groups one can determine the proportion of these groups, participating in the reaction. It was established that the chemical interaction involves up to 20% of acrylate units of EEA and up to 40% of EBA units (Figs. 4b, and 5b). This is probably due to the higher reactivity of butoxy groups, due to a large shift of the electron densities along the butyl radical compared to that of ethyl radical and, consequently, with a lower strength of C–O bond of butoxy group.

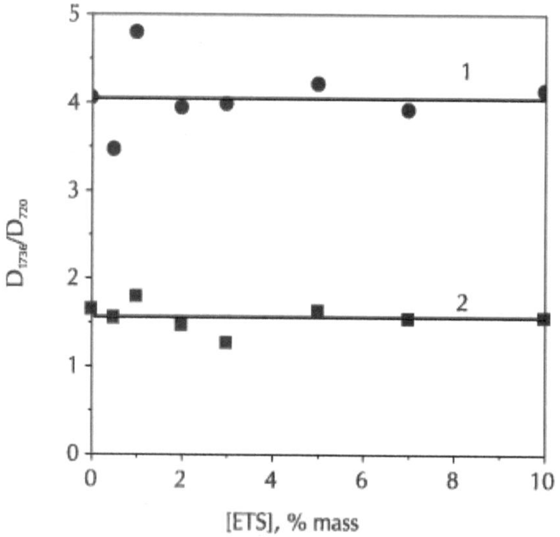

FIGURE 1 The dependence of the intensity of observed (integral) characteristic absorption bands of C=O groups stretching vibrations in the EEA (1) and EBA (2) from the modifier concentration.

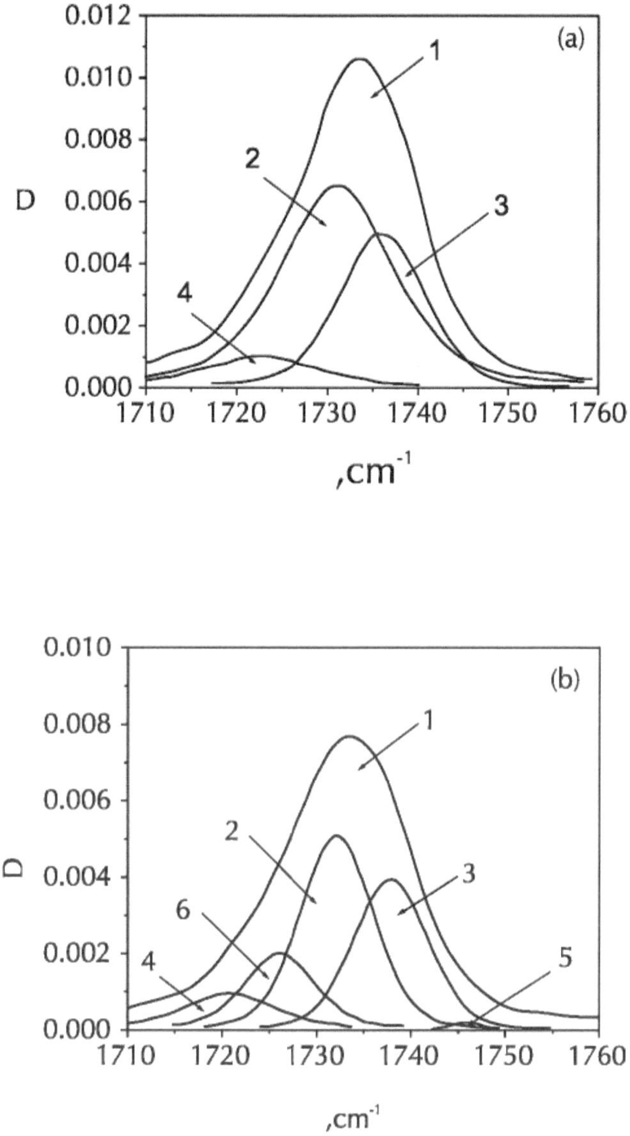

FIGURE 2 Computer decomposition of the IR – spectra ATR of the original EEA (a) and EEA, modified by ETS (b) in the region of stretching vibrations of C=O groups. Observed (integrated) absorption band of 1736 cm^{-1} (1); individual components of the spectrum: 1733 cm^{-1} (2), 1738 cm^{-1} (3), 1721 cm^{-1} (4), 1744 cm^{-1} (5), 1728 cm^{-1} (6).

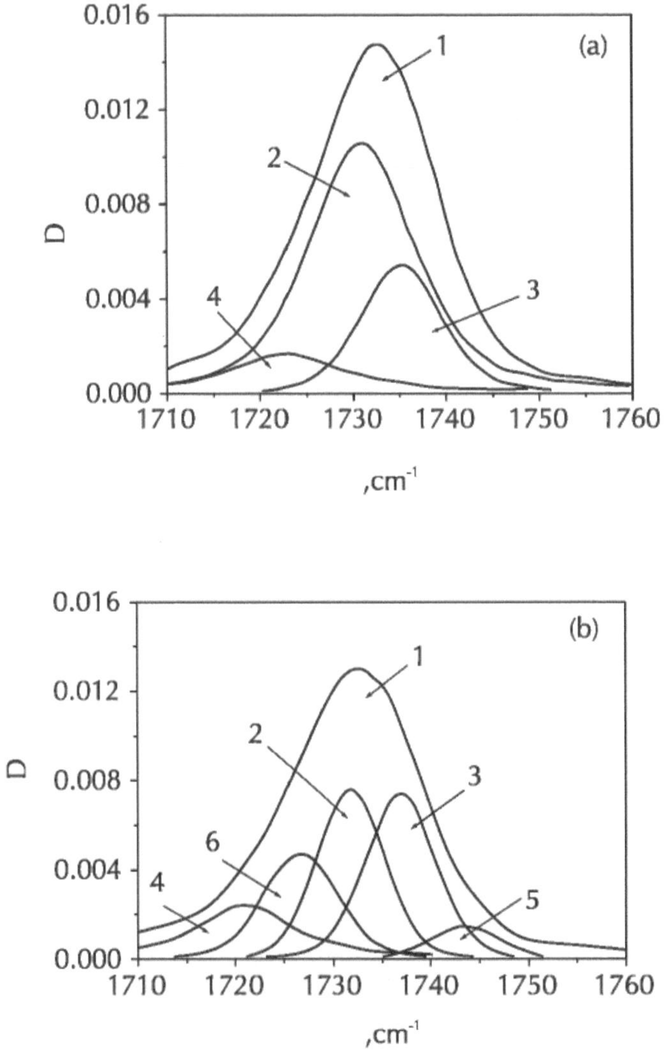

FIGURE 3 Computer decomposition of the IR – spectra ATR of the original EBA (a) and EBA, modified by ETS (b) in the region of stretching vibrations of C=O groups. Observed (integrated) absorption band of 1736 cm^{-1} (1); individual components of the spectrum: 1733 cm^{-1} (2), 1738 cm^{-1} (3), 1721 cm^{-1} (4), 1744 cm^{-1} (5), 1728 cm^{-1} (6).

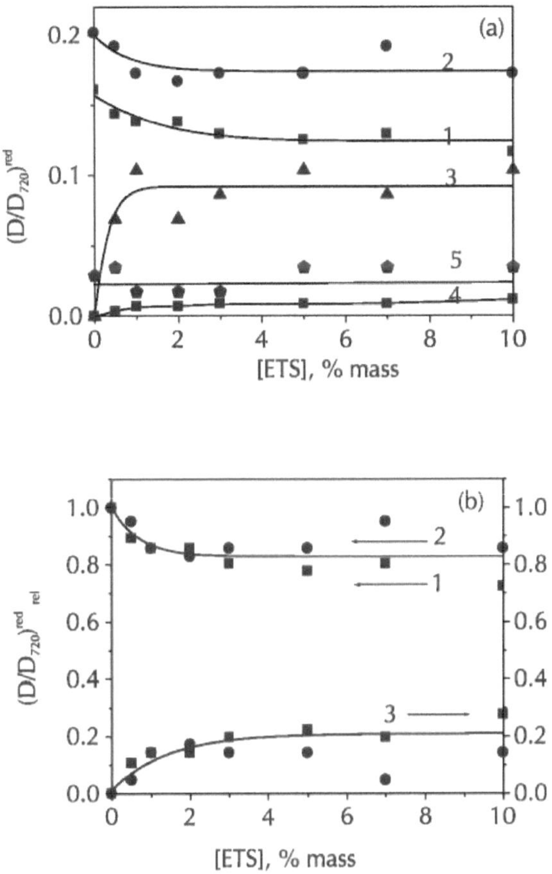

FIGURE 4 Changing of the reduced optical densities of the absorption bands of C=O groups stretching vibrations in EEA (a); the relative optical densities and the proportion of transesterification links (b): a – 1738 cm^{-1} (1), 1733 cm^{-1} (2), 1728 cm^{-1} (3), 1744 cm^{-1} (4), 1721 cm^{-1} (5); b – 1738 cm^{-1} (1), 1733 cm^{-1} (2), the percentage of the reacted acrylate comonomer units (3).

The formation of new functional groups in the polymer is confirmed by infrared spectral analysis of the received compositions. In the spectra of copolymers modified by ETS the bands related to silicone fragments in the areas of 1110–1180 cm^{-1}, 1020–1090 cm^{-1}, 780–830 cm^{-1}, characteristic for the stretching vibrations of Si–O, Si–O–Si, Si–O–C (Fig. 6) appear, while the spectra of EEA and CEBA both modified and original are the same. The difference in the nature of the observed changes of characteristic bands optical densities is possibly due to the different length of grafted organosilicon fragments.

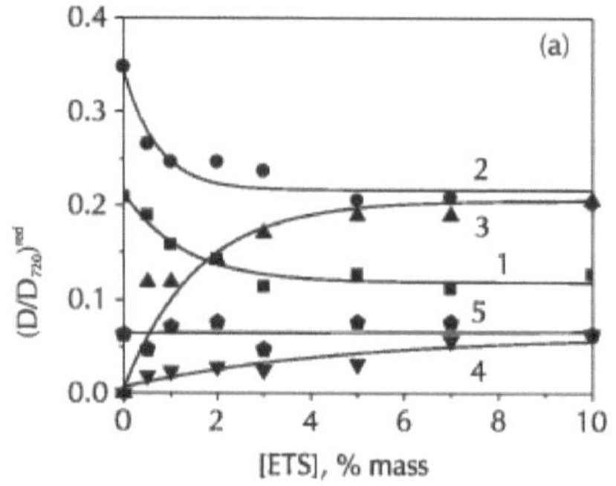

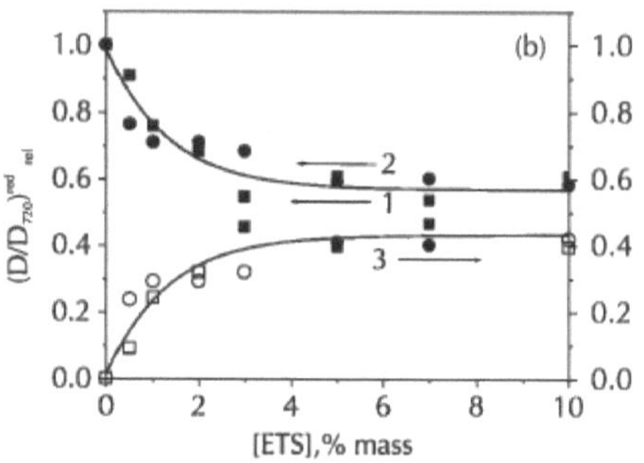

FIGURE 5 Changing of the reduced optical densities of the absorption bands of C=O groups stretching vibrations in EBA (a); the relative optical densities and the proportion of transesterification links (b): a – 1738 cm^{-1} (1), 1733 cm^{-1} (2), 1728 cm^{-1} (3), 1744 cm^{-1} (4), 1721 cm^{-1} (5); b – 1738 cm^{-1} (1), 1733 cm^{-1} (2), the percentage of the reacted acrylate comonomer units (3).

The ATR IR spectra of EEA and EBA analysis is complicated by the fact that the characteristic bands related to ester (C–O–C) and silicone fragments (Si–O–Si, Si–O–C) are in the same area, the so-called area of "fingerprint" – 1000–1100 cm^{-1}. However, there is a substantial increase in the intensities of characteristic peaks in the introduction of 1% mass. of the modifier (Fig. 6).

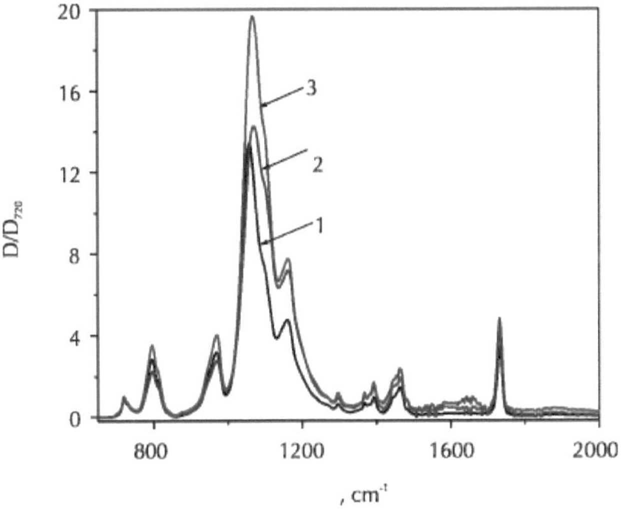

FIGURE 6 IR-spectra ATR of the initial EBA (1) and EBA modified by 1 wt.% of ETS (2), 7 wt.% of ETS (3).

In the spectra of organosilicon compounds [6] the band of 1070 cm^{-1} is pronounced. It is characteristic to the stretching vibrations of Si–O–Si groups. This band is also presents in the spectra of the modified copolymers. Introduction of the modifier into the polymer, naturally, leads to an increase in optical density of this band, and the nature of the copolymer acrylate fragment does not affect on the intensity increase, but only the concentration of the modifier (Fig. 7), as D_{1070}/D_{720} corresponds to the total content of Si–O–Si bonds, which is directly proportional to an amount of the injected additive. Reduced optical densities D_{1070}^{red} depend not only on the number of the introduced ETS, but also on the acrylate content in the copolymers, so on them can be judged on the proportion of Si–O–Si groups, accounted for per 1 mole of the polymer ester link (Fig. 8) and hence on the length of the grafted organosilicon fragment. Since the reaction involves only a part of the acrylate units, ETS due to its greater reactivity of the Si–O bond, than the C–O bond, forms a grafted siloxane chains whose length is greater, the less content of acrylate in the starting polymer. We can assume that siloxane branching in the studied modified EBA have a greater length than in the EEA, due to the different concentrations of the ester groups in the copolymers. This leads to the different amounts of siloxane fragments per 1 mole of the comonomer acrylate link.

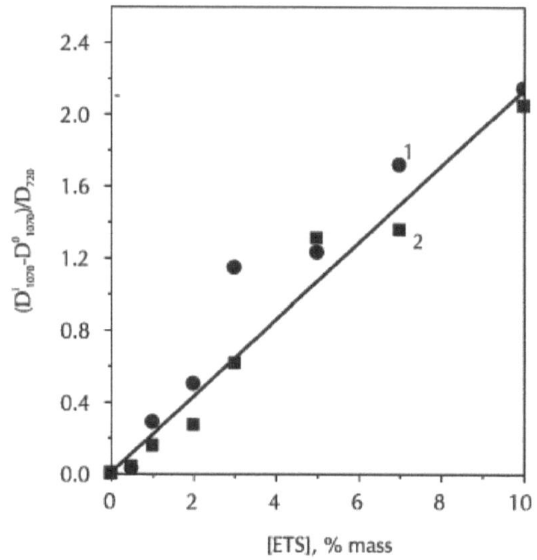

FIGURE 7 Dependence of the relative optical density of the characteristic absorption bands of Si–O–Si stretching vibrations in EEA (1) and EBA (2) from the modifier concentration.

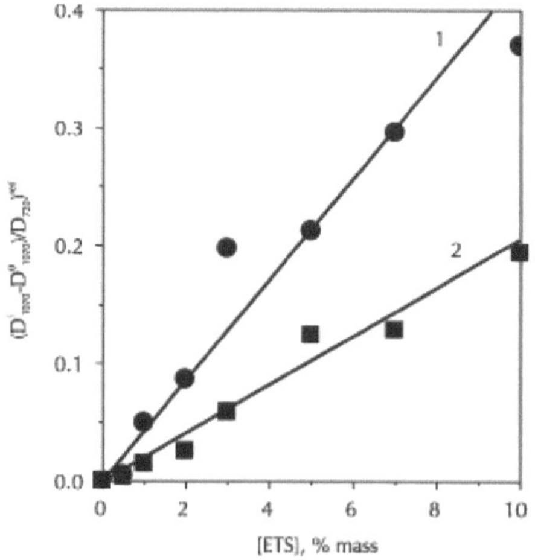

FIGURE 8 Dependence of the reduced optical density of the characteristic absorption bands of Si–O–Si stretching vibrations in EEA (1) and EBA (2) from the modifier concentration.

Our studies have allowed assuming the following scheme of the chemical reactions between ethylene-acrylate copolymers and ethoxysilane under intense shear and thermal effects:

Firstly, it is reaction (1) of tetraethoxysilane interaction with alkoxy group of copolymer acrylate fragment or with hydroxy group formed during the thermomechanodestruction of the initial polymer, which results in the grafting of siloxane link to the macromolecule. The second stage is associated with the increasing of polymerization degree of the lateral siloxane group according to the reaction (2). Since almost all polymer-analogous transformations are accompanied by the side reactions, a possible third stage is associated with interchain reactions (3) and intrachain (4) interactions that are probably leads to the fragments of nets and "cycles" formation. In addition, the reaction of polycondensation (5) of the dissolved molecules of ETS and the subsequent grafting (6) of the formed diethoxysiloxane to macromolecules of the copolymers by the reaction of transesterification analogically to the reaction 1 are possible. However, the ATR IR spectra of the treated and untreated modified polymers are identical, which suggests the absence of not grafted macromolecules of diethoxysiloxane in the polymer matrix.

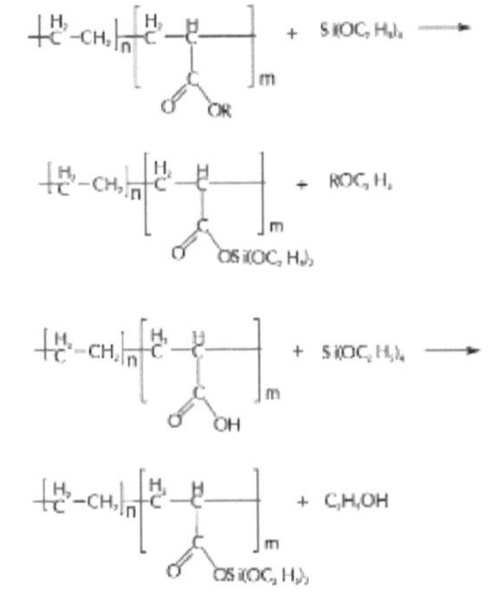

(1)

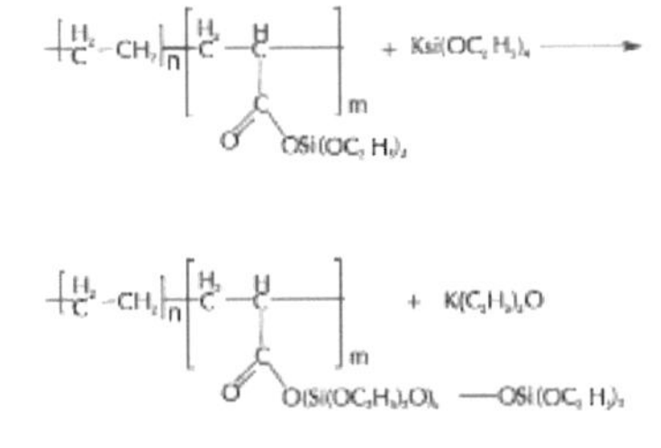

(2)

(3)

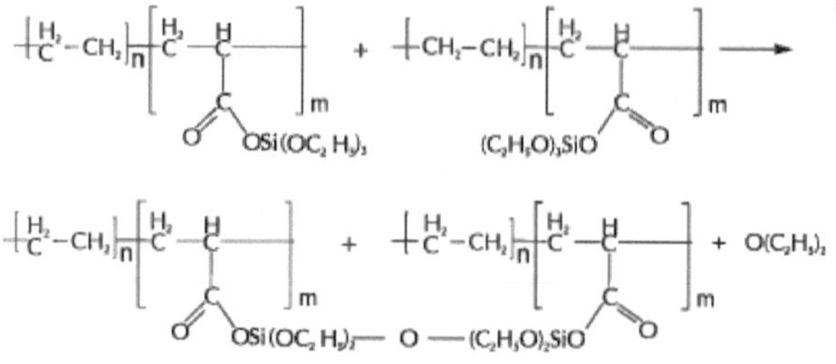

(4)

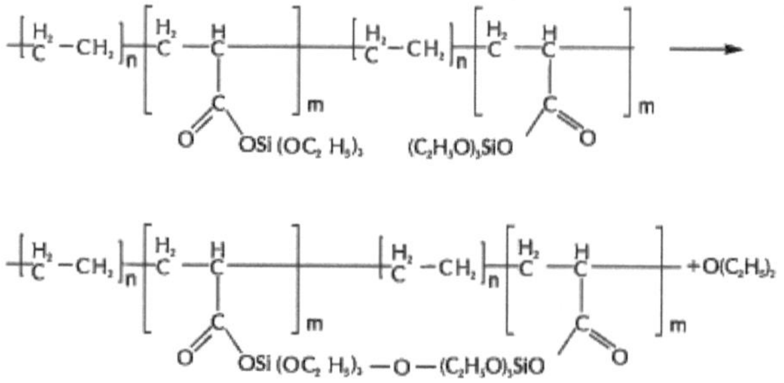

(5)

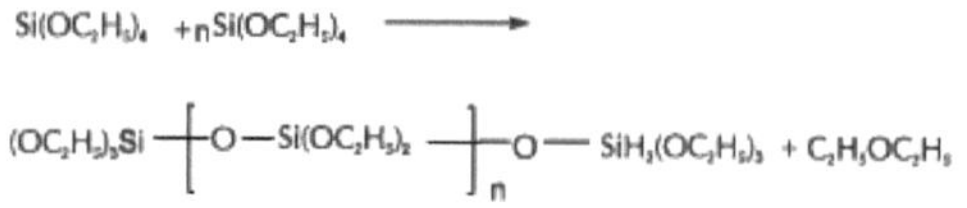

(6)

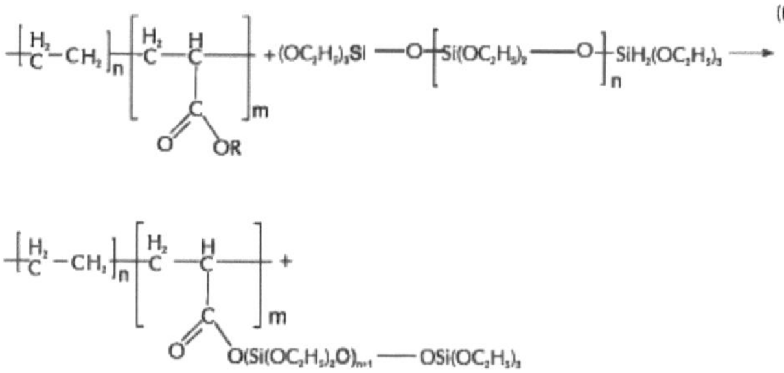

KEYWORDS

- **Ethylene copolymers**
- **Fingerprint**
- **Gaussian–Lorentzian**
- **Infrared Fourier spectrometer**

REFERENCES

1. Stoyanov, O. V.; Rusanova, S. N.; Petukhova, O. G.; *Zhurn. Prikl. Khimii,* **2000**, *71*(4), 998–1014
2. Stoyanov, O. V.; Rusanova, S. N.; Petukhova, O. G.; Remisov, A. B. *Zhurn. prikl. Khimii,* **2001**, *74*(7), 1174–1177.
3. Bounor-Legaré, V.; Ferreira, I.; Verbois, A.; Cassagnau, Ph.; Michel, A. *Polymer*, **2002**, *43*, 6085–6092.
4. Bogatyrev, V. L.; Maksakova, G. A.; Villevald, G. V.; Logvinenko, V. A. *Izv. Sib. otd. AN SSSR Ser. khim. Nauk,* **1986**, *2*(1), 104–105.
5. Tarutina, L. I.; Pozdnyakova, F. O. *Spectralny Analis Polimerov*, Khimiya, L., 1986.
6. Bellami, L. Ed.; *IK-spektry slozhnyx molekul,* Inostr.Llit., M., 1963.
7. Vasilets, L. G.; Lebedeva, E. D.; Akutin, M. S.; Kazurin V.. I. *Plast. Massy,* **1988**, *7*, 43–44.
8. Kazitsina, L. A.; Kupletskaya, N. B. *Primenenie UF-, IK- i YaMR-spectroscopii v organicheckoy khimii,* Vyssh.shkola, M, 1971.
9. Dekhant, I., Ed. *Infrakrasnaya spectroscopiya polimerov,* Khimiya, M., 1976.
10. Chalykh, A. E.; Gerasimov, V. K.; Rusanova, S. N.; Stoyanov, O. V.; Petukhova, O. G.; Kulagina, G. S.; Pisarev, S. A. *Vysokomolek. soed. A.* **2006**, *48*(10), 1801–1810.

CHAPTER 25

UPDATES ON THE ADHESION STRENGTH OF POLYOLEFIN COMPOSITIONS

E. V. SECHKO, R. M. KHUZAKHANOV, L. F. STOYANOVA, and I. A. STAROSTINA

CONTENTS

25.1 INTRODUCTION

Adhesive compositions on the base of polyolefins blends were first investigated and applied at the end of the last century [1–4]. In particular, it was shown that by mixing low-density polyethylene (high pressure, LDPE) and ethylene-vinyl acetate copolymers (EVA) in the presence of a mineral filler (talc) there is an extreme increase in the strength of an adhesive compound of the polymer mixture composition with the metal (steel) compared to raw materials [1–4]. It was shown later [5–7] that when mixing EVA with various concentrations of vinyl acetate units with each other synergistic effect is also realized – at filling talc there is an extreme increase in the strength of an adhesive compound with the metal. The reasons for this behavior of the compositions are considered in works [8, 9], but the questions about the effect of the colloidal structure of the polyolefin composition on the adhesive strength of the polymer-substrate remain unresolved. You can expect that for mixture compositions the strength of adhesive bonding will be determined by which of the copolymers – "non-adhesionnoactive" (LDPE) or "adhesionnoactive" (ethylene copolymer) will form a continuous phase in the mixture. To answer this question we chose as an object of study the system of LDPE + EVA, as they are described in detail in the literature in terms of assessing of their adhesion and physical and mechanical properties depending on the composition of the mixture [1–4, 10, 11]. In addition, for systems of LDPE-EVA the phase diagrams are constructed [12]. They give an idea about the compatibility of LDPE and EVA with different molecular characteristics and content of vinyl acetate units.

25.2 RESEARCH METHODS

As the main objects of study were selected: low-density polyethylene (LDPE) grades 15313-003, 11503-070 (GOST 16337-77), ethylene-vinyl acetate copolymers (EVA) grades 11104-030, EVA 11306-075, EVA 11507-375, EVA 11808-1750, containing various amounts of ester groups, produced by JSC "Sevilen (TU 6-05-1636-97), ethylene-vinyl acetate copolymers Evatane grades 20-20, 28-05, terpolymers of ethylene with

TABLE 1 Characteristics of the investigated polymers.

Characteristics	LDPE 11503–070	LDPE 15303–003	EVA 11104–030	EVA 11306–075	EVA 11507–375	EVA 11808–1750	EVA Evatane 20–20	EVA Evatane 28–05	EVAMA Orevac 9307	EVAMA Orevac 9305	EBA Lotryl 35 BA 320
Symbol	**LDPE 115**	**LDPE 153**	**EVA 111**	**EVA 113**	**EVA 115**	**EVA 118**	**EVA 20**	**EVA 28**	**EVAMA 14**	**EVAMA 28**	**CEBA**
VA content, %	–	–	7	14	22	29	20	28	14	28	–
MA content, %	–	–	–	–	–	–	–	–	1.5	1.5	–
BA content, %	–	–	–	–	–	–	–	–	–	–	35
Melt Flow Rate, g/10 min, T = 190°C	7.0	0.3	2.4	9.9	27.8	204	20	6.5	10.5	180	305
Melt Flow Rate, g/10 min, T = 125°C				0.85	7.25	25.6	2.23	0.53	1,1	14,06	17,48
Density, kg/m^3	918	921	925	934	936	945	940	950	940	950	930
Tensile strength, MPa	12	18	19	20	12	5	14	17	19	5	1.3
Elongation at break, %	450	660	800	890	930	830	660	830	670	800	200
Elastic modulus, MPa	168	186	110	61	32	12,5	31	17	62	9	3
Melting point, °C	104	106	103	99	91	84	82	72	90	67	67
Acidity parameter of the coverage (mJ/m^2)1/2	1.45	1.65	0.45	–3.05	–3.05	–1.75	–2.0	–1.15	3.8	4.1	1.8
Acidity parameter of the coverage (mJ/m^2)1/2 (samples with 10% talc content)	–	–	–	–	–	–	0.15	0.5	2.75	3.75	3.0

maleic anhydride and various vinyl acetate content grades Orevac 93-07 and 93-05, copolymer of ethylene with butyl acrylate grade Lotryl 35 VA 320. The main characteristics are listed in the Table 1.

The mixtures were prepared by the components mixing on the laboratory rolls. The samples of the investigated compositions were prepared by pressing on hydraulic press DV 2428, in accordance with GOST 12019-66. The dimensions of the pressed plates were set by the restrictive frameworks (100×100 mm^2 with the thickness of 0.9–1.1 mm).

The samples for adhesion testing were prepared as follows: metal plates (steel 3) were previously cleaned of rust and other raids with sandpaper and then degreased with acetone. After this, plates in the amount of 6 pieces were placed in the lower part of the mold, then was laid pressed earlier pattern of the adhesive composition and the plastic plate was laid on it. Further, the double-layer composition was pressed by the weight of the plate at 200°C for 20 min, then without removing the load the water cooling was switched on for 5 min. Then the cooling was switched off, the upper plate was raised, and the prepared samples were removed from the restrictive frameworks.

The samples of blends of LDPE 115 with EVAMA 14, EVAMA 28, EBA were pressed for 5 min and with EVA-20 and EVA-28 – for 10 min.

The samples of the three-layer composition for adhesion testing were prepared as follows: on the cleaned by sandpaper for removing oxide film and defatted by toluene metal plates liquid epoxy composition was applied by brush. Curing process was carried out at a room temperature of 22 ± 2°C for 24 h, all the rest is similar to the two-layer systems. Compounds of the composition: ED-20:TETA (100:6).

Melt flow rate of the samples was determined by capillary viscometer type IIRT according to GOST 11645-73 at the temperatures of 125°C and 190°C. Effective viscosity was determined on MRT “Monsanto”. Investigations were carried out at a temperature of 130°C in the range of shear rates $\gamma = 3.6/1226$ s^{-1}. Shear rate was calculated by the formula: $\gamma = 0.075\ (D_k)^{-3}\cdot U$, (s^{-1}), where D_k – capillary diameter; U – linear velocity of the plunger. Viscosity was calculated by the formula: $\eta = 22.98\ (D_k)^{-3}\cdot P/LU$, kPa·s, where P – pressure, L – capillary length. The dependence of log η from log γ was build according to the calculated values.

Physico-mechanical tests were carried out on breaking machine Inspekt mini. The clamps velocity was 100 mm/min.

Adhesive joint strength to steel (St-3) and to epoxy primer was evaluated by the coating flaking from the substrate size 20 × 100 mm^2 at the angle of 180° and the rate of separation of 100 mm/min on a breaking machine R-0.5 a day after forming the adhesive connection.

For obtaining the information about the colloidal structure of the mixtures of LDPE-153, LDPE-115 and various ethylene copolymers the method of selective dissolution of the components was used. The essence of the method is in the multiple extractions by a chloroform of the soluble part of the mixtures in a Soxhlet apparatus.

To estimate the parameter of acidity D at first was determined the surface free energy (SFE) and its components by measuring the contact angles of the samples surface by the test liquids. As the test liquids were used: water, dimethylformamide, glycerol, formamide, aniline, dimethylsulfoxide, saturated aqueous solutions of phenol and potassium carbonate, α-bromonaphthalene, methyleneiodide. The parameter of acidity was calculated by the method of E. Berger [13].

The structure of the pressed samples was studied on the force probe microscope "Solver P47" (Russia). The samples were subjected to etching in chloroform and then were dried at a room temperature. The samples were scanned in semi-contact mode; the height of scanning along the vertical axis – 200 nm; the diameter of the cantilever (tip) – 20 nm; the frequency of the cantilever oscillations – 150 kHz; maximum scan size – 15×15 micrometers, resolution – 256×256 pixels.

25.3 RESULTS AND DISCUSSION

As a result of studies the concentration dependences of adhesion strength (A) of the mixtures of LDPE-153 and EVA containing 7–29% of vinyl acetate groups in relation to steel and epoxy primer were obtained. The concentration dependences of physical and mechanical properties and selective dissolution of the materials were also obtained.

For information about the structure of LDPE-153 and various brands of EVA mixtures the method of selective dissolution of the mixture component in chloroform was used. It is known that LDPE and EVA with low vinyl acetate content are not soluble in chloroform under the normal conditions. The growth of the vinyl acetate content increases the solubility of EVA in chloroform, moreover EVA-118 almost completely soluble at a room temperature due to a high content of polar groups and negligible crystallinity (≈5%). Therefore, for the mixtures of LDPE and EVA with various vinyl acetate units content were chosen such temperature conditions in which complete dissolution of the copolymer – the soluble component of the mixture occurred. Evaluation of selective dissolution of the mixtures depending on the composition is a convenient way to study the concentration ranges in which the mixture components play the role of the dispersion medium, the disperse phase, or the both components have continuous phases (the area of phase inversion). However to estimate exactly the range of the continuous phase of LDPE and phase inversion region was not possible in this experiment, since it was necessary to choose a solvent that dissolves the polyethylene, but does not dissolves the EVA. However, such a solvent at a temperature of dissolution of polyethylene will melt also the EVA. Therefore, we can reliably estimate only the region of the continuous phase of EVA and the beginning of the region of phase inversion.

It was established that with decreasing of vinyl acetate units content in a series of the mixtures LDPE – EVA-118, LDPE – EVA-115, LDPE – EVA-113, LDPE – EVA-111 the region of the continuous phase formed by EVA narrows. The integrated graphics as well as the concentration dependencies of the solubility in relation to the total content of the soluble component illustrates this (Fig. 1). The solubility exceeding of more than 100% may indicate that PE is significantly more soluble in EVA, than EVA in LDPE [12] and, therefore, LDPE dissolved in EVA, also dissolves in chloroform.

As it known, the ratio of viscosities of polymer melts in mixing influences on the formation of the colloidal structure of the mixture. Dependencies of the effective viscosity for EVA-115 and EVA-118 on the shear rate have an extreme nature. Such curves are characteristic for polyolefins with a branched structure and a high content of the polar comonomer units [14]. The obtained data are consistent with the values of MFR measured by the standard method at 190°C (Table 1) – the higher the viscosity, the lower the MFR. This result allows us to use the standard value of MFR to identify the relationship between structural and rheological characteristics of the mixtures. We should expect that the higher the difference in MFR between LDPE and EVA, the wider the concentration range in which the less viscous component forms a continuous phase, i.e. is a dispersion medium. Such a relationship is shown in Table 2.

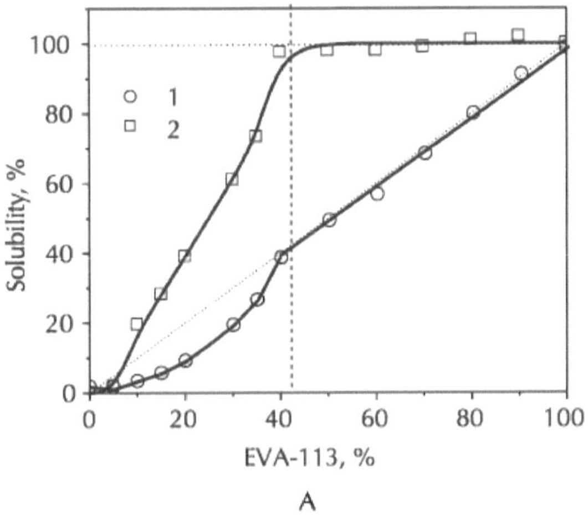

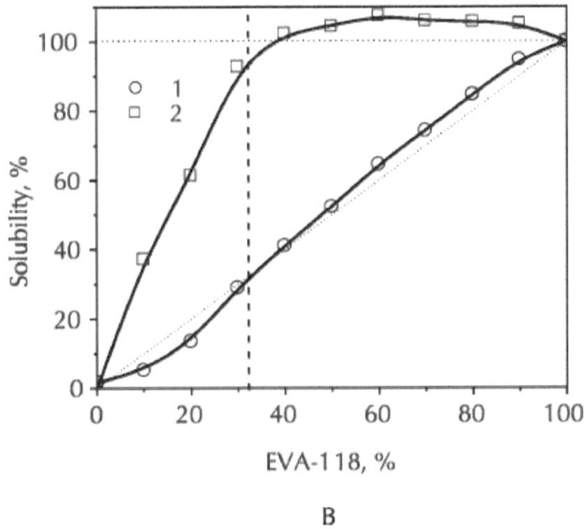

FIGURE 1 Concentration dependence of solubility for the mixtures LDPE–153 – EVA–113 (a) and LDPE–153 – EVA–118 (b): 1 – integral solubility, 2 – solubility in relation to the total content of the soluble component.

TABLE 2 Relationship of rheological and structural characteristics of the investigated mixtures.

Mixture	|ΔMFR|	The formation of a continuous phase of the copolymer at a concentration of EVA, %
LDPE – EVA–118	203.7	35
LDPE – EVA–115	27.5	40
LDPE – EVA–113	9.6	45
LDPE – EVA–111	2.1	70

Thus, these results confirm that the increase in the difference in viscosity leads to an increase in the concentration range in which the less viscous component forms a continuous phase.

Concentration dependencies of the strength of adhesive joint for the mixtures (A) to the steel and to the epoxy primer are presented in Figs. 2 and 3.

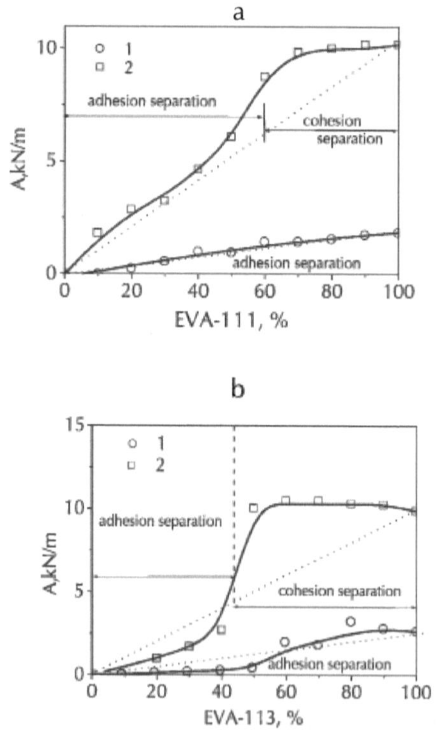

FIGURE 2 Concentration dependence of the strength of adhesive joint for the mixtures LDPE–153 – EVA–111 (a) and LDPE–153 – EVA–113 (b): 1 – adhesive joint strength to steel, 2 – adhesive joint strength towards to the cured epoxy primer.

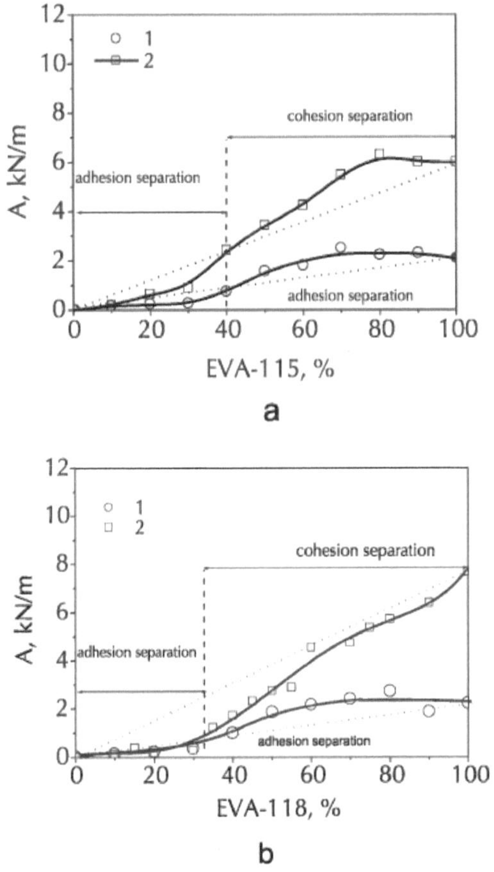

FIGURE 3 Concentration dependence of the strength of adhesive joint for the mixtures LDPE–153 – EVA–115 (a) and LDPE–153 – EVA–118 (b): 1 – adhesive joint strength to steel, 2 – adhesive joint strength towards to the cured epoxy primer.

The dependencies on the adhesive strength of the metal have the S-shaped, with the exception of the LDPE – EVA-111. For this system, the change of A is close to the additive. This result can be explained by the fact that EVA-111 forms a continuous phase only at 70% copolymer content.

Maximum values of adhesion strength for the other systems (with a positive deviation from the additive values) can be realized in the range of EVA concentrations in which EVA, like a most "adhesive active" component, forms a continuous phase.

For all systems an adhesive nature of the separation is characterized by visually. This is due to the low intensity of interphase interaction in the selected experimental conditions. The strength of adhesive joint is consists of adhesion and deformation components [15]. Adhesion in this case is a limiting factor and determines the nature of destruction.

The intensity of interphase interaction of polymer and metal may be interpreted in the framework of the adsorption theory of adhesion [16], on which a theory of acid–base interactions between the components of an adhesive compound develops [16]. From a practical point of view a convenient measure of acidity (basicity) of the adhesive compound components is a parameter of the acidity D [13], as well as a reduced parameter of the acidity $| \Delta D |$, which is a module of the difference between the parameters of the acidity of the adhesive and the adherend [17–21]. It was established by the experiment that there is a linear dependence between an adhesive strength and a reduced parameter of the acidity with correlation coefficient $k = 0.92$.

Thus A to a steel of LDPE–EVA mixtures defined by a component forming the continuous phase. While the maximum values of A is realized at the largest values of $|\Delta D|$.

The results of the adhesive joint strength assessing for the systems of polyethylene – adhesive (LDPE–EVA) – epoxy primer are generally consistent with those results for the systems of polyethylene – adhesive – steel, but have some peculiarities.

Maximum values of A are also realize for the mixtures with compositions in which EVA forms a dispersion medium (continuous phase). But firstly, there is a significant growth in A values with the increase of EVA content in the mixture. Secondly, S-shaped curve is observed only for polymer pairs LDPE – EVA-111, LDPE – EVA-113, LDPE – EVA-115. It should be noted that in the area of prevailing concentrations of LDPE in the investigated mixtures destruction has an adhesive nature, i.e., realizes along the border adhesive – epoxy primer. However, in the area of concentration range in which phase of EVA is continuous the nature of separation is cohesive.

For a mixture of LDPE – EVA-118 (Fig. 4) the transition from adhesive separation to cohesive is also occurs at the formation of the continuous phase by EVA-118. But, unlike the above-described, the growth of A with the increasing of copolymer concentration is accompanied by a negative deviation of the adhesion strength values from the additive in the whole concentration range.

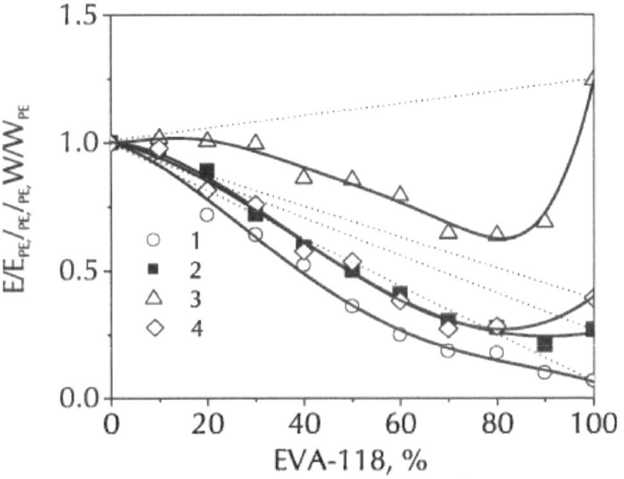

FIGURE 4 Concentration dependencies of physical and mechanical properties for a mixture of LDPE–153 – EVA–118: 1 – elastic modulus, 2 – tensile strength, 3 – elongation at break, 4 – specific work of destruction.

Cohesion nature of destruction of the systems with epoxy adherend can be explained by the fact that $|\Delta D|$ for these systems is higher (Table 1). Therefore, in these cases, the value of directly the adhesive strength at the interface boundary is higher than the cohesive strength of the adhesive and it was he who destroyed during the testing.

Negative deviation of the concentration dependence of A from the additive values for the system with EVA-118 is due to the fact that a mixture of LDPE – EVA-118 has a minimum compatibility [12]. This affects the deformation-strength properties of the compositions: the elastic modulus E, tensile strength σ, elongation at break ε and the specific work of destruction w. If the mixtures of LDPE with EVA-111 (EVA-113, EVA-115) characterized by the dependencies, close to the additive, then for a mixture of LDPE – EVA-118 is observed a negative deviation from the additive, until the appearance of a minimum (Fig. 4). This suggests a low interphase adhesion and is in a qualitative agreement with the variation of A (Fig. 3).

Thus, the adhesion of the investigated mixtures to steel and epoxy primer is due to the presence and concentration of polar vinyl acetate groups in EVA, which interact with functional groups of adherend by the acid–base mechanism [17–21].

A significant increase of A starts in the region of phase inversion. The maximum values of adhesion strength to steel, as well as to the epoxy primer, are observed in the area of prevailing values of EVA, when it forms a continuous phase. In the systems with a steel substrate an adhesive component is not sufficient for the formation of a strong adhesive joint, so the destruction has visually an adhesive character and goes over the boundary "adhesive – steel." In the case of the epoxy substrate a greater increase in the adhesive joint strength is observed. It is due to increase of differences in the parameters of acidity of the contacting surfaces during the transition from steel to epoxy primer. Cohesive strength of the material becomes less than the adhesive component, and the destruction has a cohesive character. For the investigated mixtures the adhesion to epoxy primer (fully) and to steel (in general) is consistent with the solubility graph – on the area where EVA forms a continuous phase, the adhesive strength is maximum. At the chosen conditions of adhesive contact formation the transition from the adhesive character of destruction to cohesive is determined by the substrate nature, which defines the intensity of interphase interaction.

For a more conclusive evidence of the above assumptions, we have extended the range of the studied objects (LDPE-115, EVA-20, EVA-28, EBA, EVAMA-14, EVAMA-28). To make the results more evident, the formation time of the contact was reduced, so the influence of the contact thermooxidative processes was practically excluded. Concentration dependencies of solubility for the mixtures LDPE-115 – ethylene copolymer (integral and relative to the total content of the soluble component) are the similar to the dependencies of solubility for the mixtures LDPE-153 – EVA.

Table 3 shows the relationship of MFR with structural characteristics of the mixtures at the standard measurement temperatures of 190°C and 125°C. The results presented in Table 3 agree with each other similar to those described above for LDPE-153 and EVA-111, 113, 115, 118.

TABLE 3 Relationship of rheological and structural properties of the mixtures at a MFR measurement temperature of 190°C and 125°C.

Mixture	|ΔMFR| 190°C	|ΔMFR| 125°C	The formation of a continuous phase of the copolymer at a concentration of the copolymer,%
LDPE–115 – EBA	298	16.44	40%
LDPE–115 – EVAMA–28	173	13.02	40%
LDPE–115 – EVA–20	13	1.19	50%
LDPE–115 – EVA–28	0.5	0.51	50%
LDPE–115 – EVAMA–14	3.5	0.06	40%

Thus, once again confirms the conclusion that the greater the difference in viscosity of the components, the earlier the less viscous component forms the continuous phase. However, in this group of the mixtures EVAMA-14 looks like an exception. It can be assumed that this result is due to the fact that for this composition is realized a matrix structure (both phases are completely continuous) for the considered ratios of the components in the mixture. This makes it possible to completely dissolve the soluble component, while the dispersion medium it forms at higher concentrations. This assumption is confirmed by the results obtained by the atomic force microscopy method.

Fig. 5a shows a microphotograph of the system LDPE-115 – EVAMA-14 containing 40% of EVAMA-14, i.e., at the point at which the adhesive bond strength to steel ceases to be zero (Fig. 6). There are two continuous phases: polyethylene insoluble phase and terpolymer soluble phase. We have assumed that the continuous phase of the copolymer begins to form a little bit later, when the adhesion strength value becomes additive (Fig. 6). It follows from the data of the microphotograph of the system LDPE-115 – 70% EVAMA-14 (Fig. 5b). You can see that the "islands" of LDPE-115 dispersed phase "rise" over the soluble dispersion medium formed by EVAMA-14.

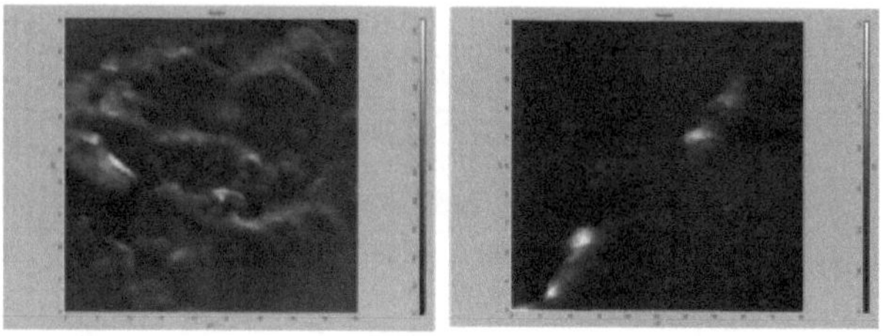

FIGURE 5 The structure of the system LDPE–115 – EVAMA–14 obtained with an atomic force microscope with EVAMA–14 content of 40% (a) и 70% (b).

With regard to adhesive strength of the other systems, for the compositions of LDPE-115 – EBA and LDPE-115 – EVAMA-28, that is, with the less strong "adhesion active" component and the largest reduced acidity parameter, the cohesion destruction character in the range when the copolymer forms a continuous phase is observed. In the region of phase inversion and the continuous phase of LDPE-115 the adhesive strength drops to zero. EVAMA-28 strength characteristics is several times higher than the characteristics of EBA and therefore higher its adhesive strength. Graphs of the concentration dependencies of A for both systems are S-shaped and lie below the additive values, as well as their physical and mechanical properties. It is due to the low interphase adhesion of incompatible components because of the high content of polar units in the copolymers.

For the both systems LDPE-153 – EVA and LDPE-115 – EBA there is a correlation between the adhesive strength and the specific work of destruction in the area of cohesive destruction with a correlation coefficient k = 0.99.

This result indicates that the nature of the adhesion strength concentration dependence and the destruction character of the system LDPE-115 – EBA is determined by the physical and mechanical properties of the compositions, which, in turn, determined by the structure of the colloidal mixture. For the system LDPE-115 – EVAMA-28 the results of adhesion and the physical and mechanical properties are in a qualitative agreement.

For the systems LDPE-115 – EVA-20 and LDPE-115 – EVA-28 in the selected conditions of adhesive contact formation the adhesive nature of separation is observed. With decreasing of polyethylene content in the composition the adhesive strength of the connection increases similarly to the above systems LDPE-153 – EVA.

Concentration dependencies of the physical and mechanical properties for the systems LDPE-115 – EVA-20 and LDPE-115 – EVA-28 are slightly different, but in terms of adhesive destruction and low values of A the limiting factor is the interphase interaction, characterized by the reduced acidity parameter, which has a low value (Table. 1).

The values of adhesive strength for the systems with EVA-20, EVA-28 and EBA are approximately equal to 2 kN/m, but for EVA an adhesion character of separation is observed, and for EBA this character is cohesion.

At the interaction with the steel EBA has a higher reduced acidity parameter (Table 1), but the low mechanical properties. In this case, the adhesive strength exceeds the cohesive, as a result the low values of stress at peeling are realized at the cohesive character of the destruction. EVA-20 and EVA-28 have high physical and mechanical properties and low reduced acidity parameter, so the cohesive strength exceeds the adhesive.

The greatest value of the strength of adhesive joint with the steel at the adhesion character of separation (Fig. 6) is the feature of the system LDPE-115 – EVAMA-14 in this group of the mixtures. This is explained by the high value of the reduced acidity parameter (Table 1), and the high strength properties of the copolymer.

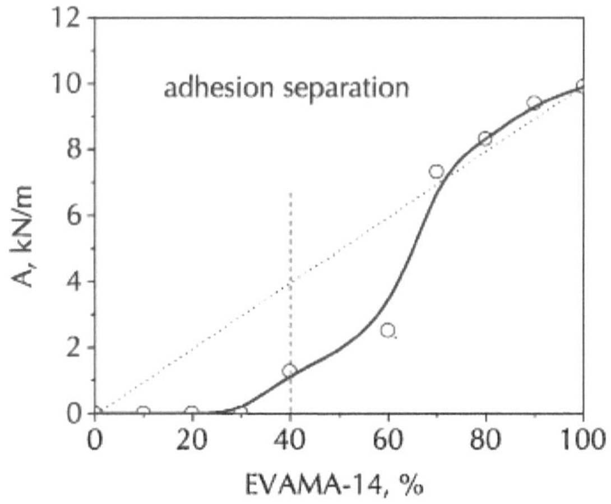

FIGURE 6 Concentration dependence of the strength of adhesive joint with the steel for a mixture LDPE–115 – EVAMA–14.

In this case, it must be assumed that the adhesive character of destruction at the values of A close to those for EVAMA-14 in the conditions of the cohesive character of separation, is due to the proximity of the adhesive and cohesive components of A, and the changing of the contact formation conditions (time increasing) will lead to a cohesive separation.

With regard to the character of the curve of concentration dependence of A for a given system, the transitional plot of the curve from zero to the high values is due to the fact that in this range a phase of LDPE as well as a phase of EVAMA-14 are continuous, and the data of atomic force microscopy method indicate that (Fig. 5).

KEYWORDS

- **Adhesion active**
- **Adhesionnoactive**
- **Inspekt mini**
- **Nonadhesionnoactive**
- **Soxhlet apparatus**

REFERENCES

1. Shubnikov, M. V.; Sayfeev, F. G.; Deberdeev, R. Ya.; Stoyanov, O. V. *Sposob polucheniya polimernoi kompozicii*. A.S.148261.
2. Kurnosov, V. V.; Dederdeev, R. Ya.; Sergeeva, Je. A.; Stoyanov, O. V. *J. Polym. Ing.* **1997**, 17(4), 282–294.
3. Sirmach, A. I.; Yansons, A. V.; Ozolnish, Ju. L. *Modificaciya Polimernykh Materialov. – Riga,* **1986**, 8–14.
4. Malers, L. Ya.; Vapkalis, A. Ju.; Yansons, A. Ya. *Modificaciya polimernykh materialov: Sb. nauch. tr. – Riga,* **1989**, 4–13.
5. Khuzakhanov, R. M.; Mukhamedsyanova, E. R.; Zaikin, A. E.; Kapitskaya, Ya. V.; Nikitina, N. N.; Stoyanov, O. V. *Vliyanie sostava sevilenovykh kompozicii na prochnost ikh adgezionnogo soedineniya so staliu Vestnik Kazan.* Tekhnol. Universiteta, 2003, Vol. 1, pp. 337–341.
6. Khuzakhanov, R. M.; Kapitskaya, Ya. V.; Mukhamedsyanova, E. R.; Dederdeev, R. Ya.; Stoyanov, O. V. *Germetiki. Tekhnologii.* **2005**, *4*, 24–27.

7. Khuzakhanov, R. M.; Kapitskaya, Ya. V.; Dederdeev, R. Ya.; Stoyanov, O. V. *Germetiki. Tekhnologii.*, **2005**, *5*, 15–16.
8. Stoyanov, O. V.; Khuzakhanov, R. M.; Stoyanova, L. F.; Gerasimov, V. K.; Chalykh, A. E.; Aliev, A. D.; Vokal, M. V. *Germetiki. Tekhnologii.* **2010**, *11*, 15–17.
9. Kapitskaya, Ya. V. Adgezionnye materialy na osnove sevilenovykh smesey: Diss... kand. tekn. nauk. Kazan, 2004, 147 s.
10. Khmelevskaya, I. O.; Bezgaev, A. F.; Trizno, M. S.; Safronova, R. F.; Straz, Ya. A. *Plast. Massy.* **1987**, **3**, 30–31.
11. Tabachnik, L. B.; Vanshtein, A. B.; Karlivan, V. P. *Plast. Massy.* **1977**, *12*, 24–26.
12. Dryz, N. I.; Chalykh, A. E.; Aliev, A. D. *Vysokomolekul. soed. – Ser. B.*, **1987**, *T.29*(2), 101–104.
13. Berger E. J. *J. Adhes. Sci. Technol.* **1990**, 4(5), 373–391.
14. http://www.komef.ru/reopolimer.pdf
15. Basin V. E. *Adgezionnaya prochnost.* – M.: Khimiya, **1981**, 208 s.
16. Kinlok, E. *Adgeziya i adgezivy*. Nauka i tekhnologiya. – M.: Mir, **1991**, 484 s.
17. Starostina, I. A.; Burdova, E. V.; Sechko, E. V.; Khuzakhanov, R. M.; Stoyanov, O. V. *Vliyanie kislotno-osnovnykh svoistv metallov, polimerov i polimernykh kompozicionnykh materialov na adgezionnoe vzaimodeistvie v metall-polimernykh sistemakh Vestnik Kazanskogo tekhnologicheskogo universiteta*, 2009, Vol. 3, 85–95.
18. Khasbiullin, R. R.; Stoyanov, O. V.; Chalykh, A. E.; Starostina, I. A. *Vliyanie kislotno-osnovnykh vzaimodeistvii na adgezionnyju sposobnost soedinenii polietilena s metallami ZhPKh.* **2001**, *T.74*(11), 1859–1862.
19. Starostina, I. A.; Burdova, E. V.; Kustovskii, V. Ya.; Stoyanov, O. V. *Rol kislotno-osnovnykh vzaimodeistvii v formirovanii adgezionnykh soedinenii polimerov s metallami Klei*. Germetiki. Tekhnologii, 2005, Vol. 10, 16–21.
20. Kustovskii, V. Ya., Starostina, I. A., Stoyanov, O. V. *Zhurnal Prikladnoi Khimii*, **2006**, *T.79(*Vup. 6), 940–943
21. Starostina, I. A.; Stoyanov, O. V.; Garipov, R. M.; Zagidullin, A. I.; Kustovskii, V. Ya.; Koltsov, N. I.; Kuzmin, M. V.; Trofimov, D. M.; Petrov, V. G. *Lakokrasochnye Materially i ikh Primeneniei*, **2007**, *5*, 32–36.

CHAPTER 26

UPDATES ON THE SURFACE FREE ENERGY PARAMETERS

A. STAROSTINA, O. V. STOYANOV, N. V. MAKHROVA, G. E. ZAIKOV, and R. YA. DEBERDEEV

CONTENTS

26.1 INTRODUCTION

According to acid–base theory, an important role in the formation of adhesion bonds are played by the interphase acid–base interactions between the adhesive and the substrate. Within this theory, the van Oss-Chaudhury-Good approach is most popular [1], in which characteristics of the acid and base interactions are nonadditive parameters γ^+ and γ^- of the surface free energy. Unfortunately, the values of γ^+ and γ^- of solids are impossible to determine if it is impossible to obtain the values of the corresponding parameters of test liquids from wettability data. The theory's authors calculated that the surfaces of most polymers (polystyrene, poly(vinyl alcohol), poly(methyl methacrylate), cellulose nitrate) are mainly basic (this fact is referred to in the scientific literature as the "basic disaster"), which prompted to introduce the term "monopole surfaces" [1]. The surface is called monopole if it is either only basic ($\gamma^- = 0$) or only acidic ($\gamma^+ = 0$). The simplicity of calculating the acid and base parameters of liquids and solids within the said hypothesis has interested many researchers, and this method is now being used quite widely (see, e.g., [2, 3]). Typically used acidic and basic monopole surfaces are those of poly(vinyl chloride) [2] and poly(methyl methacry-late) [3], respectively. However, the γ^+ and γ^- values of test liquids that are determined by this method are often inconsistent with their chemical nature. Therefore, the existence of monopole polymer surfaces requires detailed investigation. This is all the more so, as the Lewis and Bronsted theories say that there are no pure acids and pure bases and that any substance has both acid and base properties.

26.2 METHODS

Our IR spectroscopic study of polymer samples confirmed a complex and ambiguous composition of poly(methyl methacrylate) and poly(vinyl chloride). The transmission and multiple attenuated internal reflection IR spectra in the region of C=O group vibrations show not only the major band at 1740–1725 cm^{-1}, which characterizes the absorption of the ester carbonyl group, but also a sextet of bands caused by high sensitivity of the carbonyl absorption to a change in the nearest surroundings [4].

Analysis of the spectra suggested that one cannot rule out the insignificant presence of double bonds in the carbon chain at the surface (because of structural defects and destructive processes) and also the existence of various functional groups-aldehyde, ketone, and carboxyl (acidic) groups [5]. The poly(vinyl chloride) spectrum has a broad band at 3410 cm^{-1}, which is likely to be caused by hydrogen bonds involving the acidic hydro-gen of the polymer and the hydroxyl oxygen of water vapor captured during casting of the polymer film from a tetrahydrofuran solution.

Additionally, we analyzed poly(vinyl chloride) and poly(methyl methacrylate) by the quantum-chemical density functional theory method. Fig. 1 presents the results obtained.

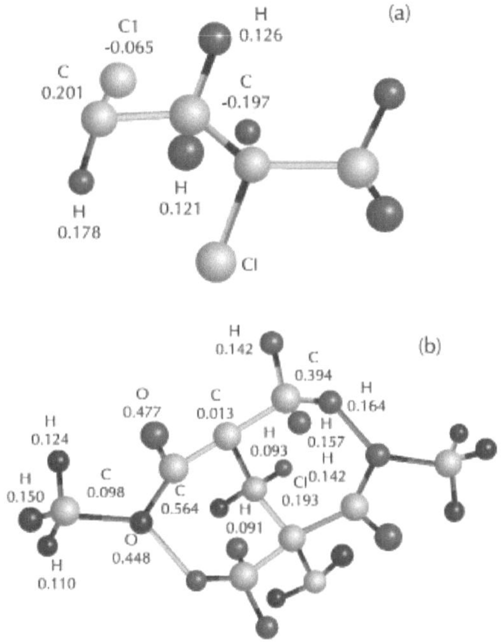

FIGURE 1 Electron density distribution in monomeric units of (a) poly(vinyl chloride) and (b) poly(methyl methacry-late).

Figure 1a shows that the poly(vinyl chloride) mac-romolecule indeed has strong electron-acceptor sites (δ_H = 0.178), which confirms the mainly acidic nature of this polymer. However, along with them, poly(vinyl chloride) also contains chlorine atoms with excess electron density (δ_{Cl}

= –0.065). Although a chlorine atom is a weak electron-donor site, it can exhibit basic properties in the presence of a strong acid. In its turn, poly(methyl methacrylate) contains mobile hydrogen atoms. Given a certain arrangement of the polymer chain, some of these atoms can be involved in the formation of an intramolecular hydrogen bond (and also intermolecular hydrogen bonds in the bulk of the polymer) (Fig. 1b).

26.3 CONCLUSION

All the above suggests that the surface of neither poly(methyl methacrylate) nor poly(vinyl chloride) can be considered monopole. Most likely, such a conclusion is valid for any polymer surface. Nonetheless, various polymer surfaces have long and successfully been used as probe surfaces, e.g., for determining the surface tension of liquids and oligomers [6]. A legitimate question arises as to whether it is possible and, if so, how to obtain reliable information on the γ^+ and γ^- values of liquids and solid surfaces. Della Volpe et al. proposed to solve this issue by solving a system of nonlinear equations [7]. We modified and simplified the method for solving such systems. The determined parameters of test liquids [8] can be used to find the sought–for values for polymers by two ways:

(1) by solving systems of nonlinear equations containing the cosines of the contact angles of an unknown polymer surface with test liquids [8]; or

(2) by applying a graphical method for determining the γ^+ and γ^- of a polymer surface.

In the first case, the system is highly overdeter-mined, there are many equisignificant solutions, and the problem of choosing the only correct solution remains open.

In the second case, the van Oss-Chaudhury-Good equation

$$W_{LS} = \gamma_L(\cos\theta + 1) = 2\sqrt{\gamma_S^{LW}\gamma_L^{LW}} + 2\sqrt{\gamma_S^+\gamma_L^-} + 2\sqrt{\gamma_S^-\gamma_L^+}$$

where W_{LS} is the adhesion work done by liquid *(L)* on solid *(S)*, is transformed to the form $y = ax + b$, and a

straight line in the coordinates $\left(A, \sqrt{\gamma_L^+\gamma_L^-}\right)$ is plotted, where

$$A = \left(\frac{W_{LS}}{2} - \sqrt{\sqrt{\gamma_S^{LW} \gamma_L^{LW}}} \right) \Big/ \sqrt{\gamma_L^-}$$

The y-intercept of the straight line is $\sqrt{\gamma_s^+}$, and the slope is $\sqrt{\gamma_s^-}$. This method gives the only value of each of the quantities γ^+ and γ^-.

Figure 2 exemplifies the graphical method for determining the surface free energy parameters of poly(vinyl chloride) and poly(methyl methacrylate). For constructing the graphs, we used the previously determined [8] surface free energy parameters of aniline, dimethyl sulfoxide, formamide, glycerol, and double distilled water. The correlation coefficients for both straight lines are 0.99. For poly(methyl methacrylate), the graphically found acid and base parameters are 0.28 and 0.06 mJ/m^2, respectively; and for poly(vinyl chloride), they are 1.54 and 0.48 mJ/m^2, respectively.

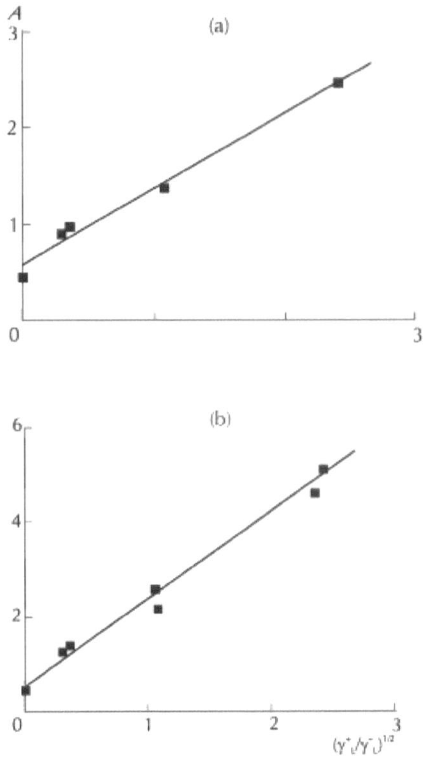

FIGURE 2 Graphical determination of the surface free energy parameters of (a) poly(vinyl chloride) and (b) poly(methyl methacrylate).

Thus, we have shown the invalidity of using poly(methyl methacrylate) and poly(vinyl chloride) polymer surfaces as probe monopole surfaces and demonstrated the possibility of correctly determining the acid and base parameters of the surface free energy of polymers by a simple and illustrative graphical method.

KEYWORDS

- **Acid–base theory**
- **Basic disaster**
- **Monopole surfaces**
- **Tetrahydrofuran**

REFERENCES

1. Van Oss, C. J.; Chaudhury, M. K.; Good, R. J., *Adv. Colloid Interface Sci.*, **1987**, *28*, 35–64.
2. Berg, J. C., *Wettability;* New York: M. Dekker, 1993.
3. Jan ‘czuk, B.; Bialopiotrowicz, T.; Zdziennicka, A. *J. Colloid Interface Sci.*, **1999**, *211*, 96–103.
4. Kazitsina, L. A.; Kupletskaya, N. B. *Primenenie UF-, IK-, YaMR- i mass-spektroskopii v organicheskoi khimii* (Applications of UV, IR, NMR, and Mass Spec-troscopy to Organic Chemistry), Mosk. Gos. Univ.: Moscow, 1979.
5. Troitskii, B. B.; Domrachev, G. A.; Anikina, L. I.; et al. *Vestn. Nizhni Novgorod. Gos. Univ.*, **2001**, *1*, 60–71.
6. Chalykh, A. E.; Busygin, V. B.; Stepanenko, V. Yu. *Vysokomol. Soedin.*, **1999**, *41*(11), 1843–1846.
7. Della Volpe, C.; Siboni, S.; Maniglio, D.; Morra, M. *J. Adhesion Sci. Technol.*, **2003**, *17*(11), 1425–1456.
8. Starostina, I. A.; Stoyanov, O. V.; Makhrova, N. V.; Deberdeev, R. Ya. *Dokl. Phys. Chem.*, **2011**, *463*(part 1), 8–9 [*Dokl. Akad. Nauk*, **2011**, *436*(3), 343–345].

CHAPTER 27

QUANTIZATION OF ENERGY INTERACTIONS

G. A. KORABLEV and G. E. ZAIKOV

CONTENTS

27.1 INTRODUCTION

Quantum concepts on atom and molecule composition constitute the foundation of modern natural science theories. Thus, in a stationary state the moment of electron motion amount equals the integral multiple from Plank's constant. This main quantum number and three other together expressly characterize the state of any atom. The repetition factor of atom quantum characteristics is expressed in spectral data for both simple and complex structures.

It is known that any complex periodic processes can be represented in the form of separate simple harmonic waves. "By Fourier theory, oscillations of any form with period T can be presented as the total of harmonic oscillations with periods T_1, T_2, T_3, T_4, etc. Knowing the periodic function form we can calculate the amplitude and phases of sinusoids with the given function as their summing up" [1, 2].

Therefore many regularities of intermolecular interactions, complex formation and nanothermodynamics are explained by the application of functional multiple quantum or wave energy characteristics of structural interactions.

For this purpose we use the notion on spatial-energy parameter (P-parameter) in this research.

27.2 INITIAL CRITERIA

In the process of interactions of oppositely charged heterogeneous systems we observe a certain compensation of volume energy of interacting structures, which leads to the resultant energy, decrease (e.g., during the hybridization of atom orbitals). But this is not a direct algebraic deduction of the corresponding energies. The comparison of multiple regularities of physical and chemical processes allows assuming that in such and similar cases the principle of adding the reciprocals of volume energies or kinetic parameters of interacting structures are fulfilled.

Lagrangian equation for the relative motion of two interacting material points with the mass m_1 and m_2 in x coordinate is as follows:

$$M_r x'' = -\frac{\partial U}{\partial x}, \text{ where } \frac{1}{m_r} = \frac{1}{m_1} + \frac{1}{m_2} \quad (1), (1a)$$

where U – mutual potential energy of material points; m_r – reduced mass. At the same time $x'' = a$ (system acceleration).

For elementary regions of interactions Δx we can take: $\frac{\partial U}{\partial x} \approx \frac{\Delta U}{\Delta x}$.

Then: $m_r a\Delta x = -\Delta U$; $\frac{1}{1/(a\Delta x)} \cdot \frac{1}{(1/m_1 + 1/m_2)} \approx -\Delta U$

or: $\frac{1}{1/(m_1 a\Delta x) + 1/(m_2 a\Delta x)} - \Delta U$

As the product $m_i a\Delta x$ by its physical sense equals the potential energy of each material point ($-\Delta U_i$),

then:
$$\frac{1}{\Delta U} \approx \frac{1}{\Delta U_1} + \frac{1}{\Delta U_2} \quad (2)$$

Thus the resultant energy characteristic of the interaction system of two material points is found by the principle of adding the reciprocals of initial energies of interacting subsystems.

"Electron with the mass m moving near the proton with the mass M is equivalent to the particle with the mass $m_r = \frac{mM}{m+M}$" [3].

Therefore modifying the Eq. (2), we can assume that the energy of atom valence orbitals (responsible for interatomic interactions) can be calculated [4] by the principle of adding the reciprocals of some initial energy components based on the equations:

$$\frac{1}{q^2/r_i} + \frac{1}{W_i n_i} = \frac{1}{P_E} \text{ or } \frac{1}{P_0} = \frac{1}{q^2} + \frac{1}{(Wrn)_i}; \; P_E = P_0/r_i \quad (3),(4),(5)$$

here: W_i – orbital energy of electrons [5]; r_i – orbital radius of i orbital [6]; $q=Z^*/n^*$ – by [7, 8], n_i – number of electrons of the given orbital, $Z^*$and n^* – nucleus effective charge and effective main quantum number, r – bond dimensional characteristics.

P_O is called a spatial-energy parameter (SEP), and P_E – effective P-parameter (effective SEP). Effective SEP has a physical sense of some averaged energy of valence orbitals in the atom and is measured in energy units, e.g. in electron-volts (eV).

The values of P_0-parameter are tabulated constants for electrons of the given atom orbital.

For SEP dimensionality:

$$[P_0]=\left[q^2\right]=[E]\cdot[r]=[h]\cdot[\upsilon]=\frac{kgm^3}{s^2}=Jm,$$

where [E], [h] and [υ] – dimensionalities of energy, Planck's constant and velocity.

The introduction of P-parameter should be considered as further development of quasi-classical concepts with quantum-mechanical data on atom structure to obtain the criteria of phase-formation energy conditions. For the systems of similarly charged (e.g., orbitals in the given atom) homogeneous systems the principle of algebraic addition of such parameters is preserved:

$$\Sigma P_E=\Sigma(P_0/r_i); \tag{6}$$

$$\Sigma P_E=\frac{\Sigma P_0}{r} \tag{7}$$

or:

$$\Sigma P_0=P_0'+P_0''+P_0'''+\ldots; \tag{8}$$

$$r\Sigma P_E=\Sigma P_0 \tag{9}$$

Here P-parameters are summed up by all atom valence orbitals.

To calculate the values of P_E-parameter at the given distance from the nucleus either the atomic radius (R) or ionic radius (r_I) can be used instead of r depending on the bond type.

Let us briefly explain the reliability of such an approach. As the calculations demonstrated the values of P_E-parameters equal numerically (in the range of 2%) the total energy of valence electrons (U) by the atom statistic model. Using the known correlation between the electron density (β) and intra-atomic potential by the atom statistic model [9], we can obtain the direct dependence of P_E-parameter on the electron density at the distance r_i from the nucleus.

The rationality of such technique was proved by the calculation of electron density using wave functions by Clementi [10] and comparing it with the value of electron density calculated through the value of P_E-parameter.

27.3 WAVE EQUATION OF P-PARAMETER

To characterize atom spatial-energy properties two types of P-parameters are introduced. The bond between them is a simple one:

$$P_E = \frac{P_0}{R}$$

where R – atom dimensional characteristic. Taking into account additional quantum characteristics of sublevels in the atom, this equation can be written down in coordinate x as follows:

$$\Delta P_E \approx \frac{\Delta P_0}{\Delta x} \text{ or } \partial P_E = \frac{\partial P_E}{\partial x}$$

where the value ΔP equals the difference between P_0-parameter of i orbital and P_{CD}-countdown parameter (parameter of main state at the given set of quantum numbers).

According to the established [4] rule of adding P-parameters of similarly charged or homogeneous systems for two orbitals in the given atom with different quantum characteristics and according to the energy conservation rule we have:

$$\Delta P_E' - \Delta P_E' = P_{E,\lambda}$$

where $P_{E,\lambda}$ – spatial-energy parameter of quantum transition.

Taking for the dimensional characteristic of the interaction $\Delta\lambda = \Delta x$, we have:

$$\frac{\Delta P_0''}{\Delta\lambda} - \frac{\Delta P_0'}{\Delta\lambda} = \frac{P_0}{\Delta\lambda} \text{ or: } \frac{\Delta P_0'}{\Delta\lambda} - \frac{\Delta P_0''}{\Delta\lambda} = -\frac{P_0\lambda}{\Delta\lambda}$$

Let us again divide by $\Delta\lambda$ term by term:

$$\left(\frac{\Delta P_0'}{\Delta\lambda}-\frac{\Delta P_0''}{\Delta\lambda}\right)\Big/\Delta\lambda=-\frac{P_0}{\Delta\lambda^2}\text{, where:}$$

$$\left(\frac{\Delta P_0'}{\Delta\lambda}-\frac{\Delta P_0''}{\Delta\lambda}\right)\Big/\Delta\lambda\sim\frac{d^2P_0}{d\lambda^2}\text{, i.e.: }\frac{d^2P_0}{d\lambda^2}+\frac{P_0}{\Delta\lambda^2}\approx 0$$

Taking into account only those interactions when $2\pi\Delta x=\Delta\lambda$ (closed oscillator), we have the following equation:

$$\frac{d^2P_0}{dx^2}+4\pi^2\frac{P_0}{\Delta\lambda^2}\approx 0$$

Since $\Delta\lambda=\frac{h}{mv}$, then:

$$\frac{d^2P_0}{dx^2}+4\pi^2\frac{P_0}{h^2}m^2v^2\approx 0$$

or

$$\frac{d^2P_0}{dx^2}+\frac{8\pi^2m}{h^2}P_0E_k=0 \qquad (10)$$

where $E_k=\frac{mV^2}{2}$ – electron kinetic energy.

Schrodinger equation for the stationery state in coordinate x:

$$\frac{d^2\psi}{dx^2}+\frac{8\pi^2m}{h^2}\psi E_k=0$$

When comparing these two equations we see that P_0-parameter numerically correlates with the value of Ψ-function:

$$P_0\approx\Psi,$$

and is generally proportional to it: $P_0\sim\Psi$. Taking into account the broad practical opportunities of applying the P-parameter methodology, we can consider this criterion as the materialized analog of Ψ-function [11,12].

Since P_0-parameters like Ψ-function have wave properties, the superposition principles should be fulfilled for them, defining the linear character of the equations of adding and changing P-parameter.

27.4 QUANTUM PROPERTIES OF P-PARAMETER

According to Planck, the oscillator energy (E) can have only discrete values equaled to the whole number of energy elementary portions-quants:

$$nE = h\nu = hc/\lambda \tag{11}$$

where h – Planck's constant, ν – electromagnetic wave frequency, c – its velocity, λ – wavelength, n = 0, 1, 2, 3…

Planck's equation also produces a strictly definite bond between the two ways of describing the nature phenomena – corpuscular and wave.

P_0-parameter as an initial energy characteristic of structural interactions, similarly to the Eq. (11), can have a simple dependence from the frequency of quantum transitions:

$$P_0 \sim \hbar(\lambda\nu_0), \tag{12}$$

where: λ – quantum transition wavelength [14]; $\hbar = h/(2\pi)$; ν_0 – kayser, the unit of wave number equaled to $2.9979 \cdot 10^{10}$ Hz.

In accordance with Rydberg equation, the product of the right part of this equation by the value $(1/n^2 - 1/m^2)$, where n and m – main quantum numbers – should result in the constant.

Therefore the following equation should be fulfilled:

$$P_0(1/n_1^2 - 1/m_1^2) = N\hbar(\lambda\nu_0)(1/n^2 - 1/m^2), \tag{12a}$$

where the constant N has a physical sense of wave number and for hydrogen atom equals $2 \cdot 10^2 Å^{-1}$.

The corresponding calculations are demonstrated in Table 1. There: r_i' = 0.5292 Å – orbital radius of 1S-orbital and r_i'' = $=2^2 \cdot 0.5292 = 2.118$ Å – the value approximately equaled to the orbital radius of 2S-orbital.

The value of P_0-parameter is obtained from the Eq. (4), e.g. for 1S-2P transition:

$$1/P_0 = 1/(13.595 \cdot 0.5292) + 1/14.394 \rightarrow P_0 = 4.7985 \text{ eVÅ}$$

The value q^2 is taken by [7, 8], for the electron in hydrogen atom it numerically equals the product of rest energy by the classical radius.

The accuracy of the correlations obtained is in the range of percentage error 0.06 (%), i.e., the Eq. (12a) is in the accuracy range of the initial data.

In the Eq. (12a) there is the link between the quantum characteristics of structural interactions of particles and frequencies of the corresponding electromagnetic waves.

But in this case there is the dependence between the spatial parameters distributed along the coordinate. Thus in P_0-parameter the effective energy is multiplied by the dimensional characteristic of interactions, and in the right part of the Eq. (12a) the Kayser value is multiplied by the wavelength of quantum transition.

In Table 1 you can see the possibility of applying the Eq. (12a) and for electron Compton wavelength ($\lambda_\kappa = 2.4261 \cdot 10^{-12}$ m), which in this case is as follows:

$$P_0 = 10^7 \hbar(\lambda_\kappa v_0) \quad (13)$$

(with the error of 0.25%).

Integral-valued decimal values are found when analyzing the correlations in the system "proton-electron" given in Table 2:

1. Proton in the nucleus, energies of three quarks $5 + 5 + 7 \approx 17$ (MeV) $\rightarrow P_p \approx 17 \text{ MeV} \cdot 0.856 \cdot 10^{-15}$ m $\approx 14.552 \cdot 10^{-9}$ eVm. Similarly for the electron $P_e = 0.511 \text{ (MeV)} \cdot 2.8179 \cdot 10^{-15}$ m (electron classic radius) $\rightarrow P_e = 1.440 \cdot 10^{-9}$ eVm.

   Therefore: $$P_p \approx 10\, P_E \quad (14)$$

2. Free proton $P_n = 938.3 \text{ (MeV)} \cdot 0.856 \cdot 10^{-15} \text{ (m)} = 8.0318 \cdot 10^{-7}$ eVm. For electron in the atom $P_a = 0.511 \text{ (MeV)} \cdot 0.5292 \cdot 10^{-5} \text{(m)} = 2.7057 \cdot 10^{-5}$ eVm.

   Then: $$3P_a \approx 10^2 P_n \quad (15)$$

The relative error of the calculations by this equation is found in the range of the accuracy of initial data for the proton ($\delta \approx 1\%$).

27.5 ON QUANTIZATION DECIMAL PRINCIPLE

From Tables 1 and 2 we can see that the wave number N is quantized by the decimal principle: $N = n10^Z$, where n and Z – whole numbers.

Other examples of electrodynamics equations should be pointed out in which there are integral-valued decimal functions, e.g. in the formula: $4\pi\varepsilon_0 c^2 = 10^7$, where ε_0 – electric constant.

In [14] the expression of the dependence of constants of electromagnetic interactions from the values of electron получено P_e-parameter was obtained:

$$k\mu_0 c = k/(\varepsilon_0 c) = P_e^{1/2} c^2 \approx 10/\alpha \quad (16)$$

where: $k = 2\pi/\sqrt{3}$; μ_0 – magnetic constant; c – electromagnetic constant; α – fine structure constant.

All the above conclusions are based on the application of rather accurate formulas. But sometimes the analogies are necessary. Apparently, such analogy means that the whole orderliness of the Universe is arranged by the method of a complex Russian nested doll: atom nucleus, atom, plasma, solar system, etc. One structure in some important case resembles the other, and many physical regularities can be the same for all subsystems, including those applicable for quantization decimal issue.

Quantum interactions in the systems “proton-electron” and “proton-positron” define the initial generation processes of the solar system as a result of thermonuclear reactions [15]. Thus, in one such reaction two protons are combined forming the deuterium nucleus ejecting the positron and neutrino. In another initial reaction two protons and electron are combined forming the deuterium nucleus and neutrino. The correlations of energy parameters of these particles (given in Table 2) are decimal, they can initiate the similar dependencies in complex macrostructures formed by them.

TABLE 1 Quantum properties of hydrogen atom parameters.

Orbitals	W_i (eV)	r_i (Å)	q_i^2 (eVÅ)	P_0 (eVÅ)	$P_0(1/n_1^2 - 1/m_1^2)$ (eVÅ)	N (Å^{-1})	λ (Å)	Quantum transition	$N\hbar\lambda v_0$ (eVÅ)	$N\hbar\lambda v_0 \cdot (1/n^2 - 1/m^2)$ (eVÅ)
1S	13.595	0.5292	14.394	4.7985	3.5989	$2\cdot10^2$	1215	1S–2P	4.7951	3.5963
1S						$2\cdot10^2$	1025	1S–3P	4.0452	3.5954
1S						$2\cdot10^2$	912	1S–nP	3.5990	3.5990
2S	3.3988	2.118	14.394	4.7985	3.5990	$2\cdot10^2$	6562	2S–3P		3.5967
2S						$2\cdot10^2$	4861	2S–4P		3.5971
2S						$2\cdot10^2$	3646	2S–nP		3.5973
1S	13.595	0.5292	14.394	4.7985		10^7	$2.4263\cdot10^{-2}$	–	4.7878	

Thus the coefficient 10 for the numbers of solar spots (R) is introduced into the Wolf formula for the groups of solar spots (g): R = k(10g + S), where S – number of separate spots, k – correction factor.

In the Eq. (12a) the wave number is multiplied by the coefficient ($1/n^2 - 1/m^2$), considering the series of quantum transitions. Thus for the initial series 1S–2P it equals 3/4. Modifying this condition to the initial values N = 10, we can assume the presence of quantum series by the formula N = 10(3 + n)/4, resulting in N = 7.5; 10; 12.5; 15; 17.5… for the series at n = 0,1,2,3,4…

Comparing with the data by Ref. [13] given in Table 3, we can see that these values of N correlate (with the accuracy from 5 up to 11%) with the duration of cycles between their maximums.

Applying these quantum numbers, the year of maximum solar activity have been calculated from the first registered ones to 2012 (Table 3).

TABLE 2 Quantum ratios of proton and electron parameters.

Particle	E (eV)	r (Å)	P = Er (eVÅ)	Ratios
Free proton	$938.3 \cdot 10^6$	$0.856 \cdot 10^{-5}$	$8.038 \cdot 10^3 = P_n$	$3P_a/P_n \approx 10^2$
Electron in the atom	$0.511 \cdot 10^6$	0.5292	$2.7042 \cdot 10^5 = P_a$	
Proton in the atom nucleus	$(5 + 5 + 7) \cdot 10^6 = 17 \cdot 10^6$	$0.856 \cdot 10^{-5}$	$145.52 = P_p$	$P_p/P_e \approx 10$
Electron	$0.511 \cdot 10^6$	$2.8179 \cdot 10^{-5}$	$14.399 = P_e$	

The relative error of the data obtained in comparison with the fixed year of solar activity maximums given in line 6 of Table 3 is rather low (from 0 to 0.07%). Thus such quantum numbers as a whole rationally reflect the cyclicity of such processes. But not only the main quantum number of the cycle can influence its duration but another number close to it, as the cycle maximums often have two activity peaks. Thus cycles No 1, 6 and 20 have two such activity peaks and respectively two quantum numbers 10 and 12.5 operate at the fixed period duration between the maximums 11.2; 11.2 and 11.0. The similar situation is in some other cases.

Therefore the error in the correlation of the average number of N cycle with the cycle duration between the maximums is under 2.2%.

Quantum number 10 is more frequently realized particularly during the last hundred years. It can be assumed that this number fulfills the most stationary conditions of such macroprocesses. In cycle No 5 the period duration between the maximums is 17.1 yr, which cannot be explained by conventional approaches. Quantum number, according to this model, in cycle No 4 operates by the same model when transiting to a new century with possible resonance coincidence of centenary and decennary cycles.

But the problem issues of the specifics and conditions of realization of such quantum numbers need to be further investigated.

TABLE 3 Years of solar cycle formation – lines 1, 2, 3, 5 – till [13].

Cycle number (N_0)	Maximum years	Wolf number W	Subsystem totals (calculated)	Cycle duration between max	Relative error, %
1	2	3	4	5	6
–12	1615.5		1600+15=1615	–	0.031
–11	1626.0		1615+10=1625	10.5	0.06
–10	1639.5		1625+15=1640	13.5	0.030
–9	1649.0		1640+10=1650	9.5	0.06
–8	1660.0		1650+10=1660	11.0	0
–7	1675.0		1660+15=1675	15.0	0
–6	1685.0		1675+10=1685	10.0	0
–5	1693.0		1685+7.5=1692.5	8.0	0.042
–4	1705.5	54	1692.5+12.5=1705	12.5	0.029
–3	1718.2	60	1705+12.5=1717.5	12.7	0.041
–2	1727.5	113	1717.5+10=1727.5	9.3	0
–1	1738.7	112	1727.5+10=1737.5	11.2	0.069
0	1750.5	92.6	1737.5+12.5=1750	11.6	0.029
1	1759.7	77.2	1750+10=1760	11.2	0.017
	1761.5	86.5	1750+12.5=1762.5		0.057
2	1769.7	115.8	1762.5+7.5=1770	8.2	0.017
3	1778.4	158.5	1770+7.5=1777.5	8.7	0.051
4	1788.1	141.2	1777.5+10=1787.5	9.7	0.034

TABLE 3 *(Continued)*

Cycle number (N_0)	Maximum years	Wolf number W	Subsystem totals (calculated)	Cycle duration between max	Relative error, %
1	2	3	4	5	6
5	1804.6	48.6	1787.5+17.5=1805	17.1	0.022
	1805.2	49.2			0
6	1816.4	48.7	1805+10=1815	11.2	0.077
	1817.2	W = 96.2	1805+12.5=1817.2		
7	1829.9	71.7	1815+15=1830	13.5	0.006
8	1837.2	146.9	1830+7.5=1837.5	7.3	0.016
9	1848.1	131.6	1837.5+10=1847.5	10.9	0.032
10	1860.1	97.9	1847.5+12.5=1860	12	0.005
11	1870.6	1405	1860+10=1870	10.5	0.032
12	1883.9	74.6	1870+12.5=1882.5	13.3	0.074
13	1894.1	87.9	1882.5+12.5=1895	10.2	0.048
14	1907.4	62.8	1895+12.5=1907.5	12.9	0.042
	1906.1	64.2	1895+10=1905		0.052
	1905.45	63.4	1895+10=1905		0.024
15	1917.6	105.4	1907.5+10=1917.5	10.6	0.01
16	1928.2	78.1	1917.5+10=1927.5	10.8	0.036
17	1937.4	119.2	1927.5+10=1937.5	9.0	0.005
18	1947.5	151.8	1937.5+10=1947.5	10.1	0
19	1957.9	201.3	1947.5+10=1957.5	10.4	0.020
20	1968.9	110.6	1957.5+10=1967.5	11.0	0.071
	1970.2	106.2	1957.5+12.5=1970		0.010
21	1978.7	138.2	1967.5+10=1977.5	10	0.061
	1980	176.3	1970+10=1980		0
22	1989.5	158.5	1980+10=1990	11	0.060
23	2000.3	120.8	1990+10=2000	10	0.015
24	2010	From 120	2000+10=2010	10	0
	2012	to 165	2000+12.5=2012.5		0.025

KEYWORDS

- **Fourier theory**
- **Proton-electron**
- **Proton-positron**
- **Spatial-energy**

REFERENCES

1. Gribov, L. A.; Prokofyeva, N. I. *Basics of Physics*. M.: Vysshaya shkola, 1992; 430p.
2. Putilov, K. A. *Course of Physics*, Vol. 1. M.: Publishing House of Technical Literature, 1954; 708p.
3. Eyring, G.; Walter, J.; Kimball, G. *Quantum Chemistry* M., F. L., 1948; 528p.
4. Korablev, G. A. *Spatial Energy Principles of Complex Structures Formation.* Brill Academic Publishers and VSP: Netherlands, Leiden, 2005; 426pp. (Monograph).
5. Fischer, C. F. *Average-Energy of Configuration Hartree-Fock Results for the Atoms Helium to Radon Atomic Data*, **1972**, *4*, 301–399.
6. Waber, J. T.; Cromer, D. T. *J. Chem. Phys.*, **1965**, 42(12), 4116–4123.
7. Clementi, E.; Raimondi, D. L. *J. Chem. Phys.*, **1963**, *38*(11), 2686–2689.
8. Clementi, E.; Raimondi, D. L. *J. Chem. Phys.*, **1967**, *47*(4), 1300–1307.
9. Gombash, P. *Statistic Theory of Atom and its Application* M.: F. L., 1951; 398p.
10. Clementi, E. *J. B.M. S. Res. Develop. Suppl.*, **1965**, *9*(2), 76.
11. Korablev, G. A.; Zaikov, G. E. *J. Appl. Polym. Sci., USA*, **2006**, *101*(3), 2101–2107.
12. Korablev, G. A.; Zaikov, G. E. *Spatial-Energy Parameter as a Materialised Analog of Wafe Function Progress on Chemistry and Biochemistry;* Nova Science Publishers, Inc.: New York, 2009; pp. 355–376.
13. Allen, K. W. *Astrophysical Values*. M.: Mir, 1977; 446p.
14. Korablev, G. A. *Exchange Spatial-Energy Interactions;* Publishing house "Udmurt University": Izhevsk, 2010; 530p. (Monograph).
15. Sagan, K.; Parker, J.; Van-Allen, J.; et al. *Solar System*. M.: Mir, 1978; 200p.

CHAPTER 28

ENERGY CONVERSION USING NANOFIBERS FOR TEXTILE SOLAR CELLS

VAHID MOTTAGHITALAB, MAEDEH SAJEDI,
and MAHMOOD SABERI MOTLAGH

CONTENTS

28.1 INTRODUCTION

In recent years, renewable energies attract considerable attention due to the inevitable end of fossil fuels and due to global warming and other environmental problems. Photovoltaic solar energy is being widely studied as one of the renewable energy sources with key significance potentials and a real alternate to fossil fuels. Solar cells are in general packed between weighty, brittle and rigid glass plates. Therefore, increasing attention is being paid to the construction of lighter, portable, robust, multipurpose and flexible substrates for solar cells. Textiles substrates are fabricated by a wide variety of processes, such as weaving, knitting, braiding and felting. These fabrication techniques offer enormous versatility for allowing a fabric to conform to even complex shapes. Textile fabrics not only can be rolled up for storage and then unrolled on site but also they can also be readily installed into structures with complex geometries.

Textiles are engaging as flexible substrates in that they have a enormous variety of uses, ranging from clothing and household articles to highly sophisticated technical applications. Last innovations on photovoltaic technology have allowed obtaining flexible solar cells, which offer a wide range of possibilities, mainly in wearable applications that need independent systems. Nowadays, entertainment, voice and data communication, health monitoring, emergency, and surveillance functions, all of which rely on wireless protocols and services and sustainable energy supply in order to overcome the urgent needs to regular battery with finite power. Because of their steadily decreasing power demand, many portable devices can harvest enough energy from clothing-integrated solar modules with a maximum installed power of 1–5 W. [1]

Increasingly textile architecture is becoming progressively of a feature as permanent or semi-permanent constructions. Tents, such as those used by the military and campers, are the best known textile constructions, as are sun shelter, but currently big textile constructions are used extensively for exhibition halls, sports complexes and leisure and recreation centers. Although all these structures provide protection from the weather, including exposure to the sun, but solar concept offers an additional precious use for providing power. Many of these large textile architectural constructions cover huge areas, sufficient to supply several kilowatts of power. Even the fabric used to construct a small tent is enough to provide a few hundred watts. In addition to textile architecture, panels made from robust solar textile fabrics could be positioned on the roofs of existing buildings. Compared to conventional and improper solar panels for roof structures lightweight and flexible solar textile panels is able to tolerate load-bearing weight without shattering.

Moreover, natural disaster extensively introduces the huge potential needs the formulation of unusual energy package based on natural source. Over the past 5 yr, more than 13 million *people* have *lost their home* and possessions because of earthquake, bush fire, flooding or other natural disaster. The victims of these disasters are commonly housed in tents until they are able to rebuild their homes. Whether they stay in tented accommodation for a short or long time, tents constructed from solar textile fabrics could provide a source of much needed power. This power could be stored in daytime and used at night, when the outdoor temperature can often fall. There are also a number of other important potential applications. The military would benefit from tents and field hospitals, especially those in remote areas, where electricity could be generated as soon as the structure is assembled.

28.2 THE BASIC CONCEPT OF SOLAR CELL

In 1839 French scientist, Edmond Becquerel found out photovoltaic effect when he observed increasing of electricity generation while light exposure to the two metal electrodes immersed in electrolytic solution [2]. Light is composed of energy packages known as photons. Typically, when a matter exposed to the light, electrons are excited to a higher level within material, but they return to their initial state quickly. When electrons take sufficient energy more than a certain threshold (band gap), move from the valance band to the conduction band holes with positive charge will be created. In the photovoltaic effect electron-hole pairs are separated and excited electrons are pulled and fed to an external circuit to buildup electricity [3] (Fig. 1).

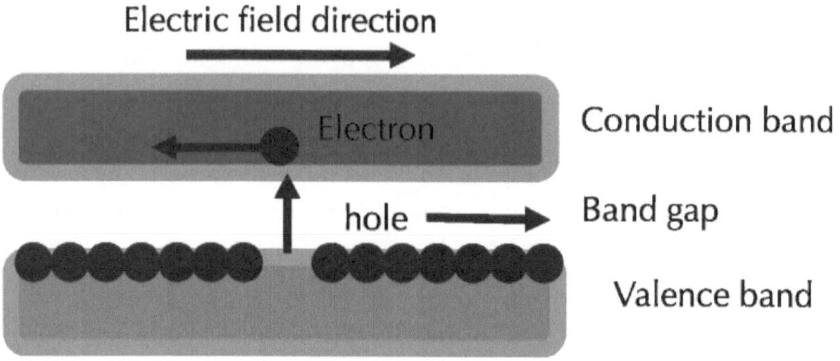

FIGURE 1 Electron excitation from valence band to conduction band.

An effective solar cell generally comprises an opaque material that absorbs the incoming light, an electric field that arises from the difference in composition between the semiconducting layers comprising the absorber, and two electrodes to carry the positive and negative charges to the electrical load. Designs of solar cells differ in detail but all must include the above features (Fig. 2).

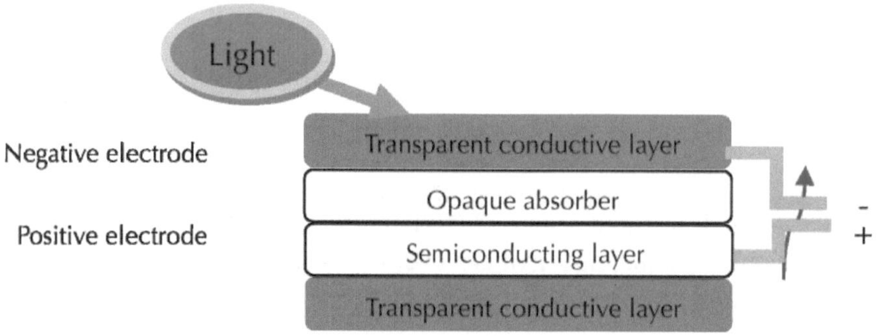

FIGURE 2 The general structure of a Solar cell.

Generally, Solar cells categorized into three main groups consisting of inorganic, organic and hybrid solar cells (Dye sensitized solar cells). Inorganic solar cells cover more than 95% of commercial products in solar cells industry.

28.2.1 ELECTRICAL MEASUREMENT BACKGROUND

All current–voltage characteristics of the photovoltaic devices were measured with a source measure unit in the dark and under simulated solar simulator source was calibrated using a standard crystalline silicon diode. The current–voltage characteristics of Photovoltaic devices are generally characterized by the short-circuit current (I_{sc}), the open-circuit voltage (V_{oc}), and the fill factor (FF). The photovoltaic power conversion efficiency (η) of a solar cell is defined as the ratio between the maximum electrical power (P_{max}) and the incident optical power and is determined by Eq. (1). [4]

$$\eta = \frac{I_{sc} \times V_{oc} \times FF}{P_{in}} \tag{1}$$

In Eq. (1), the short-circuit current (I_{sc}) is the maximum current that can run through the cell. The open circuit voltage (V_{oc}) depends on the highest occupied molecular orbital (homo) level of the donor (p-type semiconductor quasi Fermi level) and the lowest unoccupied molecular orbital (l_{umo}) level of the acceptor (n-type semiconductor quasi Fermi level), linearly. P_{in} is the incident light power density. FF, the fill-factor, is calculated by dividing P_{max} by the multiplication of I_{sc} and V_{oc} and this can be explained by the Eq. (2):

$$FF = \frac{I_{mpp} \times V_{mpp}}{I_{sc} \times V_{oc}} \tag{2}$$

In the Eq. (2), V_{mpp} and I_{mpp} represent, respectively the voltage and the current at the maximum power point (MPP), where the product of the voltage and current is maximized [4].

28.2.2 INORGANIC SOLAR CELL

Inorganic solar cells based on semiconducting layer architecture can be divided into four main categories including P–N homo junction, hetrojunction either P–I–N or N–I–P and multi junction.

The P–N homojunction is the basis of inorganic solar cells in which two different doped semiconductors (n-type and p-type) are in contact to make solar cells (Fig. 3.a). P-type semiconductors are atoms and compounds with fewer electrons in their outer shell, which could create holes for the electrons within the lattice of p-type semiconductor. Unlike p-types semiconductors, the n-type have more electrons in their outer shell and sometimes there are exceed amount of electron on n-type lattice result lots of negative charges [5].

Compared to homojunction structure amorphous silicon thin-film cells use a P–I–N hetrojunction structure, whereas cadmium telluride (CdTe) cells utilize a N–I–P arrangement. The overall picture embrace a three-layer sandwich with a middle intrinsic (i-type or undoped) layer between an N-type layer and a P-type layer (Fig. 3b). Multiple junction cells have several different semiconductor layers stacked together to absorb different wavebands in a range of spectrum, producing a greater voltage per cell than from a single junction cell which most of the solar spectrum to electricity lies in the red (Fig. 3c). Variety of semiconducting material such as single and poly crystal silicon, amorphous silicon, Cadmium–Telluride (CdTe), Copper Indium/Gallium Di Selenide (CIGS) have been employed to form inorganic solar cell based on layers configuration to enhance absorption efficiency, conversion efficiency, production and maintenance cost.

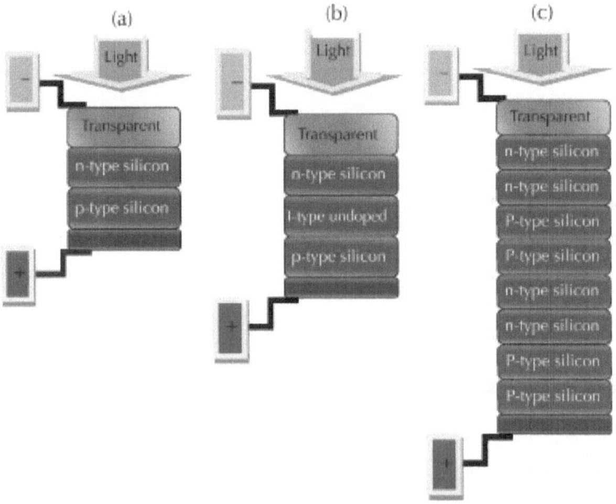

FIGURE 3 A general scheme of inorganic solar cell (a) P–N homojunction (b) P–I–N hetrojunction (c) multijunction.

28.2.3 ORGANIC SOLAR CELLS (OSCS)

Photoconversion mechanism in organic or excitonic solar cells is differing from conventional inorganic solar cells in which exited mobile state are made by light absorption in electron donor. While, light absorption creates free electron-hole pairs in inorganic solar cells [5]. It is due to law dielectric constant of organic materials and weak non-covalent interaction between organic molecules. Consequently, exciton dissociation of electron-hole pairs occurs at the interface between electron donor and electron acceptor components [6]. Electron donor and acceptor act as semiconductor p–n junction in inorganic solar cells and should be blended together to prevent electron-hole recombination (Fig. 4).

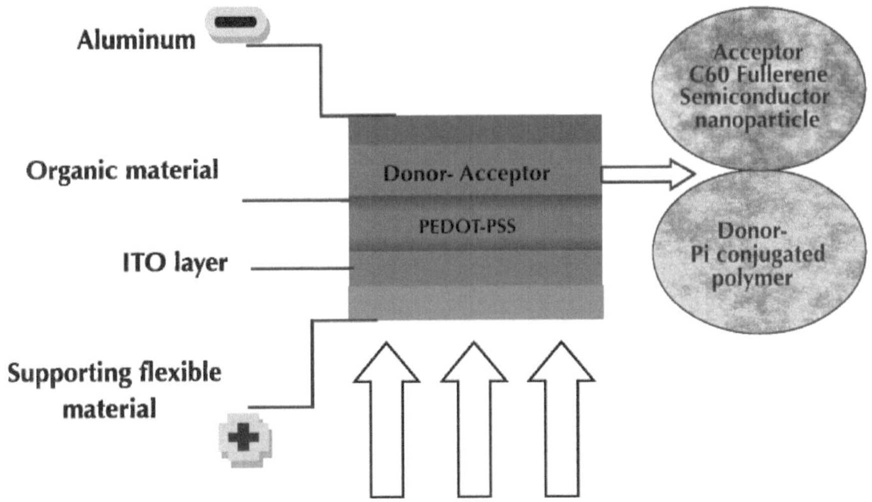

FIGURE 4 Schematic of organic solar cell.

There are two main types of PSCs including: bilayer heterojunction and bulk-heterojunction [7]. Bulk-heterojunction PSCs are more attractive due to their high surface area junction that increases conversion efficiency. This type of polymer solar cell consists of Glass, ITO, PEDOT: PSS, active layer, calcium and aluminum in which conjugated polymer are used as active layer [8]. The organic solar cells with maximum conversion efficiency about 6% still are at the beginning of development and have a long way to go to compete with inorganic solar cells. Indeed, the advantages of polymers including low-cost deposition in large areas, low weight on flexible substrates and sufficient efficiency are promising advent of new type of solar cells [9]. Conjugated small molecule attracted as an alternative approach of organic solar cells. Development of small molecule for OSCs interested because of their properties such as well-defined molecular structure, definite molecular weight, high purity, easy purification, easy mass-scale production, and good batch-to-batch reproducibility [10–12].

28.2.4 DYE-SENSITIZED SOLAR CELLS

Dye-sensitized solar cells (DSSC) use a variety of photosensitive dyes and common, flexible materials that can be incorporated into architectural elements such as windowpanes, building paints, or textiles. DSSC technology mimic photosynthesis process whereby the leaf structure is replaced by a porous titania nanostructure, and the chlorophyll is replaced by a long-life dye. The general scheme of DSSC process is shown in Fig. 5. Although traditional silicon-based photovoltaic solar cells currently have higher solar energy conversion ratios, dye-sensitive solar cells have higher overall power collection potential due to low-cost operability under a wider range of light and temperature conditions, and flexible application [13]. Oxide semiconductors materials such as TiO_2, ZnO_2 and SnO_2 have a relatively wide band gap and cannot absorb sunlight in visible region and create electron. Nevertheless, in sensitization process, visible light could be absorbed by photosensitizer organic dye results creation of electron. Consequently, excited electrons are penetrated into the semiconductor conduction band. Generally, DSSC structures consist of a photoelectrode, photosensitizer dye, a redox electrolyte, and a counter electrode. Photoelectrodes could be made of materials such as metal oxide semiconductors. Indeed, oxide semiconductor materials, particularly TiO_2, are choosing due to their good chemical stability under visible irradiation, nontoxicity and cheapness. Typically, TiO_2 thin film photoelectrode prepared via coating the colloidal solution or paste of TiO_2 and then sintering at 450°C to 500°C on the surface of substrate, which led to increase of dye absorption drastically by TiO_2 [14]. The substrate must have high transparency and low ohmic resistance to high performance of cell could be achieved. Recently many researches focused on the both organic and inorganic dyes as sensitizer regarding to their extinction coefficients and performance. Among them, B4 (N3): RuL2(NCS)2 :L=(2,2'-bipyridyl-4,4'-dicarboxylic acid) and B2(N719) : {cis-bis (thiocyanato)-bis(2,20-bipyridyl-4,40-dicarboxylato)-ruthenium(II) bis-tetrabutylammonium} due to its outstanding performance was interested.(Scheme 1). Oxide semiconductors materials such as TiO_2, ZnO_2 and SnO_2 have a relatively wide band gap and cannot absorb sunlight in visible region and create electron. Nevertheless, in sensitization process, visible light could be absorbed by photosensitizer organic dye results creation of electron.

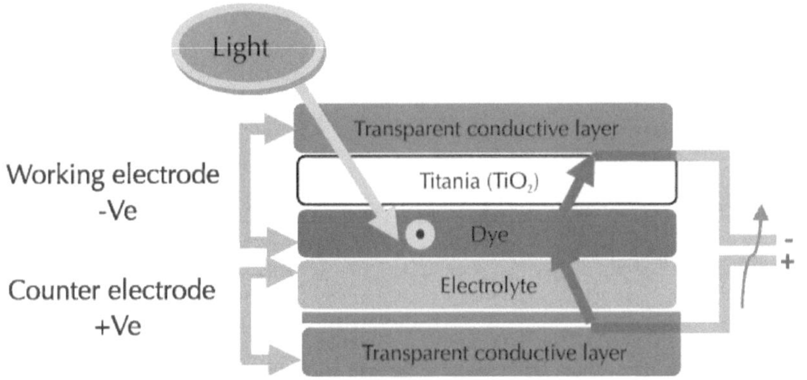

FIGURE 5 Schematic of Dye Senetesized Solar Cell (DSSC).

Consequently, excited electrons are penetrated into the semiconductor conduction band. Generally, DSSC structures consist of a photoelectrode, photosensitizer dye, a redox electrolyte, and a counter electrode. Photoelectrodes could be made of materials such as metal oxide semiconductors.

Indeed, oxide semiconductor materials, particularly TiO_2, are choosing due to their good chemical stability under visible irradiation, nontoxicity and cheapness. Typically, TiO_2 thin film photoelectrode prepared via coating the colloidal solution or paste of TiO_2 and then sintering at 450◦C to 500◦C on the surface of substrate, which led to increase of dye absorption drastically by TiO_2 [14]. The substrate must have high transparency and low ohmic resistance to high performance of cell could be achieved. Recently many researches focused on the both organic and inorganic dyes as sensitizer regarding to their extinction coefficients and performance. Among them, B4 (N3): RuL2(NCS)2 :L=(2,2'-bipyridyl-4,4'-dicarboxylic acid) and B2(N719) : {cis-bis (thiocyanato)-bis(2,20-bipyridyl-4,40-dicarboxylato)-ruthenium(II) bis-tetrabutylammonium} due to its outstanding performance was interested (Scheme 1).

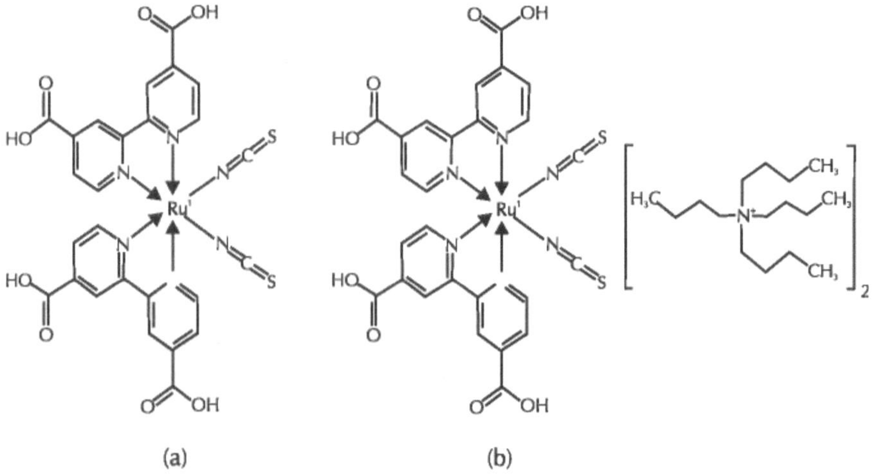

SCHEME 1 The chemical structure of (a) B4 (N3): RuL2(NCS)2 L(2,2'-bipyridyl-4,4'-dicarboxylic acid), (b) B2(N719) {cis-bis (thiocyanato)-bis(2,20-bipyridyl-4,40-dicarboxylato)-ruthenium(II) bis-tetrabutylammonium.

B2 is the most common high performance dye and a modified form of B4 to increase cell voltage. Up to now different methods have been performed to develop of new dyes with high molar extinction coefficients in the visible and near-IR in order to outperform N719 as sensitizers in a DSSC [15]. DNH2 is a hydrophobic dye, which very efficiently sensitizes wide band-gap oxide semiconductors, like titanium dioxide. DBL (otherwise known as "black dye") is designed for the widest range spectral sensitization of wide band-gap oxide semiconductors, like titanium dioxide up to wavelengths beyond 800 nm (Scheme 2).

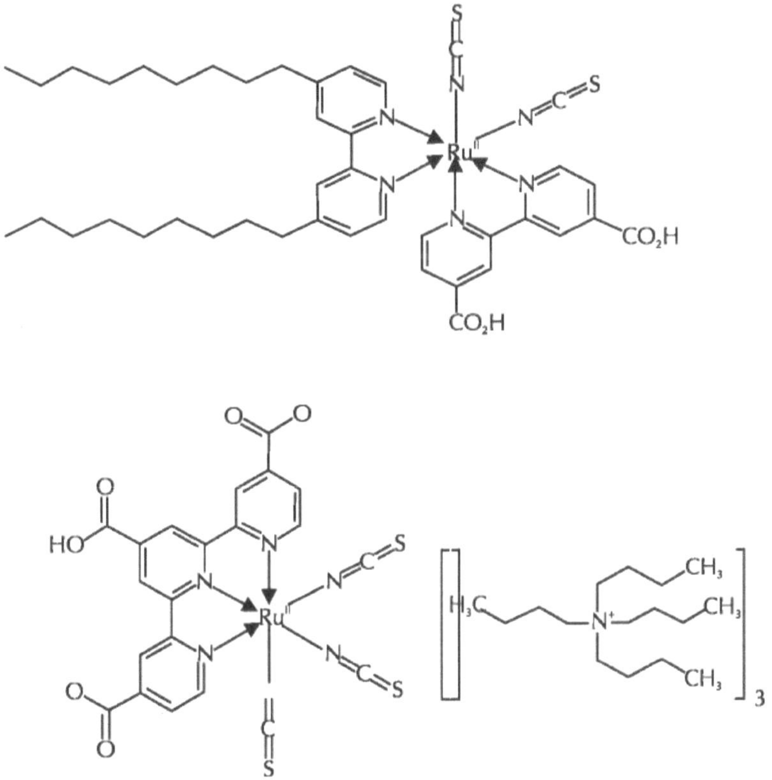

SCHEME 2 The chemical structure of (a) DNH2 (Z907) RuLL'(NCS)2, L=2,2'-bipyridyl-4,4'-dicarboxylic acid, L'= 4,4'-dinonyl-2,2'-bipyridine (b) DBL (N749) [RuL(NCS)3]: 3 TBA L= 2,2':6',2"-terpyridyl-4,4',4"-tricarboxylic acid TBA=tetra-*n*-butylammonium.

In order to continuous electron movement through the cell, the oxidized dye should be reduced by electron replacement. The role of redox electrolyte in the DSSCs is to mediate electrons between the photoelectrode and the counter electrode. Common electrolyte used in the DSSC is based on I^-/I^-_3 redox ions [16]. The mechanism of photon to current has been summarized in the following equations:

$$\text{Dye} + \text{light} \rightarrow \text{Dye}^* + e^- \quad \text{(a)}$$

$$\text{Dye}^* + e^- + TiO_2 \rightarrow e^-(\text{TiO2}) + \text{oxidized Dye} \quad \text{(b)}$$

$$\text{Oxidized Dye} + 3/2I^- \rightarrow \text{Dye} + 1/2I^{-3} \quad \text{(c)}$$

$$1/2I^-_3 + e^-(\text{counter electrode}) \rightarrow 3/2I^- \quad \text{(d)}$$

Despite of many advantages of DSSCs, still lower efficiency compared to commercialized inorganic solar cells is a challenging area. Recently, one-dimensional nanomaterials, such as nanorods, nanotubes and nanofibers, have been proposed to replace the nanoparticles in DSSCs because of their ability to improve the electron transport leading to enhanced electron collection efficiencies in DSSCs.

Subsequent sections attempts to provide fundamental knowledge to general concept of textile solar cells and their recent progress based on. Of particular interest are electrospun TiO_2 nanofibers playing the role as a key material in DSSCs and other organic solar cell, which have been shown to improve the electron transport efficiency and to enhance the light harvesting efficiency by scattering more light in the red part of the solar spectrum. A detailed review on cell material selection and their effect on energy conversion are considered to elucidate the potential role of nanofiber in energy conversion for textile solar cell applications.

28.3 TEXTILE SOLAR CELLS

Clothing materials either for general or specific use are passive and the ability to integrate electronics into textiles provides great opportunity as smart textiles to achieve revolutionary improvements in performance and the realization of capabilities never before imagined on daily life or special circumstances such as battlefield. In general, smart textiles address diverse function to withstand an interactive wearable system. Development, incorporation and interconnection of flexible electronic devices including sensor, actuator, data processing, communication, internal network and energy supply beside basic garment specifications sketch the road map toward smart textile architecture. Regardless of the subsystem functions, energy supply and storage play a critical role to propel the individual functions in overall smart textile systems.

The integration of photovoltaic (PVs) into garments emerges new prospect of having a strictly mobile and versatile source of energy in communications equipment, monitoring, sensing and actuating systems. Despite of extremely good power efficiency, most conventional crystalline silicon based semiconductor PVs are intrinsically stiff and incompatible with the function of textiles where flexibility is essential. Extensive research has been conducted to introduce the novel potential candidates for shaping the textile solar cell (TSC) puzzle. In particular, polymer-based organic solar cell materials have the advantages of low price and ease of operation in comparison with silicon-based solar cells. Organic semiconductors, such as conductive polymers, dyes, pigments, and liquid crystals, can be manufactured cheaply and used in organic solar cell constructions easily. In the manufacturing process of organic solar cells, thin films are prepared utilizing specific techniques, such as vacuum evaporation, solution processing, printing [17, 18], or nanofiber formation [19] and electrospinning [20] at room temperatures. Dipping, spin coating, doctor blading, and printing techniques are mostly utilized for manufacturing organic solar cells based on conjugated polymers [17]. Recent TSC studies revealed two distinctive strategies for developing flexible textile solar cell and its sophisticated integration.

(1) The first strategy involved the simple incorporation of a polymer PV on a flexible substrate Such as poly thyleneterphthalate (PET) directly into the clothing as a structural power source element.
(2) The second strategy was more complicated and involved the lamination of a thin anti reflective layer onto a suitably transparent textile material followed by plasma, thermal or chemical treatment. The next successive step focuses on application of a photoanode electrode onto the textile material. Subsequent procedure led to the deposition of the active material and finally evaporation of the cathode electrode complete the device as a textile PV composed of organic, inorganic and also their composites.

Regardless of many gaps need to be bridged before large-scale application of this technology, the TSC fabrication based on second strategy may be envisaged through two routes to solve pertinent issues of efficiency and stability. The solar cell architecture in first approach is founded based on knitted or woven textile substrate, however second alternative follows a roadmap to develop a wholly PV fiber for further knitting or weaving process that may form energy-harvesting textile structures in any shape and structure.

Irrespective to fabric or fiber shaped of the photovoltaic unit, the light penetration and scattering in photo anode layer needs a waveguide layer. This basic requirement naturally mimicked by polar bear hair. The optical functions of the polar bear hair are scattering of incident light into the hair, luminescence wave shift and wave-guide properties due to total reflection. The hair has an opaque, rough-surfaced core, called the medulla, which scatters incident light. The simulated synthetic core-shell fiber can be manufactured through spinning of a core fiber or with sufficient wave-guiding properties followed by finishing with an optically active, i.e. fluorescing, coating as a shell to achieve a polar bear hair' effect [21]. As described in Tributsch's original work [22], a high-energy conversion can be expected from high frequency shifts as difference between the frequency of absorbed and emitted light. According to Tributsch et al. [22] for the polar bear hair, the frequency shift is of the order of 2×10^{14} Hz.

In principle, the refractive index varies over the fiber diameter and this would provide a certain wave-guiding property of the fiber. In charge of wave-guiding property most specifically With regard to manufacturing on a larger scale, fiber morphology, crystalinity, alignment, diameter and geometry can be altered for preferred optical performance as solar energy transducer. The focus on both approached of second strategy and illuminate their opportunities and challenges is main area of interest in next parts.

28.3.1 RECENT PROGRESS OF PV FIBER

Fiber shaped organic solar cells has been subject of a few patent, project and research papers. Polymers, small molecules and their combinations were used as light-absorbing layers in previous studies. A range of synthetic substrate with various level of flexibility including optical [23], polyimide [24] and poly propylene (PP) [25] subjected to processes in which functional electrodes and light absorbing layer continuously forms on fiber scaffold. In recent PV fiber studies, conducting polymers such as P3HT in combination to small molecules nanostructure materials such as branched fullerene plays main role as photoactive material. For instance, Reference 23 introduce a light absorbing layer on optical fiber composed of poly (3-hexylthiophene) (P3HT): phenyl-C61-butyric acid methyl ester (P3HT:PCBM) (Fig. 6) . While the light was travelling through the optical fiber and generating hole-electron pairs, the 100 nm top metal electrode (which does not let the light transmit from outside) was used to collect the electrons [23].

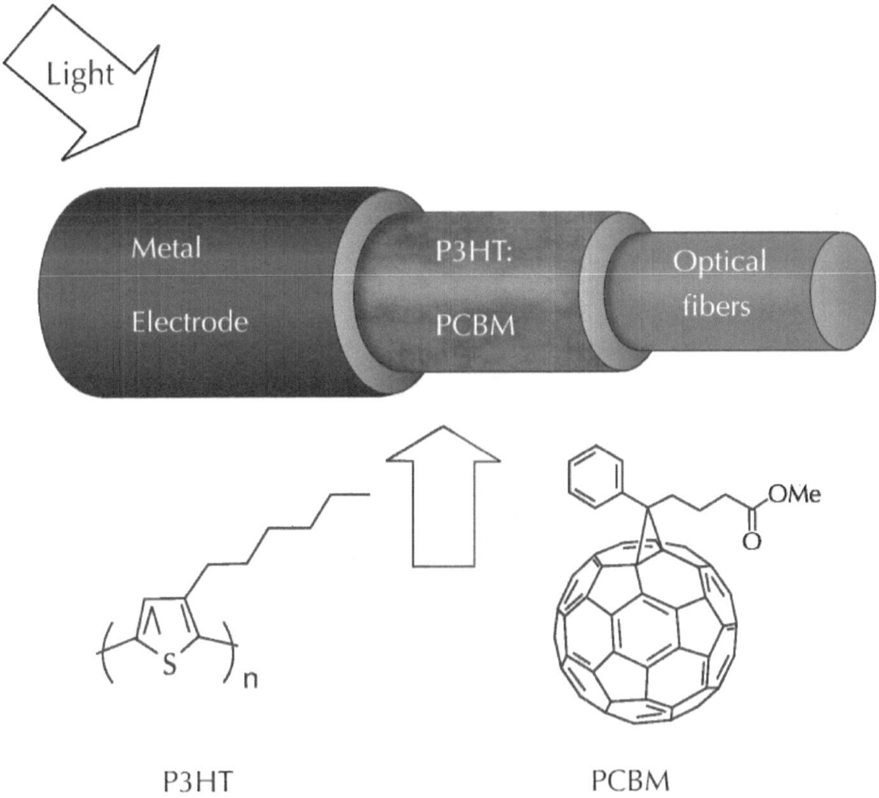

FIGURE 6 Simplified pattern of hetrojunction photo active layer in fiber solar cell.

One of the important challenges of flexible solar cell concentrated in hole collecting electrode, which is most widely used ITO as a transparent conducting material. However, the inclusion of ITO layer in flexible solar cell could not be applicable. The restrictions are mostly due to the low availability and expense of indium, employment of expensive vacuum deposition techniques and providing high temperatures to guarantee highly conductive transparent layers. Accordingly, there are some ITO-free alternative approaches, such as using carbon nanotube (CNT) layers or different kinds of poly(3,4-ethylenedioxythiophene):poly(styrenesulfonate) (PEDOT:PSS) and its mixtures [26–28], or using a metallic layer [29] to perform as a hole-collecting electrode (Scheme 3).

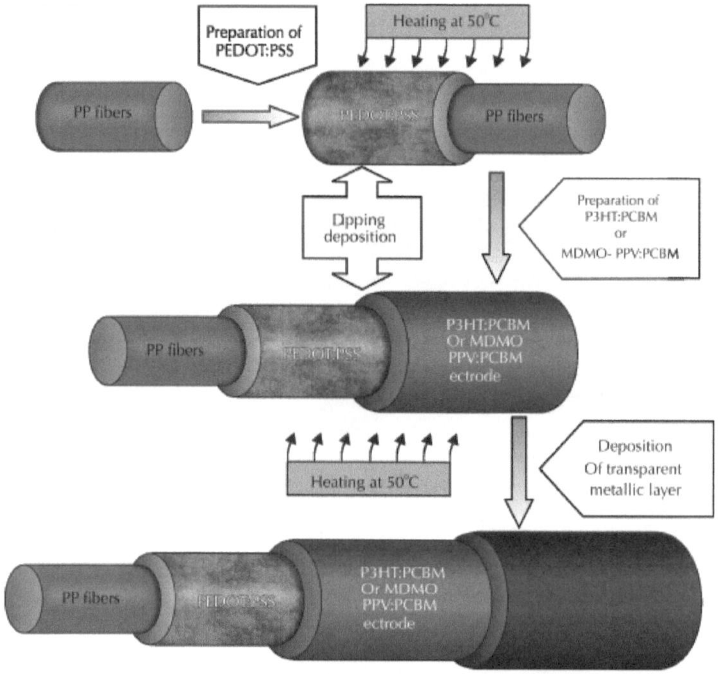

FIGURE 7 The Schematic of procedure for preparation of PV fiber.

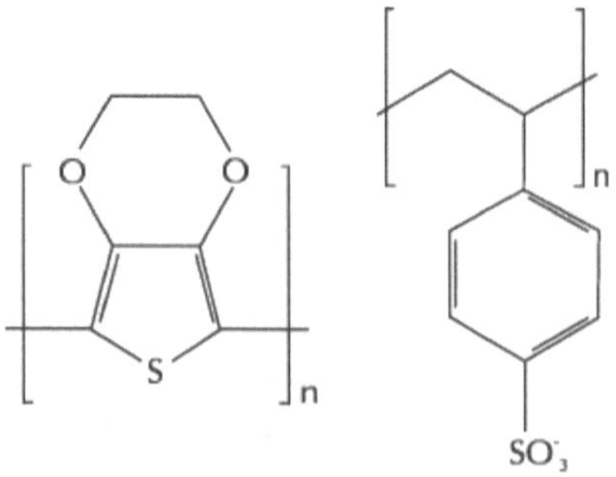

SCHEME 3 The chemical structure of poly (3,4-ethylenedioxythiophene):poly(styrene sulfonate).

The ITO free hole collecting layer was realized using highly conductive solution of PEDOT:PSS as a polymer anode that is more convenient for textile substrates in terms of flexibility, material cost, and fabrication processes compared with ITO material. Based on procedure described in reference 25 a sophisticated and simple design was presented to show how thin and flexible could be a solar cell panel.

Based on implemented pattern, the sunbeams entered into the photoactive layer with 4–10 mm^2 active area by passing through a 10 nm of lithium fluoride/aluminum (LiF/Al) layer as semi-transparent cathode outer electrode.

Table 1 gives the current density data versus voltage characteristics of the photovoltaic fibers consisting of P3HT: PCBM and MDMO–PPV: PCBM blends. Based on given results of open-circuit voltage, short-circuit current density and fill factor for two types of photoactive material, and also Eqs. (1) and (2) in Section 28.2.2, the power conversion efficiency of the MDMO-PPV: PCBM based photovoltaic fiber was higher than the P3HT: PCBM based photovoltaic fiber.

TABLE 1 Photoelectrical characteristics of photovoltaic fibers having different having different photoactive layers (P3HT: PCBM and MDMO-PPV:PCBM).

Solar cell pattern	V_{oc} (mV)	I_{sc} (mA/cm^2)	FF (%)	η (%)
PP|PEDOT:PSS| P3HT:PCBM|LiF/Al	360	0.11	24.5	0.010
PP|PEDOT:PSS| MDMO–PPV:PCBM| LiF/Al	300	0.27	26	0.021
ITO|PEDOT:PSS| MDMO–PPV:PCBM| LiF/Al	740	4.56	43.4	1.46

Comparing the solar cell characteristics of second and third pattern shows greater performance of ITO|PEDOT:PSS| MDMO-PPV:PCBM| LiF/Al rigid organic solar cell compared to PP|PEDOT:PSS| MDMO-PPV:PCBM| LiF/Al fiber solar cell. Since a same cathode, anode and photoactive material utilized in both pattern, the higher power conversion efficiency of ITO solar cell can be attributed to different wave-guide property and transparency of cell pattern in sun's ray entrance angle (i.e., LiF/AL versus ITO glass). Since using ITO is strictly restricted for PV fiber, enhancing the power conversion efficiency of needs to improving existing materials and techniques. In particular, the optical band gap of the polymers used as the active layer in organic solar cells is very important. Generally, the best bulk heterojunction devices based on widely studied P3HT: PCBM materials are active for wavelengths between 350 and 650 nm. Polymers with narrow band gaps can absorb more light at longer wavelengths, such as infra-red or near-infra-red, and consequently enhance the device efficiency. Low band gap polymers (<1.8 eV) can be an alternative for better power efficiency in the future, if they are sufficiently flexible and efficient for textile applications [30, 31]. The variety of factors influence on polymer band gaps, which can be categorized as intra-chain charge transfer, substituent effect, π-conjugation length. Systematically the fused ring low band gap copolymer composes of a low energy level electron acceptor unit coupled with a high energy level electron donor unit. The band gap of the donor/acceptor copolymer is determined by the HOMO of the donor and LUMO of the acceptor, and therefore a high energy level of the HOMO of the donor and a low energy level of the LUMO of the acceptor results in a low band gap [32].

The substituent on the donor and acceptor units can affect the band gap. The energy level of the HOMO of the donor can be enhanced by attaching electron-donating groups (EDG), such as thiophene and pyrrole. Similarly, the energy level of the LUMO of the acceptor is lowered, when electron-withdrawing groups (EWG), such as nitrile, thiadiazole and pyrazine, are attached. This will result in improved donor and acceptor units, and hence, the band gap of the polymer is decreased.

28.3.2 PV INTEGRATION IN TEXTILE

Having a complete functional textile solar cell motivates the researchers to attempt an approach for direct incorporation of photovoltaic cell elements onto the textile. The textile substrates inherently scatter most part of the incident light outward. Therefore, it was found necessary to apply a layer of the very flexible polymer PE onto the textile substrate to have a surface compatible with a layered device. The textile-PE substrate was plasma treated before application of the transparent PEDOT electrode in order to obtain good adhesion of the PEDOT layer to the PE carrier. Then screen-printing was employed for the application of the active polymer poly[2-methoxy-5-(2'-ethylhexyloxy)-p-phenylene vinylene]. (MEH-PPV). [33]

The traditional solar cell geometry was re-invented in fractal forms that allow the building of structured modules by sewing the 25–40 cm cells realized. Figure 8 shows an step by step approach for fabrication textile solar cell pattern based on polymer photo absorbing layer.

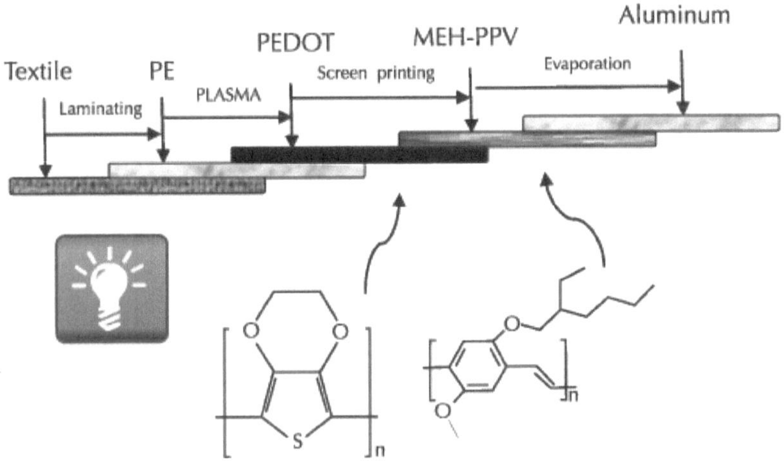

FIGURE 8 A typical fabrication procedure and key elements of textile solar cell.

The pattern designed was particularly challenging for application in solar cells and reduced the active area to 190 cm^2 (19% of the real area). The best module output power was found to be 0.27 mW with a I_{sc} = 3:8 μA, V_{oc} = 275 mV and a FF% of 25.7%. The pattern designed allows connections in different site of the cloth cell with reproducible performances within 5–10%.

28.4 NANOFIBERS AS A POTENTIAL KEY ELEMENT IN TEXTILE SOLAR CELLS

Previous sections present variety of solar cell structure and their corresponding elements and power conversion performance to indicate opportunities and challenges of producing of solar energy harvesting module based on a wholly flexible textile based photovoltaic unit. Current state of Textile Solar Cells is extremely far from commercial inorganic hetrojunction solar cells that showing around 45% conversion efficiency. Current section addresses promising potential of nanofiber 1D morphology to be utilized as solar cell elements. Of particular, enhancement of photovoltaic unit demanding properties is a great of importance. Two different strategies can be presumed including integration of functional photoanode, photo cathode, scattering layer, photoactive or acceptor – donor materials in the form of nanofiber on to textile substrate or developing fully integrated multilayer nonwoven solar cloth.

28.4.1 ELECTROSPUN NANOFIBER

Fibers with a diameter of around 100 nm are generally classified as nanofibers. What makes nanofibers of great interest is their extremely small size. Nanofibers compared to conventional fibers, with higher surface area to volume ratios and smaller pore size, offer an opportunity for use in a wide variety of applications. To date, the most successful method of producing nanofibers is through the process of electrospinning. The electrospinning process uses high voltage to create an electric field between a droplet of polymer solution at the tip of a needle and a collector plate. When the electrostatic force overcomes the surface tension of the drop, a charged, continuous jet of polymer solution is ejected. As the solution moves away from the needle and toward the collector, the solvent evaporates and jet rapidly thins and dries. On the surface of the collector, a nonwoven web of randomly oriented solid nanofibers is deposited. Material properties such as melting temperature and glass transition temperature as well as structural characteristics of nanofiber webs such as fiber diameter distribution, pore size distribution and fiber orientation distribution determine the physical and mechanical properties of the webs. The surface of electrospun fibers is important when considering end-use applications. For example, the ability to introduce porous surface features of a known size is required if nanoparticles need to be deposited on the surface of the fiber.

The conventional setup for producing a nonwoven layer can be manipulated to fabricate diverse profile and morphology including oriented [34], Core-shell [35] and hollow [36] nanofiber. Figure 9 shows the latest nanofiber profiles and its corresponding electrospinning production instrument. The variety and propagation of nanofiber products opens new horizon for development of functional profile respect to demanding application. Amongst developed techniques, coaxial electrospinning forms core-shell and/or hollow nanofiber through combination of different materials in the core or shell side, novel properties and functionalities for nanoscale devices can be found.

Increasing demands for the manufacturing of bi-component structures, in which one is surrounded by the other or the particles of one are encapsulated in the matrix of the other, at the micro or nano level, show potential for a wide range of uses. Application includes minimizing chances of decomposition of an unstable material, control releasing a substance to a particular receptor and improving mechanical properties of a core polymer by its reinforcing with another material. The electro-spinneret consists of concentric inner and outer syringe by witch two fluids are introduced to the spinneret, one in the core of the inner syringe and the other in the space between in the inner syringe and outer syringe. The droplet of the sheath solution elongates and stretches due to the repulsing between charges and form a conical shape. When the applied voltage increases, the charge accumulation reaches a certain value so a thin jet extends from the cone.

The stresses are generated in the sheath solution cause to the core liquid to deform into the conical shape and a compound co-axial jet develops at the tip of the cones (Fig. 10).

(a)

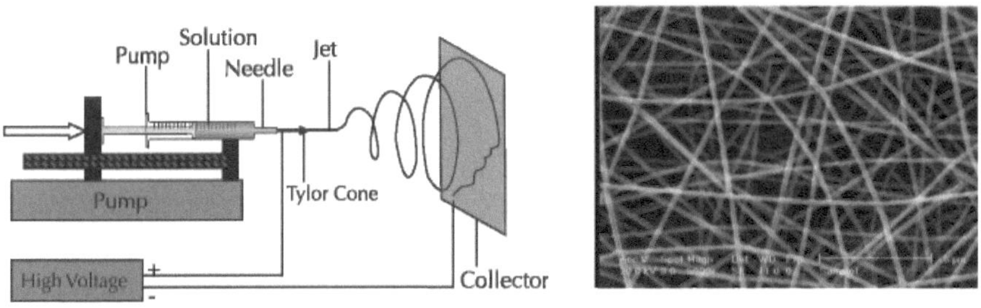

FIGURE 9 *(Continued)*

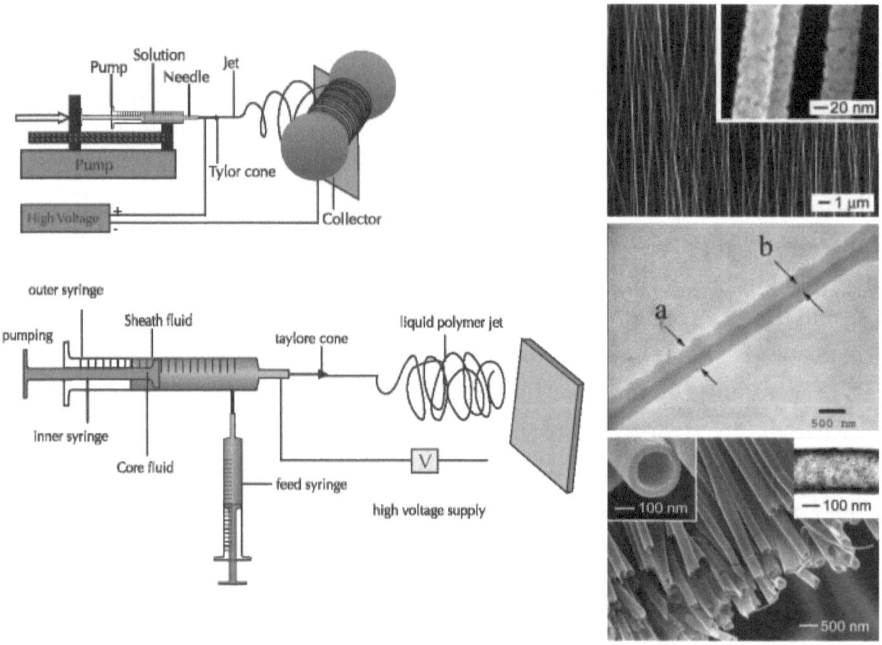

FIGURE 9 Electrospinning setup and its corresponding nanofiber profile (a) conventional nanofiber, (b) Oriented nanofiber, (c) Core-shell nanofiber, (d) Hollow nanofiber.

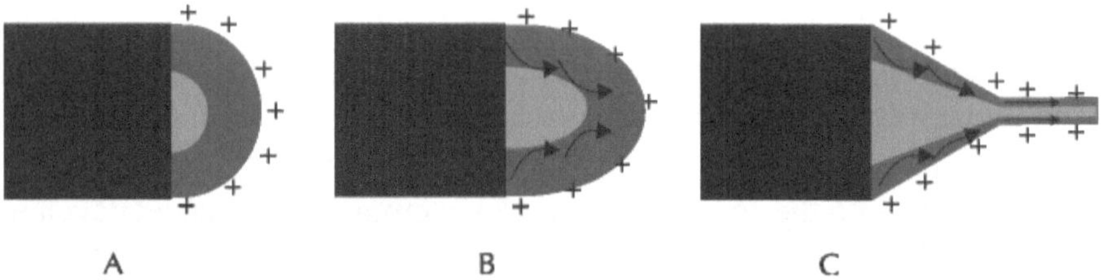

FIGURE 10 Schematic illustration of compound Taylor cone formation (A) Surface charges on the sheath solution, (B) viscous drag exerted on the core by the deformed sheath droplet, (C) Sheath-core compound Taylor cone formed due to continuous viscous drag).

28.4.2 NANOFIBERS IN DSSC SOLAR CELL

It can be presumed that the electrospun nanofiber offers high specific surface areas (ranging from hundreds to thousands of square meters per gram) and bigger pore sizes than nanoparticle or film. Meanwhile, referring to Section 28.2.3, the particle-based titanium dioxide layers have low efficiencies due to the high density of grain boundaries, which exist between nanoparticles. The 1D morphology of metal oxide fibers attracts more interest because of the lower density of grain boundaries compared to those of sintered nanoparticles [37].

28.4.2.1 ELECTROSPUN SCATTERING LAYER

Reference 38 compared the effect of TiO_2 nanofiber and nanoparticle as scattering layer and indicated the significant enhancement of all photovoltaic specefications. In another attempt ZNO nanofiber used instead of TiO_2 nanofiber to form photo anode [39].

TABLE 2 The effect of TiO2 nanofiber on cell performance for DSSC solar cell.

Solar cell pattern	V_{oc} (mV)	I_{sc} (mA/	FF	η (%)	Ref
FTO|TiO_2 NP|N719|LiI/I2/TBT|Pt|FTO	630	11.3	54	3.85	[38]
FTO|TiO_2 NF|N719|LiI/I2/TBT|Pt|FTO	660	14.3	53	4.9	[38]
FTO|ZNO NF|N719|LiI/I2/TBT|Pt|FTO	690	2.87	44	0.88	[39]
FTO|TiO_2 NF: Ag NP|N719|LiI/I2/TBT|Pt|FTO	800	7.57	55	3.3	[40]

However, the measured photocurrent density-voltage shows poorer results compared to TiO_2. The influence of Ag nanoparticle was also studied showed nearly same fill factor but lower power conversion efficiency compared to neat TiO_2 nanofiber scattering layer [40]. Recently, a composite anatase TiO_2 nanofibers/nanoparticle electrode was fabricated through electrospinning [41]. This method avoided the mechanical grinding process, and offered a higher surface area, so conversion efficiencies of 8.14% and 10.3% for areas of 0.25 and 0.052 cm^2, respectively, were reported Hybrid TiO_2 nanofibers with moderate multi walled carbon nanotubes (MWCNTs) content also can prolong electron recombination lifetimes [41]. Since MWCNTs can quickly transport charges generated during photocatalysis, the opportunity for charge recombination is reduced. Furthermore, MWCNTs decrease the agglomeration of TiO_2 nanoparticles and increase the surface area of TiO_2. These advantages make this hybrid electrode a promising candidate for DSSCs.

28.4.2.2 NANOFIBER ENCAPSULATED ELECTROLYTE

Nanofiber can be considered as promising candidate for preparation o f solid or semi solid electrolyte. This is mostly because of the inherent long-term instability of electrolyte used in DSSCs usually consists of triiodide/iodide redox coupled in organic solvents [42]. Many solid or semi-solid viscous electrolytes with low level of penetration to TiO_2 layer such as ionic liquids [43], and gel electrolytes [44] utilized to triumph over these problems. However, nanofiber with may increase the penetration of viscous polymer gel electrolytes through large and controllable pore sizes.

A few research conducted to fabricate the Electrospun PVDF-HFP membrane by electrospinning process from a solution of poly(vinylidenefluoride-co-hexafluoropropylene) in a mixture of acetone/N,N-dimethylacetamide to encapsulate electrolyte solution [45, 46]. Although the solar energy-to electricity conversion efficiency of the quasi-solid-state solar cells with the electrospun PVDF-HFP membrane was slightly lower than the value obtained from the conventional liquid electrolyte solar cells, this cell exhibited better long-term durability because of the prevention of electrolyte solution leakage.

28.4.2.3 FLEXIBLE NANOFIBER AS COUNTER ELECTRODE

Efficient charge transfer from a counter electrode to an electrolyte is a key process during the operation of dye-sensitized solar cells. One of the greatest flexible counter electrodes could be polyaniline (PAni) nanofibers on graphitized polyimide (GPi) carbon films for use in a tri-iodide reduction. These results are due to the high electrocatalytic activity of the PAni nanofibers and the high conductivity of the flexible GPi film. In combination with a dye-sensitized TiO_2 photoelectrode and electrolyte, the photovoltaic device with the PAni counter electrode shows an energy conversion efficiency of 6.85%. Short-term stability tests indicate that the photovoltaic device with the PAni counter electrode approximately preserves its initial performance [47]. The major concern for the application of alternative counter electrodes to conventional platinized TCOs in DSSCs is long-term stability. Many publications indicate that during prolonged exposure in corrosive electrolyte, catalysts will detach from the substrate and deposit onto the surface of the semiconductor photoelectrode.

TABLE 3 Current–voltage characteristics of the dye synthesized solar cells with various electrodes [47].

Solar cell pattern	V_{oc} (mV)	I_{sc} (mA/cm^2)	FF (%)	(%)
FTO|TiO_2 nanoparticle|N719|PMII/I2/TBP |Pt|FTO	820	12.61	62.3	6.44
FTO|TiO_2 nanoparticle|N719|PMII/I2/TBP |PAni|FTO	831	12.22	62.1	6.31
FTO|TiO_2 nanoparticle|N719|PMII/I2/TBP |PAni	856	11.59	58.7	5.82
FTO|TiO_2 nanoparticle|N719|PMII/I2/TBP |PAni|GPi	901	9.68	28.3	2.49

28.4.2.4 A PROPOSED MODEL FOR DSSC TEXTILE SOLAR CELL USING NANOFIBERS

Choosing proper material and structure for DSSC textile solar cell using previously mentioned nanofiber propose potential candidates for designing of an integrated photovoltaic unit. As can be seen in Fig. 11 a multilayer textile DSSC solar cell composed of a complicated pattern while nanofiber is dominant in step-by-step fabrication process.

The major concern regarding the DSSC textile solar is TiO_2 nanofiber that needs to subject to high temperature for being scattering layer. This strategy is not compatible with other textile element and it is believed that the usage of Anatase TiO_2 spinning solution provide the possibility to avoid high temperature treatment. The proposed strategy is under intensive investigation in our laboratory and future results probably mostly illuminate the opportunities and challenges.

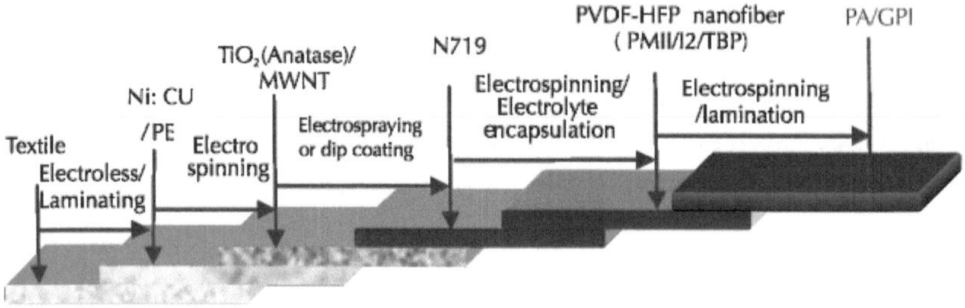

FIGURE 11 A proposed model for DSSC textile solar cell using nanofibers in successive layer.

Electroless plating is of special interest due to its advantages such as uniform deposition, coherent metal deposition, good electrical and thermal conductivity, shielding effectiveness (SE) and applicability to complex-shaped materials or nonconductors. It can be applied to almost all fiber substances and performed at any step of textile production such as yarn, stock, fabric or cloth [11–13]. Electroless copper plated fabrics are important for manufacturing of conductive textiles due to high conductivity, fast deposition and lower cost [14]. However, the fast oxidation causes this metal to lose its conductivity with time as it is exposed to the air [18]. So, there is need for this material to be coated with a protective material. Nickel could be used as coating material because of the formation of a protective oxide surface in air conditions. Therefore, the process for forming an external protective layer of nickel on the copper surface of PET fabric was done by electroless plating method. The SEM morphology at the top nickel-plated surfaces of metallic fabric are shown in Fig. 12(a, and b). These indicated that the surfaces were covered with the dense particles and evenness layers, which were clearly visible.

FIGURE 12 SEM images of the nickel-plated metallic fabric with magnifications 250×(a) and 2500×(b).

The major challenge regarding to the light transmission and outward scattering of textile material after elctroless plating turn the focus of study toward optimization of wave guide profile after textile being conductive using metallic nanoaprticles. The preliminary data of optical properties of conductive textile silk substrate after a step by step procedure outlined in table 4 that shows a semi transparent behavior with 40% reflection and 35% scattering for neat silk fabric. After electroless plating of CU:Ni shows a drastic decreasing of reflectivity. It can mostly attribute to scattering since the transmission insignificantly changes over the optical tests. However, the Tio_2 particle and nanofibers assist to increase reflectivity and as it can be expected the transmission share decreases. It means that most of the incident light might be scatters or absorbs in photo anode layers.

Experimental observation needs to be substantially carried out to find a transparent conductive layer as a flexible alternative for ITO or FTO glass.

TABLE 4 The average reflectance and transmittance of silk fabric substrate after layer-by-layer coating with electroless metallic nanoparticle, Titania nanoparticle (NP) and Titania nanofiber.

% Transmission	Reflectance	Sample
25	40	Silk
24	15	Silk|CU:Ni
22	22	Silk|CU:NI|TiO_2 NP
4	28	Silk|CU:NI|TiO_2 NP|TiO_2 NF

In spite of all opportunities proposed by DSSC solar cell technology and its bright future in energy conversion sector, but numerous challenges should be addressed for commercialized process. Ranges of transparent fabric such as Nylon, Polyester, PP and PE have been produced in large scale for various applications. While indium tin oxide (ITO) is the prevalent material of choice for conductive transparent substrtate, ITO is less desirable for emerging flexible applications since the brittleness of ITO thin films have a tendency to crack and fail under bending and flexing. Furthermore, ITO deposition requires vacuum coating techniques, and subsequent high-temperature anneal steps that are not compatible with fabric substrates and scalable to high throughput roll-to-roll or printing processes. Although using atomic layer deposition technique in low temperature showed promising potential to overcome the problem, but to best of our knowledge it is extremely expensive process for industrial scale. Therefore, if it would be inevitable to utilize semi transparent conductive textile layer, developing low band gap excitable dye critically resolve the overall issues regarding to low light transmission intensity.

28.4.3 ELECTROSPUN NONWOVEN ORGANIC SOLAR CELL

The idea of generating non-woven photovoltaic (PV) cloths using organic conducting polymers by electrospinning is quite new and has not been intensively investigated. Based on previously reported PV fiber composed of conjugated hole conducting polymer (see Section 28.3.2), a novel methodology was reported to generate a non-woven organic solar cloth. The fabrication of core-shell nanofibers has been achieved by co-electrospinning of two components such as poly(3-hexyl thiophene) (P3HT) (a conducting polymer) or P3HT/PCBM as the core and poly(vinyl pyrrolidone) (PVP) as the shell using a coaxial electrospinning set up [see Section 28.4.1] [48].

Initial measurements of the current density vs. voltage of the P3HT/PCBM solar cloth were carried out and showed current density (I_{sc}), open circuit voltage (V_{oc}) and fill factor (FF) of the fiber cloth around 3.2×10^{-6} mA/cm^2 0.12 V and 22.1%, respectively. In addition, a six order of magnitude lower photo conversion efficiency of the fiber cloth around 8.7×10^{-8} was observed that might sound disappointing. The low photovoltaic (PV) parameters of the fiber cloth could be attributed to the following factors:

a) The fiber cloth processing steps including electrospinning as well as ethanol washing were carried out under ambient conditions
(b) The thickness of the fiber cloth was ~5 μm compared to the diffusion length of the charge carriers in organic solar cells, which is only several nm.

Therefore, most of the charge carriers were lost in the fiber matrix itself. Drop casted films of the same thickness also showed similar PV parameters.

28.4.4 FUTURE PROSPECTS OF ORGANIC NANOFIBER TEXTILE SOLAR

Nanofiber revolutionize the future trend of material selection to enhance the characteristics of textile solar cell. Latest experience regarding to polymer solar cell and low band gap material needs to be considered for ambitious plans with nanofiber morphology. Figure 13 shows a schematic for layer-by-layer hetrojunction textile solar cell according to previously mentioned concerns and promising potential solar absorbing and photoactive material. The usage of anti reflective and protective materials is extremely crucial for industrial scale production. A protective layer will save the organic material from moisture and oxygen. The anti-reflective layer in solar cells can obstruct the reflection of light and also contribute to the device performance. This is an important point that should be overcome in the case of large-scale production of textile solar cell. The PEDOT:PSS combination with CNTs forms first P-N junction in the form of bi layer nanofiber. The second layer composed of core-shell or general bi-layer nanofiber of MDMO/PPV:PCBM which absorb visible spectrum. The proposed plan can be realized in a large-scale production through a needlles electrospinning setup. The continuity of process also is not beyond expectation and a range of materials as nanofiber has been already provided on given substrates.

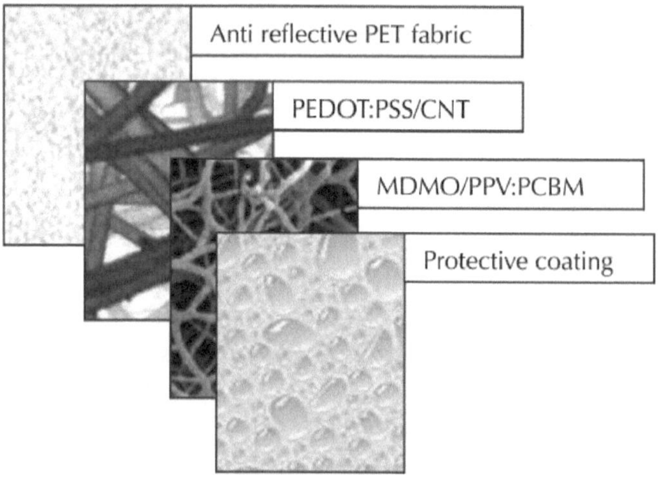

FIGURE 13 A general scheme of hetrojunction Organic nanofiber textile solar.

28.5 CONCLUDING REMARK

The multilayer solar cell energy conversion unit obeys theories of light scattering, absorption, electron excitation, charge transfer and its compensation. Each layer for specific prescribed functions needs to be intensively investigated for being applicable in diverse circumstances. The commercial solar cell products including silicon either homo or hetrojunctions, dye synthesized ll and organic solar cell subjected to demanding research and reached to high level of maturity. The huge progress in silicon solar cell and other photovoltaic system highlights developing of new generation of flexible solar cell. Current work in first part has a quick glance on variety of solar cells including inorganic, organic and hybrid structures. In overall, regardless of type of solar cell and its corresponding elements, following points have been addressed to show the effect of various parameters on cell performance:

- The wider absorption wavebands in a range of spectrum;
- The thinner the solar cell;
- The lowering the band gap;
- The higher surface area per unit mass;
- The lowering the cathode thicknesses;
- Using anti reflective and proactive coating;
- The multi junction cell to cover range of spectrum;
- Using semi-transparent wave guide material;
- The life time enhancement.

According to these concerns, the recent effort in textile solar cells has been illuminated to find substitute material and routs for integration of nanofibers in textile photovoltaic aspect. The nanofibers showed fascinating performance in DSSC and organic solar cells as scattering, active and also electrolyte layer. The observed performance is attributed to huge surface area and also ability to manipulate the morphology, crystalinity, profile and alignment. In addition making a very thin layers of functional material by a sequential multi step commercial electrospinning process persumbly propel the technology to tackle the demanding targets of solar cells including cell efficiency, stability and reliability. Incorporation of DSSC and organic solar cells into textiles is reaching encouraging performances. Stability issues need to be resolved before future commercialization can be envisaged. The mechanical stability of the devices was not limiting the function of the devices prepared. It would seem that low power conversion efficiency much more pertinent than the mechanical stability on the timescale of commercial.

KEYWORDS

- **Black dye**
- **Current–voltage**
- **Fermi level**
- **Hole collecting electrode**

REFERENCES

1. Winterhalter, C. A.; Teverovsky, J.; Wilson, P.; Slade, J.; Horowitz, W.; Tierney, E.; Sharma, V. *IEEE Trans. Inf. Technol. Biomed.,* **2005**, *9*(3), 402–406.
2. Nelson, J. *The Physics of Solar Cells*; Imperial College Press, 2003.
3. Miles, R. W.; Forbes, H. I. *Photovoltaic Solar Cells* **2005**, *51*, 1–42.
4. Günes, S.; Beugebauer, H.; Sariciftci, N. S. *Chem. Rev.,* **2007**, *107*, 1324–1338.
5. Castafier, T. M., Ed. In *Practical Handbook of Photovoltaics: Fundamentals and Applications;* Elsevier, 2003, pp 71–95.
6. Zhu, H.; Wei, J.; Wang, K.; Wu, D. *Solar Energy Materials and Solar Cells*, **2009**, *93*(9), 1461–1470.
7. Shrotriya, V.; Yao, Y.; Moriarty, T.; Emery, K.; Yang, Y. *Adv. Funct. Mater.*, **2006**, *16*, 2016–2023.
8. Krebs, F. C. *Polymer Photovoltaics A Practical Approach: SPIE*, **2008**, 1–10.
9. Cai, W.; Cao, Y. *Solar Energy Materials and Solar Cells* **2010**, *94*, 114–127.
10. Dutta, P.; Eom, S. H.; Lee, S. H. *Org. Electron.,* **2012**, *13*(2), 273–282.
11. Soa, S.; Koa, H. M.; Kima, C.; Paeka, S.; Choa, N.; Songb, K.; Leec, J. K.; Koa, J. *Solar Energy Materials and Solar Cells*, **2012**, *98*, 224–232.
12. Lina, Y.; Liua, Y.; Shia, Q.; Hua, W.; Lia, Y.; Zhan, X. *Org. Electron.,* **2012**, *13*(4), 673–680.
13. Gratzel, M. *Nature*, **2001**, *414*, 338–344.
14. Ferrazza, F. Large Size Multicrystalline Silicon Ingots. Proceedings of E-MRS 2001 Spring Meeting, Symposium E on Crystalline Silicon Solar Cells. *Solar Energy Material Solar Cells,* **2002**, *72*, 77–81.
15. Kisserwan, H. *Inorganica Chimica Acta* **2010**, *363*, 2409–2415.
16. Wei, D. *Int. J. Mol. Sci.*, **2010**, *11*, 1103–1113.
17. Günes, S.; Beugebauer, H.; Sariciftci, N. S. *Chem. Rev.*, **2007**, *107*, 1324–1338.
18. Brabec, C. J.; Dyakonov, V.; Parisi, J.; Sariciftci, N. S. *Organic Photovoltaics Concepts and Realization,* 1st ed.; Springer: New York, 2003.
19. Berson, S.; De Bettignies, R.; Bailly, S.; Guillerez, S. *Adv. Funct. Mater.,* **2007**, *17*, 1377–1384.
20. Gonzalez, R.; Pinto, N. J. *Synthetic Metals*, **2005**, *151*, 275–278.
21. Bahners, T.; Schlosser, U.; Gutmann, R.; Schollmeyer, E. *Solar Energy Materials and Solar Cells*, **2008**, *92*, 1661–1667.

22. Tributsch, H.; Goslowski, H.; Ku, U.; Wetzel, H. *Solar Energy Mater.*, **1990**, *21*, 219–236.
23. Liu, J.; Namboothiry, M. A. G.; Carroll, D. L. *Appl. Phys. Lett.*, **2007**, *90*, 063501.
24. O'Connor, B.; Pipe, K. P.; Shtein, M. *Appl. Phys. Lett.*, **2008**, *92*, 193306.
25. Bedeloglu, A.; Demir, A.; Bozkurt, Y.; Sariciftci, N. S. *Textile Res. J.*, **2007**, *80*(11), 1065–1074.
26. Ouyang, J.; Chu, C. W.; Chen, F. C.; Xu, Q.; Yang, Y. *Adv. Funct. Mater.*, **2005**, *15*, 203–208.
27. Kushto, G. P.; Kim, W.; Kafafi, Z. H. *Appl. Phys. Lett.*, **2005**, *86*, 093502.
28. Huang, J.; Wang, X.; Kim, Y.; deMello, A. J.; Bradley, D. D. C.; De Mello, J. C. *Phys. Chem. Chem. Phys.*, **2006**, *8*, 3904–3908.
29. Tvingstedt, K.; Inganäs, O. *Adv. Mater.*, **2007**, *19*, 2893–2897.
30. Perzon, E.; Wang, X.; Admassie, S.; Inganäs, O.; Andersson, M. R. *Polymer,* **2006**, *47*, 4261–4268.
31. Campos, L. M.; Tontcheva, A.; Günes, S.; Sonmez, G.; Neugebauer, H.; Sariciftci, N. S.; Wudl, F. *Chem. Mater.*, **2005**, *17*, **4031–4033**.
32. Scharber, M. S.; Mühlbacher, D.; Koppe, M.; Denk, P.; Waldauf, C.; Heeger, A. J.; Brabec, C. J. *Adv. Mater.*, **2006**, *18*, 789–794.
33. Krebs, F. C.; Biancardo, M.; Jensen, B. W.; Spanggard, H.; Alstrup, J. *Solar Energy Materials and Solar Cells,* **2006**, *90*, 1058–1067.
34. Li, D.; Wang, Y.; Xia, Y. *Adv. Mater.*, **2004**, *16*(14), 361–366.
35. Yu, J. H.; Fridrikh, S. V.; Rutledge, G. C. *Adv. Mater.*, **2004**, *16*(17), 1562–1566.
36. Li, D.; Xia, Y. *Nano Lett.*, **2004**, *4*(5), 933–938.
37. Chuangchote, S.; Sagawa, T.; Yoshikawa, S. *Appl. Phys. Lett.*, **2008**, *93*, 033310.
38. Zhao, X.; Lin, H.; Li, X.; Li, J. *Mater. Lett.*, **2011**, *65*, 1157–1160.
39. Li, S.; Zhang, X.; Jiao, X.; Lin, H. *Mater. Lett.*, **2011**, *65*, 2975–2978.
40. Li, J.; Chen, X.; Ai, N.; Hao, J.; Chen, Q.; Strauf, S.; Shi, Y. *Chem. Phys. Lett.*, **2011**, *514*, 141–145.
41. Hu, G. J.; Meng, X. F.; Feng, X. Y.; Ding, Y. F.; Zhang, S. M.; Yang, M. S. *J. Mater. Sci.*, **2007**, *42*, 7162–7170.
42. Kubo, W.; Kitamura, T.; Hanabusa, K.; Wada, Y.; Yanagida, S. *Chem. Commun.*, **2002**, 374–375.
43. Wang, P.; Zakeeruddin, S. M.; Comte, P.; Exnar, I.; Gratzel, M. *J. Am. Chem. Soc.*, **2003**, *125*, 1166–1167.
44. Wang, P.; Zakeeruddin, S. M.; Moser, J. E.; Nazeeruddin, M. E.; Sekiguchi, T.; Gratzel, M. *Nat. Mater.*, **2003**, *2*, 402–407.
45. Park, S. H.; Kim, J. U.; Lee, S. Y.; Lee, W. K.; Lee, J. K.; Kim, M. R. *J. Nanosci. Nanotechnol.*, **2008**, *8*, 4889–4894.
46. Kim, J. U.; Park, S. H.; Choi, H. J.; Lee, W. K.; Lee, J. K.; Kim, M. R. *Solar Energy Materials and Solar Cells* **2009**, *93*, 803–807.
47. Chen, J.; Lia, B.; Zheng, J.; Zhao, J.; Zhu, Z. *Electrochimica Acta* **2011**, *56*, 4624–4630.
48. Sundarrajan, S.,; Murugan, R.; Nair, A. S.; Ramakrishna, S. *Mater. Lett.*, **2010**, *64*, 2369–2372.

CHAPTER 29

EFFECT OF SAVORY ESSENTIAL OIL ON THE LEUKEMOGENESIS IN MICE

T. A. MISHARINA, E. B. BURLAKOVA, L. D. FATKULLINA, A. K. VOROBYEVA, and I. B. MEDVEDEVA

CONTENTS

29.1 INTRODUCTION

At the last time, the search for substances, including natural ones, protecting the organism from unfavorable action of the environment, particularly various carcinogenic factors, has become very important. Antioxidants are of particular interest as these biologically active substances [1]. It has been shown that a synthetic antioxidant from the class of hindered phenols [β-(4-hydroxy-3,5-ditretbutylphenyl)propionic acid (phenozan)] shows significant antitumor activity both in low and very-low doses when administered into the organism of leukemic mice [2].

It is known that many products of plant origin, i.e., herbs, spices, and their extracts, possess wide biological activity, including antioxidant and pharmacological ones [3,4]. These compounds are applied in small doses; they have low toxicity and are recommended for usage for decrease in the risk of disease caused by increased oxidation of cell components. It has been shown that natural antioxidants α-tocopherol, β-carotene, and lycopene inhibit the oxidative modification of low-density lipoproteins in vitro [5–7]. The addition of quercetin as quercetin-aglucon, rutin, and also products containing these compounds (dried apples and onions) in the feed of mice has increased the content of reduced glutathione and decreased the content of oxidized glutathione and mixed disulphide protein-glutathione in the animal liver. The antioxidant activity of plant flavonoids is caused by their ability to inhibit prooxidant enzymes, give complexes with the cations of iron and copper, and catch radicals of oxygen and nitrogen being the donor of hydrogen [8].

Among natural antioxidants of plant origin, an important place is taken by essential oils, which are a mixture of volatile compounds isolated from spice aromatic plants. The presence of antioxidant properties in many essential oils, including ones that do not contain phenol derivatives, has been proved in model experiments [9]. The number of works estimating the biological activity of essential oils in cell cultures and in vivo is far fewer. It has been shown that lemon essential oil and its separate components inhibit oxidation of human low-density lipoproteins in vitro with an efficiency close to the efficiency of synthetic phenol antioxidants [10]. Thymol, carvacrol, eugenol, and their derivatives have shown a dose dependent decrease of mitochondrial activity of cancer cells [11]; their ability to reduce the consequences of oxidative stresses superoxide dismutase and glutathione peroxidase in the liver of rats obtaining thymol or thyme oil with feed has been significantly higher than in the control group. The general antioxidant status of rats from the experimental groups has been 61–71% higher in comparison with the control [12].

In the present work the biological activity of the essential oil from summer savory (*Satureja hortensis* L.), consumed by mice in the course of their entire lives with drinking water or food was studied. The effect of savory oil on the lifetime of the AKR high cancerous line of mice and some parameters of oxidative stress in these animals have been tested. Spontaneous leukosis appears in mice of this line at the age of 6–11 months in 65–90% of cases [13]. Precisely spontaneous leukosises of mice is the closest to human leukosises according to the origin and clinical aspects and the similarity of pathological and morphological special features. Detailed study of the development kinetics of this process has been performed [14].

29.2 MATERIALS AND METHODS

Mice (120 pcs) of the AKR line at the age of 3–4 months (by the beginning of the experiment) were separated into four equal groups and placed into wire cages from stainless steel with a size of $220 \times 320 \times 500$ mm^3 (10 animals per cage). The air temperature in the room was kept at the level of 20–22°C with natural light. Both control groups of mice were given usual drinking water and standard laboratory feed (pellets), which contained wheat, corn, barley, soy oil meal, sunflower cake, fish flour, dry milk, feed yeast, limestone flour, vitamin complex, malt sprouts, and a mineral mixture (PK120 receipt, Laboratorkorm LLC, Moscow) without limitation. Mice of the first experimental group got drinking water, in which the essential oil of summer savory *Satureja hortensis* L. (Lionel Hitchen Ltd., Great Britain) was added, and standard laboratory feed also without limitation. The content of the essential oil in drinking water was 0.15 mg in 1l. This water was placed into waterers in a sufficient amount.

Mice of the second experimental group got pure drinking water without limitation and food, into which the essential oil was added. For obtaining this feed, 0.5 g of savory essential oil was mixed with 100 g of glucose;

50 mg of this mixture was added to 100 g of feed and mixed until equal distribution. Consequently, 1 g of feed contained about 2.5 μg of savory essential oil. The experiment was performed for 17 months until the natural death of the last animal. The presence of leukemia in dead animals was determined by the increased size of the thymus (above 30 mg) and spleen (above 150 mg).

Antileukemic activity of the substance under study was estimated by the survival curves (mortality curves) and the values of the average and maximum lifetime of animals in the control and experimental (savory essential oil was administered with drinking water or feed) groups. On the basis of data on lifetime, the survival curves, which represented the dependence of the portion of surviving animals on age, were plotted and nonlinear approximation of the survival curves by the Gompertz function were applied [15].

Mice were decapitated for the biochemical studies at the age of 150 and 240 days. Animal blood was taken for the investigation. The indexes of the system of lipid peroxidation (LP) in erythrocytes, i.e., the hemolysis level and the content of active products reacting with thiobarbiturc acid (TBA-AP), were determined by the method in [16]. A 5% washed erythrocyte suspension in Tris-HCl buffer (pH 7.4), diluted by a physiological salt solution in the ratio of 1 : 1, was centrifuged at 1500 rpm for 10 min and the optical density of the supernatant was determined at $\lambda = 540$ nm. Then the content of TBA products was determined spectrophotometrically in the same probe by the reaction with thiobarbituric acid on SF-103 (Russia) at λ 532 and 600 nm [16].

Structural changes in the erythrocyte membranes were studied by the method of the paramagnetic spin probes [17]. Microviscosity of various parts of the lipid component of the membrane was estimated by the time of rotating correlation of spin probes included into the membrane: 2,2,6,6-tetramethyl-4-capryloil-oxypiperidin-1-oxil (probe 1) and 5,6-benzo-2,2,6,6-tetramethyl-1,2,3,4-tetrahydro-γ-carbolin-3-oxil (probe 2), which differ in their hydrophobic properties. It is known that probe 1 is mostly localized in surface lipids and probe 2 is mostly localized in the area of deep-lying near-protein lipids of membranes [18,19]. The probes were brought into the 5% erythrocyte suspension in the form of an alcohol solution 30 min before the measuring of the samples on the ER-200D SRC EPR-spectrometer (Bruker, Germany) in the final concentration of 10^{-4} M. The time of rotating correlation of the probe (τc), which had the sense of the period of the reorientation of the probe for the angle of $\pi/2$, was counted from the obtained spectra of paramagnetic resonance (EPR). The results were expressed in relative units.

29.3 RESULTS AND DISCUSSION

Savory essential oil used in the work contained 0.5–1.7% of each of the following monoterpene hydrocarbons (α-thujene, α-pinene, camphene, β-pinene, β-myrcene, sabinene, α-phellandrene, α-terpinene), 2.1% of *p*-cymene, 14.8% of γ-terpinene, 2.8% of bornyl acetate, 18.1% of thymol, 37.8% of carvacrol, and 4% of caryophyllene. A high content of carvacrol, thymol, and γ-terpinene was respondent for the antioxidant activity of the oil [9, 20]. It was revealed earlier that the addition of thyme oil (1200 mg per 1 kg of mass) into rat feed increased the general antioxidant status of the animals and kept a high level of polyunsaturated fatty acids in cell membranes during the process of their aging [12]. Thyme and savory essential oils have a close composition of the main components; that is why we hoped that the oil of savory, which is successfully grown in central Russia, would also possess biological activity. It should be noted that savory oil doses in our work were by a factor of 100 lower than in the study by Youdim and Deans [12].

Fig. 1 represents the survival curves of the AKR line of mice in the control and in the case of administration of savory essential oil with drinking water. It is seen that the essential oil shows remarkable antileukemic action: the survival curve of the experimental group of mice is significantly shifted to the right in comparison with the control. The obtained data give evidence that the constant consumption of savory essential oil significantly increases the latent period (Fig. 1). The difference in the date of the beginning of animal mortality is marked: in the control group it began after 120 days, in the experimental groups it began after 200 (essential oil was added to water) and 250 (essential oil was added to feed) days. The initial level of mortality in the control group is higher than in the groups obtaining savory essential oil (Fig. 1). All the survival curves were approximated with the Gompertz function. It is possible that the consumption of savory essential oil has prophylactic action putting off the terms of contraction of leukemia and mass animal mortality. Thus, the average lifetime (AL) of mice increases by 47 days (20%) in the case of the consumption of savory essential oil with drinking water and by 52 days (35%) in the case of its consumption with food in comparison with the control.

However, some differences in the effect of savory essential oil added to water or feed on leukemogenesis were found. At early stages the speed of animal mortality in the experimental group administered the essential oil with water was lower than in the control group (Fig. 1), butafter death of about 40% of animals in the experimental group (280-th day) the speed of their mortality increased and later on both survival curves of mice from the control and experimental groups came close. As a result the maximum lifetimes of mice in both groups were much the same. Consequently, savory oil had remarkable effect mainly on the short-lived part of the population. In the case of consumption of savory essential oil with food, the speeds of leukemic pathological process and animal mortality were close but in the experimental group the leukemic process began to develop far later than in the control: the experimental curve is significantly shifted to the right. As a result, the maximum lifetime of mice from the experimental group was 102 days longer than in the control. It is possible that the differences in the action of savory essential oil added with water or feed is connected with the fact that the concentration of the oil in feed was ment, it is possible that antioxidant consumption at late stages of the development of the process caused the intensification of leukosis, and the animal death came earlier. The acceleration of tumor processes at the late stages of their course in case of antioxidant consumption was also noted by other authors [21]. In case of feed, savory essential oil was placed on the surface of dry feed pellets using glucose as the carrier. Such feed weakly retains volatile substances; the content of the oil in it could decrease several times after 2–3 days. Food was in cages all the time, new portions of feed were added without taking out the old ones, so mice could choose food with various contents of the essential oil, for example, with a high content in early stages of leukemia and then at later stages with a very low content of the essential oil. In case of the addition of the essential oil into feed, it is difficult to control quantitatively its consumption. In case of the addition of the oil into water, its dose is more accurate. It was shown earlier that the injection of antioxidants, for example, phenozan, inhibited lipid peroxidation processes in the blood of the AKR line of mice [22]. Savory essential oil also possesses antioxidant properties. That is why for more detailed study of its effect on the state of the organism of mice we investigated the biochemical indexes of erythrocytes (hemolysis level, the content of TBA products, microviscosity of the cell membrane) on the 120-th day (the end of the latent period) and on the 180-th day (the stage of intensive progress of leukemia).

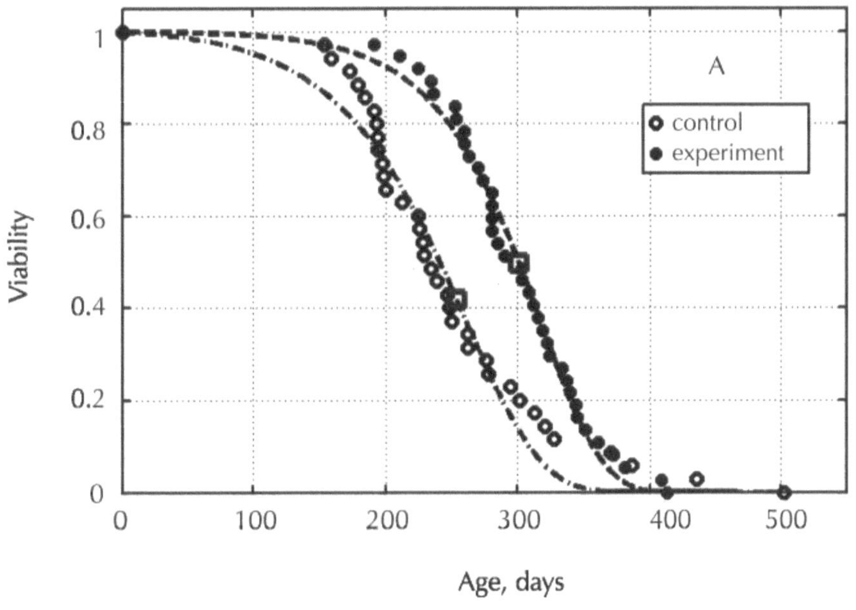

FIGURE 1 The survival curves of the AKR line of mice in the control and in the case of addition of savory oil into drinking water (experiment); approximate Gompertz function in the control and in the experiment. Selected values of the average lifetime are marked by squares.

Table 1 shows the obtained data on the hemolysis level and TBA products of lipid peroxidation in erythrocytes of the AKR line of mice aged 4 and 6 months in the control group and in the group obtaining water with savory essential oil. As is seen, in the control group of mice, the degree of hemolysis decreased with the increase of age. In the group consuming savory essential oil, the hemolysis level was 20–35% lower than in the control group and decreased with the increase of age of mice. The concentration of TBA-AP increased in the control group and decreased in case of the essential oil consumption. It should be noted that the values of the hemolysis level and TBA-AP obtained for all the groups were in the framework of the physiological standard. Consequently, the obtained data are indicative that the administration of savory essential oil caused a decrease of the level of lipid peroxidation products in erythrocytes of mice in the case of oil consumption, although all these decreases were in the framework of the physiological standard.

TABLE 1 The alteration of hemolysis level, concentration of TBA-AP, microviscosity of surface (SL) and near-protein (NPL) lipids in erythrocytes of the AKR line of mice administrated (experimental) and not administrated (control) savory essential oil with drinking water.

	Hemolysis level, %	**TBA-AP, nM/ml**	**Microviscosity of SL, τ_c 10^{-10} s**	**Microviscosity of NPL, τ_c 10^{-10} s**
Control, 4 months age	28.2 ± 0.5	36.2 ± 1.5	0.42 ± 0.04	1.20 ± 0.10
Experimental, 4 months age	20.1 ± 0.4	34.3 ± 1.8	0.41 ± 0.06	1.38 ± 0.09
Control, 6 months age	25.5 ± 0.5	41.5 ± 2.0	0.45 ± 0.05	1.21 ± 0.05
Experimental, 6 months age	17.7 ± 0.3	36.0 ± 2.1	0.40 ± 0.04	1.50 ± 0.08

The investigation of the structural characteristics of the membranes showed that with the increase of age of leukemic mice, microviscosity of lipids of erythrocyte membranes increased in the control group, while it decreased in the experimental group. The viscosity of near-protein lipids increased in both groups in the course of time. It should be noted that the changes of microviscosity with aging ranged from 10 to 20% from the control values. Therefore, the usage of essential oils prevents the development of oxidative stress in leukemic mice: lipid peroxidation parameters in blood improve and it takes place with small changes of the structural organization of erythrocyte membrane lipids.

The appearance and progress of many pathological states, including malignant growth, is connected with the failure of the structure and properties of biological membranes. The change in the physicochemical state of membranes significantly affects the processes of membrane transport, the system of trans-membrane signaling, activity of membrane-bound enzymes, etc. The work of the regulating system of lipid peroxidation in biological membranes plays a great role in regulation of cell metabolism, controlling the structure and functions of cell membranes [23, 24]. Continuous accumulation of oxidation products, which destabilize membranes and promote cell destruction, is believed to be one of the unfavorable consequences of lipid peroxidation.

Thus, it is shown that significant structural and functional changes of erythrocytes occur in blood of oncology patients: the decrease of hemolytic stability and failure of sorption ability of erythrocyte membranes and lipid peroxidation activation with the failure of lipid-protein interactions. Despite the differences in hemolysis level, the lipid peroxidation parameters and membrane structural characteristics found in our investigation did not exceed the physiological standard, and the level of deviation of these characteristics from average values was usually lower. This allows us to suppose that savory essential oil consumption stabilizes biochemical indexes of erythrocytes and can play a prophylactic role.

KEYWORDS

- **Gompertz function**
- **Leukemic mice**
- **Phenozan**
- **Spleen**
- **Thymus**

REFERENCES

1. Burlakova, E. B.; Alesenko, A. V.; Molochkina, E. M. *Bioantioksidanty v luchevom porazhenii i zlokachestvennom roste* (Bioantioxidants in Radiation Injury and Malignant Growth); Nauka: Moscow, 1975.
2. Erokhin, V. N.; Krementsova, A. V.; Semenov, V. A.; Burlakova, E. B. Effect of Antioxidant β-(4-Hydroxy-3,5-Ditertbutylphenyl) Propionic Acid (Phenosan) on the Development of Malignant Neoplasms, *Biol. Bull.*, **2007**, *34*(5), 485–491.
3. Lampe, J. W. Spicing up a Vegetarian Diet: Chemopreventive Effects of Phytochemicals, *Am. J. Clin. Nutr.*, **2003**, *78*, 579S–583S.
4. Dragland, S.; Senoo, H.; Wake, K.; Holte K.; Blomhoff R. Several Culinary and Medicinal Herbs Are Important Sources of Dietary Antioxidants, *J. Nutr.*, **2003**, *133*, 1286–1290.
5. Jessup, W.; Rankin, S. M.; De Whalley, C. V.; Hout J. R. S.; Scott J.; Leake D. S. Alfa-Tocopherol Consumption during Low-Density Lipoprotein Oxidation, *Biochem. J.*, **1990**, *265*, 399–405.
6. Oshima, S.; Ojima, F.; Sakamoto, H.; Ishiguro Y.; Terao J. Supplementation with Carotenoids Inhibits Singlet Oxygen-Mediated Oxidation of Human Plasma Low-Density Lipoprotein, *J. Agric. Food Chem.*, **1996**, *44*(10), 2306–2309.
7. Carpenter, K. L. H.; Van der Veen, C.; Hird, R.; Dennis, I. F.; Ding, T.; Mitchinson, M. J. The Carotenoids beta-Carotene, Canthaxanthin, and Zeaxanthin Inhibit Macrophage-Mediated LDL Oxidation, *FEBS Lett.*, **1997**, *401*, 262–266.
8. Meyers, K. J.; Rudolf, J. L.; Mitchell, A. E. Influence of Dietary Quercetin on Glutatione Redox Status in Mice, *J. Agric. Food Chem.*, **2008**, *56*(3), 830–836.
9. Misharina, T. A.; Terenina, M. B.; Krikunova, N. I. Antioxidant Properties of Essential Oils, *Appl. Biochem. Microbiol.*, **2009**, *45*(6), 642–647.
10. Takahashi, Y.; Inaba, N.; Kuwahara, S.; Kuki, W. Antioxidative Effect of Citrus Essential Oil Components on Human Low–Density Lipoprotein *in vitro*, *Biosci. Biotechnol. Biochem.*, **2003**, *67*(1), 195–197.
11. Mastelic, J.; Jercovic, I.; Blazevic, I.; Poljak-Blazi, M.; Borovic, S.; Ivancic-Bace, I.; Smrecki, V.; Zarkovic, N.; Brcic-Kostic, K.; Vikic-Topic, D.; Muller, N. Comparative Study on the Antioxidant and Biological Activities of Carvacrol, Thymol, and Eugenol Derivatives, *J. Agric. Food Chem.*, **2008**, *56*(11), 3989–3996.
12. Youdim, K. A.; Deans, S. G. Effect of Thyme Oil and Thymol Dietary Supplementation on the Antioxidant Status and Fatty Acid Composition of the Ageing Rat Brain, *Br. J. Nutr.* **2000**, *83*, 87–93.
13. Blandova, Z. K.; Dushkin, V.A.; Malashenko, A. M. *Linii laboratornykh myshei dlya medico-biologicheskikh issledovanii* (Laboratory Strains of Mice for Medical-Biological Investigations), Nauka: Moscow, 1983.
14. Erokhin, V. N.; Burlakova, E. B. Spontaneous Leukemia—A Model for Studying the Effects of Low and Ultralow Doses of Physical and Physicochemical Factorson Tumorigenesis, *Radiats. Biol. Radioekol.*, **2003**, *43*(2), 237–241.
15. Gompertz, B. On the Nature of the Function Expressive of the Law of Human Mortality and on a New Mode Determining Life Contingencies, *Philos. Trans. Roy. Soc. L. Ser. A*, **1825**, *115*, 513–585.

16. Yagi, K. Lipid peroxides and human diseases, *Chem. Phys. Lipids,* **1987**, *45*, 337–351.
17. Kuznetsov, A. V. *The Spin-Probe Method*; Nauka: Moscow, 1976.
18. Binyukov, V. I.; Borunova, S. F.; Goldfeld, M. G. Study of Structural Transitions in Biological Membranes by the Spin-Probe Method, *Biokhimiya*, **1972**, *36*(6), 1149–1152.
19. Goloshchapov, A. N.; Burlakova, E. B. Studies of Microviscosity and Structural Transitions in Lipids and Proteins of Cell Membranes by the Spin-Probe Method, *Biofizika, Biofizika*, **1975**, *20*, 816–821.
20. Ruberto, G.; Baratta, M. Antioxidant Activity of Selected Essential Oil Components in Two Lipid Model Systems, *Food Chem.*, **2002**, *69*(1), 167–174.
21. Hercberg, S. The History of β-Carotene and Cancers: From Observational to Intervention Studies. What Lessons Can Be Drawn for Future Research of Polyphenols? *Am. J. Clin. Nutr.*, **2005**, *81*, 218S–222S.
22. Fatkullina, L. D.; Goloschapov, A. N.; Burlakova, E. B. *Radioprotective Effect of Antioxidant Phenozan on Structural State of Biomembranes Exposed to Low Radiation Doses*. Abstr. 35th Ann. Meet. Eur. Radiat. Res., Kiev, 2006, pp 86–87.
23. Burlakova, E. B. The Role of Lipids in Information Transfer in the Cell, In *Biokhimiya lipidov i ikh rol'v obmene veshchestv* (Biochemistry of Lipids and Their Role in Metabolism); Nauka: Moscow, 1981, pp 23–25.
24. Burlakova, E. B. Preface to a Review Volume, *Biol. Membr.*, **1998**, *15*(2), 117–119.

CHAPTER 30

ANTIRADICAL PROPERTIES OF ESSENTIAL OILS FROM OREGANO, THYME, AND SAVORY

T. A. MISHARINA, E. S. ALINKINA, and L. D. FATKULLINA

CONTENTS

30.1 INTRODUCTION

In recent years, the interest in the study of properties of essential oils (EOs) of aromatic plants, which content various natural bioactive compounds, has increased significantly, as problems related to the safety of many synthetic chemicals used in various fields of human activity became urgent [1, 2]. For example, studies have shown that many EOs are antioxidants, the effectiveness of which is equal to that of synthetic antioxidants, while they and their metabolic products are safe for health. All these qualities greatly expand the boundaries of the study and application of natural EOs [3–7].

EOs are multicomponent mixtures of volatile substances with a characteristic odor inherent to the plants from which they were isolated. The main components of EOs are compounds of the terpene series: mono- and sesquiterpene hydrocarbons, aldehydes, ketones, alcohols, esters, oxides, and phenol derivatives. It is shown that EOs containing substituted phenols—eugenol, thymol, and carvacrol—have strong antibacterial and antioxidant properties [8, 9]. The two last compounds are the main components of EOs obtained from oregano, thyme, and savory. These plants belong to the family Lamiaeeae, which includes about 3500 species, among which there are a lot of aromatic ones (basil, mint, lemon balm, catnip, hyssop, rosemary, sage, marjoram, thyme, etc.). They all have a pleasant intense aroma and are widely used in cooking, and many of them have medicinal properties, known since ancient times. A large number of studies confirm the presence of antibiotic and anti-inflammatory activity in these plants and their essential oils [10, 11]. Recent studies have shown the ability of oregano and savory EOs to slow down the aging process in mammals and to exert a beneficial effect on the level of polyunsaturated fatty acids in the brain of aging animals [12].

The purpose of this work is to study the antiradical properties of EOs of thyme *(Thymus vulgare)*, oregano (*Origanum vulgare*), and savory *(Satureja hortensis)* and to compare them with a synthetic antioxidant—ionol.

30.2 METHODS

2,2-Diphenyl-1-picrylhydrazyl radical (DPPHR) and ionol (2,6-ditert-butyl-4-methylphenol) were obtained from Sigma-Aldrich (Germany).

Essential oils, used in the work, were industrial products manufactured by Lionel Hitchen Ltd. (England).

30.2.1 DETERMINATION OF THE ANTIRADICAL ACTIVITY OF EOS AND IONOL

To determine the antiradical activity towards 1 ml of 200 μM methanol solution of DPPHR, methanol solutions of antioxidants (ionol, and thyme, oregano, or savory EOs) were added to achieve the chosen concentrations, the total volume was by increased up to 2 ml. Initial concentration of DP-PHR in all reaction mixtures was 10^{-4} M (39.4 mg/l); such solutions had an optical density of about 1. We investigated four series of model reactions in which substrate concentrations varied within the following limits: from 2 to 1000 mg/l for EOs and from 1 to 100 mg/l for ionol. To obtain kinetic curves of the DPPHR reduction by antioxidants, the reaction mixtures were placed in quartz cuvettes (10 mm) with tight-fitting lids, and the optical density was recorded in an SF-2000 spectrophotometer (ZAO OKB Spectrum, Russia) at 515 nm at room temperature in the dark every 5 min for 120 min.

For the DPPHR solutions in methanol, a graph of the linear dependence of the optical density from the DPPHR concentration was constructed. According to this graph, molar absorption coefficient ε was determined to be equal to 10 010 l/mol cm (the cuvette thickness is equal to 1 cm). The concentration of the remaining radical in model reactions was calculated from the optical density values. Each series of kinetic measurements was performed three times; mathematical treatment of the results was performed using Microsoft Excel 2007 and Sigma Plot 10. The standard deviation of the mean values of three measurements does not exceed 3% (relative).

30.3 RESULTS AND DISCUSSION

The *in vivo* study of the biological activity of oregano and savory EOs, performed by us, showed that the regular use of low doses (approximately 300 ng/day) of savory oil with food or drinking water increased the lifespan of mice with the highly malignant AKP line and reduced the incidence of leukemia by 30% [12, 13]. It was found that similar systematic intake of small oregano EO doses by healthy Balb line mice increased the average lifetime by 120 days or 17%. At the same time, it was found that the intake of oil during the whole life did not cause toxic effects and had no effect on the body weight, size of immunocompetent organs, and the blood formula [14, 15]. The EO acted as a natural antioxidant, significantly reducing the content of lipid peroxidation products in the blood, liver, and brain of mice and increasing their resistance to oxidation [15]. It was found that in the brains of mice at the age of 24 months, taking oregano EO, a high content of polyun-saturated fatty acids, including the crucial docosahexaenoic acid, remained, whereas in the control group the level of these acids was reduced by 10 to 20% with increasing age [13, 15]. Such an action on the organism of mammals is associated with the presence in EOs of various biological activities, including antioxidant ones. The biological effects obtained in vivo arose our interest in a detailed study of oregano and savory EOs, including their ability to interact with free radicals, which can lead to oxidative stress and all accompanying health problems.

The studied EOs content carvacrol and thymol, which are isomeric phenols with methyl and isopropyl substituents. The similarity between their structures determines the presence of close biological activity, in particular, antioxidant activity, which has been confirmed in tests in vitro and in vivo, as well as in cell cultures [16–18]. However, slight differences in the structures of isomers lead to quantitative differences in their activity. Thus, in [16] it is shown that the antioxidant activity (AOA) of thymol and carvacrol in two lipid systems was different: in one of them, carvacrol was more active; in the other one, on the contrary, thymol was more active. At the same time, the AOA of EOs containing these phenols was higher than that of individual compounds. This confirms the fact that a property of compounds, such as AOA, is highly dependent on the structure of substances, the qualitative and quantitative composition of model systems, and the method of its evaluation [19].

To determine the antiradical activity (ARA), we chose a well-known reaction with stable DPPHR [20–22]. To compare antiradical properties, one used EC_{50} values, which are equivalent to the amount of antioxidants (EOs and ionol) necessary to reduce half of a radical. Based on kinetic curves of radical reduction in model solutions with different EOs or ionol concentrations, linear dependences of the degree of radical recovery in 30 min from the EO concentration were built. Thereon, the EO concentration at which the degree of reduction of a radical was 50% (EC_{50}) was calculated; the values obtained (g/l) are shown in Table 1. As can be seen, the EC_{50} values for all three EOs were close and ranged from 0.223 to 0.262 g/l. It should be noted that, according to [17], EC_{50} for carvacrol was 0.267 g/l; for thymol, 0.269 g/l. Thus, the EC_{50} values for EOs were smaller and differed to a greater extent than for individual phenols. This means that the antiradical activity of EOs was higher than that of carvacrol and thymol. At the same time, a comparison of the data on individual oils and phenols confirms our assumption that these two phenols determine the antiradical properties of the studied oils. According to a GLC analysis, the examined Eos vary in the content and ratio of thymol and carvacrol. Thus, in oregano EO, the total content of two phenols was the highest and amounted to 67.51%; in thyme oil, 47.47%; and in savory EO, 49.71%. The ratio of carvacrol to thymol was also different: in oregano oil, it was 15 : 1; in that of thyme, 1 : 19; and in savory EO, 1.8 : 1.

From the kinetic curves can be seen, the process of radical reduction had two stages: the first stage was fast and the second one was slow; both stages are described by linear equations of pseudo first order. The obtained Eqs. (1) and (2) are given in Table 1. The solution to Eqs. (1) and (2) for each oil and ionol enabled us to find the time of completion of the first phase and the concentration of the free radical reduced over the period. The processing results of the kinetic curves are shown in Table 1. The coefficients of x in Eq. (1) are proportional to the rate of the first fast reaction stage; as can be seen, it is maximal for oregano EO and is minimal for savory EO. For ionol the rate of the first reaction stage is two times less than that for all EOs. At the second stage, the reaction rates of all antioxidants were close (Table 1, Eq. (2)).

The content of active antiradical components in reaction 1 is given in Table 1, and as one can see this parameter ranged from 61 to 67 nmol/ml (in equivalent radical) and decreased in the order: oregano EO > ionol = thyme EO > savory EO. Given the fact that each of the EOs comprised two active antiradical components—carvacrol and thymol—with a molecular weight of 150 amu, the content of reacting phenols was calculated:

it ranged from 9.15 to 10.08 μg/ml, and that of ionol was 13.75 μg/ml. EOs (300 μg/ml) and ionol (10 μg/ml) were added to the reaction mixtures; the real total content of phenols was less, decreased in the order oregano EO > savory EO > thyme EO, and amounted to 202.5, 149.1, and 142.4 μg/ml, respectively. Given the real content of phenols, we obtain that only 5% of phenols from the oregano EO, 6.6% from the thyme EO, and 6.1% from the savory EO took part in the reaction. From these results, we can draw two conclusions. First, the antiradical activities of carvacrol and thymol were different: thymol was more active. Indeed, in the thyme EO, the thymol percentage was much higher than that of carvacrol; at the same time, the content of phenols was minimal, but their loss in the thyme EO was greater than in other oils. On the other hand, we were not able to describe quantitatively the mutual influence of thymol and carvacrol in the manifestation of their antiradical activity. This indicates that we should not equate the quantitative content of antioxidant components in a sample with its general antiradical activity. Probably, for the manifestation of antioxidant and antiradical properties in EOs, it is not only important to consider the content of individual components with high activity, but also their combination, as well as the presence of other compounds [8]. Less active EO components also contribute to the overall APA, for example, monoterpene hydrocarbons, terpinolene, α- and γ-terpinenes, owing to which EO containing no phenols can exhibit a relatively high antioxidant activity [23, 24]. In such complex multicomponent compositions as EOs, one cannot exclude the possibility of a synergistic and antagonistic interaction between individual components with each other, which, of course, cannot change but may increase or decrease the total ARA of studied EOs.

The reaction mechanism of ionol was complicated. A detailed study has shown that the reaction mechanism of DPPHR with ionol included a quick reaction of the phenol group with the radical and processes of H+ delocalization, dimerization, and complex formation, occurring between the intermediates formed, through which one molecule of ionol restored 2.8 radical molecules [20–22]. The same behavior was typical of carvacrol and thymol derivatives, which surpassed carvacrol by their ARA [17]. The reactions of the radical with EO components will be no less complex: several reactions can proceed in them simultaneously, involving both EO components and the products formed; this can both speed up and slow down the DPPHR reduction process. For a description of such systems, we used the parameters listed in Table 1. In addition to the EC_{50} values and T_{EC50}, we also used a characteristic, which links the reduction period of half a radical (T_{EC50}) and the required substrate concentration (EC_{50}). This is the value of antiradical efficiency (AE), which is calculated according to the formula proposed in [20]:

$$AE = 1 / (EC_{50} \times T_{EC50})$$

Table 1 shows that the AE values of the EOs differ by 1.5 times, and for ionol this value is by a factor of 5–10 greater than for the EOs. Note, however, that T_{EC50} of the EOs was twice less than T_{EC50} of ionol, indicating a higher reactivity of the "fast" antiradical EO components.

TABLE 1 Kinetic and physicochemical characteristics of the DPPHR reduction process by ionol and oregano, thyme, and savory EOs.

Kinetic and physico-chemical parameters of the reaction	Oregano EO	Thyme EO	Savory EO	Ionol
Eq. (1)	y = 96.84 – 0.069×, R^2=0.9352	y = 96.44 – 0.065×, R^2=0.9103	y = 97.43 – 0.060×, R^2=0,9428	y = 99.82 – 0.030×, R^2=0,9997
Eq. (2)	y = 34.82 – 0.002×, R^2=0.9860	y = 39.78 – 0.002×, R^2=0.9893	y = 41.50 – 0.002×, R^2=0.9946	y = 43.53 – 0.003×, R^2=0.9890
Content of active antiradical com–pounds that have entered into the reaction (equiva–lent DPPHR, nmol/ml)	67.12	62.47	60.92	62.49
Content of antiradicals and ionol that have entered into the reaction, μg/ml	10.08	9.38	9.15	13.75
Initial concentra–tion of EO and ionol in the model reaction, μg/ml	300	300	300	10
Total content of phenols in the EO, %	67.51	47.47	49.71	–

TABLE 1 *(Continued)*

Kinetic and physico-chemical parameters of the reaction	Oregano EO	Thyme EO	Savory EO	Ionol
Total content of carvacrol and thymol in the model reaction, μg/ml	202.5	142.4	149.1	–
Concentration of phenols, which have entered into the reaction,%	5.0	6.6	6.1	137.5
Completion time of the first stage, s	923	899	969	2077
EC_{50}, g/l	0.223	0.245	0.262	0.018
T_{EC50}, s	1080	1320	1380	2100
AE, l/g s	4.15×10^{-3}	3.09×10^{-3}	2.77×10^{-3}	2.65×10^{-2}

Based on these results, it can be concluded that, although the ARA of EOs is highly dependent on the composition, the quantitative content of antioxidant components in a sample is not always proportional to its antiradical activity. Of great importance is the ratio of components, through which the synergistic effects contributing to higher ARA of multicomponent mixtures in comparison with individual compounds can be manifested. Thus, despite differences in the quantitative ratio of the main components, the three EOs studied are close in their properties and are practically as good as synthetic antioxidant ionol, which enables their use as effective natural antioxidants. The definition of kinetic characteristics and the elucidation of the mechanisms of synergistic and antagonistic action of antioxidants are very important, as it allows one to expand the circle of affordable and highly effective agents for various industries, including the food, pharmaceutical, and cosmetics industries.

KEYWORDS

- **Balb line mice**
- **Carvacrol**
- **Docosahexaenoic acid**
- **Thymol**

REFERENCES

1. Kahl, R.; Kappus, H., *Z. Lebensmitt. Unters. Forsc.*, **1993**, *196*(2), 329–338.
2. Cozzi, R.; Ricordy, R.; Aglitti, T.; Gatta, V.; Petricone, P.; De Salvia, R. *Carcinogenesis*, **1997**, *18*, 223–228.
3. Murcia, M. A.; Egea, I.; Romojaro, F.; Parras, P.; Jimeanez, A. M.; Martianez-Tomea, M. *J. Agric. Food Chem.*, **2004**, *52*(7), 1872–1881.
4. Sacchetti, G.; Maietti, S.; Muzzoli, M.; Scaglianti, M.; Manfredini, S.; Radice, M.; Bruni, R. *Food Chem.*, **2005**, *91*, 621–632.
5. Wei, A.; Shibamoto, T. *J. Agric. Food Chem.*, **2007**, *55*(5), 1737–1742.
6. Zhelyazkov, V. D.; Cantrell, C. L.; Tekwani, B. *J. Agric. Food Chem.*, **2008**, *56*(2), 380–385.
7. El-Ghorab, A.; Shaaban, H. A.; El-Nassry, K. F.; Shibamoto, T. *J. Agric. Food Chem.*, **2008**, *56*(13), 5021–5025.
8. Misharina, T. A.; Terenina, M. B.; Krikunova, N. I. *Appl. Biochem. Microbiol.*, **2009**, *45*(6), 710–716.
9. Milos, M.; Makota, D. *Food Chem.*, **2012**, *131*, 296–299.
10. Miguel, M. G. *Flavour Fragr. J.*, **2010**, *25*(1), 291–312.
11. Ultee, A.; Bennink, M. H. J.; Moezelaar, R. *Appl. Environ. Microbiol.*, **2002**, *68*(4), 1561–1568.
12. Burlakova, E. B.; Erokhin, V. N.; Misharina, T. A.; Fatkullina, L. D.; Semenov, V. A.; Terenina, M. B.; Vorob'eva, A. K.; Goloshchapov, A. N. *Biol. Bull.*, **2010**, *37*(6), 612–618.
13. Misharina, T. A.; Burlakova, E. B.; Fatkullina, L. D.; Terenina, M. B.; Vorob'eva, A. K.; Erokhin, V. N.; Goloshchapov, A. N. *Biomed. Khim.*, **2011**, *57*(6), 604–614.
14. Burlakova, E. B.; Misharina, T. A.; Fatkullina, L. D.; Terenina, M. B.; Krikunova, N. I.; Erokhin, V. N.; Vorob'eva, A. K. *Dokl. Biochem. Biophys.*, **2011**, *437*(3), 409–412.
15. Burlakova, E. B.; Misharina, T. A.; Vorob'eva, A. K.; Alinkina, E. S.; Fatkullina, L. D.; Terenina, M. B.; Krikunova, N. I. *Dokl. Biochem. Biophys.*, **2012**, *444*(6), 676–679.
16. Ruberto, G.; Baratta, M. T. *Food Chem.*, **2000**, *69*, 167–174.
17. Mastelic, J. *J. Agric. Food Chem.*, **2008**, *56*(14), 3989–3996.
18. Slamenova, D.; Horvathova, E.; Wsolova, L. *Neoplasma*, **2008**, *55*(5), 394–399.

19. Misharina, T. A.; Alinkina, E. S.; Fatkullina, L. D.; Vorob'eva, A. K.; Medvedeva, I. B.; Burlakova, E. B. *Appl. Biochem. Microbiol.*, **2012**, *48*(1), 117–123.
20. Brand-Williams, W.; Cuvelier, M. E.; Berset, C. *Lebenm. Wiss. Technol.*, **1995**, *28*(1), 25–30.
21. Sanchez-Moreno, C.; Larrauri, J. A.; Saura-Calixto, F. *J. Sci. FoodAgric.*, **1998**, *76*(1), 270–276.
22. Huang, D.; Ou, B.; Prior, R. L. *J. Agric. Food Chem.*, **2005**, *53*(6), 1841–1856.
23. Sacchetti, G.; Maietti, S.; Muzzoli, M.; Scaglianti, M.; Manfredini, S.; Radice, M.; Bruni, R. *J. Agric. Food Chem.*, **2005**, *91*(3), 621–632.
24. Wang, H. F.; Yih, K. H.; Huang, K. F. *J. Food Drug Anal.*, **2010**, *18*(1), 24–33.

CHAPTER 31

ALPHA-TOCOPHEROL AND THE LIPID PEROXIDATION REGULATION IN MURINE TISSUES UNDER ACUTE IRRADIATION

N. V. KHRUSTOVA, YE. V. KUSHNIREVA, and L. N. SHISHKINA

CONTENTS

31.1 INTRODUCTION

One of the basic natural antioxidants, α-tocopherol (the most biologically active form of vitamin E) plays the significant role in regulation of peroxidation processes [1–3]. Besides the familiar role as lipid-soluble chain-terminating inhibitor of lipid peroxidation the structural role of α-tocopherol (TP) in membranes was also proposed [2]. TP may enhance curvature stress or counteract similar stresses induced by other lipids such as lysoforms of phospholipids [2]. In this connection the interrelation between the TP content and the relative content of lysoforms of phospholipids would be paid a special attention in this research. The biological consequences of the radiation action are due to the level of the antioxidant activity (AOA) of the tissue lipids, the activities of the antioxidant defense enzymes and the lipid peroxidation (LPO) intensity in organs and tissues of animals and also their ability to repair and normalization after the radiation injury [4–7]. Besides, it's very important to analyze the interrelation between α-tocopherol content and the lipid composition of the murine tissues in order to prognosticate the mechanism of the inhibitory effectiveness of antioxidants in the presence of phospholipids and sterols, because in model systems they affect the activity of nature and synthetic antioxidants [8–11]. The AOA level itself is connected with the antioxidants content in the tissue lipids [4], thus, the consequence of the acute irradiation may be resulted from the TP content and its inhibitory effectiveness in the various tissues of mice.

The aim of this work is to analyze the interrelations between the α-tocopherol content and parameters of the lipid composition in the tissues of mice with the different antioxidant status (liver, spleen, blood erythrocytes) in norm and under the acute X-radiation at the dose of 5 Gy.

31.2 EXPERIMENTAL PART

The experiments were carried out on 43 mice SHK (males, 17–23 g) in spring. In April mice were divided into three groups: the first served as control group (n = 5 mice), the single injection *per os* of the sunflower oil (0.2 ml per 20 g of the mouse mass) was done to the second group (n = 8 mice), the same injection plus the single acute X-irradiation at the dose of 5 Gy was applied to the third group (n = 30 mice). After one month (May) the decapitation of mice and the lipid investigations of different tissues were done for age control group 1, group 2 under the sunflower oil injection and group 3 under the oil injection plus irradiation. The biological consequences of the single injection *per os* of the sunflower oil and the radiation action were investigated on lipids of three types of tissue – liver, blood erythrocytes and spleen, which are characterized by the different level of the lipid AOA.

The single acute X-irradiation at the dose of 5 Gy was performed by a RUT-200-20-3 apparatus in the special plastic containers per 10 mice at the following conditions: current strength – 15 mA; filter 0.5 mm Cu; the dose rate is 44 cGy/min; the total irradiation duration 11 min. The survival of mice in this experiment was 10 ± 5.5% and this value is in agreement with literature data [6]. After decapitation of mice, blood was collected in test tubes treated by the 5% solution of the sodium citrate. The erythrocytes were separated from the different blood components by centrifugation. The liver and spleen were placed in ice-cooled weighting bottles immediately after decapitation.

Lipids were isolated by the method of Blay and Dyer in the Kates modification [12]. The AOA of lipids was determined by using the methyl oleate oxidation model [4], as the ratio of the differences in the induction period of methyl oleate oxidation in presence and absence of lipids to the concentration of the added lipid. Preliminary methyl oleate was purified by the vacuum distillation. The lipid concentration in methyl oleate was 20 mg/ml. Oxidation was carried out in a temperature-controlled chamber at a constant temperature 37±0.1°C by blowing air through at a rate providing oxidation in a kinetic range. In detail the lipid AOA analysis was described in [13]. The peroxide content in lipids was determined iodometrically. The antiperoxide activity (APA) of lipids, that is their ability to decompose peroxides, was evaluated as the ratio of the difference in the peroxide concentrations of methyl oleate in the absence and the presence of lipids to the amount of the added lipid [14].

The qualitative and quantitative composition of phospholipids (PL) was determined by thin-layer chromatography [15]. It was used type G or H silica gel (Sigma, USA), glass plates 9×12 cm and mixture of solvents chloroform : methanol : glacial acetic acid : water (25:15:4:2) as a mobile phase. All the solvents were specially pure or chemically pure grade. The development of chromatograms was performed by iodine vapour. For the color reaction for phosphorus we used ammonium molybdate and ascorbic acid (Serva, FRG) and perchloric acid of chemically pure grade. The amount of inorganic phosphorus was judged from the optical density of the solution at the 810 nm wavelength with Beckman Du-50 (USA). The sterol content was determined spectrophotometrically by the method presented in Ref. [16]. In addition to quantitative analysis of the different fractions of PL and sterols the following parameters of lipid composition were also evaluated: the molar ratio of [sterols]/[PL]; the PL proportion in the total lipid composition (% PL); the ratio of the phosphatidyl choline to phosphatidyl ethanolamine content in PL (PC/PE) and the ratio of the sums of the more easily to the more poorly oxidizable fractions of PL (∑EOPL/∑POPL). The last value was calculated by the formula: (PI + PS + PE + CL+ PA)/(LPC + SM + PC), where PI is phosphatidylinositol, PS is phosphatidylserine, CL is cardiolipin, PA is phosphatidic acid, LPC are lysoforms of PL, SM is sphingomielin.

The TP content in lipids is determined by fluorescence technique [17]. Fluorescence intensity is measured by the help of a Hitachi M-850 (Japan) spectrofluorimeter. The wavelength of the exiting was 295 nm, the fluorescence spectra were measured at the 323 nm wavelength. The TP content (μg) in samples was determined by calibration curves, which was done by TP ("Sigma," USA) and was calculated as a proportion in the total lipid (%TL) composition.

The content of the LPO products which have reacted with 2-thiobarbituric acid (TBA-reactive substances, TBA-RS) was determined spectrophotometrically by the method described in [18] with adding 10 μl of 0.01% 4-methyl-2, 6-ditert.butylphenol (BHT) solution in ethanol. Protein was determined according [19].

Under the performance of the investigated parameters mice in the control and experimental groups were divided into subgroups per 1–4 animals.

The data were processed by the commonly used variatiomal statistic method [21]. The variability of indices was evaluated as the ratio of mean square error of average mean to average mean for group expressed as a percentage. Results were presented in the form of the averaged arithmetic values with an indication of the mean square errors of the mean arithmetic value (M ± m).

31.3 RESULTS

As known, the lipid antioxidant activity and the LPO intensity have the season variability [4, 6, 20]. The influence of the season on the radiation injury of mice was also noted [4, 6]. Earlier it has shown that although the α-tocopherol content contributes in the tissue antioxidant status supply there is no direct correlation between the AOA and TP content [22]. The α-tocopherol content and the answer of the LPO regulatory system (values of the AOA, APA, TBA-reactive substances) within one month after actions in the liver, spleen and blood components were investigated in the age control and experimental groups. The analysis of these data, presented in Figs. 1 and 2, allows us to reveal the following regularities.

In May the organs of the intact mice SHK (males) in the age control group (in spring the tissue lipids possess a low level of antioxidants) have the liver lipids curve as an upper bound for the AOA values (Fig. 1, curve 1) and as a lower bound for the TBA-reactive substances content in tissues (Fig. 2, curve 1).

The values of the AOA and [TBA-RS] for the spleen and blood components have more or less the same limits for the variations under the actions (area I in Figs. 1, and 2), and the TP content for these lipids in the age control group is approximately alike and it is 4 times higher than one in the liver lipids. Besides, variations of the TP content in the liver and spleen lipids are almost in the same interval (0.12–0.52%) after the single injection *per os* of the sunflower oil (group 2) and the combined action of the same injection and acute irradiation (group 3), but for group 2 we see the increasing of the TP content in the liver lipids and its decreasing in the spleen lipids. The TP content in the blood erythrocytes is increasing 4 times after the sunflower oil injection and 4.1 times after the combined action of the sunflower oil injection and the acute irradiation in comparison with the age control group. The lipids from the spleen and blood erythrocytes are characterized by the prooxidant activity (Fig. 1) in the age control and experimental groups under the actions, as it is often found in the autooxidation reactions [23]. The prooxidant effect was also found for the liver lipids of the group 2 (Fig. 1, grey triangles) and the group 3 (Fig. 1, black triangular) within one month after the sunflower oil injection and the acute irradiation of mice. In a group of the age control the liver lipids possess both the antioxidant and antiperoxide activities (the APA mean value is 0.52±0.08 mmol/g×g of lipids). The antiperoxide activity of the liver lipids of mice in groups 2 and 3 increases 1.6 and 1.9 times as compared with control correspondingly.

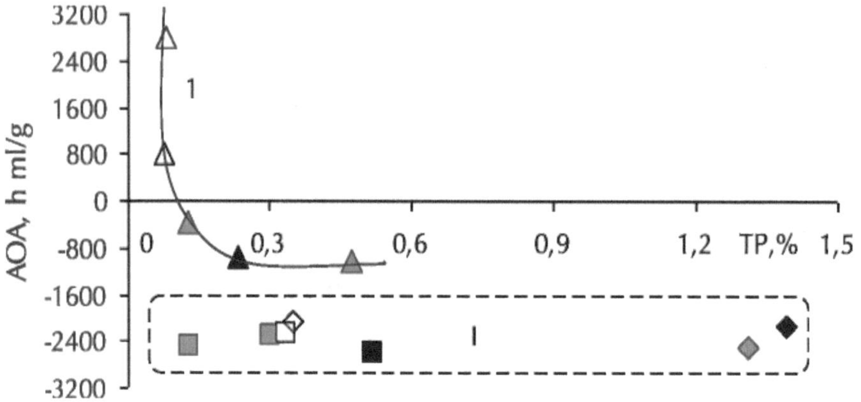

FIGURE 1 Relation between AOA of lipids and α-tocopherol content in the different tissue lipids of mice SHK in the control and experimental groups: spleen – squares, blood erythrocytes – rhombus, liver – triangles, for groups of mice (males) within one month after the experiment beginning: white labels – the age control group, grey labels – the group of mice after the single injection of the sunflower oil, black labels – the group of mice after the combined action of the oil injection and the X-irradiation at the dose of 5 Gy.

The LPO intensity is usually evaluated by the TBA-reactive substances content in a complex biological system [24, 25]. The tissues of the intact mice SHK (males, spring) may be presented by following consequence in accordance with their LPO intensity: blood plasma > spleen > liver (Fig. 2). The single injection of the sunflower oil affects the quantity of TBA-reactive substances in the tissues of mice: it's decreasing in the liver and blood plasma, but it's increasing in the spleen achieving the same level as in the blood plasma, so that the consequence of the diminution of the TBA-RS content is seen as: blood plasma ≈ spleen > liver. The experimental data for the liver show that the injection of the sunflower oil not only decreases the TBA-RS amount but has such effect that the TBA-RS value becomes the same for groups 2 and 3 (Fig. 2, grey and black triangles). It can be stipulated by the low intensity LPO in liver. Besides, as obtained for group 3, the TBA-reactive substances quantity increases in the spleen lipids – it's increasing in the spleen and decreases in the blood plasma after the injection of the sunflower oil and the acute irradiation of mice at the dose of 5 Gy.

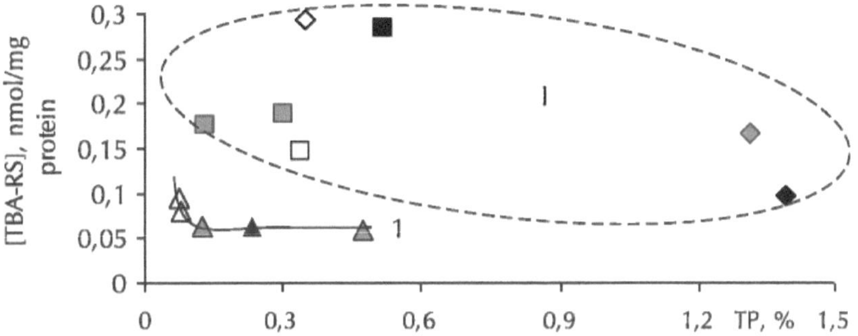

FIGURE 2 Interrelation between the TBA-reactive substances content in the liver, spleen and blood plasma and the TP content in the lipids of the spleen (squares), blood erythrocytes (rhombus) and liver (triangles) for groups of mice (males) within one month after the experiment beginning: white labels – the age control group, grey labels – the group of mice after the single injection of the sunflower oil, black labels – the group of mice after the combined action of the oil injection and the X-irradiation at the dose of 5 Gy.

The lipid composition of the different tissues was studied to form a true notion of interrelations between the LPO parameters in lipids within one month after the action of the investigated factors. As to the PL proportion in the total lipid composition (%PL) in tissues for the intact mice, it has the same value (43±2%) for the blood erythrocytes and spleen lipids (the TP content is also the same for these tissues lipids) and it is higher in the liver lipids (75±5%). The tendency of increasing of TP content with decreasing %PL was observed for the spleen and liver lipids in control and after the actions. The variability of %PL after the actions lies in the range from 48% to 80% for the liver lipids and from 20% to 58% for the spleen lipids (the variability of the TP content is equal in lipids of these tissues, as was mentioned above). The difference was found in a character of changes of the PL share in lipids of the different tissues: %PL decreased more within one month after the sunflower oil injection than after the combined action of the oil injection and irradiation in the blood erythrocytes and liver lipids, whereas in the spleen lipids %PL increased after the sunflower oil injection and decreased after the combined action as compared with the age control group.

It's noted that the PL fractions can be divided into two series. In the first one the relative content of the PL fraction in the age control and experimental groups of mice has its own limits of the variability and their limits in tissues are in the different ranges, which are not overlap; in the second one these ranges are overlapping. In the first series of the PL fractions there are SM, PA proportion in the total lipid composition and molar ration of [sterols]/[PL]. So, the variability of the SM share is in the range from 4.5% to 7.5% in the liver PL and from 9% to 13% in the spleen and blood erythrocytes PL. For the relative content of PA these ranges are from 0.25% to 0.85% in the liver PL, from 0.87% to 3.3% in the spleen PL and from 5.2% to 13.2% in the blood erythrocytes PL. In accordance with the diminution of the [sterols]/[PL] molar ratio lipids of the investigated tissues are following consequence: blood erythrocytes > spleen > liver. Besides, as seen from Fig. 3, these ratio values in lipids of the various tissues locate in the different areas in the dependence on the TP content, because in lipids of all investigated tissues the TP injection doesn't strongly change the [sterols]/[PL] molar ratio in comparison with the age control groups; a tendency of increasing the [sterols]/[PL] molar ratio with the TP growth in the spleen and blood erythrocytes lipids for group 3 is also noted (Fig. 3).

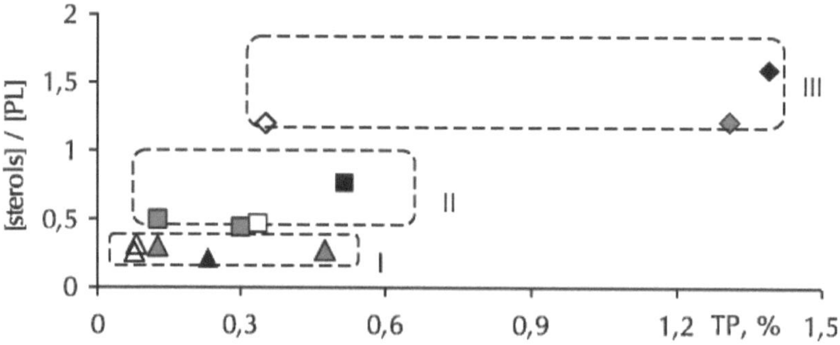

FIGURE 3 Interrelation between the molar ratio of sterols to phospholipids and the TP content in the lipids of the spleen (squares), blood erythrocytes (rhombus) and liver (triangles), for groups of mice (males) within one month after the experiment beginning: white labels – the age control group, grey labels – the group of mice after the single injection of the sunflower oil, black labels – the group of mice after the combined action of the oil injection and the X-irradiation at the dose of 5 Gy.

Among the second series there are such PL fractions as PE, PI + PS, CL and lysoforms of PL. For these PL fractions we can't select the different areas of the lipid composition parameters for the various tissues lipids. In the spleen, liver and blood erythrocytes PL the relative content of the PE is in the range from 21% to 29%, PI + PS – from 1.8% to 7.9%, CL – from 1.2% to 5.3% and LPC – from 3.8% to 11.7% in PL of all tissues. The ratio of the sums of the more easily to the more poorly oxidizable fractions of PL for the liver and spleen lipids has the same value of ∑EOPL/∑POPL = 0.59 ± 0.006 for the intact mice; under the actions (groups 2 and 3) this ratio is decreasing and its variations are from 0.55 to 0.44. In the blood erythrocytes the value of ∑EOPL/∑POPL = 0.52 ± 0.04 in the age control group of mice and in the group 2 after the oil injection; after the combined action of the oil injection and acute irradiation this ratio was increased up to the value 0.87 ± 0.05.

The relative content of PC, a main fraction of PL, has a similar share in the age control group for the blood erythrocytes (45%) and spleen PL (43%), and a higher value for liver PL (54%). The PC share does not change in the liver and blood erythrocytes PL, but it increases in the spleen PL after the sunflower oil injection (group 2). Under the additional action of the radiation (group 3) the PC share has a little enlargement in the liver and spleen PL, but for certain reduction in the blood erythrocytes PL from 45% to 35%. The relative content of PE, a second main fraction of PL, has the substantial changes for all tissues both after the sunflower oil injection and the combined action of the sunflower oil injection and irradiation. However, in group 2 the PE share increases in the spleen and blood erythrocytes PL and decreases in the liver PL. Under the combined action the PE share hardly decreases in the spleen PL from 24% to 23% and in the liver PL from 27% to 25%, but increases considerably in the blood erythrocytes PL from 21% to 27.5%. As a result, in accordance with the reduction of their PC/PE in the PL tissues of the age control group of mice may be presented by the following consequence: blood erythrocytes > liver > spleen. After the sunflower oil injection (group 2) this consequence is as follows: liver > blood erythrocytes > spleen, but for the group 3: liver > spleen > blood erythrocytes. For this group an inverse correlation between the value of PC/PE and the TP content is observed (Fig. 4. curve 1, the correlation coefficient R = –0.98 ± 0.02), while a direct correlation was found for the spleen lipids in the experimental and control groups (Fig. 4. curve 2, R= 0.91 ± 0.05).

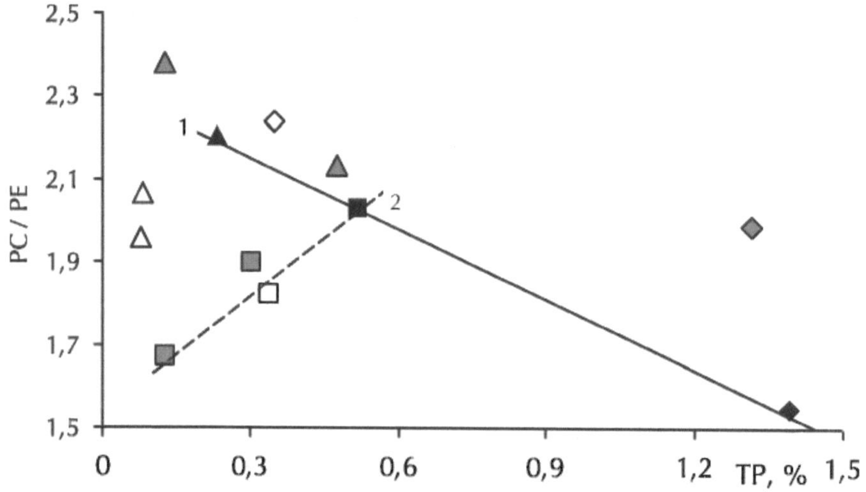

FIGURE 4 Relation between the ratio of the phosphatidyl choline to phosphatidyl ethanolamine content and the TP content in lipids of the different tissues of mice SHK: spleen – squares, erythrocytes – rhombus, liver – triangles, for groups of mice (males) within one month after the experiment beginning: white labels – the age control group, grey labels – the group of mice after the single injection of the sunflower oil, black labels – the group of mice after the combined action of the oil injection and the X-irradiation at the dose of 5 Gy.

There are some interesting features in the lipid composition answer to the sunflower oil injection of mice (group 2). For example, the share of the cardiolipin and phosphatidic acid sum is the same for the liver and spleen PL in group 2 (CL+PA= 3.45± 0.15%). The relative content of the phosphatidylinositol and phosphatidylserine sum for PL of all tissues (PI + PS = 3.7 ± 1.2%) is lower in the group 2 than in the control group 1 (PI + PS = 6.5 ± 1.5%).

The lysoforms fraction share in PL depends on the group and tissues of mice (Fig. 5), but in the blood erythrocytes PL the LPC content is the same (8.2 ± 0.1% PL) for groups 2 and 3.

A dependence between the share of LPC and α-tocopherol content in the spleen lipids is a direct correlation for the average values in 1, 2 and 3 groups (Fig. 5, curve 1, R= 0.94 ± 0.04); for the liver lipids this dependence is like the saturation curve (Fig. 5, curve 2), where the saturation level (7.7 ± 0.3%PL) for the LPC share in the liver PL is close to the analogical parameter in the blood erythrocytes PL for groups 2 and 3 (Fig. 5).

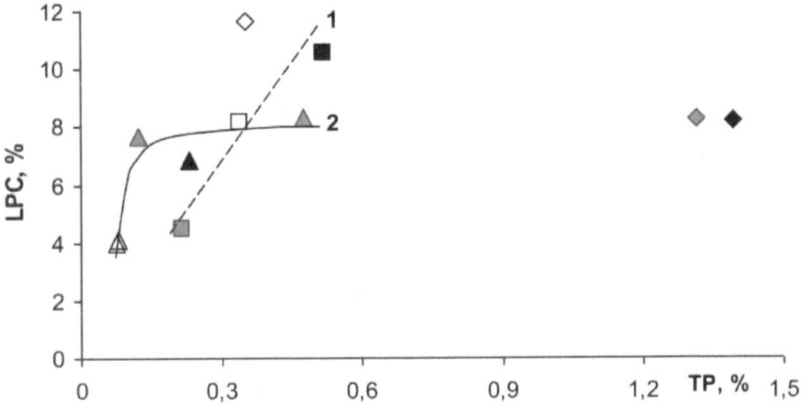

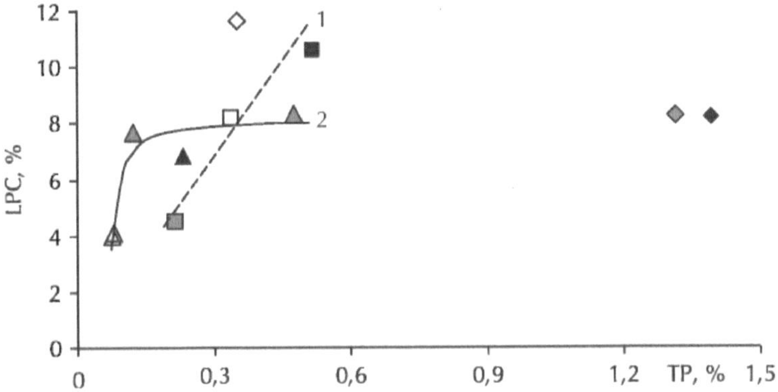

FIGURE 5 Interrelation between the relative content of lysoforms of phospholipids and the TP content in the lipids of the spleen (squares), blood erythrocytes – rhombus, liver – triangles, for groups of mice (males) within one month after the experiment beginning: white labels – the age control group, grey labels – the group of mice after the single injection of the sunflower oil, black labels – the group of mice after the combined action of the oil injection and the X-irradiation at the dose of 5 Gy.

It is need to note that experiments were performed on mice SHK in spring, when the tissue lipids possess a low level of antioxidants [6, 13]. In case of the low level of antioxidants, the less value of the α-tocopherol content in the total lipids is in the liver lipids (0.08%), and 4 times higher values of the TP content are observed in the spleen (0.34%) and blood erythrocytes (0.35%) lipids in the age control group of mice.

Several parameters are more or less the same for the spleen and blood erythrocytes lipids in norm: the TP content, AOA value, %PL in the total lipid composition, the PC share and the value of the ∑EOPL/∑POPL ratio. Obviously, in this case we can say that the TP content in the total lipids of the blood erythrocytes of the control mice may be taken as one of markers of the antioxidant status of the organism. And what happens within one month if we artificially by the oil injection increase the level of tocopherol (once more, in norm we had the low level of antioxidants)? May be the some lipid parameters in the spleen and blood erythrocytes would have an equal response? Yes, after the single injection of the sunflower oil the intensity of the lipid peroxidation and AOA values in the blood erythrocytes and spleen lipids within one month achieve the same level (Figs. 1, and 2). It seems that such similarity in the behavior after the sunflower oil injection is due to the fact of the strong blood system dependence on diet, and the spleen lipids in this case represent nourishment system. Also, the interrelation between the SM content in PL and sterols content in the total lipid composition demonstrates interesting dependences in experimental groups of mice (Fig. 6).

The share of sphingomielin fraction in PL has the similar values in the spleen and blood erythrocytes after the sunflower oil injection (12.5 ± 1%), and after the oil injection and irradiation it has the same response (10.5 ± 1%), although in norm the average share of SM in PL are 8.6% for the blood erythrocytes and 11.2% for the spleen. The sterols content in the total lipid composition has a near level in these tissues (12.5 ± 1.5%) after the sunflower oil injection, despite the difference in the control groups. The SM share exhibits the direct correlation on the sterols content in the spleen lipids (Fig. 6, curve 1, the correlation coefficient $R = 0.87 \pm 0.04$) and the inverse correlation in the blood erythrocytes lipids in the experimental and age control groups (Fig. 6, curve 2, $R = -0.95 \pm 0.04$).

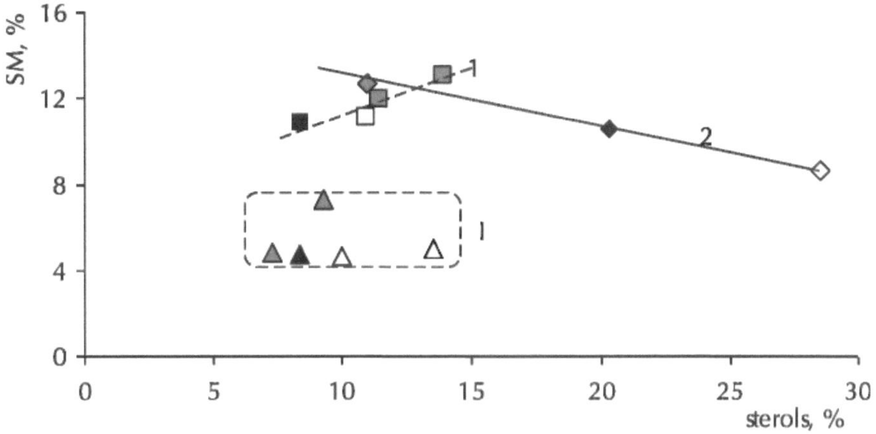

FIGURE 6 Relation between the relative content of sphingomielin and the sterols content in the lipids of the different tissues: spleen – squares, blood erythrocytes – rhombus, liver – triangles, for groups of mice (males) within one month after the experiment beginning: white labels – the age control group, grey labels – the group of mice after the single injection of the sunflower oil, black labels –the group of mice after the combined action of the oil injection and the X-irradiation at the dose of 5 Gy.

Is the TP content higher in lipids of all tissues within one month after the sunflower oil injection? – Not. In the spleen lipids the level of TP decreases. Does the α-tocopherol injection affect all parameters of lipids within one month? – No, there is an exception – the oil injection has no influence on the [sterols]/[PL] molar ratio in all tissues lipids. Another question arises – is there special organ, in the lipid parameters of which we can see correlations between the TP content and the effects under the oil injection and irradiation? Yes, in our research it is the spleen. The fraction of lysoforms of phospholipids, the ratio of the phosphatidyl choline to phosphatidyl ethanolamine, the [SM]/[sterols] ratio are increasing in accordance with an increasing of the TP content in the spleen lipids in norm, under the sunflower oil injection and under the combined action with the acute irradiation, whereas the relative contents of PC and PE are decreasing. It was also observed, that the some liver lipid parameters are independent to the TP injection and the combined action of the TP injection and acute irradiation: the PC share in the liver PL (54%) is independent to the studied actions, the sum of CL+PA (3.4%) in the liver PL is independent to the injection. However, it must be noted, that it is the case of the low level of antioxidants and high level of phospholipids (75%) in the liver total lipids in norm.

KEYWORDS

- **Lysoform**
- **Phosphatidyl choline**
- **Phosphatidyl ethanolamine**
- **Sunflower oil injection**

REFERENCES

1. Burlakova, E. B.; Krashakov, S. A.; Khrapova, N. G. Kinetic peculiarities of tocopherols as antioxidants. *Chem. Phys. Rep*., **1995**, *14(*10), 151–182.
2. Brigelius-Flohe, R.; Traber, M. G. *FASEB J.*, *13*, **1999**, 1145–1155.
3. Atkinson, J.; Epand, R. F.; Epand, R. M. *Free Rad. Biol. Med.*, **2008**, *44*, 739–764.
4. Burlakova, E. B.; Alesenko, A. V.; Molochkina, E. M.; Pal'mina, N. P.; Khrapova, N. G. *Bioantioxidants in the Radiation Damage and the Tumor Growth*. Nauka: Moscow, 1975, 214p (in Russian).
5. Vigo-Perfley, C., Ed. *Membrane Lipid Oxidation*. CRC Press: Boca Raton, Ann Arbor, Boston, 1991, III, 300p.
6. Shishkina, L. N.; Burlakova, E. B. In *Chemical and Biological Kinetics. New Horizons. Vol. 2. Biological Kinetics;* Burlakova, E. B.; Varfolomeev, S. D., Eds.; VPS Leden: Boston, 2005, 334–364.
7. Halliwell, B.; Gutteridge, J. M. C. *Free Radicals in Biology and Medicine*, 4th ed.; Oxford, New York, 2007, 851p.
8. Boguslavskaya, L. V.; Burlakova, E. B.; Kol'tsova, E. A.; Maksimov, O. B.; Khrapova, N. G. *Biophysics*, **1990**, *35*, 928–932 (in Russian).
9. Burlakova, E. B.; Mazaletskaya, L. I.; Sheludchenko, N. I.; Shishkina, L. N. *Russ. Chem. Bull.*, **1995**, *44*, 1014–1025.
10. Mazaletskaya, L. I.; Sheludchenko, N. I.; Shishkina, L. N. *Appl. Biochem. Microbiol.*, **2010**, *46*, 135–139.
11. Mazaletskaya, L. I.; Sheludchenko, N. I.; Shishkina, L. N. In *Chemical and Biochemical Kinetics: New Perspectives;* Zaikov, G., Ed.; Nova Science Publishers: New Jersey, 2011, 167–176.
12. Kates, M. *Technique of Lipidology: Isolation, Analysis and Identification of Lipids*. Elsevier: Amsterdam, 1972.
13. Shishkina, L. N.; YE.Kushnireva, V.; Smotryaeva, M. A. *Oxidation Commun.*, **2001**, *24(2)*, 276–286.

14. V. A. Men'shov, Shishkina, L. N.; Kishkovskii, Z. N. *Appl. Biochem. Microbiol.*, **1994**, *30*, 359–369.
15. Findley, J. B. C.; Evans W. H.; Eds. *Biological Membranes. A Practic Approach;* Mir: Moscow, 1990, 424p. (Russian version).
16. Sperry, W. M.; Webb, M. *J. Biol. Chem.*, **1950**, *187*, 97–109.
17. Duggan, D. E. *Arch. Biochem. Biophys.*, **1959**,*84*, 116–122.
18. Asakawa, T.; Matsushita, S. *Lipids*, **1980**, *15*, 137–140.
19. Itzhaki, R.; Gill, D. M. *Anal. Biochem.*, **1964**, *9*, 401–410.
20. Kozlov, M. V.; Urnysheva, V. V.; Shishkina, L. N. *J. Evolut. Biochem. Phys.*, **2008**, *44*, 470–475.
21. Lakin, G. F. *Biometry*, 3rd ed.; Moscow, 1990 (in Russian).
22. Aristarhova, S. A.; Burlakova, E. B.; Khrapova, N. G. *Biophysics*, **1974**, *19*(4), 688–691 (in Russian).
23. Shishlina, L. N.; Khrustova, N. V. *Biophysics*, **2006**, *51*, 292–298.
24. Lefevre, G.; Beljean-Leymarie, M.; Beyerle, F.; Bonnefont-Rousselot, D.; Cristol, J. P.; Therond, P.; Torreiles, J. *Ann. Biol. Clin. (Paris)*, **1998**, *56*, 305–319.
25. Meagher, E. A.; Fitzgerald, G. A. *Free Radic. Biol. Med.*, **2000**, *28*, 1745–1750.

CHAPTER 32

THE BIOCOMPATIBILITY OF POLYMERS BASED ON 2-HYDROXYETHYL METHACRYLATE FOR CREATION OF NEW EMBOLIZATION MATERIALS POSSESSED OF TRANSPORT FUNCTION

E. V. KOVERZANOVA, S. V. USACHEV, K. Z. GUMARGALIEVA, M. I. TITOVA, and L. S. KOKOV

CONTENTS

32.1 INTRODUCTION

The technique of synthesis of new embolization materials based on 2-hydroxyethyl methacrylate with the use of dimethacrylate glycols with different quantity oxyethyl fragments between acrylic groups as cross-linking agents is offered. An accessible and reliable method of an estimating the biocompatibility of a polymeric material with blood cells is developed. It is shown that synthesized embolization materials have a regular superamolecular structure, have a strengthened hemostatic effect at biocompatibility preservation.

Since the middle of the 70s yr of the last century, there has been steadily increasing interest in polymers for biomedical cross applications, which are widely and successfully used in medicine and pharmacology. Polymers for biomedical applications must satisfy such requirements as: occlusion to the body, resistance to a biological environment, etc. [1, 2]. One of the representatives from of a large class of such polymers are three-dimensional (cross-linked) hydrogels based on 2-hydroxyethyl methacrylate (HEMA). The main advantages of such gels are large porosity, swelling in aqueous solutions (up to 65%), non-toxic properties, elasticity and resistance to external biological environment [3]. They are the properties, which determine the high biocompatibility of these materials with body tissues, enabling to apply them in the target embolization of vessels. [4]. Thus, the embolization material (EM) based on poly-HEMA has been successfully used for the treatment of certain tumor (ischemia of liver tumors, kidney, myomatous nodes, etc.). The high porosity of the hydrogels based on *poly*-HEMA has a positive effect on the material compatibility with the tissues body tissues and promotes a growth of connective fibrous tissue in the pores of the hydrogel, which leads to a stable fixation in the blood vessel [3]. Another obvious advantage of the porosity is a possibility of using EM for delivery and controlled release of drugs [5].

Traditionally, ethyleneglycol dimethacrylate (EGDMA) has been using as the cross-linking agent. A possibility of using cross-linking agents different in qualitative or quantitative structure of allows modifying physical and chemical properties of the resulting synthesized hydrogels, the functional characteristics being preserved. In order to expand the range of EM based on *poly*-HEMA for application in various pathologies and with a possibilities of speed regulation of the organization of blood clot, ingrowing of tissues into polymeric material, and also saturation by medications we have developed methods of syntheses of *poly*-HEMA with use *tri*-ethyleneglycol dimethacrylate (*tri*-EGDMA) and *tetra*-ethyleneglycol dimethacrylate (*tetra*-EGDMA) as cross-linking agents.

The development and application of the EM based on *poly*-HEMA in clinical practice primarily requires the development of simple, available and reliable methods of assessing the biocompatibility of a polymeric material (PM) with blood cells, as well as studying the genesis of its haemostatic effect. Analysis of changes in parameters of donor blood during the interaction with the newly obtained materials, synthesized with different types of cross-linking agents should include a obligatory program of research: that is the definition of the number of leukocytes, erythrocytes, the levels of Hb, platelets and the assessment of the degree of hemolysis of the blood

The present study has been aimed at investigating the possibility of using *tri*-EGDMA and *tetra*-EGDMA as cross-linking agents, as well as the research of the morphological structure of the polymers, their biocompatibility and hemostatic effect.

32.2 MATERIALS AND METHODS

Cylindrical emboli have been synthesized in aqueous solution under the action of the initiating system: ammonium persulfate – *N, N, N', N'*-tetramethyl-ethylenediamine at 20°C [6], with a constant ratio of monomer: cross-linker 46,7:1 mol/mol. As the cross-linking agents were used EGDMA, *tri*-EGDMA or *tetra*-EGDMA. The polymerization has been carried out in tubes of a length of 12–15 cm and a diameter of 1 mm. After polymerization the blanks of hydrogels have been cut into pieces 1 cm long and washed in water at 80°C for several h with the replacement of one to remove unreacted monomer. Washed emboli have been placed in sterile containers with isotonic sodium chloride solution and stored at room temperature before further research.

The surface morphology and pore structure of cylindrical emboli has been investigated with the use of a scanning electron microscope Jeol

YSM 5300 LV (Japan). Contrast samples have been coated with gold on the installation Jeol YFC-1100 E "Ion sputtering device." The thickness of the sputtering was about 40 Å.

To assess the preservation of blood cells 0.1 ml of the solution in which EM was stored (0.08 g of PM in 10 ml) has been added to 0.1 ml of blood. In stained blood smears for Romanowsky-Giemsa have been studied the morphology of leukocytes, erythrocyte and platelets, which characterize the degree of integrity and intactness of the blood during its contact with the PM. Increased hemolysis (20%), leukopenia, anemia and sharp thrombocytopenia, as well as the morphological features of damaged blood cells (leukocytes, erythrocytes and platelets) is an indicator of biological incompatibility of the PM with blood, precluding a possibility of its application in clinical practice [3].

32.3 RESULTS AND DISCUSSIONS

The process of three-dimensional polymerization is of microheterogeneous nature [7]. Primary gel structures appear at the earliest stages of polymerization, and the concentration of such microgels, ceasing to increase since some moment. The reactionary system consists of a set of local microreactors in which the polymerization reaction proceeds though is at different stages. Simultaneously with chemical processes of polymerization and formation of cross-link occurs in microreactors. The physical processes of interlacing of both formed and growing chains also takes place. The polymerization occurs on the surface of individual microreactors (globule) building-up the new macromolecules chains. As a result of increasing globules their contact and aggregation is occurring. Further polymerization takes place in zones of contact and in the space out of these zones. The processes described above define structural and morphological features of EM being formed. The resulting polymer has a complex heterogeneous structure and contains both dense globular formation and flexible weak links between them. For EM, formed by *poly*-HEMA-EGDMA, the size of the globules varies in the range of 2–10 microns, the packaging is friable, and the linear-chain-like aggregates of globules organize multi-strands, leading to the organization of macro-cell structure. The latter, in turn, create macropores up to several tens of microns (Fig. 1). The material based on *poly*-HEMA-*tri*-EGDMA virtually do not exhibit communication between globules, packing is more dense, the size of the globules varies in the range of 2–10 microns (Fig. 2). The structure of *poly*-HEMA-*tetra*-EGDMA is more regular, the bond between the globules is absent, and there appear "unloaded" chain-like aggregates. The size of the globules varies in the range of 1–2 microns (Fig. 3). Similar to *poly*-HEMA-EGDMA, this polymer tends to create a macro-cell structure, but without voluminous voids.

FIGURE 1 The structure of *poly*-HEMA-EGDMA.

FIGURE 2 The structure of *poly*-HEMA-*tri*-EGDMA.

FIGURE 3 The structure of *poly*-HEMA-*tetra*-EGDMA.

Previously we have shown the following [2, 8]:

- under the contact of the donor blood with *poly*-HEMA-EGDMA used after several removals of residual low molecular weight substances by water washing out, have been observed minimal damage of blood cells and for these indicators this material is suitable for application in endovascular surgery;
- Hb levels varied in the range $11.5 \div 12$ g/L;
- hemolysis of the blood did not exceed 11 mg%;
- the number of erythrocytes was not less than $3.5 \times 10^{12}\ L^{-1}$, at the average, this reduction amounted to $10 \div 15\%$ relative to the level of erythrocyte of the donor blood, with no observed morphological changes of erythrocytes being observed;
- number of leukocytes decreased in the range the $15 \div 18\%$, i.e., from $(6.2 \pm 0.4) \times 10^9$ to $(5.1 \pm 0.3) \times 10^9\ L^{-1}$, $p < 0.05$;
- morphological preservation of white blood cells (mean, 4 of damaged cells to 200 cells), at the average.

Figure 4–7 shows the micrographs of the original donor blood (4) and the blood after the interaction with the medium in which EM has been stored (5, 6, 7). The photomicrographs present the unchanged blood cells, mainly erythrocytes, leukocytes, and platelets. The interaction of the donor blood with a liquid medium of *poly*-HEMA-EGDMA exhibit no signs of hemolysis. The morphology of blood is good, there are small aggregates of platelets, some neutrophilic leukocytes, erythrocytes and their aggregates (Fig. 5).

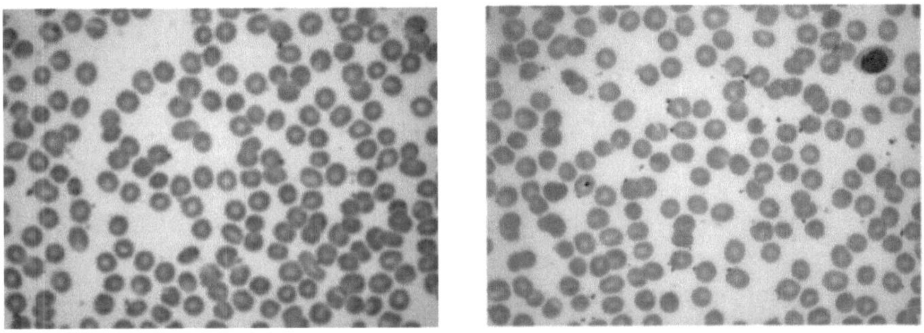

FIGURE 4 The morphology of cells donor blood.

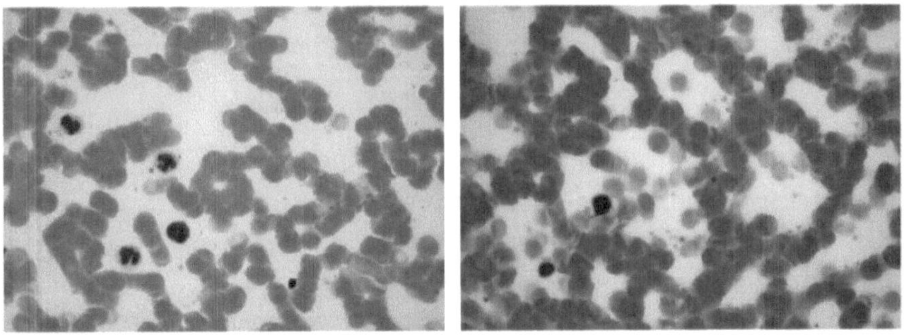

FIGURE 5 The morphology of cells after interaction with isotonic sodium chloride solution, in which was kept embolization material *poly*-HEMA-EGDMA.

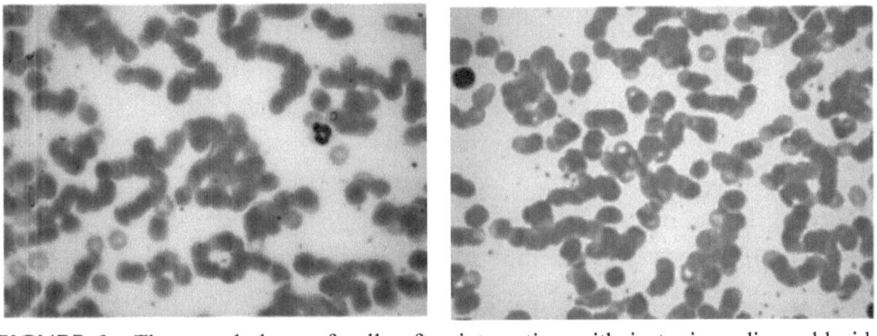

FIGURE 6 The morphology of cells after interaction with isotonic sodium chloride solution, in which was kept embolization material *poly*-HEMA-*tri*-EGDMA.

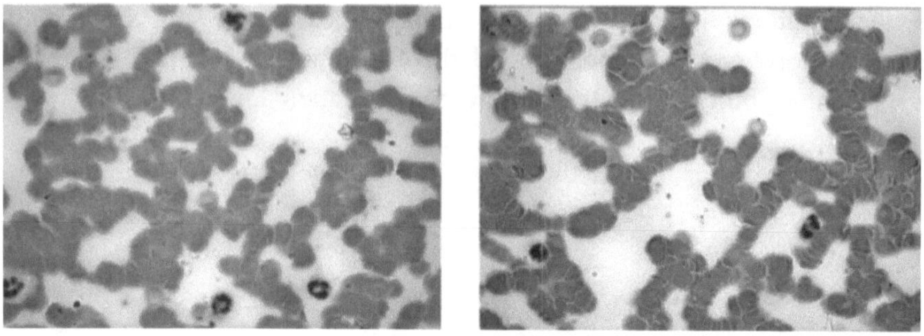

FIGURE 7 The morphology of cells after interaction with isotonic sodium chloride solution, in which was kept embolization material *poly*-HEMA-*tetra*-EGDMA.

The contact of the liquid medium of *poly*-HEMA-EGDMA with the donor blood continued to cause thrombocytopenia due to increased formation of platelet aggregates, which reflects a specific property of *poly*-HEMA associated with the release of residual monomer at a concentration of 10^{-5} g/g, thus accelerating the process of platelet aggregation [2, 8]. The number of platelet decreased, by 3 times from $(22.1 \pm 0.5)\times10^{9}$ to $(7.5 \pm 0.3)\times10^{9}$ L^{-1} at the average.

Thus, the emboli from the *poly*-HEMA-EGDMA are hemostatically active material, which primarily activates the adhesive-aggregative properties of platelets via desorption of residual monomer in this concentration. Activation of plasma hemostatic factors is secondary character and develops in 1–3 days and stored post embolization period of 10–15 days, depending on the speed and volume of injected EM in the blood vessel when the X-ray endovascular occlusion.

In testing the newly synthesized EM with cross-linking agents, *tri*-ethyleneglycol dimethacrylate or *tetra*-ethyleneglycol dimethacrylate on the proposed scheme the following results has been obtained. The contact of the donor blood with a liquid medium in which the emboli from *poly*-HEMA-*tri*-EGDMA have been stored also causes a minimal damage to blood cells, the preservation of blood remaining good (Fig. 6). Leukocytes (neutrophils), stab neutrophil, platelet macroform are observed in the visible region. The level of Hb in this test varied in the range of $11.6\div12$ L^{-1}. Hemolysis of blood did not exceed 10 mg%. The number of erythrocytes did not decrease below $(3.35\pm0.6)\times10^{12}$ L^{-1} and the average was $10\div14\%$

relative to the level of erythrocytes of the donor blood. There were no signs of morphological damage to cells.

The number of leukocytes decreased in the range of 15÷20%, i.e., of $(6.2\pm 0.6)\times 10^9$ to $(5\pm 0.4)\times 10^9\ L^{-1}$, $p < 0.05$. Morphological preservation of the white blood was good. The increased formation of platelet aggregates results in accelerated thrombocytopenia, the number of platelet decreased by 1.5÷2.5 times from $(21.5\pm 0.7)\times 10^9$ to $(8.6\pm 0{,}2)\times 10^9\ L^{-1}$, $p < 0.01$, respectively. However, unlike the *poly*-HEMA-EGDMA the aggregation properties of erythrocytes increased and instantly organized into micro- and macroaggregates capable of adhesion to leukocytes and platelets.

Under the contact the donor blood with the liquid medium in which the emboli from the *poly*-HEMA-*tetra*-EGDMA have been stored there was a slight damage of blood cells (damaged cells have not been identified), the preservation of the formed elements was satisfactory (Fig. 7). In the visible region is observed densely organized aggregations of erythrocytes, neutrophils, leukocyte with the platelet adjacent to its surface. A characteristic feature of such contact is contiguity of platelets or their groups on the aggregations of erythrocytes. The level of Hb varied within 11÷12 L^{-1}, hemolysis of blood did not exceed 12÷13 mg%. The number of erythrocytes did not decrease below $(3.1\pm 0.4)\times 10^{12}\ L^{-1}$, and the average was 15÷16% relative to the level of erythrocytes of the donor blood. In addition, an increasing erythrocytes aggregation activity has been observed, which is reflected in the organization of macroaggregates in the form of a erythrocytic clot. The number of leukocytes is reduced in the range of 20%, i.e., from $(6.1\pm 0.5)\times 10^9$ to $(5.05\pm 0.3)\times 10^9\ L^{-1}$, $p < 0.05$.

Table 1 presents the data on changes in of the cellular elements of the donor blood under contact with EM, which was synthesized with different cross-linking agents.

Morphological preservation of white blood cells was also satisfactory. Apparently, an increased formation of platelet aggregates causes the activation of thrombocytopenia with a decreasing number of platelet by 3 times, from $(21.5\pm 0.7)\times 10^9$ to $(7.1\pm 0.3)\times 10^9\ L^{-1}$.

It should be noted that the genesis of hyper-coagulative activity under the interaction of both *poly*-HEMA-*tri*-EGDMA and *poly*-HEMA-*tetra*-EGDMA can be specified as an increase in the aggregative properties of erythrocytes with the formation of macroaggregates and erythrocytic clots with participation of platelets, which instantly increases the area of obturation in the vessel.

KEYWORDS

- **Biocompatibility**
- **Dimethacrylate glycols**
- **Hemostatic effect**
- **Leukopenia**
- **Poly-HEMA**
- **Romanowsky-Giemsa**

REFERENCES

1. Jenkins, M., Ed. *Biomedical Polymers. The Scientific World;* Moscow, 2011; p 256 (in Russian).
2. Gumargalieva, K. Z.; Zaikov, G. E. *Biodegradation and Biodeterioration of Polymer;* Nova Science Publishers: New York, 1998; p 409.
3. Horak, D.; Gumargalieva, K. Z.; Zaikov, G. E. *Hydrogel in Endovascular Embolization. Chemical Reaction in Liquid and Solid Phase;* Nova Science Publishers: New York, 2003; pp 11–59.
4. Horak, D.; Galibin, I. E.; Adamyan, A. A.; et al. *J Mater Sci: Mater Med.*, **2008**, *19*, 1265–1274.
5. **Koverzanova, E. V.; Usachev, S. V.; Gumargalieva**, K. Z.; **Kokov, L. S.** *J. Balkan Tribol. Assoc.*, **2011**, *17*(2), 275–280.
6. Adamyan, A. A.; Kokov, L. S.; Titova, M. I., et al. RF Patent No. 61120, 2007 (in Russian).
7. Korolev, G. V.; Mogilevich, M. M. *Three-dimensional Radical Polymerization. Network and Hyperbranched Polymers*. Himizdat: St. Petersburg, 2006, p 344 (in Russian).
8. Horak, D.; Sitnikov, A.; Guseinov, E.; et al. *Polym. Med.*, **2002**, *32*(3–4), 48–62.

CHAPTER 33

INTERACTION AND STRUCTURE FORMATION OF GELATIN TYPE A WITH THERMO AGGREGATES OF BOVINE SERUM ALBUMIN

Y. A. ANTONOV, and I. L. ZHURAVLEVA

CONTENTS

33.1 INTRODUCTION

Intermacromolecular interactions are very important in natural biological systems [1–3] as well as in biotechnological applications [4–6]. For example, many molecules including biopolymers participate in biological functions as a molecular assembly or tissue: the self-assembly of the bacterial flagella, antigen-antibody reactions, the high activity and selectivity of enzymes, etc. is accurately achieved by intermacromolecular interactions [1]. The binding of proteins to nucleic acids, which are natural polyelectrolyte, is an integral step in gene regulation [1, 7, and 8]. A weak intermacromolecular interactions are responsible for a dramatically changes in thermodynamic compatibility of biopolymers [9–11]. Intermacromolecular interactions can be utilized for isolation of proteins [4, 5, and 12] and enzymes [13], enzyme immobilization [3, 14], encapsulation [15] and drug delivery [14, 16]. The most intriguing type of complexes is that containing two proteins [1]. When two or more protein binds together, often to carry out their biological function [2]. Proteins might interact for a long time to form part of a protein complex, [17] or a protein may interact briefly with another protein just to modify it [18]. Therefore understanding of the effect of protein structure on protein–protein interactions, for example, of smooth and skeletal muscle proteins permit the manipulation of protein side chains in order to enhance gelation properties. Of particular interest would be studies on the mechanism of complexation as well as on the molecular characteristic of resulting complexes. These studies have also aided in the understanding of biological systems. This application note highlights the use of the Malvern Zetasizer Nano ZS for characterization of interbiopolymer complexes [19]. However, data are lacking for understanding of how structural and aggregation properties of the interacting proteins affect structure formation, of complexes.

This work studies the effect of the limited thermo aggregation of BSA on the structure development in the complex forming water-acid gelatin-BSA system above the temperature of the conformation transition of gelatin, using dynamic light scattering (DLS), fluorescence measurements, and circular dichroism.

It is known [20, 21], that thermoaggregation of BSA in water at moderate temperatures changes thermodynamic properties of the protein decreasing significantly thermodynamic compatibility of BSA with other proteins, for example, with ovalbumin. Since the limited thermoaggregation of the globular protein can lead to changes its conformation state, and electrical properties, as well as activity of the protein in its saturated solutions, we assumed that such thermomodification of BSA molecules can affect on the molecular and structural properties of the inter protein complexes.

BSA or plasma albumin is a well-known globular protein (M_W = 67 kDa) that has the tendency to aggregate in macromolecular assemblies. Its three-dimensional structure is composed of three domains, each one formed by six helices. A 17 disulphide bonds are located in BSA molecule. The most common molecular form is ellipsoid (4.1 nm   14.1 nm) [22]. Gelatin is a protein derived by partial hydrolysis of collagen. The protein is unique in that it is made up of triplets of amino acids, gly-X–Y. The X and Y can be any amino acid but the most common are proline and 4-hydroxyproline, which have a five-member ring structure. Helix-coil transition of gelatin and gelation of gelatin solutions has been studied extensively in past [23]. Gelatin is well known, widely used in industry for its textural and structuring properties [24], and capacity to form interpolymer complexes with polyelectrolyte (see, e.g., [5, 25, 26]. Gelatin-BSA mixtures are well known and used in biomedical and controlled release areas [27].

33.2 MATERIALS AND METHODS

The BSA Fraction V, pH 5 (Lot A018080301), was obtained from Across Organics Chemical Co. (protein content = 98–99%; trace analysis, Na < 5000 ppm, CI < 3000 ppm, no fat acids were detected). The isoelectric point of the protein is about 4.8–5.0, and the radius of gyration at pH 5.3 is equal to 30.6 Å [22]. The water used for solution preparation was distilled three times. Most measurements were performed at pH 5.3. The extinction of 1% BSA solution at 279 nm was $A^{1cm}{}_{279}$= 6.70, and that value is very close to the tabulated value of 6.67 [28]. The gelatin sample used is gelatin type A 200 Bloom PS 8/30 (Lot 09030) produced by SBW Biosystems, France. The Bloom number, weight average molecular mass and the isoelectric point of the sample, as reported by the manufacturer, are, respectively, 207, 99.3 kDa and 8–9. Since the commercial sample contained traces of peptides and various substances regarded as impurities, an additional purification by washing with deionised water for 3 h at 5°C, was used. The major characteristics of the purified gelatin were described in a previous paper [29]. To prepare molecularly dispersed gelatin solutions, deionised water was gradually added to the gelatin, and stirred first at 60°C for 20 min and then at 40°C for 1 h. The required pH value of the solutions (5.4) was adjusted by addition of 0.1–0.5 M NaOH or HCl. The resulting solutions were centrifuged at 50,000 g for 1 h at 40°C to remove insoluble particles.

The ternary water-gelatin-BSA systems containing gelatin in coil conformation and native BSA sample were prepared by mixing solutions of each biopolymer at 40°C. All measurements were performed after previous holding of biopolymer solutions and their mixtures during 12–15 h. Previous experiments shown that longer store of solutions before measurement do not change the results of measurements.

Solutions of BSA aggregates with the middle size 100 nm were obtained by heating 5 g of 5 wt% solution of the native BSA sample at 57°C for 10 min and subsequent dilution to the desired concentrations.

33.2.1 DLS

Determination of Intensity-size distribution, and volume–size distribution functions, as well as zeta potentials of BSA, gelatin, and BSA+gelatin particles were performed, by the Malvern Zetasizer Nano instrument (England), using a rectangular quartz capillary cell. The concentration of gelatin in the water-gelatin-BSA mixtures was kept 0.25% (w/w). The Intensity-size distribution functions were determined for mixtures with different BSA/gelatin weight ratio (q). For each sample the measurement was repeated 3 times. The samples of native BSA and gelatin were filtered before measurement through DISMIC-25cs (cellulose acetate) filters (sizes hole of 0.22 μm). Subsequently the samples were centrifuged for 30 seconds at 4,000 g to remove air bubbles, and placed in the cell housing. Solutions of BSA aggregates were used without filtration. The detected scattering light intensity was processed by Malvern Zetasizer Nano software.

33.2.2 CD MEASUREMENTS

CD measurements of BSA solution alone and in the presence of gelatin were performed using a Chiroscan Applied Photophysics instrument. Far-UV CD spectra were measured in 1-mm cells at 20°C and 40°C. The solutions were scanned at 50 nm min^{-1} using 2 s as the time constant with a sensitivity of 20 mdeg and a step resolution of 0.1.

33.2.3 FLUORESCENCE MEASUREMENTS

Fluorescence emission spectra between 280 and 420 nm were recorded on a RF 5301 PC Spectrofluorimeter (Shimadzu, Japan) at 25°C and 40°C with the excitation wavelengths set to 250, 270, and 290 nm, slit widths of 3 nm for both excitation and emission, and an integration time of 0.5 s. The experimental error was approximately 2%.

33.2.4 ROSENBERG'S METHOD

The method of Middaugh et al, [30] (hereafter called Rosenberg's method) has been used to quantify the solvent quality. The method consists in determining the dependence of protein solubility in the given aqueous solvent on the concentration of PEG in the Solvent (1) Protein (2) PEG (3) system. Extrapolation of this dependence to $C_{PEG} = 0$ gives the value for the effective activity of the protein in its saturated solution ($\log C_{biopolymer}$). Evidence for the validity of this extrapolation includes (a) the experimentally observed linearity of log solubility versus PEG concentration plots, (b) the extrapolation of such plots to correct activities in the situation where protein activities can be experimentally determined, and (c) the independence of the extrapolated activities on protein concentration over a wide range. A more detailed analysis [31] makes it possible to relate the activity to the value of the second virial protein coefficient characterizing the protein-solvent interaction. The data obtained represents the activity of BSA in its saturated solution and gives – in the case of water as a solvent – an indication of the hydrophily of the biopolymer. In thermodynamic terms, Rosenberg's methods relate the activity of the protein to the second virial coefficient A_{12} that characterizes the water (1) biopolymer (2) interaction [31].

Experimentally, the method consists of preparing binary solutions of PEG and BSA at the required temperature, and pH. After mixing for 1h, a separation of phases is established by means of centrifugation at 50,000 g for 30 min. The weight concentration of BSA in the supernatant is determined by determination of absorption at 279 nm.

33.3 RESULTS AND DISCUSSION

33.3.1 CHARACTERIZATION OF THE NATIVE BSA AND THERMALLY AGGREGATED BSA SAMPLE BY DLS

The intensity size distributions and the volume size distributions functions for 0.25 wt% solutions native BSA and the BSA^{TA} samples preheated at different temperatures and pH 5.4 are shown in Fig. 1. Our preliminary experiments shown that the intensity size distributions for BSA does not depends on the concentration, at least in the range from 0.1 wt% till 0.5 wt% (data not shown). One can see from Fig. 1 that approximately 90% of all BSA particles have the average radius 3.7 nm that is in accordance with known data [19]. Preheating the BSA sample at 54°C do not leads to appreciable changes in the intensity size distributions and the volume size distributions functions. At 55°C a small part of the protein undergo to aggregation (Fig. 1b). At the temperature 57°C all BSA molecules undergo to the limited aggregation with the average radius of particles 110 nm. At 58°C the average radius of particles increases up to 300 nm. Further increase the temperature leads to the beginning of thermal denaturation and irreversible precipitation of BSA [32].

The intensity size distributions for BSA and BSA^{TA} samples at pH 5.4 and different concentrations of the protein and at 40°C (before beginning of thermo aggregation of BSA) is presented in Fig. 2.

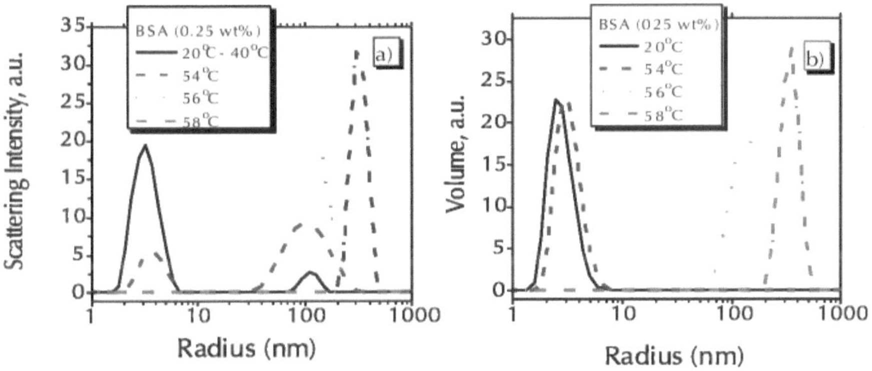

FIGURE 1 The intensity size distributions function (a) and the volume size distributions function (b) for 0.25 wt% solutions native BSA and the BSA^{TA} sample preheated at different temperatures. pH 5.4; 40°C.

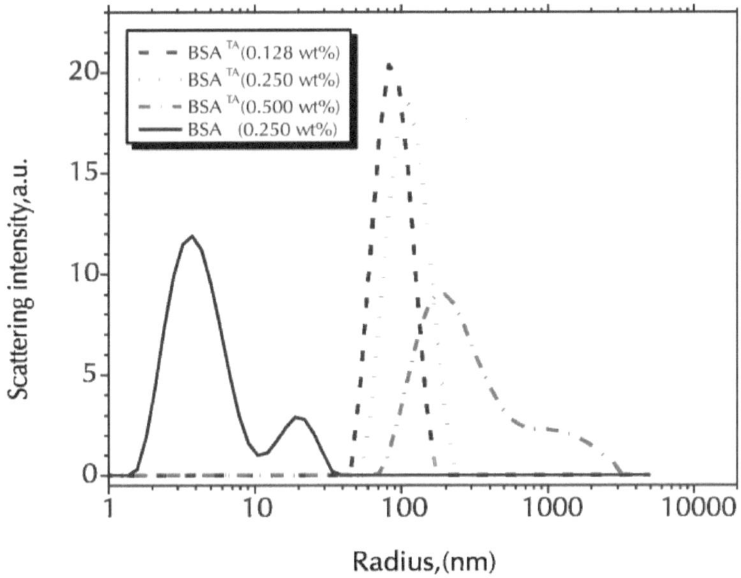

FIGURE 2 The intensity size distributions functions for BSA and BSA^{TA} samples at different concentrations of the protein. pH 5.4;40°C.

On the contrary to the native BSA, the average size of BSA^{TA} and the width of it's the intensity size distribution function are sensitive to the protein concentration. The average radius increases (from 84 nm to 110 nm) when the BSA^{TA} concentration changes from 0.128 wt% to 0.25 wt%, and it reach 180 nm at the concentration of 0.5 wt%. Since thermo aggregation of BSA at the concentrations higher then 0.25 wt% leads to formation of aggregates with a larger polydispersity, all subsequent experiments with BSA^{TA} sample were performed with the 0.25 wt% solutions preheated at 56.5°C for 10 min. as it was described in the experimental part. At this temperature the secondary and the local tertiary structures of BSA doesn't undergo an appreciable changes according to the data of circular dichroism, fluorescent spectroscopy, and differential scanning microcalorymetry [32].

33.3.2 LIMITED THERMO AGGREGATION AS A FACTOR DETERMINING MOLECULAR SIZES OF BSA-GELATIN COMPLEXES AND STABILITY BSA STRUCTURE IN COMPLEX

The intensity size distribution functions in water-BSA-gelatin and water-BSA^{TA}-gelatin systems at different q values are shown correspondingly in Figs. 3 and 4.

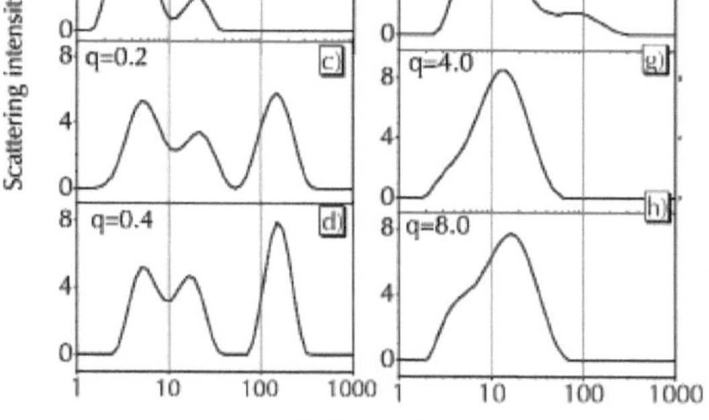

FIGURE 3 The intensity size distributions function for gelatin (0.25 wt%) (a), BSA (0.25 wt%) (b), and their mixtures at the concentration of gelatin 0.25 wt%) and different q values (c–h). pH 5.4; 40°C. The overlap concentrations (C*) for BSA and gelatin are 1.25 wt% and 3.0 wt% correspondingly [38].

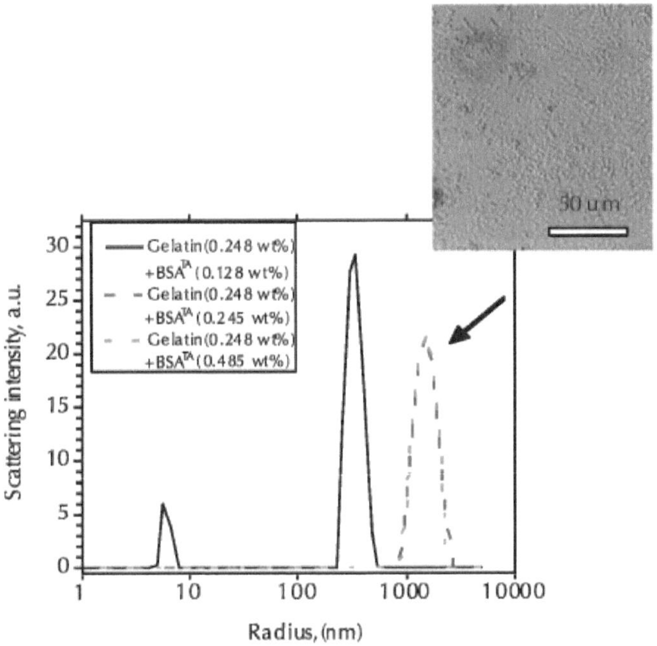

FIGURE 4 The intensity size distributions function for gelatin (0.25 wt%) (a), BSA (0.25 wt%) (b), and their mixtures at the concentration of gelatin 0.25 wt%) and different q values (c–h). pH 5.4; 40°C. The overlap concentrations (C*) for BSA and gelatin are 1.25 wt% and 3.0 wt% correspondingly [38].

The gelatin sample was strongly polydisperse. The main peak (about 50%) has the average radius 140 nm. Small parts of the low molecular weight fractions with the average sizes 3.8 nm (27%) and 9 nm (23%) correspondingly were also detected. As can be seen from Fig. 3, the presence of small amount of BSA in the gelatin solution (at q = 0.2 and q = 0.4) do not effect on the scattering intensity and position of the main gelatin peak. At a higher q values (1.0) the intensity of this peak decrease considerably. At q = 4 this peak completely disappears and the new dominant peak appears with the average radius 14 nm. At a higher content of BSA in the mixture (at q = 8) the appearance of shoulder was detected with the average radius ≅ 4 nm. The shoulder reflect "free" BSA molecules, and it becomes larger at q = 8. The main conclusion from the data presented in Fig. 2 is that above the temperature of the conformational transition intermacromolecular interaction of gelatin with native BSA leads to collapse gelatin macromolecules and formation compact (30 nm in radius) BSA-gelatin complexes. In the presence of high ionic strength (0.25, NaCl), when electrostatic interactions were suppressed the size distribution function of gelatin becomes insensitive to the presence of BSA (data not presented). Taking into account that the molecular weight of gelatin is 243,000 Da and that of BSA is equal to 67,000 Da, we can roughly evaluate the "molar" ratio BSA/gelatin in the water insoluble complex. A simple calculation shows that this ratio is ~6/1.

The behavior of gelatin in the presence of BSA^{TA} is completely different (Fig. 4). The interaction of gelatin with BSA^{TA} leads to formation of the large complex particles. At q = 0.5 the average size of the particlesis 356 nm that is 3 times higher then those of the binary water-gelatin and water-BSA^{TA}. At a higher q values (q = 1.0) the average size of the particles sharply increases up to 1,500 nm.

An addition of NaCl in the ternary water-gelatin-BSA^{TA} system results in the partial dissociation of the complex aggregated, and the absence of free BSA molecules (Fig. 5). On the other hand, an addition of NaCl in the in the initial binary solutions of BSA and gelatin (0.25 M NaCI) leads to the partial dissociation of the complex aggregated, and the absence of free BSA molecules (Fig. 6).

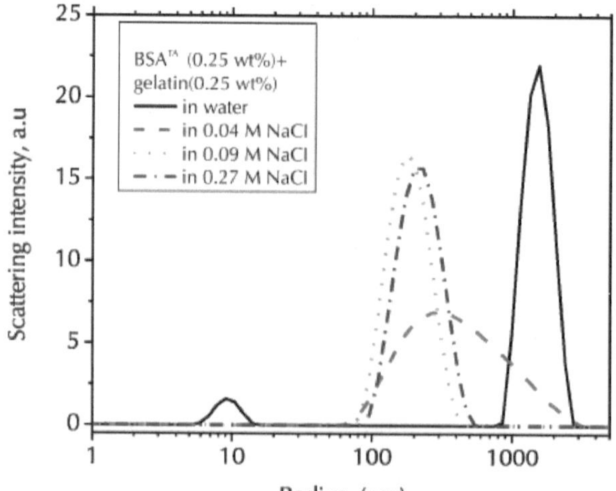

FIGURE 5 The effect of an addition NaCI in the ternary gelatin (0.25 wt%)–BSA^{TA} (0.25 wt%) system on the intensity size distributions functions. 5.4; 40°C. The overlap concentrations (C*) for BSA and gelatin are 1.25 wt% and 3.0 wt% correspondingly [38].

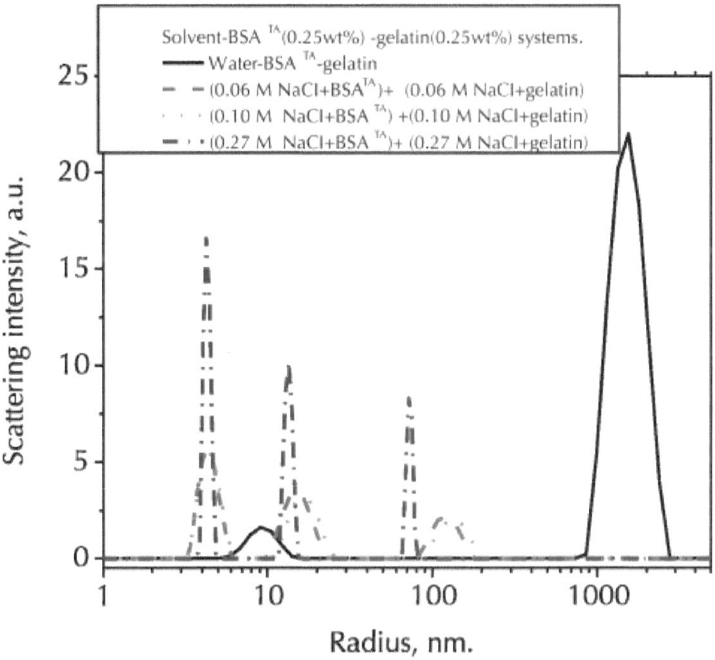

FIGURE 6 The effect of an addition of NaCl in the binary water-gelatin and water-BSA^{TA} system on the intensity-size distribution functions of the ternary water-gelatin (0.25 wt%)–BSA^{TA} (0.25 wt%) system, obtained by mixing of the binary salt solutions of gelatin and BSA^{TA}.

These facts shows that with one hand the complexes are formed via electrostatic interaction, rather than through hydrogen bonds formation or hydrophobic interaction. The role of salt is to "soften" the interactions, which is equivalent to making the electrostatic binding constant smaller. On the other hand, non electrostatic forces play an important role in stabilization of the formed complexes (Fig. 6).

The question is arises, what is the reason of the great difference in structure formation of the native and BSA^{TA} samples with gelatin? We consider three possible reasons for such behaviour of BSA: (1) Change in the solvent quality of BSA molecules after their limited thermoaggregation. (2) Possible changes in structure and in the charge of the thermally aggregated BSA alone and in the presence of gelatin compared with native BSA sample. (3) Steric reasons affecting on complex formation.

Figure 7 shows the effect of PEG on the solubility of BSA in water before and after thermally induced aggregation at pH 5.4. The dependences obtained proved to be rectilinear. Extrapolation of this dependence to $C_{PEG} = 0$ gives the value for the effective activity of the protein in its saturated solution ($\log C_{biopolymer}$) [30].

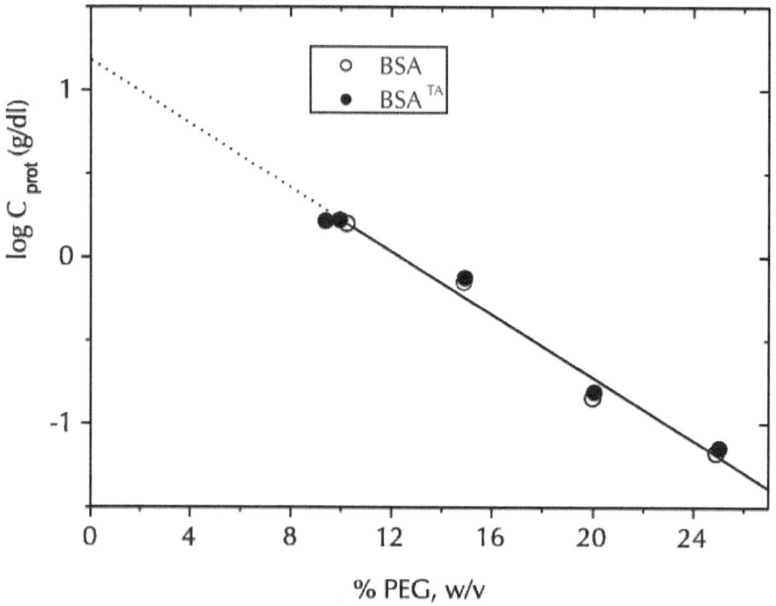

FIGURE 7 Effect of PEG on the solubility of BSA and BSATA samples (40°C, pH 5.4). The solid line is a least-square fit of the data to a straight line. The dotted line is a linear extrapolation of the solubility to zero PEG concentration.

The results obtained show that thermally induced aggregation of BSA does not affect appreciably on the activity of saturated BSA solutions, and therefore solvent quality of the BSA before and after the limited aggregation is almost the same.

It is known that at a relatively high temperature, polyelectrolyte can effect on the secondary, and the local tertiary structures of globular proteins, elaborating part of the charged functional groups of the protein from inside of globule to the globule surface [32]. Such structural changes can affect on electrostatic interaction of the oppositely charged proteins. In order to examine such possibility, the CD and fluorescence spectra of BSA samples were obtained in the presence of gelatin. The data obtained are shown in Figs. 8 and 9.

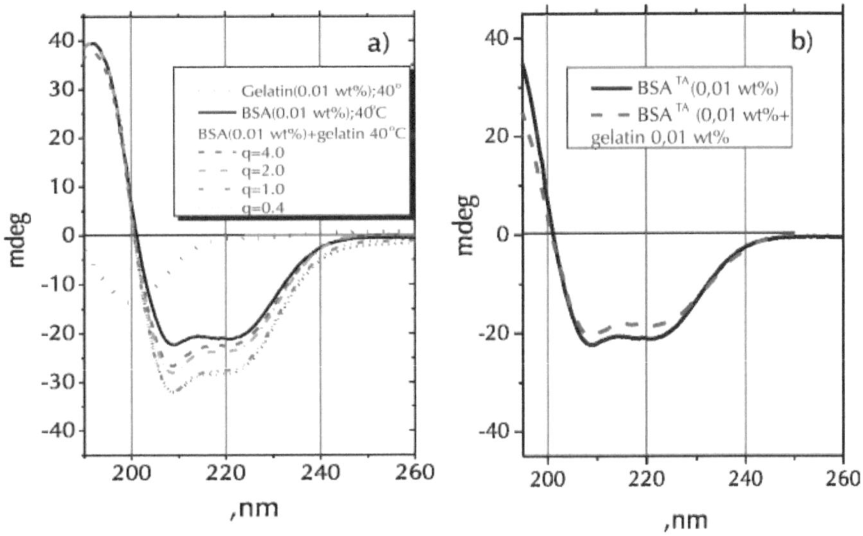

FIGURE 8 Comparison of the circular dichroism spectra of 0.01 wt% gelatin, 0.01 wt% BSA, and the water-gelatin (var)-BSA (0.01 wt%) mixtures at different q values after subtraction of the gelatin spectrum (a); comparison of the circular dichroism spectra of 0.01 wt% BSA^{TA} and 0.01 wt% BSA^{TA} + 0.01 wt% gelatin after substraction of the gelatin spectrum (b). 40°C. pH =5.4.

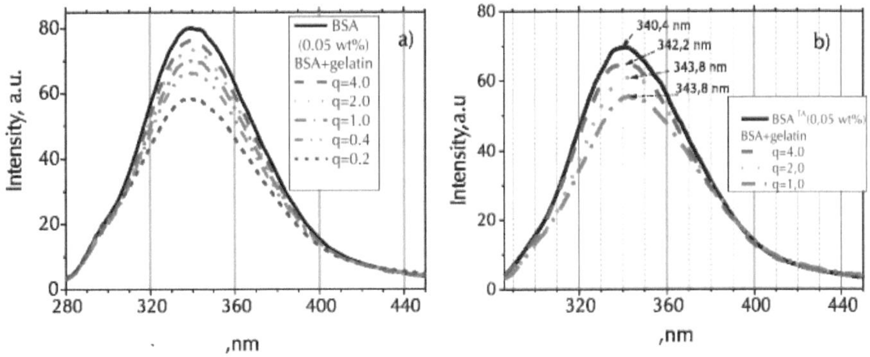

FIGURE 9 Fluorescence spectra of 0.05 wt% BSA and BSA-gelatin systems at different q values (a); Fluorescence spectra of 0.05 wt% BSA^{TA} and BSA^{TA}–gelatin systems at different q values (b). Exc λ=270 nm. 40°C. pH 5.4.

Figure 8a shows far-UV CD spectra at 40°C for BSA and gelatin in water. Gelatin at +40°C shows the coil conformation. The triple helix can partly reform when the temperature of the sample falls below the gelation temperature [33]. At 40°C BSA contains 51.4% of the alpha helical structure. The average helix content, f_H (in%) for BSA molecules was calculated according to [34, 35]. The presence of gelatin in the BSA solution at q=2 leads to significant increase in the negative band at 222 nm and significant increase in the negative maximum at 209 nm as shown in Fig. 8a. At higher gelatin content in the mixture (at q = 1.0) this effect was more pronounced. Such increase in the negative band at 222 nm corresponds to an increase in the content of alpha-helical structures from 51.4% to 62%. Introduction of NaCl in the water results in full insensitivity of CD spectra of gelatin to the presence of BSA in all the *q* range studied (data not shown). In contrary of the native BSA, the presence of gelatin in the BSA^{TA} sample results in an appreciable decrease in the negative band at 222 nm and decrease in the negative maximum at 209 nm as shown in Fig. 8b. Increase gelatin content in the mixture from q = 1 to q = 2.0 and 4.0 do not effect on CD spectra of the BSA^{TA} sample (data not shown). Figure 9 shows the effect of gelatin on the spectra of fluorescence intensity of the native BSA sample (Fig. 9a) and the BSA^{TA} sample (Fig. 9b) at the excitation wavelength (λex) = 270 nm.

The wavelength of maximum emission (λ_{max}) for the both BSA samples was about 339–340 nm, regardless of λ_{exc} in the range of 250–290 nm. The presence of gelatin in the native BSA sample leads to a visible decrease in fluorescence intensity. The decrease in the intensity was also observed at the λ_{exc} = 250 nm and λ_{exc} = 290 nm (data not shown) although to a lesser degree than at λ_{exc} = 270 nm. The fluorescence quenching was especially remarkable at a highest content of gelatin in the BSA solution (at q = 0.4). At the same time we did not observe any appreciable shift the maximum of the fluorescence intensity at q values from 0.4 to 4.0 that indicates the absence of an appreciable unfolding of BSA molecules in the presence of gelatin molecules. Moreover, CD data shows even on the significant ordering of the secondary structure of the native BSA in the presence of gelatin.

In contrast to the system BSA-gelatin containing native BSA sample, the results of fluorescence analysis for the system containing BSA^{TA} sample shows a shift in the wavelength of the maximum of the intrinsic fluorescent emission with quenching of fluorescence, indicating changes in protein conformation [34]. In proteins that contain all three aromatic amino acids, fluorescence is usually dominated by the contribution of the tryptophan residues, because both their absorbency at the wavelength of excitation and their quantum yield of emission are considerably greater than the respective values for tyrosine and phenylalanine. With proteins containing tryptophan, the changes in intensity and shifts in wavelength are usually observed upon unfolding. It is well known that, in general, the fluorescence quantum yield of tryptophan decreases when its exposure to solvent increases [36, 37]. The emission maximum is usually shifted from shorter wavelengths to about 350 nm upon protein unfolding, which corresponds to the fluorescence maximum of pure tryptophan in aqueous solution. According to this, we take into account the fluorescence quenching in terms of local rearrangements of the tryptophan surroundings.

The apparent wavelength of maximum emission (λ_{max}) for pure BSA is 339 nm (λ_{max} for pure tryptophan in aqueous solution is 354 nm). Comparing these values indicates that the tryptophan residues of the native protein are only partially located on the surface and exposed to water. Significant parts are located inside the globule. An excess of gelatin results in a decrease in the fluorescence intensity and in a small shift of λ max from 339 to 344 nm. These facts indicate a partial unfolding of BSA with an additional exposure of the hydrophobic tryptophan residues to the surface.

If the partial destabilization of the secondary structure of BSA^{TA} and an additional exposition of the hydrophobic tryptophan residues to the surface in the presence gelatin takes place it may increase the charge of the interacting molecules and effect on the structure of the complexes formed. Really, the z- average potential on the BSA (z^{za}_{BSA}) is equal to 9.27±0.8 mV), whereas $z^{za}_{BSA}{}^{TA} = -12.31 \pm 0.8$. At the compositions corresponding to the maximal binding of BSA by gelatin z average negative potential of the complex particles gelatin-BSA ($z^{za}_{BSA-Gel}$) ≅ 0, whereas the values of the same parameter for the BSA^{TA} sample is –6.16. Therefore, the charges of the interacting BSA and BSA^{TA} molecules and the complex particles formed are different. These factors can affect on the sizes of complexes gelatin with BSA.

On the other hand, it is reasonably to suggest that formation of the large aggregates of BSA^{TA} sample leads to the steric difficulty in their interaction with gelatin with subsequent collapse of the later. In this case it is more probable that formation of the complex particles results from the simple joining of the BSA^{TA} to gelatin without significant changes of the sizes of interacting biopolymers. The relatively higher negative charge of the complex BSA^{TA}/gelatin particles then that of BSA-gelatin particles is also do not promote to collapse of the gelatin molecules.

Figure 10 shows schematically the main structural changes observed during interaction BSA or BSA^{TA} with gelatin in coil conformation. The interaction of BSA with gelatin leads to the collapse of the coil. The collapsed gelatin molecule wraps six BSA molecules. BSA^{TA} interacts with gelatin molecules results in formation the large macromolecular assembly. The interaction leads to the partial unfolding globular protein and formation of the charged complex particles.

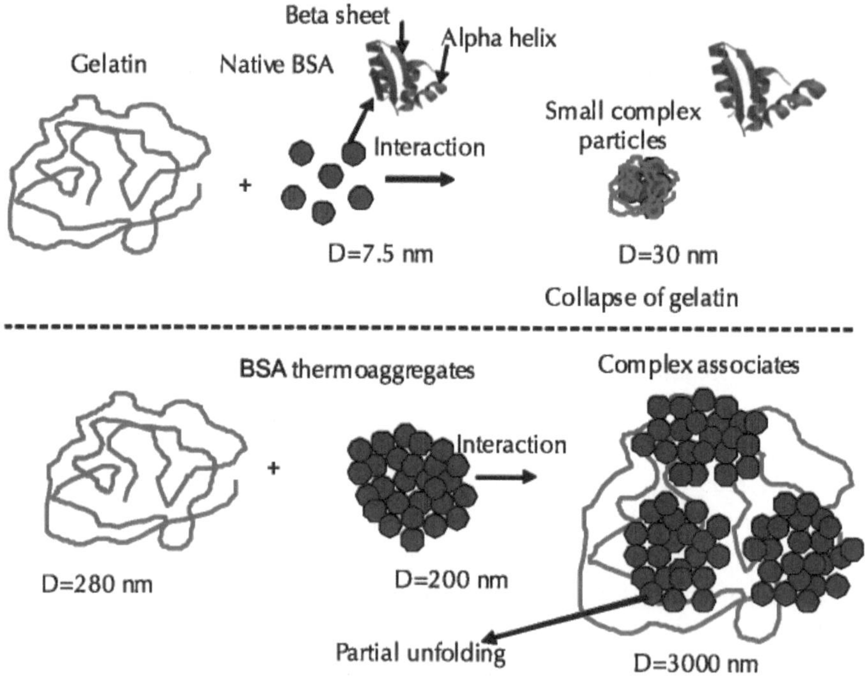

FIGURE 10 Schematic representation of the main structural changes observing during interaction of the native and thermally aggregated BSA with gelatin in coil conformation.

KEYWORDS

- **Bacterial flagella**
- **Interbiopolymer complexes**
- **Malvern Zetasizer Nano ZS**
- **Polyelectrolyte**

REFERENCES

1. Cantor, C. R.; Schimmel, P. R. In *Biophysical Chemistry*. Freeman WH & Co: San Francisco, 1980; pp 849–886.
2. Dusenbery, D. B. *Sensory Ecology: How Organisms Acquire and Respond to Information*. Freeman WH & Co: New York, 1992.
3. Xia, J.; Dubin, P. L. In *Macromolecular Complexes in Chemistry and Biology*; Dubin, P. L.; Davis, R.M.; Schultz, D.; Thies, C. Eds.; Springer-Verlag: Berlin, 1994 ; pp 247–271.
4. Dubin, P. L.; Gao, J.; Mattison, K. *Sep. Purif Methods,* **1994**, *23*(1), 1–16.
5. Tolstoguzov, V. B. In *Functional Properties of Food Macromolecules;* Hill, S. E.; Ledward, D. A.; Mitchel, J. R., Eds.; Aspen Publishers Inc.: Gaithersburg, M. D., 1998; pp 253–277.
6. Antonov, Y. A.; Moldenaers, P. *Biomacromolecules*, **2009,** *10*(12), 3235–3245.
7. Record, M. T., Jr; Anderson, C. F.; Lohman, T. M. *Q. Rev. Biophys.*, **1978**, *11*, 103–178.
8. Von Hippel, P. H.; Bear, D. G.; Morgan, W. D.; McSwiggen, J. A. *Annu. Rev. Biochem.*, **1984**, *53*, 389–446.
9. Antonov, Y. A.; Lefebvre, J.; Doublier, J. L. *J. Appl. Polym. Sci.,* **1999**, *71*, 471–482.
10. Antonov, Y. A.; Soshinsky, A. A. *Int. J. Biol. Macromol.*, **2000**, *27*(4), 279–285.
11. Antonov, Y. A.; Dmitrochenko, A. P.; Leontiev, A. L. *Int. J. Biol. Macromol.,* **2006**, *38*(1), 18–24.
12. Serov, A. V.; Antonov, Y. A.; Tolstoguzov, V. B. *Nahr*, **1985**, *29*(1), 19–30.
13. Kiknadze, E. V.; Antonov, Y. A. *Appl. Biochem. Microbiol.* (Russia, Engl. Transl), **1998**, *34*(5), 462–465.
14. Ottenbrite, R. M.; Kaplan, A. M. *Ann. N.Y. Acad. Sci.*, **1985**, *446*, 160–168.
15. Magdassi, S.; Vinetsky, Y. In *Microencapsulation of Oil-In-Water Emulsions by Proteins;* Benita, S., Ed.; Marcel Dekker Inc.: New York, 1997; pp 21–33.
16. Regelson, W. *Interferon,* **1970**, *6*, 353–359.
17. Denning, D.; Patel, S.; Uversky, V.; Fink, A.; Rexach, M. *Proc. Natl. Acad. Sci. USA,* **2003**, *100*, 2450–2455.
18. St. Stout, T. J.; Foster, P. G.; Matthews, D. J. *Curr. Pharm. Des.*, **2004**, *10*, 1069–1082.

19. Kayitmazer, B.; Shaw, D.; Dubin, P. *Characterization of Protein-Polyelectrolyte Complexes*. Zetasizer Nano Application Note. http://www.malvern.com/common/downloads/campaign/MRK513-02.pdf
20. Polyakov, V. I.; Popello, I. A.; Grinberg, V. Y.; Tolstoguzov, V. B. *Nahr,* **1986**, *30*(1), 81–88.
21. Polyakov, V. I.; Popello, I. A.; Grinberg, V. Y.; Tolstoguzov, V. B. *Food Hydrocoll.,* **1997**, *11*(2), 171–180.
22. Peters, T., Jr. *Albumin: An Overview and Bibliography*, 2nd ed. Miles Inc Diagnostics Division: Kankakee, I. L., 1992.
23. Djabourov, M.; Leblond, J.; Papon, P. *J. Phys. Fr.*, **1988**, *49*, 319–332.
24. Ward, A. G.; Courts, A. *The Science and Technology of Gelatin.* Academic Press: London-New York-San Francisco, 1977.
25. Kaibara, K.; Okazaki, T.; Bohidar, H. B.; Dubin, P. *Biomacromology*, **2000**, *1*, 100–107.
26. Bowman, W. A.; Rubinstein, M.; Tan, J. S. *Macromology,* **1997**, *30*(11), 3262–3270.
27. Migneault, I.; Dartiguenave, C.; Bertrand, M.; Karen, J.; Waldron, C. *BioTechniques,* **2004**, *37*, 790–802.
28. Kirschenbaum, D. M. *Anal. Biochem.,* **1977**, *81*(2), 220–246.
29. Antonov, Y. A.; Van Puyvelde, P.; Moldenaers, P. *Biomacromoloy,* **2004**, *5*, 276–283.
30. Middaugh, C. R.; Tisel, W. A.; Haive, R. N.; Rosenberg, A. *J. Biol. Chem.,* **1979**, *254*, 367–370.
31. Polyakov, V. I.; Popello, I. A.; Grinberg, V. Y.; Tolstoguzov, V. B. *Nahr*, **1985**, *29*, 323–333.
32. Antonov, Y. A.; Wolf, B. A. *Biomacromol*ogy, **2005**, *6*(6), 2980–2989.
33. Zhang, Z.; Li, G.; Shi, B. *J. Soc. Leather Technol. Chem.,* **2006**, *90*, 23–28.
34. Woody, R. W. *Methods Enzymol.*, **1995**, *246*, 34–71.
35. Creighton, T. E. *Protein Structure*, 2nd ed. Oxford University Press: Oxford, 1997.
36. Campbell, I. D.; Dwek, R. A. *Biological Spectroscopy*. The Benjamin Cummings Publishing Co.: Menlo Park, 1984.
37. D'Alfonso, L.; Collini, M.; Baldini, G. *Biochemistry*, **2002**, *241*, 326–333.
38. Lefebvre, J. *Rheol. Acta,* **1982**, *21*, 620–625.

CHAPTER 34

EXPRESS ASSESSMENT OF CELL VIABILITY IN BIOLOGICAL PREPARATIONS

L. P. BLINKOVA, Y. D. PAKHOMOV, O. V. NIKIFOROVA, V. F. EVLASHKINA, and M. L. ALTSHULER

CONTENTS

34.1 INTRODUCTION

Lyophilized cells in biological preparations exist in a form of artificial anabiosis that in our opinion may be paralleled with natural forms of dormancy caused by environmental stresses (physical, chemical, biological). Activity of bacterial cultures, including probiotic preparations is commonly evaluated as a number of colony forming units (CFU/ml) after plating serial dilutions of bacterial cultures in normal saline. Microbiologists believe that a colony is not always formed from a single cell and real number of cells is somewhat greater. Plating and other culture-based methods also fail to reveal cells that are not readily culturable at the moment of sampling. Such cells may be dormant, sublethally injured or exist in viable but nonculturable state [1–4]. We tested the commercially available kit for differentiating between viable and dead for accuracy when viability of probiotics is assessed.

34.2 MATERIALS AND METHODS

We studied five samples of commercially available colibacterin (*Escherichia coli* M17) produced by different manufacturers and with different storage period after expiration date. These include:

(1) Perm'RIVS, batch # 572–2 and expiration date 11.1982.
(2) ImBiO, batch #170–3and expiration date06.2001.
(3) Mechnikov Biomed, batch #40–3and expiration date03.2008.
(4) Mechnikov Biomed, batch #240–3and expiration date05.2009.
(5) Mechnikov Biomed, batch #270–3and expiration date07.2009.

Samples were resuspended in 1 ml per dose of normal saline. Aliquot was taken from the sample, diluted 10-fold and immediately stained with Live/Dead staining kit (Baclight™). For our study we used a kit with separate dyes. Equal volume of each dye was mixed and diluted 10-fold. From this mixture 3 μl was added to 100 μl of a sample. Viable and dead cells emit different colors, therefore viable/dead cells ratio can easily be evaluated as well as in terms of numbers in given volume. As described in the instructions, cells that emitted green fluorescence were considered viable and red fluorescing cells – dead. Cells that emitted yellow or orange fluorescence (where present) were counted as viable, but damaged during the processes of sample preparation and visualization, by the DNA binding dyes or the exciting light.

Total cell counts were made in a counting chamber after further 10-fold or 20-fold dilution depending on the sample.

CFU/ml value was assessed by plating on nutrient agar immediately after rehydration and after 48–72 h.

Colibacterin features			Biological parameters					
Manufacturer	Batch number, expiration date	Vessel, capacity	Total cell counts		% of viable cells (Live/Dead) Total viable counts		CFU/ml (0 h)	% of viable cells unable to form colonies
			0 h	72 h	0 h	72 h		
Perm'RIVS	572–2	Ampoules,	2.11×10^{10}	3.72×10^{10}	52.2%	95.91%	2.89×10^{7}	99.73%
	11.1982	3 doses			1.1×10^{10}	3.56×10^{10}		
ImBiO	170–3	Ampoules,	3.11×10^{10}	4×10^{10}	86.7%	96.99%	5.55×10^{9}	79.5%
	06.2001	3 doses			2.7×10^{10}	3.88×10^{10}		
Mechnikov	40–3	Vials,	1.21×10^{10}	3610^{10}	90.3%	97.66%	1.58×10^{9}	85.5%
Biomed	03.2008	5 doses			1.09×10^{10}	3.52×10^{10}		
Mechnikov	240–3	Vials,	1.44×10^{10}	2.58×10^{10}	88.2%	94.98%	1×10^{10}	21.3%
Biomed	05.2009	5 doses			1.27×10^{10}	2.45×10^{10}		
Mechnikov	270–3	Vials,	1.32×10^{10}	3.56×10^{10} (48 h)	91.3%	99.05%	1.16×10^{10}	4.13%
Biomed	07.2009	5 doses				(48 ч)		
					1.2×10^{10}	3.53×10^{10}		

34.3 RESULTS AND DISCUSSION

We propose an approach to visual assessment of viable bacteria in probiotics and other biological preparations using commercially available fluorescence staining kit Live/Dead (Baclight™). Using this method we assessed viability of a model strain of E. coli M17 in several samples of colibacterin probiotics preparations produced by different manufacturers (Mechnikov Biomed; ImBiO; Perm' RIVS) with different storage period after their expiration date (1982–2009), kinds of vessel (ampoules and vials). According to the results of fluorescent staining viability of studied preparations varied from 52.2%, for the oldest sample, to 91.3% for the one that expired in July 2009. Results of direct counting of viable and dead cells were compared with CFU/ml values. Using Live/Dead it is possible to assess at least two sample preparations per hour. We showed that 4–99% of lyophilized E. coli M17 cells didn't form colonies immediately after rehydration while remaining viable in normal saline. Further incubation of serial decimal dilutions of the sample in normal saline for 48–72 h resulted in increase of number of colony forming units by 3–4 orders of magnitude. Such increase occurred in tubes where initially only some individual colonies have formed. Such increase is clearly impossible if only due to division of initially culturable cells. Apparently processes similar to delayed multiplication of some rehydrated cells may occur in the intestines of warm–blooded animals. These cells might exist in some form of anabiosis, or be sublethally injured which requires more extended recovery period before first division is possible. It should be noted that presence of yellow fluorescing cells increased with time of storage (data not shown). It is probably due to the increase of internal damage in dry cells with time. In our study we might have slightly overestimated total numbers of viable cells, because some of them could be structurally intact, but too severely damaged internally. In contrast some cells considered as dead might be able to repair their membranes and become able to form colonies. However, we have shown that real viability of even long expired preparations may be far greater then revealed by traditional plating procedures. In fact it can be very close to the initial values.

KEYWORDS

- **Fluorescence**
- **Live/Dead staining kit**
- **Lyophilized**
- **Probiotic**

REFERENCES

1. Aersten, A.; Michielis, C. W. *Crit. Rev. Microbiol.,* **2004**, *30*, 263.
2. Kaprelyants, A. S.; Kell, D. B. *Appl. Environ. Microbiol.,* **1993**, *59*(10), 3187.
3. Ganesan, B.; Mark, R.; Stuart, M. R.; Weimer, B. C. *Appl. Environ. Microbiol.*, **2007**, *73*(8), 2498.
4. Oliver, J. D. *J. Microbiol.*, **2005**, *43*(S), 93.

CHAPTER 35

A NEW APPROACH TO THE CREATION OF BIOCOMPATIBLE MAGNETICALLY TARGETED NANOSYSTEMS FOR A SMART DELIVERY OF THERAPEUTIC PRODUCTS

A. V. BYCHKOVA, M. A. ROSENFELD, V. B. LEONOVA, O. N. SOROKINA, and A. L. KOVARSKI

CONTENTS

35.1 INTRODUCTION

Magnetic nanoparticles (MNPs) have many applications in different areas of biology and medicine. MNPs are used for hyperthermia, magnetic resonance imaging, immunoassay, purification of biologic fluids, cell and molecular separation, tissue engineering [1–6]. The design of magnetically targeted nanosystems (MNSs) for a smart delivery of drugs to target cells is a promising direction of nanobiotechnology. They traditionally consist on one or more magnetic cores and biological or synthetic molecules, which serve as a basis for polyfunctional coatings on MNPs surface. The coatings of MNSs should meet several important requirements [7]. They should be biocompatible, protect magnetic cores from influence of biological liquids, prevent MNSs agglomeration in dispersion, provide MNSs localization in biological targets and homogenity of MNSs sizes. The coatings must be fixed on MNPs surface and contain therapeutic products (drugs or genes) and biovectors for recognition by biological systems. The model which is often used when MNSs are developed is presented in Fig. 1.

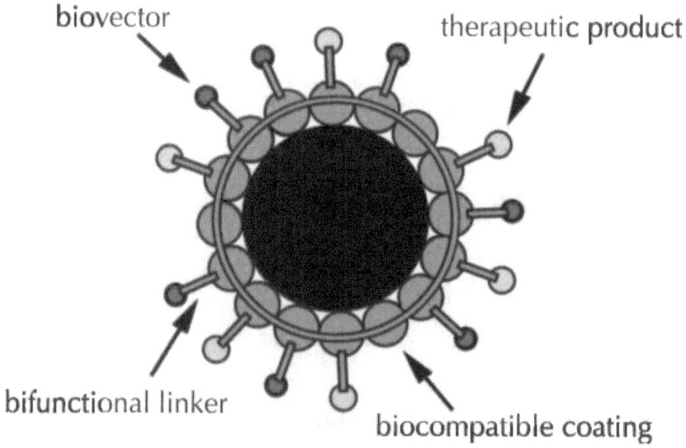

FIGURE 1 The classical scheme of magnetically targeted nanosystem for a smart delivery of therapeutic products.

Proteins are promising materials for creation of coatings on MNPs for biology and medicine. When proteins are used as components of coatings it is of the first importance that they keep their functional activity [8]. Protein binding on MNPs surface is a difficult scientific task. Traditionally bifunctional linkers (glutaraldehyde [9–10], carbodiimide [11–12]) are used for protein cross-linking on the surface of MNPs and modification of coatings by therapeutic products and biovectors. Authors of the study [9] modified MNPs surface with aminosilanes and performed protein molecules attachment using glutaraldehyde. In the issue [10] bovine serum albumin (BSA) was adsorbed on MNPs surface in the presence of carbodiimide. These works revealed several disadvantages of this way of protein fixing which make it unpromising. Some of them are clusters formation as a result of linking of protein molecules adsorbed on different MNPs, desorption of proteins from MNSs surface as a result of incomplete linking, uncontrollable linking of proteins in solution (Fig. 2). The creation of stable protein coatings with retention of native properties of molecules still is an important biomedical problem.

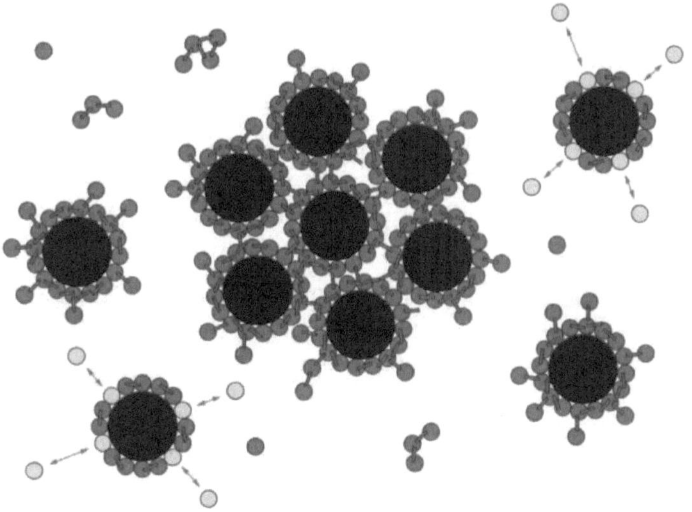

FIGURE 2 Nonselective linking of proteins on MNPs surface by bifunctional linkers leading to clusters formation and desorption of proteins from nanoparticles surface.

It is known that proteins can be chemically modified in the presence of free radicals with formation of cross-links [13]. The goals of the work were to create stable protein coating on the surface of individual MNPs using a fundamentally novel approach based on the ability of proteins to form interchain covalent bonds under the action of free radicals and estimate activity of proteins in the coating.

35.2 MATERIALS AND METHODS

35.2.1 MAGNETIC SORBENT SYNTHESIS

Nanoparticles of magnetite Fe_3O_4 were synthesized by co-precipitation of ferrous and ferric salts in water solution at 4°C and *in the alkaline medium*:

$$Fe^{2+} + 2Fe^{3+} + 8OH^- \rightarrow Fe_3O_4\downarrow + 4H_2O$$

A 1.4 g of $FeSO_4{\cdot}7H_2O$ and 2.4 g of $FeCl_3{\cdot}6H_2O$ ("Vekton," Russia) were dissolved in 50 ml of distilled water so that molar ratio of Fe^{2+}/Fe^{3+} was equal to 1:2. After filtration of the solution 10 ml of 25 mass% NH_4OH ("Chimmed," Russia) was added to it on a magnetic stirrer. 2.4 g of PEG 2 kDa ("Ferak Berlin GmbH," Germany) was added previously in order to reduce the growth of nanoparticles during the reaction. After the precipitate was formed the solution (with 150 ml of water) was placed on a magnet. Magnetic particles precipitated on it and supernatant liquid was deleted. The procedure of particles washing was repeated for 15 times until neutral pH was obtained. MNPs were stabilized by double electric layer with the use of US-disperser ("MELFIZ," Russia). To create the double electric layer 30 ml of 0.1 M phosphate-citric buffer solution (0.05 M NaCl) with pH value of 4 was introduced. MNPs concentration in hydrosol was equal to 37 mg/ml.

35.2.2 PROTEIN COATINGS FORMATION

Bovine serum albumin ("Sigma-Aldrich," USA) and thrombin with activity of 92 units per 1 mg ("Sigma-Aldrich," USA) were used for protein coating formation. Several types of reaction mixtures were created: "A1-MNP-0," "A2-MNP-0," "A1-MNP-1," "A2-MNP-1," "A2-MNP-1-acid," "T1-MNP-0," "T1-MNP-0" and "T1-0-0." All of them contained:

(1) A 2.80 ml of protein solution (“A1” or “A2” means that there is BSA solution with concentration of 1 mg/ml or 2 mg/ml in 0.05 M phosphate buffer with pH 6.5 (0.15 M NaCl) in the reaction mixture; “T1” means that there is thrombin solution with concentration of 1 mg/ml in 0.15M NaCl with pH 7.3),
(2) 0.35 ml of 0.1 M phosphate-citric buffer solution (0.05 M NaCl) or MNPs hydrosol (“MNP” in the name of reaction mixture means that it contains MNPs),
(3) 0.05 ml of distilled water or 3 mass% H_2O_2 solution (“0” or “1” in the reaction mixture names correspondingly).
Hydrogen peroxide interacts with ferrous ion on MNPs surface with formation of hydroxyl-radicals by Fenton reaction:

$$Fe^{2+} + H_2O_2 \rightarrow Fe^{3+} + OH^{\cdot} + OH^{-}$$

“A2-MNP-1-acid” is a reaction mixture, containing 10 μl of ascorbic acid with concentration of 152 mg/ml. Ascorbic acid is known to form free radicals in reaction with H_2O_2 and generate free radicals in solution but not only on MNPs surface.

The sizes of MNPs, proteins and MNPs in adsorption layer were analyzed using dynamic light scattering (Zetasizer Nano S “Malvern,” England) with detection angle of 173° at temperature 25°C.

35.2.3 STUDY OF PROTEINS ADSORPTION ON MNPS

The study of proteins adsorption on MNPs was performed using ESR-spectroscopy of spin labels. The stable nitroxide radical used as spin label is presented in Fig. 3. Spin labels technique allows studying adsorption of macromolecules on nano-sized magnetic particles in dispersion without complicated separation processes of solution components [14]. The principle of quantitative evaluation of adsorption is the following. Influence of local fields of MNPs on spectra of radicals in solution depends on the distance between MNPs and radicals [14–16]. If this distance is lower than 40 nm for magnetite nanoparticles with the average size of 17 nm [17] ESR spectra lines of the radicals broaden strongly and their intensity decreases to zero. The decreasing of the spectrum intensity is proportional to the part of radicals, which are located inside the layer of 40 nm in thickness around MNP. The same happens with spin labels covalently bound to protein macromolecules. An intensity of spin labels spectra decreases as a result of adsorption of macromolecules on MNPs (Fig. 4). We have shown that spin labels technique can be used for the study of adsorption value, adsorption kinetics, calculation of average number of molecules in adsorption layer and adsorption layer thickness, concurrent adsorption of macromolecules [18–20].

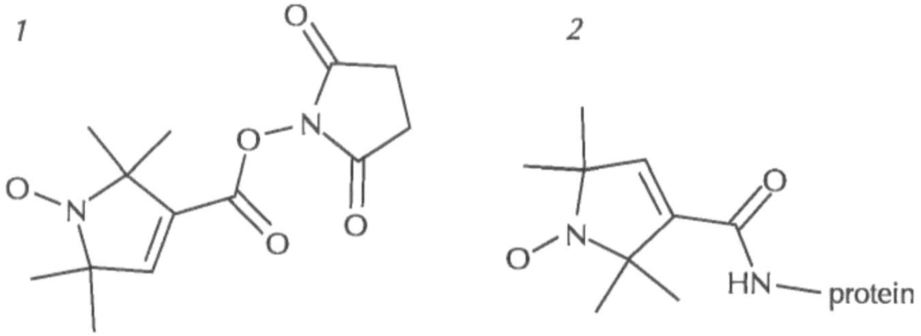

FIGURE 3 The stable nitroxide radical used for labeling of macromolecules containing aminogroups (*1*) and spin label attached to protein macromolecule (*2*).

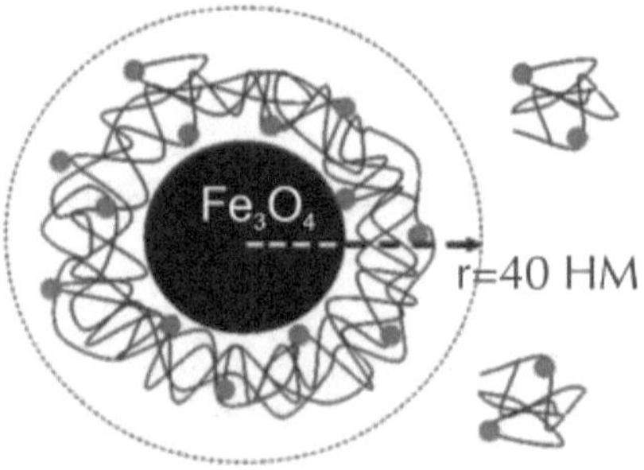

FIGURE 4 Magnetic nanoparticles and spin-labeled macromolecules in solution.

The reaction between the radical and protein macromolecules was conducted at room temperature. A 25 μl of radical solution in 96% ethanol with concentration of 2.57 mg/ml was added to 1 ml of protein solution. The solution was incubated for 6 h and dialyzed. The portion of adsorbed protein was calculated from intensity of the low-field line of nitroxide radical triplet I_{+1}.

The method of ferromagnetic resonance was also used to study adsorption layer formation.

The spectra of the radicals and magnetic nanoparticles were recorded at room temperature using Bruker EMX 8/2.7 X-band spectrometer at a microwave power of 5 mW, modulation frequency 100 kHz and amplitude 1 G. The first derivative of the resonance absorption curve was detected. The samples were placed into the cavity of the spectrometer in a quartz flat cell. Magnesium oxide powder containing Mn^{2+} ions was used as an external standard in ESR experiments. Average amount of spin labels on protein macromolecules reached 1 per 4–5 albumin macromolecules and 1 per 2–3 thrombin macromolecules. Rotational correlation times of labels were evaluated as well as a fraction of labels with slow motion (τ>1 ns).

35.2.4 COATING STABILITY ANALYSIS AND ANALYSIS OF SELECTIVITY OF FREE RADICAL PROCESS

In our previous works it was shown that fibrinogen (FG) adsorbed on MNPs surface forms thick coating and micron-sized structures [18]. Also FG demonstrates an ability to replace BSA previously adsorbed on MNPs surface. This was proved by complex study of systems containing MNPs, spin-labeled BSA and FG with spin labels technique and ferromagnetic resonance [20]. The property of FG to replace BSA from MNPs surface was used in this work for estimating BSA coating stability. 0.25 ml of FG ("Sigma-Aldrich," USA) solution with concentration of 4 mg/ml in 0.05 M phosphate buffer with pH 6.5 was added to 1 ml of the samples "A1-MNP-0," "A2-MNP-0," "A1-MNP-1," "A2-MNP-1." The clusters formation was observed by dynamic light scattering.

The samples "A2-MNP-0," "A2-MNP-1," "T1-MNP-0," "T1-MNP-1" were centrifugated at 120000 g during 1 hour on «Beckman Coulter» (Austria). On these conditions MNPs precipitate, but macromolecules physically adsorbed on MNPs remain in supernatant liquid. The precipitates containing MNPs and protein fixed on MNPs surface were dissolved in buffer solution with subsequent evaluation of the amount of protein by

Bradford colorimetric method [21]. Spectrophotometer CF-2000 (OKB "Spectr," Russia) was used.

Free radical modification of proteins in supernatant liquids of "A2-MNP-0," "A2-MNP-1" and the additional sample "A2-MNP-1-acid" were analyzed by IR-spectroscopy using FTIR-spectrometer Tenzor 27 ("Bruker," Germany) with DTGS-detector with 2 cm^{-1} resolution. Comparison of "A2-MNP-0," "A2-MNP-1" and "A2-MNP-1-acid" helps to reveal the selectivity of free radical process in "A2-MNP-1."

35.2.5 ENZYME ACTIVITY ESTIMATION

Estimation of enzyme activity of protein fixed on MNPs surface was performed on the example of thrombin. This protein is a key enzyme of blood clotting system, which catalyzes the process of conversion of fibrinogen to fibrin. Thrombin may lose its activity as a result of free radical modification and the rate of the enzyme reaction may decrease. So estimation of enzyme activity of thrombin cross-linked on MNPs surface during free radical modification was performed by comparison of the rates of conversion of fibrinogen to fibrin under the influence of thrombin contained in reaction mixtures. 0.15 ml of the samples "T1-MNP-0," "T1-MNP-1" and "T1-0-0" was added to 1.4 ml of FG solution with concentration of 4 mg/ml. Kinetics of fibrin formation was studied by Rayleigh light scattering on spectrometer 4400 ("Malvern," England) with multibit 64-channel correlator.

35.3 RESULTS AND DISCUSSION

ESR spectra of spin labels covalently bound to BSA and thrombin macromolecules (Fig. 5) allow obtaining information about their microenvironment. The spectrum of spin labels bound to BSA is a superposition of narrow and wide lines characterized by rotational correlation times of 10^{-9} s and $2{\cdot}10^{-8}$ s respectively. This is an evidence of existence of two main regions of spin labels localization on BSA macromolecules [22]. The portion of labels with slow motion is about 70%. So a considerable part of labels are situated in internal areas of macromolecules with high microviscosity. The labels covalently bound to thrombin macromolecules are characterized by one rotational correlation time of 0.26 ns. These labels are situated in areas with equal microviscosity.

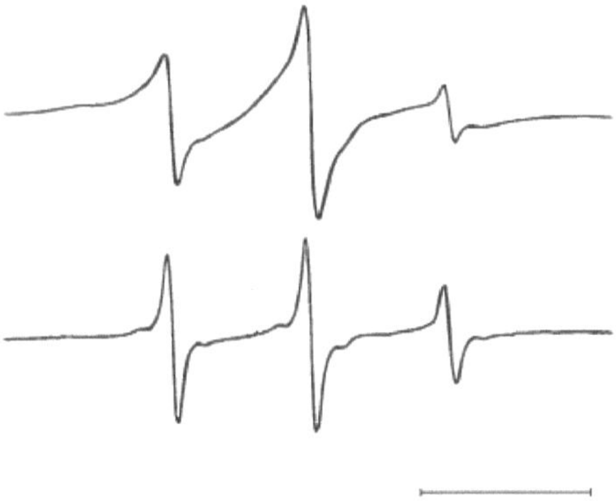

FIGURE 5 ESR spectra of spin labels on BSA (*1*) and thrombin (*2*) at 25°C.

The signal intensity of spinlabeled macromolecules decreased after introduction of magnetic nanoparticles into the solution *that testifies to the* protein adsorption on MNPs (Fig. 6). Spectra of the samples "A1-MNP-0" and "T1-MNP-0" consist of nitroxide radical triplet, the third line of sextet of Mn^{2+} (the external standard) and ferromagnetic resonance spectrum of MNPs. Rotational correlation time of spin labels does not change after MNPs addition. The dependences of spectra lines intensity for spinlabeled BSA and thrombin in the presence of MNPs on incubation time are shown in Table 1. Signal intensity of spinlabeled BSA changes insignificantly. These changes correspond to adsorption of approximately 12% of BSA after the sample incubation for 100 min. The study of adsorption kinetics allows establishing that adsorption equilibrium in "T1-MNP-0" takes place when the incubation time equals to 80 min and ~41% of thrombin is adsorbed. The value of adsorption *A* may be estimated using the data on the portion of macromolecules adsorbed and specific surface area calculated from MNPs density (5,200 mg/m^3), concentration and size. Hence BSA adsorption equals to 0.35 mg/m^2 after 100 min incubation. The dependence of thrombin adsorption value on incubation time is shown in Fig. 7. Thrombin adsorption equals to 1.20 mg/m^2 after 80 min incubation.

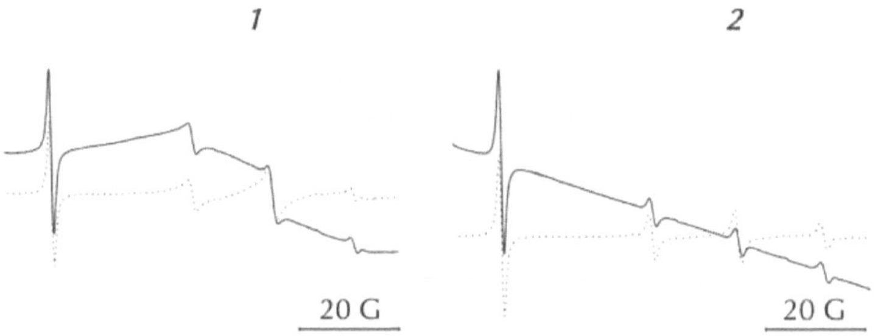

FIGURE 6 ESR spectra of spin labels on BSA (*1*) and thrombin (2) macromolecules before (dotted line) and 75 min after (solid line) MNPs addition to protein solution at 25°C. External standard – MgO powder containing Mn^{2+}.

TABLE 1 The dependence of relative intensity of low-field line of triplet I_{+1} of nitroxide radical covalently bound to BSA and thrombin macromolecules, and the portion N of the protein adsorbed on incubation time t of the samples "A1-MNP-0" and "T1-MNP-0."

	Spin-labeled BSA		Spin-labeled thrombin	
***t*, min.**	**I_{+1}, rel. units**	***N*,%**	**I_{+1}, rel. units**	***N*,%**
0	0.230 ± 0.012	0 ± 5	0.25 ± 0.01	0 ± 4
15	–	–	0.17 ± 0.01	32 ± 4
35	0.205 ± 0.012	9 ± 5	0.16 ± 0.01	36 ± 4
75	0.207 ± 0.012	10 ± 5	0.15 ± 0.01	40 ± 4
95	–	–	0.15 ± 0.01	40 ± 4
120	0.200 ± 0.012	13 ± 5	0.14 ± 0.01	44 ± 4

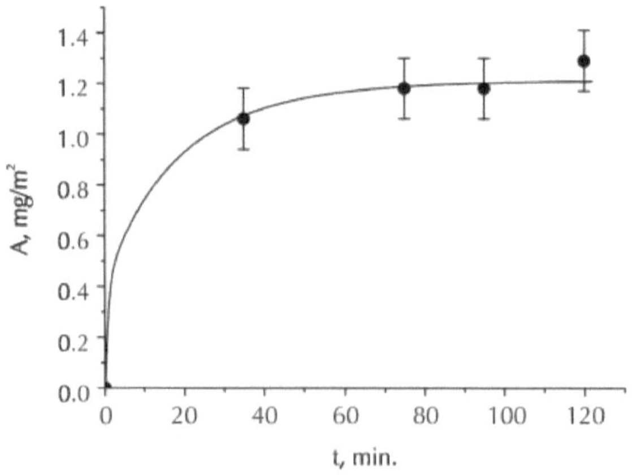

FIGURE 7 Kinetics of thrombin adsorption on magnetite nanoparticles at 25°C. Concentration of thrombin in the sample is 0.9 mg/ml, MNPs – 4.0 mg/ml.

The FMR spectra of the samples "A1-MNP-0," "T1-MNP-0" and MNPs are characterized by different position in magnetic field (Fig. 8). The center of the spectrum of MNPs is 3254 G, while the center of "A1-MNP-0" and "T1-MNP-0" spectra is 3253 G and 3449 G, respectively. Resonance conditions for magnetic nanoparticles in magnetic field of spectrometer include a parameter of the shift of FMR spectrum $|M_1| = \frac{3}{2}|H_1|$, where H_1 is a local field created by MNPs in linear aggregates, which form in spectrometer field. $H_1 = 2\sum_1^{\infty} \frac{2\mu}{(nD)^3}$, where D is a distance between MNPs in linear aggregates, μ is MNPs magnetic moment, n is a number of MNPs in aggregate [23]. Coating formation and the thickness of adsorption layer influence on the distance between nanoparticles decrease dipole interactions and particles ability to aggregate. As a result the center of FMR spectrum moves to higher fields. This phenomenon of FMR spectrum center shift we observed in the system "A1-MNP-0" after FG addition [20]. The spectrum of MNPs with thick coating becomes similar to FMR spectra of isolated MNPs. So the similar center positions of FMR spectra of MNPs without coating and MNPs in BSA coating point to a very thin coating and low adsorption of protein in this case. In contrast according to FMR center position the thrombin coating on MNPs is thicker than albumin coating. *This result is consistent with the data* obtained by ESR spectroscopy.

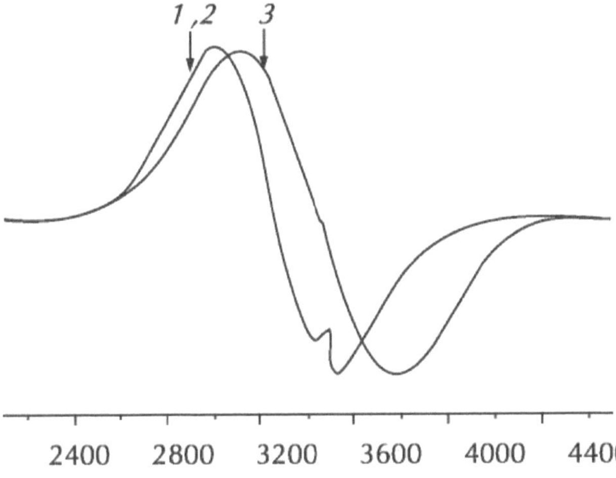

FIGURE 8 FMR spectra of MNPs (*1*), MNPs in the mixture with BSA (the sample "A1-MNP-0") after incubation time of 120 min (*2*) and MNPs in the mixture with thrombin (the sample "T1-MNP-0") after incubation time of 120 min (*3*).

FG ability to replace BSA in adsorption layer on MNPs surface is demonstrated in Fig. 9. Initially there is bimodal volume distribution of particles over sizes in the sample "A2-MNP-0" that can be explained by existence of free (unadsorbed) BSA and MNPs in BSA coating. After FG addition the distribution changes. Micron-sized clusters form in the sample that proves FG adsorption on MNPs [18]. In the case of "A2-MNP-1" volume distribution is also bimodal. The peak of MNPs in BSA coating is characterized by particle size of maximal contribution to the distribution of ~23 nm. This size is identical to MNPs in BSA coating in the sample "A2-MNP-0." It proves that H_2O_2 addition does not lead to uncontrollable linking of protein macromolecules in solution or cluster formation. Since MNPs size is 17 nm, the thickness of adsorption layer on MNPs is approximately 3 nm.

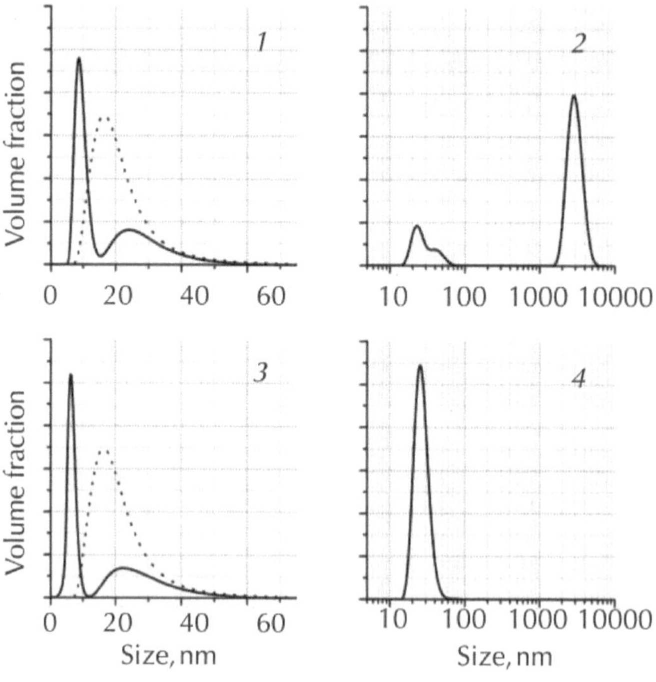

FIGURE 9 Volume distributions of particles in sizes in systems without (*1*, *2*) and with (*3*, *4*) H_2O_2 ("A2-MNP-0," "A2-MNP-1") incubated for 2 h before (*1*, *3*) and 20 min after (*2*, *4*) FG addition. Dotted line is the volume distribution of nanoparticles in sizes in dispersion.

After FG addition to "A2-MNP-1" micron-sized clusters do not form. So adsorption BSA layer formed in the presence of H_2O_2 keeps stability. This stability can be explained by formation of covalent bonds between protein macromolecules [13] in adsorption layer as a result of free radicals generation on MNPs surface. Stability of BSA coating on MNPs was demonstrated for the samples "A1-MNP-1" and "A2-MNP-1" incubated for more than 100 min before FG addition. Clusters are shown to appear if the incubation time is insufficient.

The precipitates obtained by ultracentrifugation of "A2-MNP-0," "A2-MNP-1," "T1-MNP-0" and "T1-MNP-1" were dissolved in buffer solution. The amount of protein in precipitates was evaluated by Bradford colorimetric method (Table 2). The results showed that precipitates of systems with H_2O_2 contained more protein than the same systems without H_2O_2. Therefore in the samples containing H_2O_2 the significant part of protein molecules does not leave MNPs surface when centrifuged while in the samples "A2-MNP-0" and "T1-MNP-0" the most of protein molecules leaves the surface. This indicates the stability of adsorption layer formed in the presence of free radical generation initiator and proves cross-links formation.

TABLE 2 The amount of protein in precipitates after centrifugation of the samples "A2-MNP-0," "A2-MNP-1," "T1-MNP-0" and "T1-MNP-1" of 3.2 ml in volume.

Sample name	Amount of protein in precipitates, mg
"A2-MNP-0"	0.05
"A2-MNP-1"	0.45
"T1-MNP-0"	0.15
"T1-MNP-1"	1.05

Analysis of content of supernatant liquids obtained after ultracentrifugation of reaction systems containing MNPs and BSA, which differed by H_2O_2 and ascorbic acid presence ("A2-MNP-0," "A2-MNP-1" and "A2-MNP-1-acid") allows evaluating the scale of free radical processes in the presence of H_2O_2. As it was mentioned above in the presence of ascorbic acid free radicals generate not only on MNPs surface but also in solution.

So both molecules on the surface and free molecules in solution can undergo free radical modification in this case. From Fig. 10 we can see that the IR-spectrum of "A2-MNP-1-acid" differs from the spectra of "A2-MNP-0" and "A2-MNP-1," while *the spectra of* "A2-MNP-0" and "A2-MNP-1" *almost* have no differences. The IR-spectra differs in the region of 1200–800 cm^{-1}. The changes in this area are explained by free radical oxidation of amino acid residues of methionine, tryptophane, histidine, cysteine, and phenylalanine. These residues are sulfur-containing and cyclic ones which are the most sensitive to free radical oxidation [13, 24]. The absence of differences in "A2-MNP-0" and "A2-MNP-1" proves that cross-linking of protein molecules in the presence of H_2O_2 is selective and takes place only on MNPs surfaces.

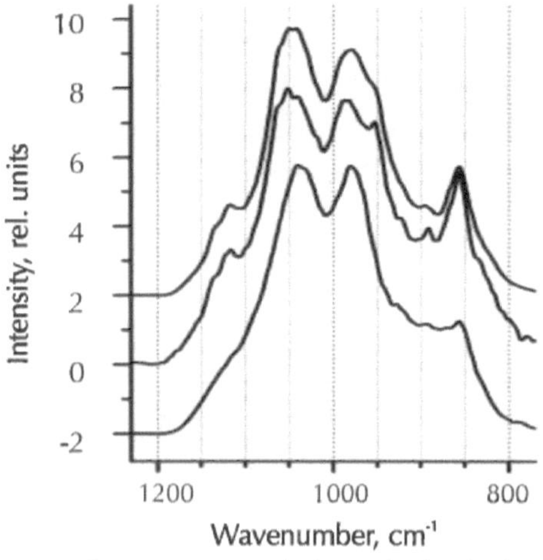

FIGURE 10 IR-spectra of supernatant solutions obtained after centrifugation of the samples "A2-MNP-0" (*1*), "A2-MNP-1" (*2*) and "A2-MNP-1-acid" (*3*).

When proteins are used as components of coating on MNPs for biology and medicine their functional activity retaining is very important. Proteins fixed on MNPs can lose their activity as a result of adsorption on MNPs or free radical modification, which is cross-linking and oxidation but it was shown that they do not lose it. Estimation of enzyme activity of thrombin cross-linked on MNPs surface was performed by comparison of the rates of conversion of fibrinogen to fibrin under the influence of thrombin contained in reaction mixtures "T1-MNP-0," "T1-MNP-1" and "T1-0-0" (Fig. 11). The curves for the samples containing thrombin and MNPs which differ by the presence of H_2O_2 had no fundamental differences that illustrates preservation of enzyme activity of thrombin during free radical cross-linking on MNPs surface. Fibrin gel was formed during ~15 min in both cases. Rayleigh light scattering intensity was low when "T1-0-0" was used and small fibrin particles were formed in this case. The reason of this phenomenon is autolysis (self-digestion) of thrombin. Enzyme activity of thrombin, one of serine proteinases, decreases spontaneously in solution [25]. So the proteins can keep their activity longer when adsorbed on MNPs. This way, the method of free radical cross-linking of proteins seems promising for enzyme immobilization.

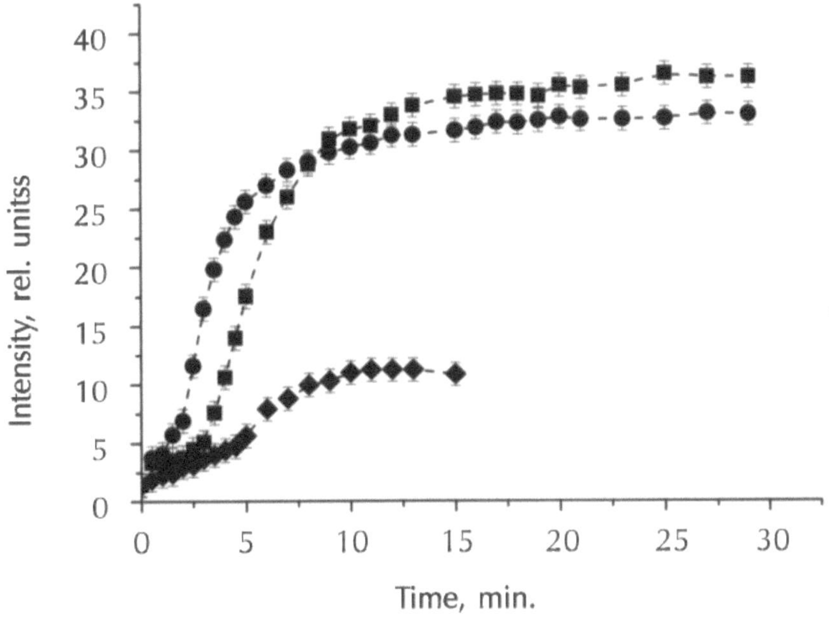

FIGURE 11 Kinetics curves of growth of Rayleigh light scattering intensity in the process of fibrin gel formation in the presence of "T1-MNP-0" (*1*), "T1-MNP-1" (*2*) and "T1-0-0" (*3*).

KEYWORDS

- **A2-MNP-1-acid**
- **Fenton reaction**
- **Fibrinogen**
- **Localization**

REFERENCES

1. Gupta, A. K.; Gupta, M. Synthesis and surface engineering of iron oxide nanoparticles for biomedical applications. *Biomaterials,* **2005**, *26*, 3995–4021.
2. Vatta, L. L.; Sanderson, D. R.; Koch, K. R. Magnetic nanoparticles: Properties and potential applications. *Pure Appl. Chem.*, **2006**, *78*, 1793–1801.
3. Lu, A. H.; Salabas, E. L.; Schüth, F. Magnetic Nanoparticles: Synthesis, Protection, Functionalization, and Application. *Angew. Chem. Int. Ed.*, **2007**, *46*, 1222–1244.
4. Laurent, S.; Forge, D.; Port, M.; Roch, A.; Robic, C.; Elst, L. V.; Muller, R. N. Magnetic Iron Oxide Nanoparticles: Synthesis, Stabilization, Vectorization, Physicochemical Characterizations, and Biological Applications. *Chem. Rev.*, **2008**, *108*, 2064–2110.
5. Pershina A. G., Sazonov A. E., Milto I. V. Application of magnetic nanoparticles in biomedicine. *Bull. Siberian Med.* [in Russian], **2008**, *2*, 70–78.
6. Trahms, L. Biomedical Applications of Magnetic Nanoparticles. *Lect. Notes Phys.*, **2009**, *763*, 327–358.
7. Bychkova, A. V.; Sorokina, O. N.; Rosenfeld, M. A.; Kovarski, A. L. Multifunctional biocompatible coatings on magnetic nanoparticles. *Uspekhi Khimii (Russian journal)*, **2012**, in Press.
8. Koneracka, M.; Kopcansky, P.; Antalik, M.; Timko, M.; Ramchand, C. N.; D.Lobo, Mehta, R. V.; Upadhyay, R. V. Immobilization of proteins and enzymes to fine magnetic particles. *J. Magn. Magn. Mater.*, **1999**, *201*, 427–430.
9. Xu, L.; Kim, M.-J.; Kim, K.-D.; Choa, Y.-H.; Kim, H.-T. Surface modified Fe_3O_4 nanoparticles as a protein delivery vehicle. *Colloids Surf. A: Physicochem. Eng. Aspects*, **2009**, *350*, 8–12.
10. Peng, Z. G.; Hidajat, K.; Uddin, M. S.; Adsorption of bovine serum albumin on nanosized magnetic particles. *J. Colloid Interface Sci.*, **2004**, *271*, 277–283.
11. Šafárík, I.; Ptáčková, L.; Koneracká, M.; Šafáríková, M.; Timko, M.; Kopčanský, P. Determination of selected xenobiotics with ferrofluid-modified trypsin. *Biotechnol. Lett.*, **2002**, *24*, 355–358.
12. Li, F.-Q.; Su, H.; Wang, J.; Liu, J.-Y.; Zhu, Q.-G.; Fei, Y.-B.; Pan, Y.-H.; Hu, J.-H. Preparation and characterization of sodium ferulate entrapped bovine serum albumin nanoparticles for liver targeting. *Int. J. Pharm.*, **2008**, *349*, 274–282.
13. Stadtman, E. R.; Levine, R. L. Free radical-mediated oxidation of free amino acids and amino acid residues in protein. *Amino Acids*, **2003**, *25*, 207–218.

14. Bychkova, A. V.; Sorokina, O. N.; Shapiro, A. B.; Tikhonov, A. P.; Kovarski, A. L. Spin Labels in the Investigation of Macromolecules Adsorption on Magnetic Nanoparticles. *Open Colloid Sci. J.*, **2009**, *2*, 15–19.
15. Abragam, A. *The Principles of Nuclear Magnetism;* Oxford University Press: New York, 1961.
16. Noginova, N.; Chen, F.; Weaver, T.; Giannelis, E. P.; Bourlinos, A. B.; Atsarkin, V. A. Magnetic resonance in nanoparticles: between ferro- and paramagnetism. *J. Phys.: Cond. Matter*, **2007**, *19*, 246208–246222.
17. Sorokina, O. N.; Kovarski, A. L.; Bychkova, A. V. Application of paramagnetic sensors technique for the investigation of the systems containing magnetic particles. In *Progress in Nanoparticles Research*; Frisiras, C. T., Ed.; Nova Science Publishers: New York, 2008, 91–102.
18. Bychkova, A. V.; Sorokina, O. N.; Kovarskii, A. L.; Shapiro, A. B.; Leonova, V. B.; Rozenfel'd, M. A. Interaction of fibrinogen with magnetite nanoparticles. *Biophysics (Russian J)*, **2010**, *55*(4), 544–549.
19. Bychkova, A. V.; Sorokina, O. N.; Kovarski, A. L.; Shapiro, A. B.; Rosenfeld, M. A. The Investigation of Polyethyleneimine Adsorption on Magnetite Nanoparticles by Spin Labels Technique. *Nanosci. Nanotechnol. Lett. (ESR in Small Systems)*, **2011**, *3*, 591–593.
20. Bychkova, A. V.; Rosenfeld, M. A.; Leonova, V. B.; Lomakin, S. M.; Sorokina, O. N. Surface modification of magnetite nanoparticles with serum albumin in dispersions by free radical cross-linking method. *Russian Colloid J.*, **2012**, in print.
21. Bradford, M. M. A rapid and sensitive method for the quantitation of microgram quantities of *protein utilizing* the *principle* of *protein-dye binding. Anal. Biochem.*, **1976**, *72*, 248–254.
22. Antsiferova, L. I.; Vasserman, A. M.; Ivanova, A. N.; Lifshits, V. A.; Nazemets, N. S. *Atlas of Electron Paramagnetic Resonance Spectra of Spin Labels and Probes* [in Russian]; Nauka: Moscow, 1977.
23. Dolotov, S. V.; Roldughin, V. I. Simulation of ESR spectra of metal nanoparticle aggregates. *Russian Colloid J*, **2007**, *69*, 9–12.
24. Smith, C. E.; Stack, M. S.; Johnson, D. A. Ozone effects on inhibitors of human neutrophil proteinases. *Arch. Biochem. Biophys.*, **1987**, *253*, 146–155.
25. Blomback, B. Fibrinogen and fibrin – proteins with complex roles in hemostasis and thrombosis. *Thromb. Res.*, **1996**, *83*, 1–75.

CHAPTER 36

MICROPROPAGATION OF HYPERICUM PERFORATUM L. FOR MEDICINAL AND ORNAMENTAL PURPOSES

Y. A. GURCHENKOVA, A. E. RYBALKO, and L. G. KHARUTA

CONTENTS

36.1 INTRODUCTION

The family Hypericaceae includes 47 genera and 850 species, mostly of tropical origin. Only two kinds are found in the territory Russia, one of them is represented by one species, growing in the Far East, and the second genus – has more than 50 species.

Hypericum – tree, shrub, rarely herbaceous plants. Various species of the genus Hypericum have used as a tool for the treatment of ulcers, burns, wounds, abdominal pain and bacterial diseases since ancient times. In recent in clinical trials have been used to treat depression and viral diseases. Antidepressant activity due to the presence hypericins, hyperphorins and flavonoids. Types of Hypericum maculatum Crantz., H. perforatum L. are several pharmacopoeias as components of the drug Herba Hyperici, however, little attention was paid to the secondary metabolites and their dynamics. Despite this, many authors found that plants H. maculatum synthesize and accumulate hypericin and flavonoids, in amounts similar that found in H. perforatum. In vitro culture was used for the reproduction of Hypericum L, the views of scientists should focus on the reproduction of H. perforatum L. Recently the method of an induction of indirect regeneration Hypericum triquetrifolium Tur. has been described Ref. [2]. In the given research callus was induced from cotyledon explants of 35 days old aseptic seedlings. Applied semisolid MS supplemented with IAA (0.5 mg l^{-1}) combined with BAP (2 mg l^{-1}). At level reduction cytokinins on a surface of callus developed meristemoids.

We investigated the effect of PGR on the micropropagation in vitro for a quick increase in the number of shoots from the axillary buds H. perforatum L. For this purpose, nodal segments were cultured on MS medium.

36.2 MATERIALS AND METHODS

The plant material selected in July 2011 among the wild plants in the area of the city Sochi. Sterilization of plant tissue was determined by standard methods. The nutrient medium used in the experiment consisted in MS salts formula, with 0.1 mg/l NAA, 20 g/l sucrose, 100 mg/l and MS vitamins, solidified with 6 g/l agar and pH adjusted to 5.5. Test of different concentrations of BAP and kinetin (0.1; 1,0 и 2.0 mg/l).

36.3 RESULTS

We investigated the effect of plant growth regulators on the micropropagation in vitro for a quick increase in the number of shoots from axillary buds H. perforatum L. For this purpose, nodal segments were cultured on MS medium. Sterilization of plant tissue was determined by standard methods. Test of different concentrations of BAP and kinetin (0.1; 1.0 and 2.0 mg/l) on the background of a standard dose of naphthalene acetic acid (0.1 mg/l). In epy flasks with 1 and 2 mg/l BAP accrued green embryogenic callus for allocation medicinal a substance. For mass propagation used test tubes in diameter of 20 mm. Within 8 weeks in them accrued green callus, which it was possible to divide into 40–45 clusters in size of 3 mm. Kinetin promoted regeneration of stems.

36.4 DISCUSSION

Similar researches on micropropagation of the given region are extremely insufficient. Are not grown up ornamental cultivars of Hypericum. Workings out on micropropagation endemic kinds of the North Caucasus are required. As along with H. perforatum also rare species – H. androsaemum L. and H. xylosteifolium (Spach) N. Robson.

KEYWORDS

- **Callus**
- **Hypericaceae**
- **Hypericum**
- **Epy flasks**

REFERENCES

1. Pretto, F. R.; Santarem, E. R. Callus formation and plant regeneration from Hypericum perforatum leaves. *Plant Cell, Tissue and Organ Culture*, **2000**, *62*, 107–113.

2. Oluk, E. A.; Orhan, S.; Karaka, O.; Qakir, A.; Goniiz, A. High efficiency indirect shoot regeneration and hypericin content in embryogenic callus of Hypericum triquetrifolium Turra. *African J. Biotechnol.*, **2010**, *9*(15), 2229–2233.
3. Murashige, T.; Skoog, F. A revised medium for rapid growth and bioassays with tobacco culture. *Physiologia Plantarum*, **1962**, *15*(4), 473–479.

CHAPTER 37

ACTIVITY OF LIPOSOMAL ANTIMICROBIC PREPARATIONS CONCERNING *STAPHYLOCOCCUS AUREUS*

N. N. IVANOVA, G. I. MAVROV, S. A. DERKACH, and E. V. KOTSAR

CONTENTS

37.1 INTRODUCTION

The skin of patients with atopic dermatitis planted on different microorganisms, whose number is much bigger than the skin of healthy people. For example, *Staphylococcus aureus* (*S. aureus*) sow from the skin of patients with atopic dermatitis in 80–100% of cases. Skin diseases of microbial etiology in most cases basic drug treatment and prevention are antibiotics. Widespread use of antibiotics has negative consequences, one of which is the emergence of pathogens with resistance to penicillin, gentamicin, tetracycline, methicillin, lincomycin, sulfonamides, as well as a new generation of antibiotics: quinolones, cephalosporins. Consequently, the problem of prevention and treatment of infectious diseases is urgent. One of the ways of its solution is the introduction of new chemotherapeutic agents into medical practice.

It is known that nanoparticles and liposomal forms of medicines allow significantly improving the efficacy, reducing toxicity, therapeutic dose and qualitatively changing the nature of their actions. Thus, in the work Ref. [4] it was shown that, despite the low concentrations, the efficiency of liposomal benzyl penicillin to inhibit growth of bacterial biofilms of *S. aureus* was higher than the intact benzyl penicillin. According to the above-mentioned issues, the purpose of the study was to investigate the efficiency of antimicrobial agents in liposomal form relatively to Staphylococcus aureus.

37.2 MATERIALS AND METHODS

The strain ATCC 25923 *S. aureus* was taken from SE "Mechnicov Institute of Microbiology and Immunology AMSU." We have also used the following items: egg lecithin (Ukraine, "Biolek"), DMSO (Russia), the mixture of negatively charged lipids that were obtained by the original technology of Dr. Nina Ivanova, substance of benzoyl peroxide ("Aldrich," USA), lincomycin (JSC "Darnitsa," Ukraine), Mueller–Hinton agar (HiMedia Laboratories Pvt. Limited, India), meat-peptone broth (MPB).

37.2.1 THE RECEIVING OF LIPOSOMES

The substance of benzoyl peroxide (BP) dissolved in chloroform due to its poor solubility in aqueous solutions and added to an alcohol or chloroform solution of lipids in the ratio of BP: Lipids 1:10, 1:20. The liposomes were obtained by evaporating the lipids and antibiotics on a rotary vacuum evaporator (Switzerland). Next mixture was suspended in sterile buffered saline. Liposomes prepared in the extruder EmulsiFlex-C5 (Canada "Avestin"), punching with compressed air (10 cycles) to achieve a constant optical density on spectrophotometer (DU-7 Spectrophotometer Beckman, USA) at temperatures that above the phase transition temperature of any of the lipid components was present. The average size of liposomes was 160-180 nm, concentration of lipids in the liposomes was 2%, ratio LN : lipids and BP : lipids was 1:20 [2].

Cleaning of switched antibiotics in liposomes (Ls) from those that are not involved in making liposomes using ultracentrifuge (MSE-Superspeed Centrifuge 65, England) for an hour at 105000g. The output of liposomes was determined spectrophotometrically at 450 nm.

The determination of minimum inhibitory concentration (MIC) of antibiotics was performed by microtiter method. Antimicrobial agents were diluted by serial dilutions of meat-peptone broth (MPB) in flat-bottomed plates, they were also added to the culture of Staphylococcus aureus, and were incubated for 24 h at 34°C. The control was culture *S. aureus* without antimicrobial agents. After that mixture sowed on solid nutrient medium Mueller-Hinton agar for calculation of amount of colony forming particles and MIC definitions. MIC was considered to be the lowest concentration, which retards the growth of *S. aureus* during the incubation period.

37.3 RESULTS AND DISCUSSION

Among the antibiotics we stopped on lincomycin. It has a bacteriostatic effect on a wide range of microorganisms, with increasing doses of lincomycin it has a bactericidal effect. Antimicrobial mechanism of action of lincomycin is the inhibition of protein synthesis in the cells of microorganisms.

The drug is active with respect to Gram-positive aerobic and anaerobic microorganisms, including *Staphylococcus spp*.

BP has a wide spectrum of antimicrobial activity. It is active against the bacteria, in the case it is also resistant to antibiotics [3, 4].

In the first phase of our study the minimum inhibitory concentrations (MIC) of lincomycin and BP in Ls on the basis of egg lecithin were used, which is a soft lipid and traditionally used in the creation of liposomal forms of drugs. As a result, the definition MIC of lincomycin and its liposomal preparations were found and also Ls received on the basis of egg lecithin and lincomycin are more effective than epy solution of lincomycin under the action of *S. aureus* planktonic cells (Fig. 1b). MIC of Ls this composition decreased in 3 times in comparison with the MIC of the lincomycin solution.

Negatively charged liposomes received on the basis of polar lipids and lincomycins were the most effective. The using of negatively charged liposomes that contained lincomycin reduced the MIC of the lincomycin solution in 7 times concerning planktonic cells of *S. aureus* (Fig. 1c).

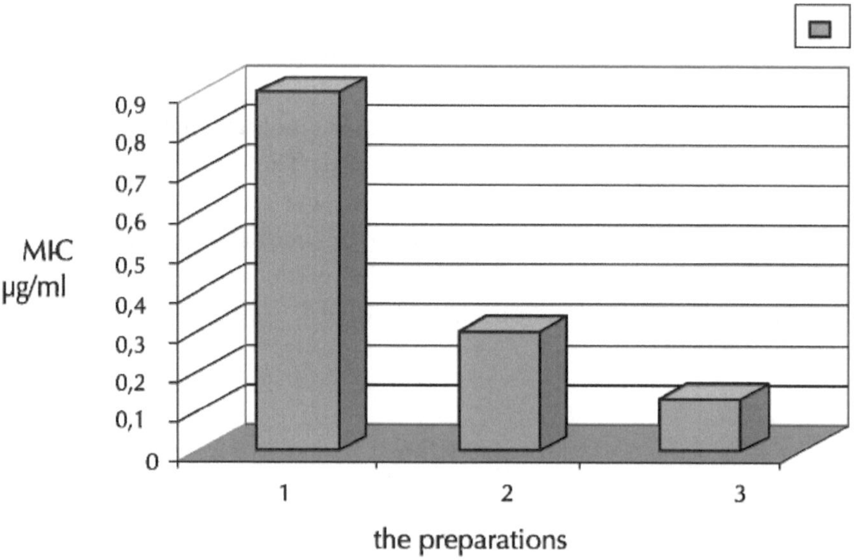

FIGURE 1 The minimum inhibitory concentration of the liposomal lincomycin: (a) the control (lincomycin solution), (b) liposomal lincomycin on the basis of egg lecithin, (c) negatively charged liposomal lincomycin.

Liposomal BP on the basis of egg lecithin was also more effective against *S. aureus* in comparison with the MIC of intact BP. At definition MIC ofliposomal forms BP on the basis of egg lecithin it has been found that its MIC made 31 μg/ml that was in 14 times less MIC BP dissolved in DMSO.

37.4 CONCLUSION

1. It is found that minimally inhibitory concentration of liposomal antibiotic solution with the neutral charge (lecithin liposomes) decreased in 3 times in comparison with minimally inhibitory concentration of the lincomycin solution concerning *S. aureus*.
2. The using of negatively charged liposomes on the basis of polar lipids with the antibiotic, strengthens efficiency of antibiotic action in greater degree than using lecithin liposomes. Negatively charged liposomes containing lincomycin reduced its minimally inhibitory concentration in 7 times concerning *S. aureus*.
3. The minimally inhibitory concentration of the liposomal antimicrobic preparation benzoyl peroxide with the neutral charge (lecithin liposomes) decreased in 14 times in comparison with minimally inhibitory concentration of benzoyl peroxide concerning *S. aureus*.
4. The received results enable to predict the using of liposomal forms of antimicrobic substances for increase of pharmacological efficiency in treatment of Staphylococcus infections.

KEYWORDS

- **Cephalosporins**
- **Lecithin liposomes**
- **Lincomycin**
- **Quinolones**

REFERENCES

1. Kim, H. J.; Jones, M. N. *J. Liposome Res.*, **2004**, *4.-C.*, 123–139.
2. Sorokoumova, G. M.; Selishcheva A. A.; Kaplun A. P. *Educat. Tools Bioorg. Chem.*, **2000**, 105 p (in Russian).
3. Tanghetti E. A.; Popp K. F.: *Dermatol. Clin.,* **2009**, *27*, 17–24.
4. Tanghetti E. A. *Cutis*., **2008**, *82*, 5–11.

CHAPTER 38

POLYELECTROLYTE MICROSENSORS AS A NEW TOOL FOR METABOLITES' DETECTION

L. I. KAZAKOVA, G. B. SUKHORUKOV, and L. I. SHABARCHINA

CONTENTS

38.1 INTRODUCTION

Design of micro and nanostructured systems for in-situ and in-vivo sensing become an interesting subject nowadays for biological and medical-oriented research [1–3]. Miniaturization of sensing elements opens a possibility for non-invasive detection and monitoring of various analytes exploiting cell and tissues residing reporters. Typical design of such sensor is nano- and microparticles loaded with sensing substances enable to report on presence of analyte by optical means [4–9]. The particles containing fluorescent dye can be use as a sensor for relevant analytes such as H^+, Na^+, K^+, and Cl^- et al. [10–13]. The fluorescent methods are most simple and handy among the possible ways of registration. They provide high sensitivity and relative simplicity of data read-out. For analysis of various metabolites it is necessary to use the enzymatic reactions to convert analyte to optically detectable compound [14, 15]. In order to proper functioning all components of sensing elements (fluorescence dyes, peptides, enzymes) are to be immobilized in close proximity of each other. That "tailoring" of several components in one sensing entity represents a challenge in developing of a generic tool for sensor construct. One approach to circumvent problem has been introduced by the PEBBLE (Photonic Explorers for Bioanalyse with Biologically Localized Embedding) system [5, 16]. PEBBLE is a generic term to describe use co-immobilization of sensitive components in inert polymers, substantially polyacrylamide, by the microemulsion polymerization technique [17]. This technique is useful for fluorescent probe, but to our mind, is too harsh for peptides and enzymes capsulation due to organic solvents involved in particle processing. Multilayer polyelectrolyte microcapsules have not this shortcoming as they are operated fully in aqueous solution at mild condition. These capsules are fabricated using the Layer-by-Layer (LbL) technique based on the alternating adsorption of oppositely charged polyelectrolytes onto sacrificial colloidal templates [18, 19]. Immobilization of one or more enzymes within polyelectrolyte microcapsules can be accomplished by the coprecipitation of these enzymes into the calcium carbonate particles, followed by particle dissolution in mild condition leaving a set of protein retained in capsule [15, 20, 21]. A fluorescence dye can be included in polyelectrolyte capsules as well. Thus, the multilayer polyelectrolyte encapsulation technique microcapsules allows in principle combining enzyme activity for selected metabolite and registration ability of dyes in one capsule. In this work we demonstrate urea detection using capsules containing urease and pH sensitive dye.

The concentration of urea in biological solutions (blood, urine) is a major characteristic of the condition of a human organism. Its value may suggest a number of acute and chronic diseases: myocardial infarction, kidney and liver dysfunction. The measurement of urea concentration is a routine procedure in clinical practice. Urease based enzymatic methods are most widely used for urea detection and use urease enzyme as a reactant. There are multiple urease-based methods, which differ from each other by the manner of monitoring the enzymatic reaction. Despite of high specificity, reproducibility and extremely sensitivity for urea as urea is the only physiological substrate for urease, these methods are all laborious and time-consuming since freshly prepared chemical solutions and calibration are required daily. They all are lacking in-situ live monitoring what makes them inappropriate for analysis in vivo as residing sensors. Embedding of urease into polyelectrolytes microcapsules can help to solve these problems [22, 23]. The encapsulated urease completely preserves its activity at least 5 days at the fridge storage [22].

Aim of this work was to demonstrate a particular example of a sensor system, which combines catalytic activity for urea and at the same time, enabling monitoring enzymatic reaction by optical recording. The proposed sensor system is based on multilayer polyelectrolyte microcapsules containing urease and a pH-sensitive fluorescent dye, which translates the enzymatic reaction into a fluorescently registered signal.

38.2 EXPERIMENTAL DETAILS

38.2.1 MATERIALS

Sodium poly(styrene sulfonate) (PSS, MW = 70,000) and poly(allylamine hydrochloride) (PAH, MW = 70,000), calcium chloride dihydrate, sodium carbonate, sodium chloride, ethylenediaminetetraacetic acid (EDTA), TRIS, maleic anhydride, sodium hydroxide (NaOH) and Bromocresol purple were purchased from Sigma-Aldrich (Munich, Germany). Urease (Jack bean, Canavalia ensiformis) was purchased from Fluka. SNARF-1 dextran (MW = 70,000) was obtained from Invitrogen GmbH (Molecular Probes #D3304, Karlsruhe, Germany). All chemicals were used as received. The bidistilleted water was used in all experiments.

38.2.2 PREPARATION OF SNARF-1 DEXTRAN AND SNARF-1 DEXTRAN/UREASE CONTAINING $CACO_3$ MICROPARTICLES

The preparation of loaded $CaCO_3$ microspheres was carried out according to the coprecipitation-method [20, 21]. To prepare the $CaCO_3$ microspheres loaded with SNARF-1 dextran were used: 1.6 ml H_2O, 0.5 ml 1M $CaCl_2$, 0.5 ml 1M Na_2CO_3 and 0.4 ml SNARF-1 dextran solution (1 mg/ml).

To prepare the $CaCO_3$ microspheres contained different ratio of SNARF-1 dextran and urease were used:

Sample I: 0.6 ml H_2O, 0.5 ml 1M $CaCl_2$, 0.5 ml 1M Na_2CO_3, 0.4 ml SNARF-1 dextran solution (1 mg/ml) and 1 ml urease (3 mg/ml);

Sample II: 0.8 ml H_2O, 0.5 ml 1M $CaCl_2$, 0.5 ml 1M Na_2CO_3, 0.2 ml SNARF-1 dextran solution (1 mg/ml) and 1 ml urease (3 mg/ml).

The solutions were rapidly mixed and thoroughly agitated on a magnetic stirrer for 30 s at 4°C. After the agitation, the precipitate was separated from the supernatant by centrifugation (250× g, 30 s) and washed three times with water. The procedure resulted in highly spherical microparticles containing SNARF-1 dextran or SNARF-1 dextran and urease with an average diameter ranging from 3.5–4 μm.

38.2.3 FABRICATION OF SNARF-1 DEXTRAN LOADED MICROCAPSULES

Microcapsules were prepared by alternate layer-by-layer (LbL) deposition of oppositely charged polyelectrolytes poly(allylamine hydrochloride) (PAH, MW = 70,000) and poly(styrene sulfonate) (PSS, MW = 70,000) onto CaCO3 particles containing SNARF-1 dextran or SNARF-1 dextran and urease to give the following shell architecture: (PSS/PAH)4PSS. Short ultrasound pulses were applied to the sample prior to the addition of each polyelectrolyte in order to prevent particle aggregation. The decomposition of the $CaCO_3$ core was achieved by treatment with EDTA (0.2 M, pH 7.0) followed by triple washing with water. The microcapsules were immediately subjected to further analysis or stored as suspension in water at 4°C.

38.2.4 SPECTROSCOPIC STUDY

All spectroscopic studies were carried out with UV-vis spectrophotometer *Varian Cary 100* at constant agitation and thermostatic control at 20°C.

The SNARF-1 dextran concentrations in different capsules samples were estimated by matching absorption intensity of supernatant after co-precipitation of the dye with $CaCO_3$ to intensity of calibrated of SNARF-1 dextran concentrations in free solution. The average content of SNARF-1 dextran per capsule was calculated to be: (1) for SNARF-1 dextran $CaCO_3$ microparticales – 1 pg; (2) for SNARF-1 dextran/urease $CaCO_3$ microparticles: ***Sample I*** – 0.6 pg, ***Sample II*** – 0.2 pg.

The amount of active urease immobilized into the polyelectrolyte microcapsules was determined under assumption that the enzyme retains its activity while encapsulated. Free urease had 100 U/mg according to the data sheets. The activity of free and encapsulated enzyme were determined from the decomposition of urea into two ammonia molecules and CO_2 using a pH-sensitive dye Bromocresol purple [24]. The urease aliquot solutions were added to a reaction mixture contained a necessary amount of urea and 0.015 mM Bromocresol, whose pH was apriory brought up to 6.2. The reaction kinetics was recorded as a change in the optical absorption of the dye at 588 nm to obtain the linear calibration plot. Then, the known number of microcapsules containing urease and SNARF-1 dextran was added to the reaction solution. The revealed activity of enzyme was compared with amount of free urease.

38.2.5 SPECTROFLUORIMETRIC STUDY

All spectrofluorimetric studies of SNARF-1 dextran and SNARF-1 dextran/urease were carried out with the spectrofluorimeter *Varian Cary Eclipse*, at constant agitation, thermostatic control at 20°C, λ_{exc}=540 nm, slit width: excitation at 10 nm and emission at 20 nm. The microcapsule suspensions were used at concentration 2×10^6 capsules/ml, which was estimated with the cytometer chamber. All solutions were prepared on bidistilleted water.

TRIS-maleate buffer solutions for pH setting were prepared by adding appropriate quantity of 0.2 NaOH to 0.2 M TRIS and maleic anhydride mixture and diluted to 0.05 M concentration.

38.2.6 CONFOCAL LASER SCANNING MICROSCOPY

Confocal images were obtained by Leica Confocal Laser Scanning Microscope TCS SP. For capsules visualization 100× oil immersion objective was used throughout. 10 μl of the SNARF-1 dextran/urease capsules suspension was placed on a coverslip. To this suspension 10 μl of 0.1 mol/l urea is added. After about 20 min confocal images were obtained. The red fluorescence emission was accumulated at 600–680 nm after excitation by the FITC–TRIC–TRANS laser at 543 nm.

38.3 RESULTS AND DISCUSSIONS

Degradation of urea ($CO(NH_2)_2$) is catalyzed by urease and results in the shift of the medium pH into the alkaline range.

$$CO(NH_2)_2 + H_2O = CO_2 + 2\ NH_3$$

Monitoring of the urea degradation can be done by using SNARF-1 as pH-sensitive dye to follow changes of the pH in the enzyme driven reaction. In order to fabricate the sensing microcapsule the urease and SNARF-1 bearing dextran were simultaneously co-precipitated to form $CaCO_3$ spherical particles 3.5–4 μm in size, containing both components urease and fluorescent dye SNARF-1 coupled dextran (MW=70,000) [15]. Then the particles were coated by standard layer-by-layer protocol with nine alternating layers of oppositely charged polyelectrolytes PSS and PAH. The formed shell had a $(PSS/PAH)_4PSS$ architecture. After the dissolution of $CaCO_3$ with the EDTA solution the obtained capsule samples contain an enzyme and a fluorescent dye in its cavity.

The dye and enzyme concentrations inside the polyelectrolyte capsule are predetermined essentially at the stage of formation of the $CaCO_3$/SNARF-1 dextran/urease conjugate microparticles. Obviously, the amount of both components of urease and SNARF-1 dextran in the capsules and their ration should play an important role while functioning of entire sensing system is concerned. However, the final composition of co-precipitated particles and later capsules in fabricated samples may be different, though the same initial concentration of components used while preparing co-precipitating particles. It depends on a number of factors: adsorption, capturing and distribution of the components among the $CaCO_3$ particles, the size and the number of the particles yielded [25]. These parameters might vary from one experiments to another and therefore, it makes problematic to obtain two capsule samples with exactly the same content of encapsulated substances while relying on single capsule detection. Yet, it is imperative to observe this condition to reproduce the efficiency of any sensor. Thus, we always run experiments with at least two samples of capsules in parallel produced independent and having the same parameters at preparation. One of major problem on single particle/capsule detecting is deviation of fluorescent intensity from one particle to another due to uneven fluorescent distribution over population of capsules. To avoid this bottleneck and to obtain a sensor whose reliability and efficiency would not depend on the concentration of the reacting and registering substances in single capsule we opted the SNARF-1 fluorescent dye for the present study (Fig. 1).

The emission spectrum of SNARF-1 undergoes a pH-dependent wavelength shift from 580 nm in the acidic medium to 640 nm in alkaline environment. The ratio $R = I_{580nm}/I_{640nm}$ of the fluorescence intensities from the dye at two emission wavelengths allows to determine the pH value according to the ratiometric method. Dual emission wavelength monitoring is well-established method eliminating a number of fluorescence measurement artifacts, including photobleaching, sample's size thickness variation, measuring instrument stability and non-uniform loading of the indicator [26]. It becomes very important particularly if one is using not an ensemble of capsules but only single or few capsules in the analysis, e.g., in the experiments with cells.

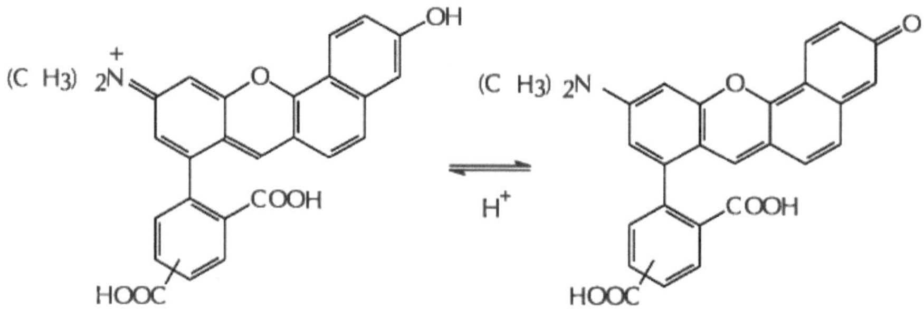

FIGURE 1 Structure of the protonated and deprotonated forms of the SNARF-1 dye.

In order to verify a feasibility of fluorescence based urea sensing on two component co-encapsulation we fabricated two polyelectrolyte capsule samples of the $(PSS/PAH)_4PSS$ shell architecture with different content of dye and urease. The first sample (***Sample I***) contained in average 0.6 pg SNARF-1 dextran per capsule, while the content of the SNARF-1 dextran in the other sample was 0.2 pg per capsule (***Sample II***). The concentration of active urease in samples was opposite 0.2 and 0.6 pg/capsule respectively what gives an average SNARF-1 dextran/urease ratio of 3:1 and 1:3 in these investigated samples.

Spectrofluoremetric studies were carried out to determine the correlation between the fluorescence intensity of the SNARF-1 dextran/urease capsules and the pH of the medium. Both the capsule samples were stored for 10 min in the 0.05 M TRIS-maleate buffer at pH in the range 5.5–9. The excitation wavelength was 540 nm. The capsules fluorescence spectra of the first sample are shown in Fig. 2. The spectra obtained for both samples were similar. The encapsulated dye is capable to provide information of the medium acidity in a reasonably wide range of pH. It is seen that fluorescence spectra are characteristic for every pH value. This fact can be used for calibration regardless amount of dye per capsules in studied samples.

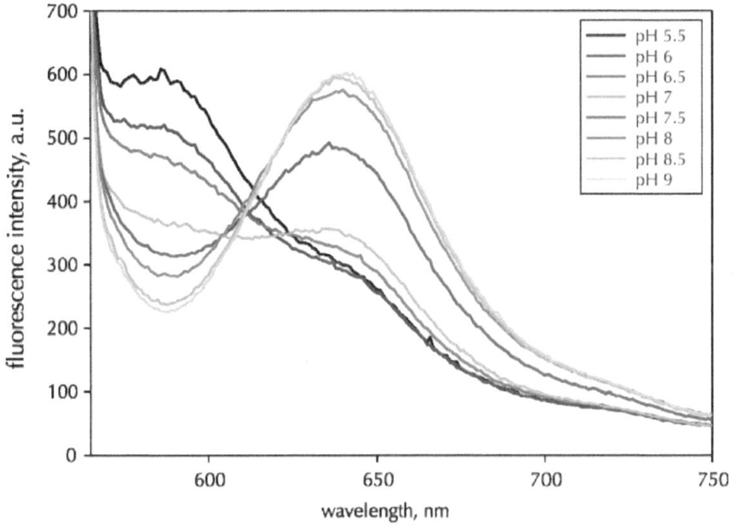

FIGURE 2 Fluorescence spectra of the SNARF-1 dextran/urease capsules in the 0.05 M TRIS-maleate buffer at pH in the range 5.5–9.

The ratio between fluorescence intensity and pH can be described by the following equation according to [26]:

$$pH = pK_a - \log\left(\frac{R - R_{min}}{R_{max} - R} * \frac{I_{640nm}(B)}{I_{640nm}(A)}\right) \quad (1)$$

where $R = I_{580nm}/I_{640nm}$; R_{min} and R_{max} – are the minimal and maximal R values in the titration curve (Fig. 3, curves 3,4); $I_{640nm}(A)$ and $I_{640nm}(B)$ – fluorescence intensities at 640 nm for the protonated and deprotonated forms of the dye, i.e., in the acidic and alkaline media, respectively. The R_{min} and R_{max} meanings depend on the experimental conditions. In this study they were determined to be:

Sample 1: $R_{min} = 0.41$, $R_{max} = 1.96$, $I_{640nm}(B)/I_{640nm}(A)=2$;

Sample 2: $R_{min} = 1.06$, $R_{max} = 2.14$, $I_{640nm}(B)/I_{640nm}(A)= 1.69$.

To yield of the pK_a value the data were plotted as the log of the $[H^+]$ versus the $\log\{(R-R_{min})/(R_{max}-R)*(I_{640nm}B)/I_{640nm}A\}$. In this form, the data gave a linear plot with an intercept equal to the pK_a (Fig.4, curves 3,4). As follows from this data pK_a for sample 1 is equal 7.15, for the sample 2 – 7.25. Thus, the pK_a value differed on 0.1 for different samples whereas the concentration of dye in them differed in 3 times. However dependence of fluorescence intensity for the two samples was linear in the smaller interval of values.

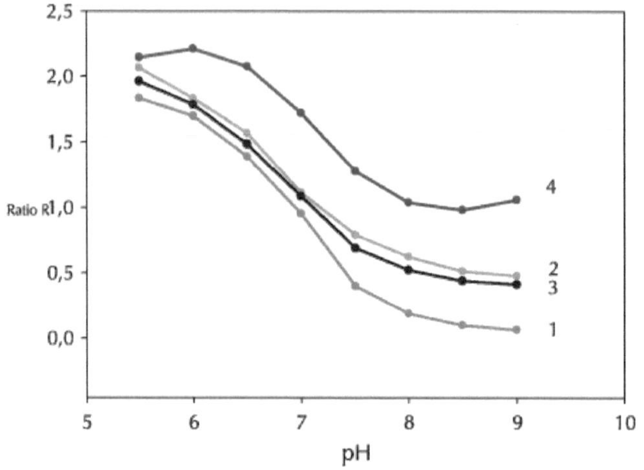

FIGURE 3 Change of fluorescence intensity ratio –R at 580 and 640 nm for: curve 1– SNARF-1 dextran water solution; curve 2 – containing SNARF-1 dextran capsules; curve 3 – SNARF-1 dextran/urease capsules (sample I, 0.6 pg dye/capsule; curve 4 – SNARF-1 dextran/urease capsules (sample II, 0.2 pg dye/capsule) in 0.05M TRIS-maleate buffer at pH in range 5.5–9.

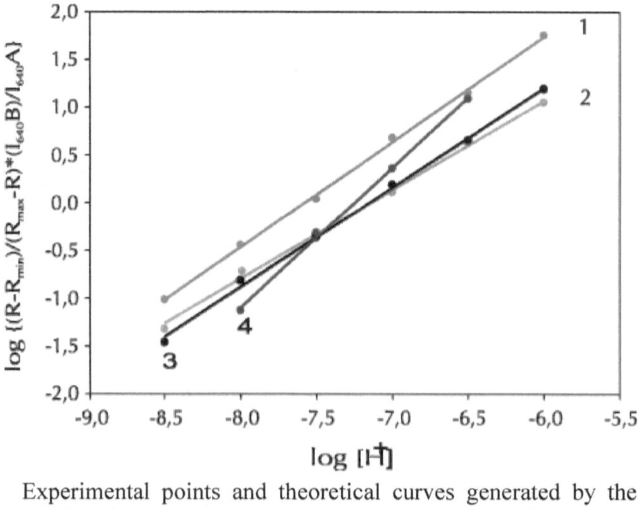

FIGURE 4 Experimental points and theoretical curves generated by the ratiometric method to determine the pK_a values: curve 1 – for SNARF-1 dextran water solution, pK_a=7.58; curve 2 – for containing SNARF-1 dextran capsules, pK_a=7.15; curve 3 – for SNARF-1 dextran/urease capsules (sample I, 0.6 pg dye/capsule), pK_a=7.15; curve 4 – for SNARF-1 dextran/urease capsules (sample II, 0.2 pg dye/capsule), pK_a=7.25.

In Fig. 3, the curves 1, 2 shows the ratio R dependences on pH value for free fluorescent dye in comparison with SNARF-1 dextran capsules. The calculation of pK_a values (Fig. 4, the curves 1, 2) has demonstrated that for encapsulated dye it is less, than for the free dye solution. It is reasonable to assume that this effect is the result of the interaction between SNARF-1 dextran and polyelectrolyte shell, notably with PAA, because they have opposite charges.

Fig. 5 illustrates the effect the urea in concentration from 10^{-6} to 10^{-2} mol/l produces on fluorescence spectra of capsules containing SNARF-1 dextran and urease. Urea concentration dependence is reflected spectroscopically by apparent pH change in the course of enzymatic reaction inside the capsules. The ammonium ions generated via enzymatic reaction in capsule interior effect on pH shift what is recorded by SNARF-1 on the plot (Fig. 5). The fluorescent spectra were measured at the 30 min time point after adding urea solutions at these concentrations to the SNARF-1 dextran/urease capsule's samples. Our particular attention was paid to the kinetics of the change at the fluorescence intensity ratio at 580 nm to 640 nm R (Fig. 6) and its relevance to amount of SNARF/urease. Parameter R was plotted versus time and as one can see on curves at high concentration of the urea substrate (10^{-3} M) level off at about 15–20 min after the beginning of the of the enzymatic reaction, while at low concentrations of urea (10^{-5} M) the time needed for flattening out spectral characteristics reaches 25–30 min. Remarkably, there are no substantial changes for samples at variable concentrations of the dye and urease inside the capsule at least at studied range of 0.2–0.6 pg per capsule. SNARF-1 dextran indicates only the course of the enzymatic reaction, therefore the time needed for the R parameter curve to level off correlates with the time needed to reach the equilibrium in the enzymatic reaction urea/urease occurring inside the capsule. Presumably, it takes about few min to equilibrate concentration of urea and its access to urease. We assume rather fast diffusion of urea through the multilayers due to small molecular size of the molecules [27].

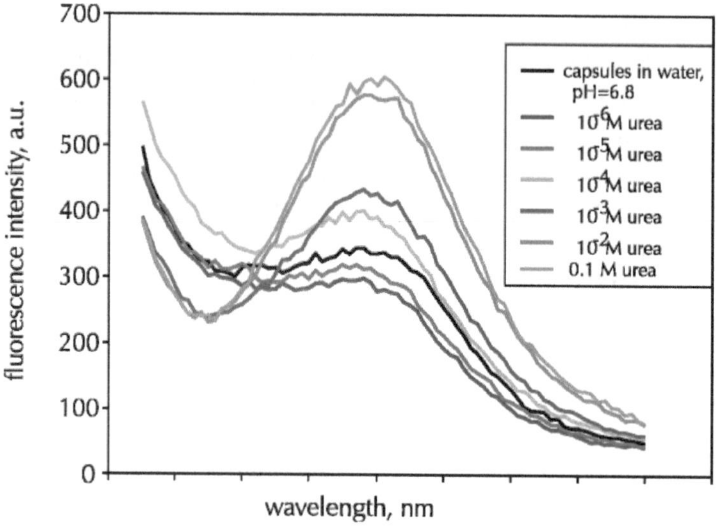

FIGURE 5 SNARF-1 dextran/urease capsule fluorescence spectra in water in the presence of urea from 10^{-6} to 0.1 M.

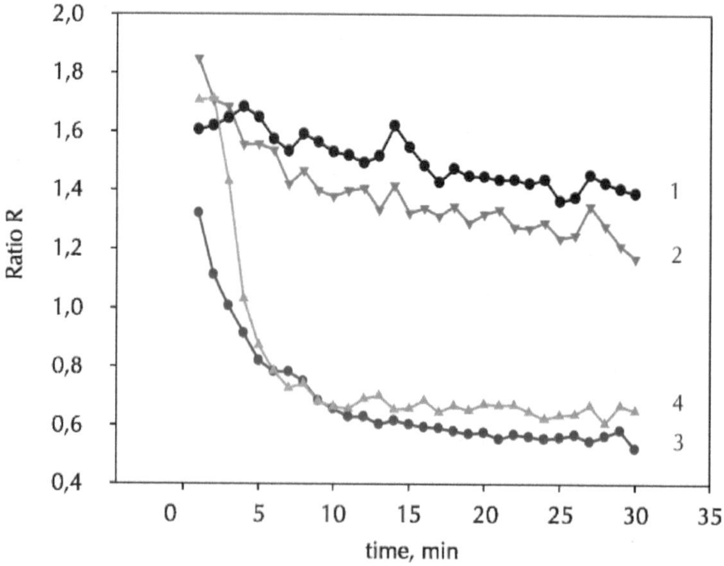

FIGURE 6 Kinetics of the change of the SNARF-1 dextran/urease capsule fluorescence intensity expressed through the fluorescence intensity ratio R at 580 and 640 nm in the presence of 10^{-5} M (curve 1 – sample I, curve 2 – sample II) and 10^{-3} M urea (curve 3 – sample I, curve 4 – sample II).

To calculate the apparent pH caused by urea concentration the values of R_{min}, R_{max}, pK_a and $I_{640nm}(B)/ I_{640nm}(A)$ were used. R_{max} was determined as a fluorescence intensity ratio at 580 nm and 640 nm in the capsules stored in bidistilled water (pH = 6.4). R_{min} is the ratio of the fluorescence intensity spectrum related to the minimal R-value at 580 nm and 640 nm in the SNARF-1 dextran/urease capsules when "high" concentration of urea (0.1 M) was added and 30 min after the enzymatic reaction begun. The pK_a value were assumed to be 7.15 and 7.25 for sample 1 and 2, respectively, that was in accord with the experiment with buffer solutions (Figs. 3, and 4). From the values obtained a calibration curve of the pH dependence inside the capsules on the urea concentration present in the solution was plotted. The calibration curve is presented in Fig. 7.

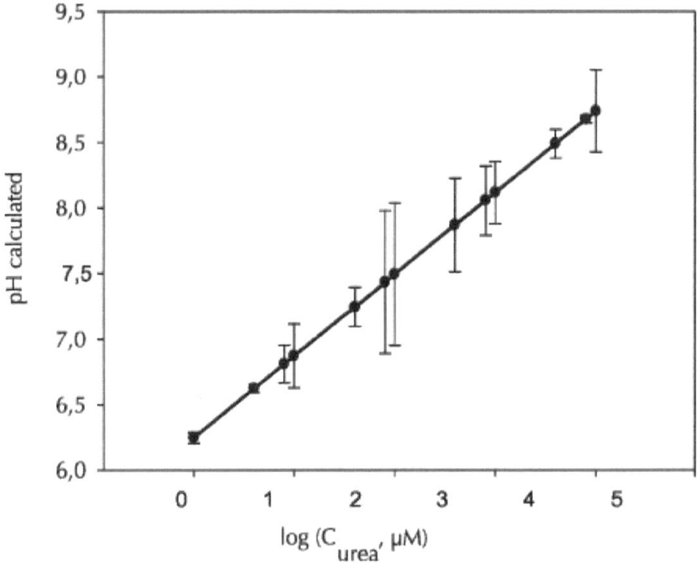

FIGURE 7 Calibration curve for detecting urea using the SNARF-1 dextran/urease capsules in water solutions.

Thus, to determine the urea concentration in the solution it is necessary to obtain the following three spectra: (1) of the capsules without substrate (urea); (2) of the capsules at "high" concentration of urea in the solution, e.g., 0.1 M as used for our calibration plot; (3) of the capsules in the investigated sample studied. These data will suffice to calculate the values of

R_{min}, R_{max} and $I_{640nm}(B)/ I_{640nm}(A)$, which are characteristic of each particular sample of the sensing capsules. Using Eq. (1) one then can calculate the pH as apparently reached in the capsules in the course of urea degradation and to compare its value with the calibration curve in Fig. 7.

It is worth to notice, that this calibration curve is obtained for SNARF-1 dextran/urease capsule in pure water without substantial contamination of any salt, which could buffer the systems and spoil truly picture for urea detection. We carried out experiments to build a similar calibration curve in the presence of the 0.001 M TRIS-maleate buffer (were used solutions with the pH 6.5 and 7.5) but it resulted in overwhelming effect of pH buffering. Buffering the solution eliminates the pH change caused in a course of enzymatic reactions. Thus, it sets a limit for detection of urea concentration using SNARF-1-dextran/urease capsules. However the calibration in conditions of particular experimental system is reasonable at salt free solution assumption. Summarizing, one can state the presented in Fig. 7 calibration curve as suitable for estimation of urea concentrations in-situ in water solutions.

Feasibility studies on single capsule detection of urea presence were carried out using Confocal Fluorescent Microscopy. CLSM image of SNARF-1 dextran/urease capsules in absence and at 0.1M urea added to the same capsules are presented on Fig. 8. Small distinctions in the form and the sizes of capsule population are connected with non-uniformity of SNARF-1 dextran/urease $CaCO_3$ particles received in co-precipitation process what is rather often observed for calcium carbonate templated capsules containing proteins [20].

The Table 1 shows the increase of Mean energy of individual capsules before (F_{low}) and after the addition 0.1 M urea solution (F_{high}) to water capsule suspension. The red fluorescence emission was accumulated at 600–680 nm after excitation by the FITC–TRIC–TRANS laser at 543 nm.

The images have been processed with Leica Confocal Laser Scanning Microscope TCS SP software to quantify the effect. The image areas corresponding to location of 10 selected capsules are set off in different colors – ROI_{1-10} (Region Of Interest). The value of average intensity of a luminescence is defined as parameter Mean Energy by formula:

$$I_{Mean}^2 = \frac{1}{N_{Pixel}} \sum_{Pixel} I_i^2$$

where, I_{mean} – the average image energy of ROI areas; N_{pixel} – the total number of pixels that are included in the calculation; I_i – the energy correspond for particular pixel.

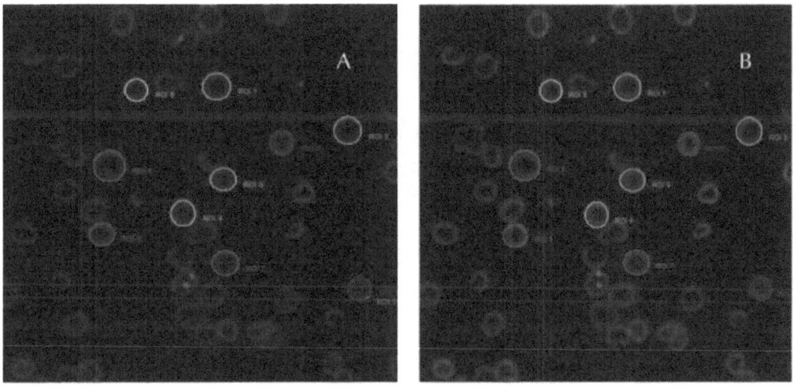

FIGURE 8 Confocal fluorescence microscopy images of $(PSS/PAH)_4PSS$ capsules loaded with SNARF-1 dextran (MW= 70kDa) and urease enzyme in water (image A) and 0.1 M urea concentrations (image B).

The energy meanings of individual capsules are presented in the Table 1.

TABLE 1 The change of mean fluorescence intensity of 10 selected capsules in presence of 0.1 M urea.

ROI	Mean Energy, I_{low}	Mean Energy, I_{high}	I_{high}/I_{low}
ROI 1	2947.95	6467.42	2.1939
ROI 2	4635.87	10316.34	2.2253
ROI 3	2424.13	5694.92	2.3493
ROI 4	4021.83	11161.52	2.7752
ROI 5	1666.76	3727.05	2.2361
ROI 6	3177.11	7237.68	2.2781
ROI 7	2237.61	5932.76	2.6514
ROI 8	3192.70	7158.54	2.2422
ROI 9	4100.76	8944.83	2.1813
ROI 10	2758.57	6127.32	2.2212

Although, value of fluorescence intensity is seen to be different for each capsule the more than double increase of integrated intensity is well pronounced in all monitored capsules upon addition of urea solution. The distribution of energy relation therefore remains almost constant for each capsule. These data on single capsules are in good agreement with data presented on Fig. 5 obtained on entire capsule population. Indeed, integrated area under spectra for black (no urea) and light blue (0.1M urea) with spectral range of 600–680 nm is about twice in difference. This fact demonstrates the principal applicability to use single capsule for carrying out analysis of urea presence.

KEYWORDS

- **Layer-by-Layer**
- **Miniaturization**
- **PEBBLE**
- **Tailoring**
- **Varian Cary Eclipse**

REFERENCES

1. Arregui, F. J. *Sensors Based on Nanostructured Materials*, 2009.
2. Fehr, M.; Okumoto, S.; Deuschle, K.; Lager, I.; Looger, L. L.; Persson, J.; Kozhukh, L.; Lalonde, S.; Frommer, W. B. *Biochem Soc Trans.*, **2005**, *33*(1), 287–290.
3. Vo-Dinh, T.; Griffin, G. D.; Alarie, J. P.; Cullum, B.; Sumpter, B.; Noid, D. *Summary Nanomed.*, **2009**, *4*(8), 967–979.
4. Sukhorukov, G. B.; Rogach, A. L.; Garstka, M.; Springer, S.; Parak, W. J.; Muñoz-Javier, A.; Kreft, O.; Skirtach, A. G.; Susha, A. S.; Ramaye, Y.; Palankar, R.; Winterhalter, M.; *Small, 3*(6), 944–955.
5. Lee, Y. E.; Smith, R.; Kopelman, R. *Annu. Rev. Anal. Chem. (Palo Alto Calif.)*, **2009**, *2*, 57–76.
6. Sukhorukov, G. B.; Rogach, A. L.; Zebli, B.; Liedl, T.; Skirtach, A. G.; Köhler, K.; Antipov, A. A.; Gaponik, N.; Susha, A. S.; Winterhalter, M.; Parak, W. J. *Small*, **2005**, *1*(2), 194–200.
7. De Geest, B. G.; De Koker, S.; Sukhorukov, G. B.; Kreft, O.; Parak, W. J.; Skirtach, A. G.; Demeester, J.; De Smedt, S. C.; Hennink, W. E. *Soft Matter*, **2009**, *5*, 282.

8. Peteiro-Cartelle, J.; Rodríguez-Pedreira, M.; Zhang, F.; Rivera Gil, P.; del Mercato, L. L.; Parak, W. J. *Nanomedicine*, **2009**, *4*(8), 967–979.
9. Sailor, M. J.; Wu, E. C. *Adv. Funct. Mater.*, **2009**, *19*(20), 3195–3208.
10. Nayak, S.; McShane, M. J. *Sens. Lett.*, **2006**, *4*, 433–439.
11. Kreft, O.; Muñoz Javier, A.; Sukhorukov, G. B.; Parak, W. J. *J. Mater. Chem.*, *42*, 4471–4476.
12. Brown, J. Q.; McShane, M. J. *IEEE Sens. J.*, **2005**, *5*, 1197–1205.
13. del Mercato, L. L.; Abbasi, A. Z.; Parak, W. J. *Small*, **2011**, doi: 10.1002/smll.201001144.
14. Brown, J. Q.; McShane, M. J. *Biosens. Bioelectron.*, **2005**, *21*, 1760–1769.
15. Stein, E. W.; Volodkin, D. V.; McShane, M. J.; Sukhorukov, G. B. *Biomacromolecules*, **2006**, *7*, 710–719.
16. Brasuel, M.; Aylott, J. W.; Clark, H.; Xu, H.; Kopelman Hoyer, R. M.; Miller, T. J.; Tjalkens, R.; Philbert, M. *Sens. Mater.*, **2002**, *14*, 309–338.
17. Xu, H.; Aylott, J. W.; Kopelman, R. *Analyst*, **2002**, *127*, 1471–1477.
18. Donath, E.; Sukhorukov, G. B.; Caruso, F.; Davis, S. A.; Möhwald, H. *Angew. Chem., Int. Ed.*, **1998**, *37*, 2202–2205.
19. Sukhorukov, G. B.; Donath, E.; Davis, S.; Lichtenfeld, H.; Caruso, F.; Popov, V. I.; Möhwald, H. *Polym. Adv. Technol.*, **1998**, *9*, 759–767.
20. Petrov, A. I.; Volodkin, D. V.; Sukhorukov, G. B. *Biotechnol. Prog.* **2005**, *21*, 918–925.
21. Sukhorukov, G. B.; Volodkin, D. V.; Gunther, A. M.; Petrov, A. I.; Shenoy, D. B.; Möhwald, H. *J. Mater. Chem.*, **2004**, *14*, 2073–2081.
22. Lvov, Y.; Antipov, A. A.; Mamedov, A.; Möhwald, H.; Sukhorukov, G. B. *Nano Lett.*, **2001**, *1*, 125.
23. Lvov, Y.; Caruso, F. *Anal. Chem.*, **2001**, *73*, 4212.
24. Paddeu, S.; Fanigliulo, A.; Lanzin, M.; Dubrovsky, T.; Nicolini, C. *Sens. Actuators*, **1995**, *25*, 876–882.
25. Halozana, D.; Riebentanz, U.; Brumen, M.; Donath, E. *Colloids Surf. A: Physicochem. Eng. Aspects*, **2009**, *342*, 115–121.
26. Whitaker, J. E.; Haugland, R. P.; Prendergast, F. G. *Anal Biochem.* **1991**, *194*(2), 330–344.
27. Antipov, A. A.; Sukhorukov, G. B.; Leporatti, S.; Radtchenko, I. L.; Donath, E.; Möhwald, H. *Colloids Surf. A: Physicochem. Eng. Aspects*, **2002**, *198–200*, 535–541.

CHAPTER 39

SELECTION OF MEDICAL PREPARATIONS FOR TREATING LOWER PARTS OF THE URINARY SYSTEM

Z. G. KOZLOVA

CONTENTS

39.1 INTRODUCTION

Stones – a metabolism illness due to various endogenous or exogenous causes and often of a hereditary nature characterized by the urino-formation of stones in the urinary system.

There are people in all age groups who suffer from irretention of urine. Thirty percent of women suffer from this in one form or another, i.e., unable to regulate the functioning of the urinary bladder. This problem may be solved by strengthening the wall of the urinary bladder, decreasing the inflammatory process in the urinary tract and strengthening the connective tissue.

Antioxidants (AO) are nutrient substances (vitamins, microelements etc.), which human organisms require constantly. They serve to maintain a balance between free-radicals and AO forces.

Modern medicine uses AO to improve people's health and as a prophylaxis. Therefore, in recent times, the AO properties of various compounds are being widely studied. The most prevalent source of AO is considered to be vegetative objects on the basis of which medicinal preparations and BAAs are prepared.

As a criterion for evaluating the quality of a medicinal preparation, we took its Antioxidant activity (AOA), i.e., concentration of natural AO in it, which was measured on a model chain reaction of liquid-phase oxidation of hydrogen by molecular oxygen.

Set task: Quantitatively measure AO content in investigated preparations since they constitute a vegetation composition and evaluate their effectiveness in improving the quality of life.

The following are medicinal preparations and BAAs investigated for treating illnesses of the urethra (cystitis, enuresis and urine stones): Cyston (India), Urotractin (Italy), Promena (USA), Spasmex (Germany), Urolyzin (Russia), Contrinol (USA), Tonurol (Anti-Enuresis) (Russia), Blemaren (Germany).

39.2 CHARACTERISTICS OF PREPARATIONS

CYSTON (India) – Complex therapy for stones in the bladder, crystallization, infection of the urethra, podagra.

Each tablet contains: Didymocarpus pedicellata R. Br., Saxifraga Ligulata Wall, Rubia cordifolia L.; Cyperus scariosus R.Br.; Achyranthes aspera L.; Onosma bracteatum Wall.; Vernonia cinerea (L.) Less. BAA.

CONTRINOL (USA) – A Mixture of eastern and western medicinal plants used for normal functioning of the urethra. It strengthens the connective tissue.

Each capsule contains: Horsetail Herb, White Poplar Bark, Dogwood Berry and Schizsandra Berry. BAA.

PROMENA (USA) – provides important support for the prostate, possesses anti-inflammation properties, normalizes functioning of the urethra and strengthens the immune system.

Each capsule contains: Vitamin E, Vitamin C, Zinc, Vitamin A, Parsley Leaf, Echinacea, Pumpkin Seed, Gravel Root, Corn silk, Bee Pollen. BAA.

SPASMEX (Germany), TROSPIYA CHLORED (PRO. MED. CS PRAHA a.s.) – quarter amine is safer to use because of its unique chemical structure. Lowers tone of the smooth muscles of the urinary bladder, reduces detrusion of the bladder.

Each capsule contains: Dry birch bark extract, Irish Moss, Origanum vulgare L.

UROLYZIN (Russia) – source of Arbutin and Flavonoids used as a diuretic and antiseptic.

Content: Extracts of Folium Betula, Polygonum avicular L., Orthosiphon stamineus Benth, Sprout vaccinium myrtillus L., Fructus Aronia melanocarpa Elliot, Fructus Sorbus aucuparia L., Burdock Root. BAA.

TONUROL (ANTI-ENURESIS) (Russia) – supports urethra organs, strengthens bladder wall, helping in case of incontinence and has antiseptic and anti-inflammation properties.

Capsule content: Equisetum arvense L., White Poplar Bark, Hypericum perforatum L. flowers, Dogwood Berry, Schisandra chinensis Baill (Turez), Matricaria recutita (L.) extract.

BLEMAREN (Germany) – for prophylactic and treatment of stones in the bladder. Burbling tablets.

39.3 METHOD OF EXPERIMENT

Chain reactions of oxidation can be used for quantitative characterization of the properties of inhibitors (antioxidants). The investigated samples were analyzed by means of a model chain reaction of initiating oxidation of cumene [1–4]. Initiated oxidation of cumene in the presence of studied AO proceeds in accordance with the following scheme:

Initiation of chain Origination of $RO_2^{\cdot}$ radicals, initiation rate W_i

Continuation of chain $RH + RO_2^{\cdot} \xrightarrow{k_3};\rightarrow_{O_2} ROOH + RO_2$

Break of chain:

$2RO_2^{\cdot}$

molecular products

$H{+}RO_2^{\cdot}\ RO_2H{+}In^{\cdot}$

$RO_2^{\cdot}$ + Inmolecular products

(We use the widely accepted numeration of rate constants of elementary reaction of inhibited oxidation.)

In accordance with this scheme, each independent inhibiting group of AO breaks two chains of oxidation.

to determine the initial concentration of the inhibitor (more exactly, the concentration of inhibiting groups) taken for the reaction from the experimentally determined value of the period of induction τ. In the right member of Eq. (1), we have: W_i, the standard given rate of initiation; f, the inhibiting coefficient, equal to 2; n, the number of inhibiting groups in an antioxidant molecule; $[InH]_0$, the initial AO concentration.

The constant of inhibiting rate k_7, determining the anti-radical activity of AO and being its qualitative characteristic, is found from relation (2), using the known constant of chain continuation rate k_3, concentration [RH] for hydrocarbon, experimentally determined period of induction τ and quantity of absorbed oxygen ΔO_2.

Cumene (isopropyl benzene) was used as oxidizing hydrocarbon and azo-bis-isobutyronitrile as initiator, which forms free-radicals upon thermal decay. Initiating rate was determined from the following formula:

$$W_i = 6.8 \times 10^{-8} [AIBN] \text{ mole} / 1 \times s,$$

where [AIBN] (AZO-bis-isobutyronitrile) is the initiator concentration in mg per ml of cumene.

The period of induction τ is determined by plotting the dependence of the quantity of absorbed oxygen ΔO_2 in the reaction against time t. The end of the kinetic curve is a linear portion representing non-inhibited reaction, i.e., the portion after expending AO. The AO expenditure time τ is determined graphically on the kinetic dependence of oxygen absorption by the point of intersection of two straight lines: one of the lines is the line that the kinetic curve assumes after AO is consumed and the other is a tangent to the kinetic curve, the tangent of the angle of inclination of which is one half the tangent of the angle of the first. The greater the amount of AO in the sample the greater the period of induction τ.

The method is direct and based on the use of the chain reaction of liquid-phase oxidation of hydrocarbon by molecular oxygen.

The method is functional, i.e., the braking of the oxidizing reaction is determined only by the presence of AO in the system being analyzed. Other possible components of the system (not AO) do not exert a significant effect on the oxidizing process, which enables to analyze AO in complex systems, avoiding a stage of separation.

The method is very sensitive, exact and informative.

The method is absolute, i.e., does not require calibration, is simple to apply and does not require complex equipment [1–5].

39.3 RESULTS AND DISCUSSION

The following medicinal preparations were taken for investigation: Cyston, Spasmex, Urotractin, Urolyzin, Contrinol, Promena, Tonurol (Anti-Enuresis), Blemaren. The AOA of these preparations was determined.

As an example, the Fig. 1 shows the kinetic dependences of oxygen absorption in a model reaction of initiated cumene oxidation in the absence of antioxidant (straight line 1) and in the presence of Urotractin (curve 2), Spasmex (curve 3), Promena (curve 4), Contrinol (curve 5), Cyston (curve 6), Urolyzin (curve 7).

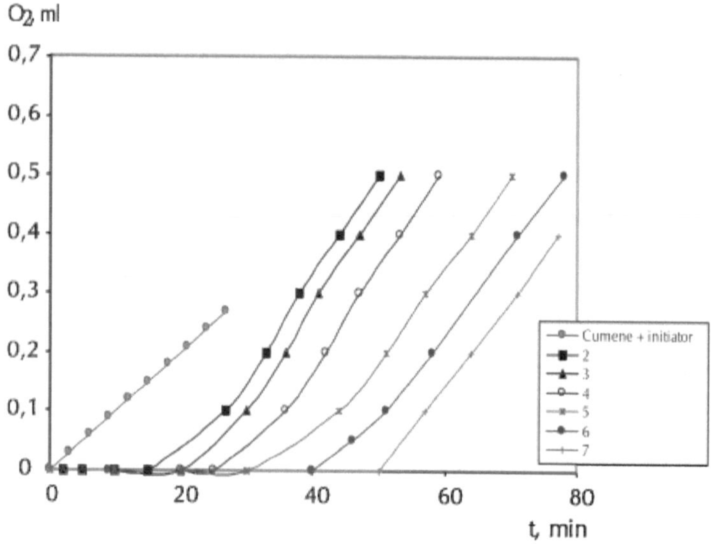

FIGURE 1 Kinetic Dependences of Oxygen Absorption. 1 ml of hydrocarbon, 1 mg of initiator, t = 60°C.

1 – hydrocarbon (cumene) + initiator AZO-bis-IZOBUTYRONITRILE, 1 mg),

2 – with Urotractin added (6.6 mg), $\tau = 20$ min,

3 – with Spasmex added (6.0 mg), $\tau = 25$ min,

4 – with Promena added (7.5 mg), $\tau = 30$ min,

5 – with Contrinol added (20 mg), $\tau = 38$ min,

6 – with Cyston added (9.6 mg), $\tau = 45$ min,

7 – with Urolyzin added (14.6 mg), $\tau = 50$ min.

It can be seen from the Fig. 1 that, in the absence of the additive, hydrocarbon oxidation proceeds at constant rate (straight line 1). When the preparation is added, the oxidation rate at the beginning is strongly retarded but begins to increase after a certain period of time. This is indicative of the presence of antioxidant in the additive.

The rise in reaction rate is due to the expenditure of antioxidant. When it is used up, the reaction proceeds at the constant rate of an uninhibited reaction. The time of antioxidant (τ) is determined graphically by the intersection of two straight lines on the kinetic curve. One of these is the straight-line portion of the kinetic curve after AO has been used up. The other is the tangent to the kinetic curve whose inclination angle is one-half the tangent angle of the first.

Data on the antioxidant content of investigated preparations are presented in the Table 1. These data are illustrated by the diagram.

TABLE 1 Concentration of Antioxidants in Studied Preparations.

№	Preparations	Antioxidant Concentration (M/kg)
1	Cyston	9.6×10^{-3}
2	Spasmex	8.5×10^{-3}
3	Promena	8.2×10^{-3}
4	Urolyzin	7.0×10^{-3}
5	Urotractin	6.2×10^{-3}
6	Tonurol (Anti–Enuresis)	5.4×10^{-3}
7	Contrinol	3.9×10^{-3}
8	Blemaren	–

Results of analysis show that AOA is in the interval of values (3.9–9.6) $\times 10^{-3}$ M/kg, which correlates with the values determined earlier for AO in plants: Basilicum, Hypericum perforatum L. -8.8×10^{-3} M/kg, Coriandrum sativum L. -3.9×10^{-3} M/kg, Valeriana officinalis L. -7.7×10^{-3} M/kg, Eleutherococcus Senticosus max. -9.4×10^{-3} M/kg. Thus, the effectiveness of certain BAAs may be due to their AO content (more than 10^{-3} M/kg of dry substance) and their inclusion in complex therapy of corresponding illnesses may be justified. The close values of AO concentration between the studied preparations and medicinal plants give basis for assuming they're having a positive effect on the human organism [6].

KEYWORDS

- **Antioxidants**
- **Azo-bis-isobutyronitrile**
- **Cumene**
- **Oxygen absorption**

REFERENCES

1. Tsepalov, V. F. In *Investigation of Synthetic and Natural Antioxidants In Vitro and In Vivo;* Moscow, 1992 (in Russian).
2. Kharitonova, A. A.; Kozlova, Z. G.; Tsepalov, V. F.; Gladyshev, G. P. *Kinetika i Kataliz. J.*, **1979**, *20*(3), 593–599 (in Russian).
3. Tsepalov, V. F.; Kharitonova, A. A.; Kozlova, Z. G. In *Bioantioxidant*; Moscow, 1998, p 94 (in Russian).
4. Tsepalov, V. F.; Kharitonova, A. A.; Kozlova, Z. G.; Bulgakov, V. G. In *Pishchevye Ingredienty*; Moscow, 2000; pp 7–8 (in Russian).
5. Tsepalov, V. F. *Zavodskaya Lab.*, **1964**, *1*, 111 (in Russian).
6. Kozlova, Z. G.; Eliseyeva, L. G.; Nevolina, O. A.; Tsepalov, V. F. *Content of Natural Antioxidants in Spice-aromatic and Medicinal Plants*. Theses of Report at International Scientific Conference: Populations Quality of Life-Basis and Goal of Economic Stabilization and Growth, Oryol, 23.09–24.09.1999; pp 184–185 (in Russian).

CHAPTER 40

COMMENTARY: MODERN SCIENTIFIC EDUCATION

N. I. KOZLOWA, A. E. RYBALKO, K. P. SKIPINA, and L. G. KHARUTA

CONTENTS

40.1 INTRODUCTION

The deep shocks connected with serious intervention in the unique nature of the Sochi Black Sea Coast, for long years will have negative consequences. In the created conditions the requirement for the ecologists capable competently to solve most different problem at creation of projects on restoration of ecologically safe environment, necessary for successful development of tourism and restoration of the resort industry raises. It is necessary to find ecologically comprehensible technical decisions connected with an active urbanization of territory, to keep the sites of the National park which has been not mentioned by creation of Olympic objects, to restore and create anew places of dwelling rare and vanishing species of representatives of flora and fauna. Experts IOC and our ecologists testify that in the conditions of the rough building limited to control terms of delivery in operation of objects, is very difficult to watch observance of positions ecologically a sustainable development of the Sochi region.

The requirement for the decision of the arisen environmental problems of region considerably will increase after end of all actions connected by carrying out Olimpiad-2014 when the period of search of the measures connected with transformation of region in vacation spot and tourism of world value will inevitably come, and also with development of the Sochi region as all-the-year-round resort. The success of these measures will be substantially provided by participation in the planned actions of highly skilled ecologists not only on a design stage, but also at realization of the best in every respect projects, which should combine economic and ecological expediency. Only such approach will lead to occurrence of successful vacation spots, routes of ecological tourism, biodiversity preservation in region of Games.

In the developed conditions of natural increase of requirement of ecological shots the Sochi institute of RPFU, expanding preparation of experts of a biological direction, begins training on two new specialty having an ecological orientation – "Biology" and "Ecology and Wildlife Management."

The new status of institute (from 2011 year Sochi branch RPFU is transformed to the Sochi institute RPFU) gives possibility to raise vocational training level. Now the decision on creation of the scientifically-educational center (SEC), *carrying out carrying out of researches and a professional training of the higher scientific qualification* that will allows to carry out scientific researches and retraining of experts of the directions claimed in region is accepted. All scientific institutions with which the institute had creatively productive relations will take part in center work.

The major qualifying characteristics of the scientifically-educational center are high scientific level of carried out researches, high productivity of preparation of scientific shots, participation in preparation of students on a scientific profile of the scientifically-educational center, use of results of scientific researches in educational space of the developing biological direction connected with opening of new specialty – Biology, Ecology and wildlife management, Veterinary science and Veterinary and sanitary examination.

Within the limits of SEC implementation of possibility of improvement of professional skill of experts, in perspective area of a science – biotechnologies is planned. Researches on preservation of rare plants on which the Sochi Black Sea Coast is the most capacious enclave of Russia and to creation of possibility of use of these valuable plants in practice of the most various areas of plant growing on the basis of methods of biotechnology of new generation will be strengthened. The cellular engineering in a combination to modern methods of plant virology is an effective basis of introduction in culture of rare kinds, and also plants perspective for use in decorative gardening.

Ongoing research of students and teachers devoted to improving the educational process, the inclusion of biotechnology and to attract resources, academic institutions with which the Institute cooperates [1, 2, 4, 8, 15], improving the legal protection of the environment [3, 14]. A number of papers devoted to the use of biotechnology in breeding of many rare plants for conservation and utilization in the economy [7, 9–16].

Scientific interests of teachers of chair the physiology directed on working out of methods of introduction in culture of disappearing representatives of local flora and their accelerated microclonao of reproduction, along with search of optimum ways of increase of productivity of educational process, are the integral making preparation of the qualified experts. In laboratory of physiology of plants graduates of chair are prepared for the vigorous activity under the successful decision of similar problems under condition of development of a corresponding infrastructure. Therefore within the limits of created SEC there is a real possibility of participation of students in scientifically-practical activities on the accelerated reception of the improved landing material necessary for increasing requirement of the developing infrastructure of landscape gardening.

Themes of the degree works executed in laboratory of physiology of plants:

– Application of methods of biotechnology in preservation of a biodiversity of hand bells

The Caucasian biospheric reserve;

– Preservation of a specific variety Campanulaceae the Western Caucasus Biotechnology methods;

– Influence of plant growth regulator on growth and development of plants Gerbera the multiplied;

– Features of micropropgation of a carnation acantolimonoides (Vanishing species);

– Reproduction conditions zephyrantes in culture in vitro;

– Micropropagation of lilies;

– High-quality features of reproduction of tulips methods of culture of fabrics;

– Studying of conditions of micropropagation of a potato of a cultivar Lugovsky;

– Working out of modes of regeneration virus free plants of Lisiantus (*Eustoma grandiflorum*) in meristem culture;

- Determination of the primary structure and characteristics of the new plant lipid transfer protein dill Anethum geaveolens;
- Assessment of the regeneration potential of different varieties of durum and soft wheat;
- Biotechnological process for the preparation of totally labeled with stable isotopes of lipid-transporting protein of lentils;
- Evergreen oaks Sukhumi subtropical arboretum.

Has received development the scientific direction connected with researches in area and problems of protection of an environment, in creation of model of the favorable environment in vacation spots and dwellings. Participation of students in these researches has come to the end with protection of degree works on problems of influence of anthropogenic changes of environment on health and social potential of the population of the Black Sea coast of Caucasus.

Results of scientific researches of students and teachers will be used for work on prospects of development of a sanatorium complex of Sochi which practical value is difficult for overestimating. On their basis it is necessary to define criteria of preservation and optimization of ecosystem, defining recreational possibilities of an environment. It can serve as a starting point on a way of maintenance steady and long-term development of territory of the Sochi region of the Black Sea coast of Krasnodar territory, especially with toughening of ecological requirements in connection with the forthcoming Winter Olympic Games in Sochi in 2014.

The themes of degree works executed under the guidance of science officers of National park:

– Ecological estimation of Navaginsky range of a firm field waste of Sochi;

– Dynamics of distribution of a HIV-infection in G. Sochi;

– Influence of anthropogenous loadings on a condition of water pool Red Glades;

– Receptions of preservation of the Nature sanctuary «the Site from the sandy seaside "Vegetation" between bases of rest "Chernomorets" and "Energy" in Imeretinsky lowland;

– Birds of the Sochi Black Sea Coast;

– Birds of Imeretinsky lowland.

Thus, active participation of students of chair in expeditions under the account of a livestock of wild animals in places of dwelling and a condition of flora of the Sochi Black Sea Coast becomes a basis for the future teamwork of the Sochi institute with the Sochi national park and the Caucasian biospheric reserve within the limits of SEC.

Creative cooperation of the Sochi institute and integration of educational resources of chair of physiology with Institute medical primatology the Russian Academy of Medical Science on problems of ecological physiology also began with the moment of opening of chair in 2000. The works of students connected with ecology of the person, and their direct participation in the scientific program of laboratories of Institute medical primatology the Russian Academy of Medical Science in the field of virology, microbiology, immunology, studying of behavior of primacies have scientific and practical value. Participation of students under the guidance of visible scientists in priority directions of a medical science has allowed giving a scientific basis of a choice of the future trade to young scientists.

Now by request of Olympic committee the project on creation of an ecological track in places of growth of wild orchids of the Sochi Black Sea Coast is developed and there are begun works on its registration. The project of preservation rare and vanishing species of plants within the limits of the project on biodiversity preservation in region of Olympic games 2014 is submitted to consideration.

Thus, all above-stated opens prospects of development of activity SOC of the Sochi institute and reception of positive results of its activity on education of scientific shots in modern conditions of cooperation of high schools and scientific institutions.

KEYWORDS

- **Black Sea**
- **Livestock**
- **Olympic objects**
- **Russian Academy of Medical Science**

REFERENCES

1. Skipina, K. P., Rybalko, A. E. *Training of Personnel for Introduction of Biotechnological Methods of Preservation of a Biodiversity in the Suburb of Sochi*. Proceeding The Moscow International Scientific and Practical Conference (Moscow, on March, 15–17th, 2010) M: Joint-Stock Company "Ekspo-biochim-tehnologies", D. I. Mendeleyev University of Chemistry and Technology of Russia, 2010, p 374.
2. Kozlowa, N. I.; Rybalko, A. E.; Skipina, K. P.; Kharuta, L. G. *Biotechnological Education Problems in Subdivision of Sochi Institute Russian University of People's Friendship in View of Preparation to Olympic Games-2014*. Proceeding of the Moscow International Scientific and Practical Conference (Moscow, on March, 21–23st, 2011) M: Joint-Stock Company "Ekspo-biochim-tehnologies," D. I. Mendeleyev University of Chemistry and Technology of Russia, 2011, p 372.
3. Rybalko, A. A. *Features of Ecolaw Training under Implementation of Biotechnological Methods for Preservation of Biological Diversity in Sochi Black Sea Region*. Ibid., p 373.

4. Rybalko, A. E.; Tkachenko, V. P.; Rybalko, A. A. *Biotechnology for Landscape Construction in Subtropical Zone – Element for Training of Skilled Personnel for Landscape Construction*. Ibid., p 371.
5. Bogdanov, I. V.; Finkina, E. I.; Balandin, S. V.; Rybalko, A. E.; Ovchinnikova, T. V. *Biotechnological Uniform Stable Isotope Labeling of Lentil Lipid Transfer Protein*. Ibid.
6. Pavlova, L. E.; Melnikova, D.N.; Finkina, E. I.; Balandin, S. V.; Rybalko, A. E.; Ovchinnikova, T. V. *New Lipid Transporting Fiber from Fruits Citrus natsudaidai*. Ibid.
7. Rybalko, A. A.; Titova, S. M.; Rybalko, A. E.; Bogatyreva, S. N.; Kharuta, L. G. *Micropropagation in Vitro of Plants of Family Gentiana and Others Endemic of Plants of Northern Caucasus*. Biotechnology and Ecology of Big Cities. Sergey Varfolomeev, D.; Zaikov, G. E.; Krylova, L. P., Eds.; Nova Science Publishers, Inc., 2011, pp 61–69.
8. Kozlowa, N. I.; Rybalko, A. E.; Skipina. K. P.; Kharuta. L. G. *Training of Personnel for Instillation of Bbiotechnological Methods of Biovariety Conservation in the Suburbs of Sochi*. Ibid. , pp 145–150.
9. Maevskij, S. M.; Gubaz, S. L.; Rybalko, A. A. *Introduction in Culture In Vitro Three Herbs of the North Caucasus (Mentha Longifolia L., Origanum Vulgare L., Thymus Vulgaris L.)*. Proceeding of the Moscow International Scientific and Practical Conference "Pharmaceutical and Medical Biotechnology" (Moscow, on March, 20–22th, 2012) M: Joint-Stock Company «Expo-biohim-technologies», D. I. Mendeleyev University of Chemistry and Technology of Russia, D. I. Mendeleyev University of Chemistry and Technology of Russia, 2012, pp 456–457.
10. Gurchenkova, Y. A.; Rybalko, A. E.; Kharuta, L. G. *Micropropagation of Hypericum for Medicinal and Ornamental Purposes*. Ibid., pp 456–457.
11. Matskiv, A. O.; Arakeljan, M. A.; Rybalko, A. E. *Introductions in Culture In Vitro Rare Bulbous Plants of the Sochi Black Sea Coast (Scilla, Muscari, Galanthus)*. Ibid., p 455.
12. Urevich, I. A.; Petikjan, E. V.; Rybalko, A. E. *Eustoma - New Raw Materials for Pharmaceutics (Eustomoside, Eustoside and Eustomorusside)*. Ibid., pp 452–453.
13. Averyanova, E. A.; Rybalko, A. E.; Skipina, K. P.; Kharuta, L. G. *Wild-growing Kolchidsky Orchids and their Preservation as Objects of Education and Producers of Medicinal Substances*. Ibid., p 453.
14. Rybalko, A. A. *New Approaches to Teaching of the Ecological Right in Connection with an Ecological Condition of a Biodiversity of the Sochi Black Sea Coast*. Ibid., pp 436–437.
15. Kozlowa, N. I.; Rybalko, A. E.; Skipina, K. P.; Kharuta, L. G. *Cooperation of High Schools and Scientific Institutions on a Way of Education of Scientific Shots to Modern Conditions*. Ibid., pp 435–436.
16. Shevlyakova, L. A.; Orlova, G. L.; Rybalko, A. A. *Study of Introduction in Culture In Vitro Rare Pharmaceutical Plant Blackstonia Perfoliata (L). Huds*. (Working Out of Technology of Reception of Pharmaceutical Raw Materials). Ibid., pp 451–452.
17. Zagorodnuk, E. D.; Rybalko, A. E. *Working Out of a Technique of Micropropagation of Pharmaceutical Plant Lysimachia Vulgaris L*. Ibid., pp 450–451.

CHAPTER 41

RECENT ADVANCES IN APPLICATION OF METAL-ORGANIC FRAMEWORKS (MOFS) IN TEXTILES

M. HASANZADEH and B. HADAVI MOGHADAM

CONTENTS

41.1 INTRODUCTION

Recently the application of nanostructured materials has garnered attention, due to their interesting chemical and physical properties. Application of nanostructured materials on the solid substrate such as fibers brings new properties to the final textile product [1]. Metal-organic frameworks (MOFs) are one of the most recognized nanoporous materials, which can be widely used for modification of fibers. These relatively crystalline materials consist of metal ions or clusters (named secondary building units, SBUs) interconnected by organic molecules called linkers, which can possess one, two or three dimensional structures [2–10]. They have received a great deal of attention, and the increase in the number of publications related to MOFs in the past decade is remarkable (Fig. 1).

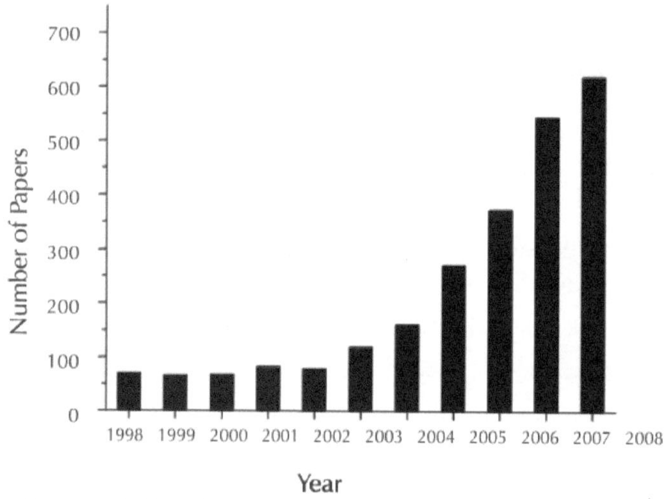

FIGURE 1 Number of publications on MOFs over the past decade, showing the increasing research interest in this topic.

These materials possess a wide array of potential applications in many scientific and industrial fields, including gas storage [11, 12], molecular separation [13], catalysis [14], drug delivery [15], sensing [16], and others. This is due to the unique combination of high porosity, very large surface areas, accessible pore volume, wide range of pore sizes and topologies, chemical stability, and infinite number of possible structures [17, 18].

Although other well-known solid materials such as zeolites and active carbon also show large surface area and nanoporosity, MOFs have some new and distinct advantages. The most basic difference of MOFs and their inorganic counterparts (e.g., zeolites) is the chemical composition and absence of an inaccessible volume (called dead volume) in MOFs [10]. This feature offers the highest value of surface area and porosities in MOFs materials [19]. Another difference between MOFs and other well-known nanoporous materials such as zeolites and carbon nanotubes is the ability to tune the structure and functionality of MOFs directly during synthesis [17].

The first report of MOFs dates back to 1990, when Robson introduced a design concept to the construction of 3D MOFs using appropriate molecular building blocks and metal ions. Following the seminal work, several experiments were developed in this field such as work from Yaghi and O'Keeffe [20].

In this review, synthesis and structural properties of MOFs are summarized and some of the key advances that have been made in the application of these nanoporous materials in textile fibers are highlighted.

41.2 SYNTHESIS OF MOFS

MOFs are typically synthesized under mild temperature (up to 200°C) by combination of organic linkers and metal ions (Fig. 2) in solvothermal reaction [2, 21].

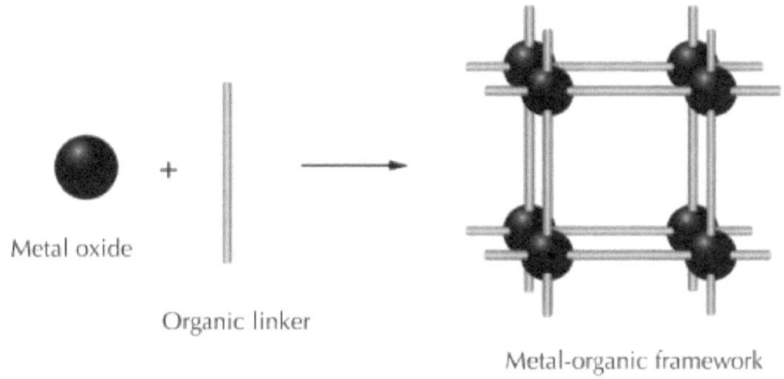

FIGURE 2 Formation of metal organic frameworks.

Recent studies have shown that the character of the MOF depends on many parameters including characteristics of the ligand (bond angles, ligand length, bulkiness, chirality, etc.), solubility of the reactants in the solvent, concentration of organic link and metal salt, solvent polarity, the pH of solution, ionic strength of the medium, temperature and pressure [2, 21].

In addition to this synthesis method, several different methodologies are described in the literature such as ball-milling technique, microwave irradiation, and ultrasonic approach [22].

Post-synthetic modification (PSM) of MOFs opens up further chemical reactions to decorate the frameworks with molecules or functional groups that might not be achieved by conventional synthesis. In situations that presence of a certain functional group on a ligand prevents the formation of the targeted MOF, it is necessary to first form a MOF with the desired topology, and then add the functional group to the framework [2].

41.3 STRUCTURE AND PROPERTIES OF MOFS

When considering the structure of MOFs, it is useful to recognize the secondary building units (SBUs), for understanding and predicting topologies of structures [3]. Figure 3 shows the examples of some SBUs that are commonly occurring in metal carboxylate MOFs. Figure 3(a–c) illustrates inorganic SBUs include the square paddlewheel, the octahedral basic zinc acetate cluster, and the trigonal prismatic oxo-centered trimer, respectively. These SBUs are usually reticulated into MOFs by linking the carboxylate carbons with organic units [3]. Examples of organic SBUs are also shown in Fig. 3(d–f).

It should be noted that the geometry of the SBU is dependent on not only the structure of the ligand and type of metal utilized, but also the metal to ligand ratio, the solvent, and the source of anions to balance the charge of the metal ion [2].

A large number of MOFs have been synthesized and reported by researchers to date. Isoreticular metal-organic frameworks (IRMOFs) denoted as IRMOF-n (n = 1 through 7, 8, 10, 12, 14, and 16) are one of the most widely studied MOFs in the literature. These compounds possess cubic framework structures in which each member shares the same cubic topology [3, 21]. Figure 4 shows the structure of IRMOF-1(MOF-5) as simplest member of IRMOF series.

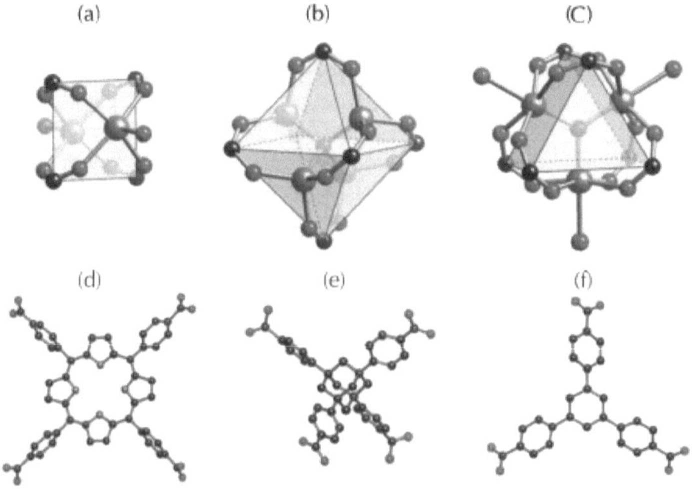

FIGURE 3 Structural representations of some SBUs, including (a–c) inorganic, and (b–f) organic SBUs. (Metals are shown as blue spheres, carbon as black spheres, oxygen as red spheres, nitrogen as green spheres).

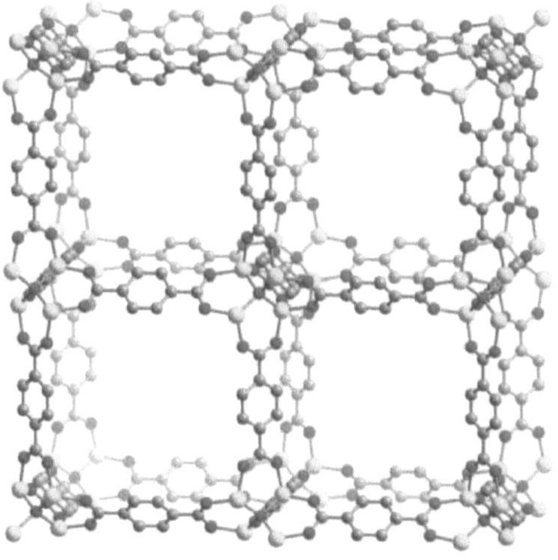

FIGURE 4 Structural representation of IRMOF-1 (Yellow, gray, and red spheres represent Zn, C, and O atoms, respectively).

41.4 APPLICATION OF MOFS IN TEXTILES

41.4.1 INTRODUCTION

There are many methods of surface modification, among which nanostructure based modifications have created a new approach for many applications in recent years. Although MOFs are one of the most promising nanostructured materials for modification of textile fibers, only a few examples have been reported to data. In this section, the first part focuses on application of MOFs in nanofibers and the second part is concerned with modifications of ordinary textile fiber with these nanoporous materials.

41.4.2 NANOFIBERS

Nanofibrous materials can be made by using the electrospinning process. Electrospinning process involves three main components including syringe filled with a polymer solution, a high voltage supplier to provide the required electric force for stretching the liquid jet, and a grounded collection plate to hold the nanofiber mat. The charged polymer solution forms a liquid jet that is drawn towards a grounded collection plate. During the jet movement to the collector, the solvent evaporates and dry fibers deposited as randomly oriented structure on the surface of a collector [23–28]. The schematic illustration of conventional electrospinning setup is shown in Fig. 5.

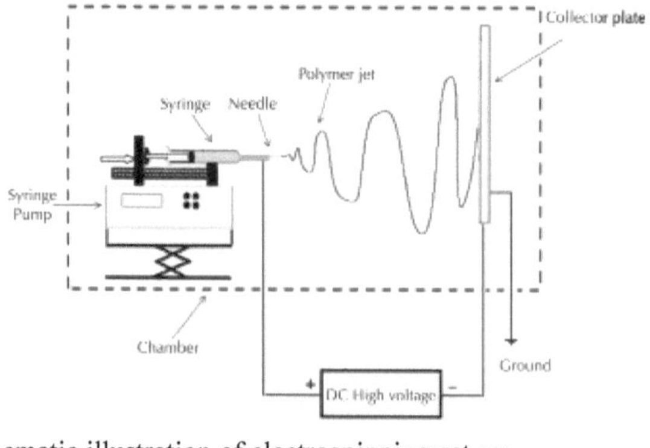

FIGURE 5 Schematic illustration of electrospinning set up.

At the present time, synthesis and fabrication of functional nanofibers represent one of the most interesting fields of nanoresearch. Combining the advanced structural features of metal-organic frameworks with the fabrication technique may generate new functionalized nanofibers for more multiple purposes.

While there has been great interest in the preparation of nanofibers, the studies on metal-organic polymers are rare. In the most recent investigation in this field, the growth of MOF (MIL-47) on electrospun polyacrylonitrile (PAN) mat was studied using in situ microwave irradiation [18]. MIL-47 consists of vanadium cations associated to six oxygen atoms, forming chains connected by terephthalate linkers (Fig. 6).

It should be mentioned that the conversion of nitrile to carboxylic acid groups is necessary for the MOF growth on the PAN nanofibers surface.

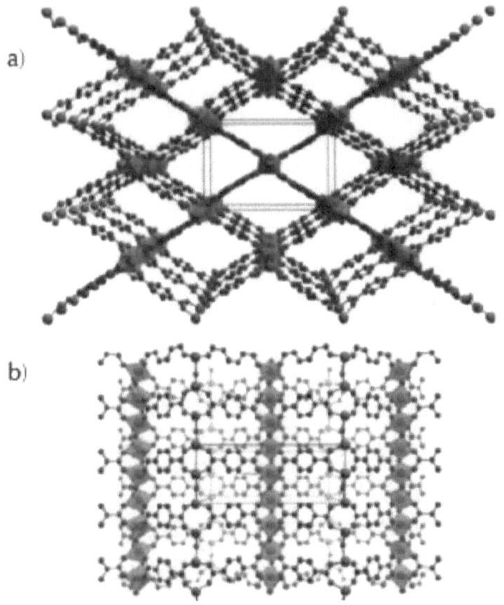

FIGURE 6 MIL-47 metal-organic framework structure: view along the b axis (a) and along the c axis (b).

The crystal morphology of MIL-47 grown on the electrospun fibers illustrated that after only 5 s, the polymer surface was partially covered with small agglomerates of MOF particles. With increasing irradiation time, the agglomerates grew as elongated anisotropic structures (Fig. 7) [18].

FIGURE 7 SEM micrograph of MIL-47 coated PAN substrate prepared from electrospun nanofibers as a function of irradiation time: (a) 5 s, (b) 30 s, (c) 3 min, and (d) 6 min.

It is known that the synthesis of desirable metal-organic polymers is one of the most important factors for the success of the fabrication of metal-organic nanofibers [29]. Among several novel microporous metal-organic polymers, only a few of them have been fabricated into metal-organic fibers.

For example, new acentric metal-organic framework was synthesized and fabricated into nanofibers using electrospinning process [29]. The two dimensional network structure of synthesized MOF is shown in Fig. 8. For this purpose, MOF was dissolved in water or DMF and saturated MOF solution was used for electrospinning. They studied the diameter and morphology of the nanofibers using an optical microscope and a scanning electron microscope (Fig. 9). This fiber display diameters range from 60 nm to 4 μm.

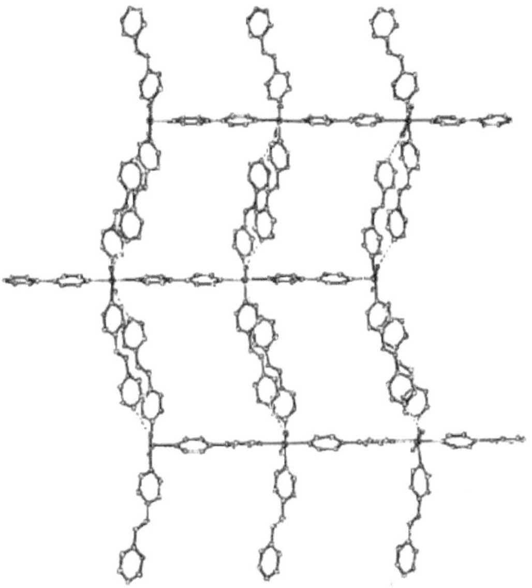

FIGURE 8 Representation of polymer chains and network structure of MOF.

FIGURE 9 SEM micrograph of electrospun nanofiber.

In 2011, Kaskel et al. [30], reported the use of electrospinning process for the immobilization of MOF particles in fibers. They used HKUST-1 and MIL-100(Fe) as MOF particles, which are stable during the electrospinning process from a suspension. Electrospun polymer fibers containing up to 80 wt% MOF particles were achieved and exhibit a total accessible inner surface area. It was found that HKUST-1/PAN gives a spider web-like network of the fibers with MOF particles like trapped flies in it, while HKUST-1/PS results in a pearl necklace-like alignment of the crystallites on the fibers with relatively low loadings.

41.4.3 ORDINARY TEXTILE FIBERS

Some examples of modification of fibers with metal-organic frameworks have verified successful. For instance, in the study on the growth of $Cu_3(BTC)_2$ (also known as HKUST-1, BTC=1,3,5-benzenetricarboxylate) MOF nanostructure on silk fiber under ultrasound irradiation, it was demonstrated that the silk fibers containing $Cu_3(BTC)_2$ MOF exhibited high antibacterial activity against the gram-negative bacterial strain *E. coli* and the gram-positive strain *S. aureus* [1]. The structure and SEM micrograph of $Cu_3(BTC)_2$ MOF is shown in Fig. 10.

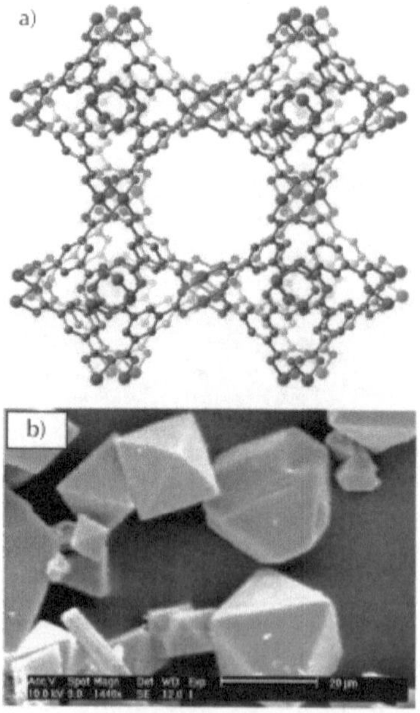

FIGURE 10 (a) The unit cell structure and (b) SEM micrograph of the $Cu_3(BTC)_2$ metal-organic framework (Green, gray, and red spheres represent Cu, C, and O atoms, respectively).

$Cu_3(BTC)_2$ MOF has a large pore volume, between 62% and 72% of the total volume, and a cubic structure consists of three mutually perpendicular channels [32].

The formation mechanism of $Cu_3(BTC)_2$ nanoparticles upon silk fiber is illustrated in Figure 11. It is found that formation of $Cu_3(BTC)_2$ MOF on silk fiber surface was increased in presence of ultrasound irradiation. In addition, increasing the concentration cause an increase in antimicrobial activity [1]. Figure 12 shows the SEM micrograph of $Cu_3(BTC)_2$ MOF on silk surface.

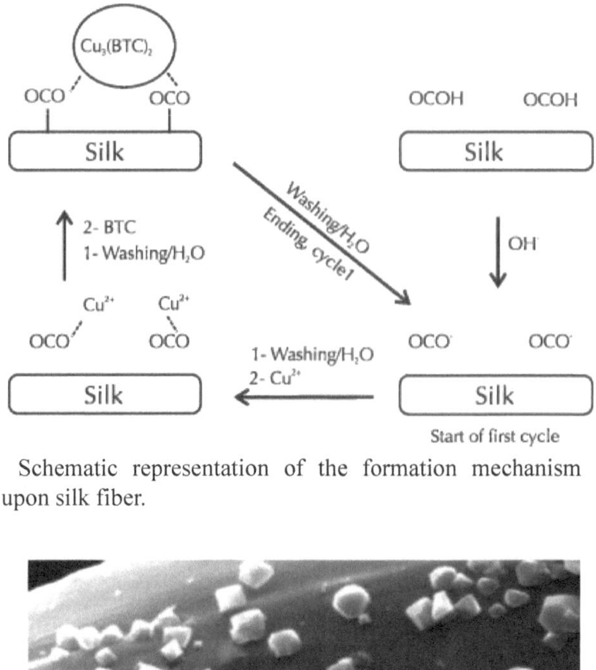

FIGURE 11 Schematic representation of the formation mechanism of $Cu_3(BTC)_2$ nanoparticles upon silk fiber.

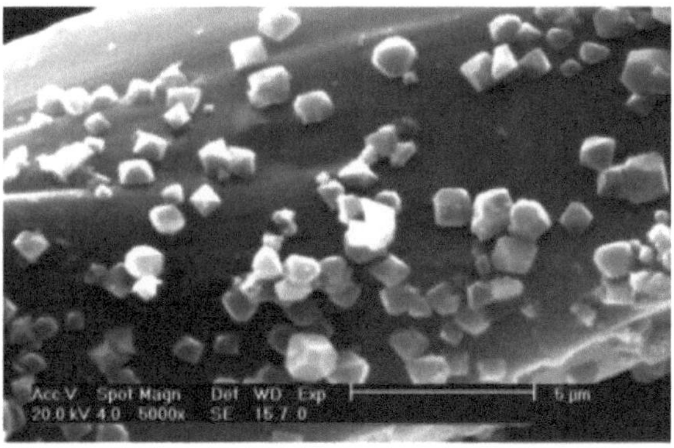

FIGURE 12 SEM micrograph of $Cu_3(BTC)_2$ crystals on silk fibers.

The FT-IR spectra of the pure silk yarn and silk yarn containing MOF (CuBTC-Silk) are shown in Fig. 13. Owing to the reduction of the C=O bond, which is caused by the coordination of oxygen to the Cu^{2+} metal center (Fig. 11), the stretching frequency of the C=O bond was shifted to lower wavenumbers (1654 cm^{-1}) in comparison with the free silk (1664 cm^{-1}) after chelation [1].

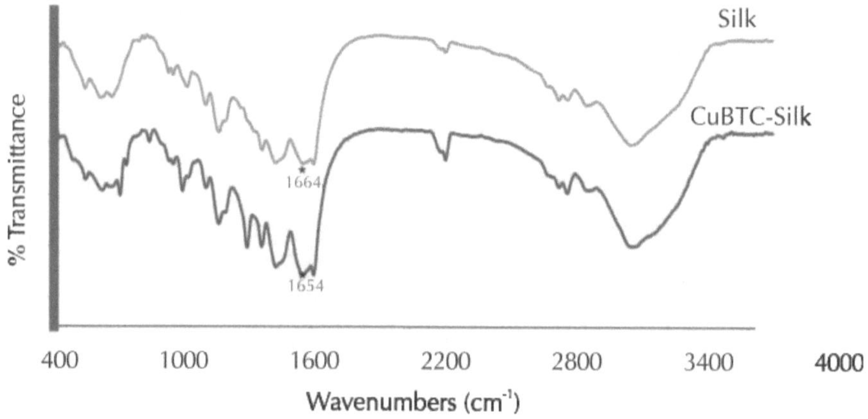

FIGURE 13 FT-IR spectra of the pure silk yarn and silk yarn containing $Cu_3(BTC)_2$.

In another study, $Cu_3(BTC)_2$ was synthesized in the presence of pulp fibers of different qualities [33]. The following pulp samples were used: a bleached and an unbleached kraft pulp, and chemithermomechanical pulp (CTMP).

All three samples differed in their residual lignin content. Indeed, owing to the different chemical composition of samples, different results regarding the degree of coverage were expected. The content of $Cu_3(BTC)_2$ in pulp samples, *k*-number, and single point BET surface area are shown in Table 1. *k*-number of pulp samples, which is indicates the lignin content indirectly, was determined by consumption of a sulfuric permanganate solution of the selected pulp sample [33].

TABLE 1 Some characteristics of the pulp samples.

Pulp sample	MOF content[a] (wt.%)	*k*-number[b]	Surface area[c] (m^2 g^{-1})
CTMP	19.95	114.5	314
Unbleached kraft pulp	10.69	27.6	165
Bleached kraft pulp	0	0.3	10

[a]Determined by thermogravimetric analysis.
[b]Determined according to ISO 302.
[c]Single point BET surface area calculated at p/p_0=0.3 bar.

It is found that CTMP fibers showed the highest lignin residue and largest BET surface area. As shown in the SEM micrograph (Fig. 14), the crystals are regularly distributed on the fiber surface. The unbleached kraft pulp sample provides a slightly lower content of MOF crystals and BET surface area with 165 m^2 g^{-1}. Moreover, no crystals adhered to the bleached kraft pulp, which was almost free of any lignin.

FIGURE 14 SEM micrograph of $Cu_3(BTC)_2$ crystals on the CTMP fibers.

KEYWORDS

- **Electrospinning**
- **Electrospun polyacrylonitrile**
- **Metal-organic framework**
- **Solvothermal reaction**

REFERENCES

1. Abbasi, A. R.; Akhbari, K.; Morsali, A. Dense coating of surface mounted CuBTC metal-organic framework nanostructures on silk fibers, prepared by layer-by-layer method under ultrasound irradiation with antibacterial activity. *Ultrasonics Sonochem.*, **2012**, *19*, 846–852.
2. Kuppler, R. J.; Timmons, D. J.; Fang, Q.-R.; Li, J.-R.; Makal, T. A.;Young, M. D.; Yuan, D.; Zhao, D.; Zhuang, W.; Zhou, H.-C. Potential applications of metal-organic frameworks. *Coord. Chem. Rev.*, **2009**, *253*, 3042–3066.
3. Rowsell, J. L .C.; Yaghi, O. M. Metal-organic frameworks: A new class of porous materials. *Microporous Mesoporous Mater.*, **2004**, *73*, 3–14.
4. An, J.; Farha, O. K.; Hupp, J. T.; Pohl, E.; Yeh, J. I.; Rosi, N. L. Metal-adeninate vertices for the construction of an exceptionally porous metal-organic framework. *Nat. Commun.*, **2012**, doi:10.1038/ncomms1618.
5. Morris, W.; Taylor, R. E.; Dybowski, C.; Yaghi, O. M.; Garcia-Garibay, M. A. Framework mobility in the metal-organic framework crystal IRMOF-3: Evidence for aromatic ring and amine rotation *J. Mol. Struct.*, **2011**, *1004*, 94–101.
6. Kepert, C. J. Metal-organic framework materials. in 'Porous Materials.' In *Porous Materials*, Bruce, D. W.; O'Hare, D.; Walton, R. I., Eds.; John Wiley & Sons: Chichester, 2011.
7. Rowsell, J. L. C.; Yaghi, O. M. Effects of functionalization, catenation, and variation of the metal oxide and organic linking units on the low-pressure hydrogen adsorption properties of metal-organic frameworks. *J. Am. Chem. Soc.*, **2006**, *128*, 1304–1315.
8. Rowsell, J. L. C.; Yaghi, O. M. Strategies for hydrogen storage in metal–organic frameworks. *Angew. Chem. Int. Ed.*, **2005**, *44*, 4670–4679.
9. Farha, O. K.; Mulfort, K. L.; Thorsness, A. M.; Hupp, J. T. Separating solids: purification of metal-organic framework materials. *J. Am. Chem. Soc.*, **2008**, *130*, 8598–8599.
10. Khoshaman, A. H. *Application of Electrospun Thin Films for Supra-Molecule Based Gas Sensing*. M.Sc. thesis, Simon Fraser University, 2011.
11. Murray, L. J.; Dinca, M.; Long, J. R. Hydrogen storage in metal-organic frameworks. *Chem. Soc. Rev.*, **2009**, *38*, 1294–1314.
12. Collins, D. J.; Zhou, H.-C. Hydrogen storage in metal-organic frameworks. *J. Mater. Chem.*, **2007**, *17*, 3154–3160.

13. Chen, B.; Liang C.; Yang J.; Contreras D. S.; Clancy Y. L.; Lobkovsky E. B.; Yaghi O. M.; Dai S. A microporous metal-organic framework for gas-chromatographic separation of alkanes. *Angew. Chem. Int. Ed.*, **2006**, *45*, 1390–1393.
14. Lee, J. Y.; Farha, O. K.; Roberts, J.; Scheidt, K. A.; Nguyen, S. T.; Hupp, J. T. Metal-organic framework materials as catalysts. *Chem. Soc. Rev.*, **2009**, *38*, 1450–1459.
15. Huxford, R. C.; Rocca, J. D.; Lin, W. Metal-organic frameworks as potential drug carriers. *Curr. Opin. Chem. Biol.*, **2010**, *14*, 262–268.
16. Suh, M. P.; Cheon, Y. E.; Lee, E. Y. Syntheses and functions of porous metallosupramolecular networks. *Coord. Chem. Rev.*, **2008**, *252*, 1007–1026.
17. Keskin, S.; Kızılel, S. Biomedical applications of metal organic frameworks. *Ind. Eng. Chem. Res.*, **2011**, *50*, 1799–1812.
18. Centrone, A.; Yang, Y.; Speakman, S.; Bromberg, L.; Rutledge, G. C.; Hatton, T. A. Growth of metal-organic frameworks on polymer surfaces. *J. Am. Chem. Soc.*, **2010**, *132*, 15687–15691.
19. Wong-Foy, A. G.; Matzger, A. J.; Yaghi, O. M. Exceptional H_2 saturation uptake in microporous metal-organic frameworks. *J. Am. Chem. Soc.*, **2006**, *128*, 3494–3495.
20. Farrusseng, D. *Metal-organic Frameworks: Applications from Catalysis to Gas Storage;* Wiley-VCH: Weinheim, 2011.
21. Rosi, N. L.; Eddaoudi, M.; Kim, J.; O'Keeffe, M.; Yaghi, O. M. Advances in the chemistry of metal-organic frameworks. *CrystEngComm*, **2002**, *4*, 401–404.
22. Zou, R.; Abdel-Fattah, A. I.; Xu, H.; Zhao, Y.; Hickmott, D. D. Storage and separation applications of nanoporous metal-organic frameworks, *CrystEngComm*, **2010**, *12*, 1337–1353.
23. Reneker, D. H.; Chun, I. Nanometer diameter fibers of polymer, produced by electrospinning, *Nanotechnology*, **1996**, *7*, 216–223.
24. Shin, Y. M.; Hohman, M. M.; Brenner, M. P.; Rutledge, G. C. Experimental characterization of electrospinning: The electrically forced jet and instabilities. *Polymer,* **2001**, *42*, 9955–9967.
25. Reneker, D. H.; Yarin, A. L.; Fong, H.; Koombhongse, S. Bending instability of electrically charged liquid jets of polymer solutions in electrospinning, *J. Appl. Phys.*, **2000**, *87*, 4531–4547.
26. Zhang, S.; Shim, W. S.; Kim, Design of ultra-fine nonwovens via electrospinning of Nylon 6: Spinning parameters and filtration efficiency, J. *Mater. Des.*, **2009**, *30*, 3659–3666.
27. Yördem, O. S.; Papila, M.; Menceloğlu, Y. Z. Effects of electrospinning parameters on polyacrylonitrile nanofiber diameter: An investigation by response surface methodology. *Mater. Des.*, **2008**, *29*, 34–44.
28. Chronakis, I. S. Novel nanocomposites and nanoceramics based on polymer nanofibers using electrospinning process—A review. *J. Mater. Process. Technol.*, **2005**, *167*, 283–293.
29. Lu, J. Y.; Runnels, K. A.; Norman, C. A new metal-organic polymer with large grid acentric structure created by unbalanced inclusion species and its electrospun nanofibers. *Inorg. Chem.*, **2001**, *40*, 4516–4517.
30. Rose, M.; Böhringer, B.; Jolly, M.; Fischer, R.; Kaskel, S. MOF processing by electrospinning for functional textiles. *Adv. Eng. Mater.*, **2011**, *13*, 356–360.

31. Basu, S.; Maes, M.; Cano-Odena, A.; Alaerts, L.; De Vos, D. E.; Vankelecom, I. F. J. Solvent resistant nanofiltration (SRNF) membranes based on metal-organic frameworks. *J. Membr. Sci.*, **2009**, *344*, 190–198.
32. Hopkins, J. B. *Infrared Spectroscopy of H_2 Trapped in Metal Organic Frameworks*. Thesis, B. A.; Oberlin College Honors, 2009.
33. Küsgens, P.; Siegle, S.; Kaskel, S. Crystal growth of the metal–organic framework $Cu_3(BTC)_2$ on the surface of pulp fibers. *Adv. Eng. Mater.*, **2009**, *11*, 93–95.

CHAPTER 42

CONCLUSION CALCULATED DEPENDENCIES, LINK GEOMETRICAL AND OPERATIONAL PARAMETERS SCRUBBERS

R. R. USMANOVA and G. E. ZAIKOV

CONTENTS

42.1 STATUS OF THE ISSUE, THE RELEVANCE

To studying vortex devices in many papers, extensive experimental data. However, many important problems of analysis and design of vortex devices have not yet found a systematic review.

Existing research in this area show a strong sensitivity of the output characteristics of the regime and design of the device. This indicates a qualitatively different flow hydrodynamics at different values of routine-design parameters.

Thus, it is important to consider the efficiency of fluid flow and vortex devices, receipt and compilation dependencies between regime-design parameters of the machine. Creating effective designs is the actual problem.

42.2 STATEMENT OF THE PROBLEM, THE ASSUMPTIONS

One of the most common devices for dust cleaning equipment considered centrifugal machines. Due to their widespread use simplicity of design, reliability, and low capital cost.

Consider the mechanism of dust-gas cleaning scrubber is something dynamic [1]. Capture dust scrubber is based on the use of centrifugal force. Dust particle flows at high velocity tangentially enters the cylindrical part of the body and makes a downward spiral. The centrifugal force caused by the rotational motion flow, dust particles are moved to the sides of the device (Fig. 1).

When moving in a rotating curved gas flow dust under the influence of centrifugal force and resistance.

Analysis of the swirling dust and gas flow in the scrubber will be carried out under the following assumptions:

1. Gas considered ideal incompressible fluid and, therefore, its potential movement.
2. Gas flow is axisymmetric and stationary.
3. Circumferential component of the velocity of the gas changes in law

$$w_\varphi = const \cdot \sqrt{r}$$

This law is observed in the experiments [2, 3], will provide a simple solution that is convenient for the quantitative analysis of the particles.

4. Particle does not change its shape over time and diameter, it does not happen any crushing or coagulation. Deviation of the particle shape on the field is taken into account the coefficient K.
5. Wrapping a strong flow of gas is viscous in nature. A turbulent fluctuation of gas is not taken into account, which is consistent with the conclusion of [4]: turbulent diffusion of the particles has no significant impact on the process of dust removal.
6. Not considered force Zhukovsky, Archimedes, severity, since these forces by orders of magnitude smaller than the drag force and centrifugal [5, 6, 3].
7. Concentration dust is small, so we cannot consider the interaction of the particles.
8. Neglect the uneven distribution of the axial projection of the radial velocity of the gas, which is in accordance with the data of [7]. Axial component of the velocity of the particles changes little on the tube radius.

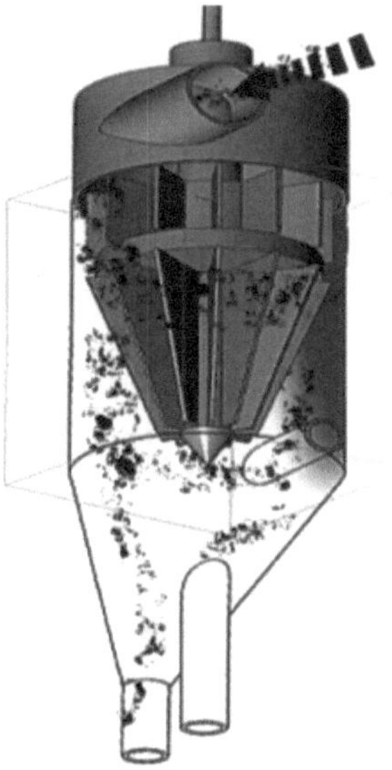

FIGURE 1 The trajectory of the particles in a dynamic scrubber.

The rotation of the purified stream scrubber creates a field of inertial forces, which leads to the separation of a mixture of gases and particles. Therefore, to calculate the trajectories of the particles it is necessary to know their equations of motion and aerodynamics of the gas flow. In accordance with the assumption of a low concentration of dust particles on the influence of the gas flow can be neglected. Consequently, we can consider the motion of a single particle in the velocity field of the gas flow. Therefore, the task to determine the trajectories of particles in the scrubber is decomposed into two parts:

(1) Determination of the velocity field of the gas flow;
(2) Integration of the equations of motion of a particle for a calculation of the velocity field of the gas.

The assumption of axial symmetry of the problem (with the exception of the mouth) allows for consideration of the motion of the particles using a cylindrical coordinate system.

The greatest difficulty is to capture fine dust, for which the strength of the resistance with sufficient accuracy is given by Stokes. By increasing the ratio of dust cleaning machine grows [8], so the calculation parameters scrubber with low dust content (by assumption 6) guarantees a minimum efficiency.

42.3 DERIVE THE EQUATION OF MOTION OF A PARTICLE

To calculate the trajectories of the particles need to know their equations of motion. Such a problem for some particular case is solved by the author [9].

We introduce a system of coordinates OXYZ. Its axis is directed along the OZ axis of symmetry scrubber (Fig. 2). Law of motion of dust particles in the fixed coordinate system OXYZ can be written as follows:

$$m\frac{d\overrightarrow{v_p'}}{dt} = \overrightarrow{F_{st}} \tag{1}$$

where m – mass of the particle; $d\overrightarrow{v_p'}$ – velocity of the particle; and $\overrightarrow{F_{st}}$ – aerodynamic force.

For the calculations necessary to present the vector Eq. (1) motion in scalar form. Position of the particle will be given by its cylindrical coordinates (r; φ; z). Velocity of a particle is defined by three components: U_p–tangential, V_p – radial and W_p – axial velocity.

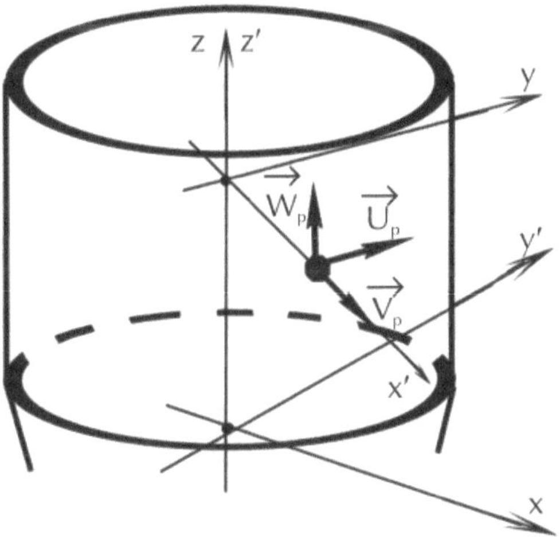

FIGURE 2 The velocity vector of the particle.

We take a coordinate system O'X'Y'Z', let O'X' passes axis through the particle itself, and the axis O'Z' lies on the axis OZ. Adopted reference system moves forward along the axis OZ W_p speed and rotates around an angular velocity

$$\omega(t)=\frac{U_p}{r_p} \tag{2}$$

The equation of motion of a particle of mass $m=\frac{1}{6}\pi\rho_p d_p^3$ coordinate system O'X'Y'Z' becomes:

$$m\frac{d\overrightarrow{v_p'}}{dt}=\overrightarrow{F_{st}}-m\overrightarrow{a_0'}+m\left[\overrightarrow{r_p'}\cdot\vec{\omega}\right]+m\left[\overrightarrow{r_p'}\cdot\vec{\omega}\right]+m\left[\vec{\omega}\cdot\left[\overrightarrow{r_p'}\cdot\vec{\omega}\right]\right]+2m\left[\overrightarrow{v_p'}\cdot\vec{\omega}\right]$$

or

$$\frac{d\overrightarrow{v_p'}}{dt}=\frac{1}{m}\overrightarrow{F_{st}}-\overrightarrow{a_0'}+\left[\overrightarrow{r_p'}\cdot\overrightarrow{\omega}\right]+\left[\overrightarrow{\omega}\cdot\left[\overrightarrow{r_p'}\cdot\overrightarrow{\omega}\right]\right]+2\left[\overrightarrow{v_p}\cdot\overrightarrow{\omega}\right] \quad (3)$$

where $\overrightarrow{a_0'}$ – translational acceleration vector of the reference frame; $\overrightarrow{dv_p}$ –velocity of the particle; r_p' – the radius vector of the particle; $\left[\overrightarrow{r_p'}\cdot\overrightarrow{\omega}\right]$ – acceleration due to unevenness of rotation; $\left[\overrightarrow{\omega}\cdot\left[\overrightarrow{r_p'}\cdot\overrightarrow{\omega}\right]\right]$ – the centrifugal acceleration; $2\left[\overrightarrow{v_p}\cdot\overrightarrow{\omega}\right]$ – Coriolis acceleration.

The first term on the right-hand side of Eq. (3) is the force acting c gas flow on the particle, and is given by Stokes,

$$F_{st}=3\pi\mu_g d_p\left[\overrightarrow{v_g-v_p}\right] \quad (4)$$

$\overrightarrow{\mu_g}$ – dynamic viscosity of the gas.

The second term (3) is defined as,

$$\frac{dW_p}{dt}\overrightarrow{e_z}=\frac{dW_p}{dt}\overrightarrow{e_{z'}}$$

Convert the remaining terms:

$$\left[\overrightarrow{r_p'}\cdot\overrightarrow{\omega}\right]=\left[\overrightarrow{r_p'}\cdot\frac{d\overrightarrow{\omega}}{dt}\right]=\left[\overrightarrow{r_p'}\cdot\frac{d}{dt}\left(\frac{U_p}{r_p}\overrightarrow{e_{z'}}\right)\right]=-r_p\left(\frac{1}{r_p}\frac{dU_p}{dt}-\frac{U_p}{r_p^2}V_x\right)\overrightarrow{e_y'}=\left(-\frac{dU_p}{dt}+\frac{U_pV_p}{r_p}\right)e_y$$

$$\left[\overrightarrow{\omega}\cdot\left[r_p^{-1}\cdot\overrightarrow{\omega}\right]\right]=\frac{U_p^2}{r_x}\left[\overrightarrow{e_z}\cdot\left[\overrightarrow{e_x}\cdot\overrightarrow{e_z}\right]\right]=-\frac{U_p^2}{r_p}\left[\overrightarrow{e_z}\cdot\overrightarrow{e_v}\right]=\frac{U_p^2}{r_p}e_x$$

$$2\left[\overrightarrow{v_p}\cdot\overrightarrow{\omega}\right]=2v_{x'}\left[\overrightarrow{e_{x'}}\cdot\overrightarrow{\omega}\right]=2v_{x'}\frac{U_x}{r_p}\left[\overrightarrow{e_{x'}}\cdot\overrightarrow{e_{z'}}\right]=\left(-2\frac{U_pV_p}{r_p}\right)e_{y'}\sqrt{a^2+b^2}$$

где $\overrightarrow{e_+},\overrightarrow{e_v},\overrightarrow{e_z}$ – vectors of the reference frame and used the fact that $\overrightarrow{r_p}=\overrightarrow{e_x}\cdot r_+\cdot v_x=V_p$

Substituting these expressions in the equation of motion (3),

$$m\frac{d\overrightarrow{v_p'}}{dt}=\overrightarrow{F_{st}}-m\overrightarrow{a_0'}+m\left[\overrightarrow{r_p'}\cdot\overrightarrow{\omega}\right]+m\left[\overrightarrow{r_p'}\cdot\overrightarrow{\omega}\right]+m\left[\overrightarrow{\omega}\cdot\left[\overrightarrow{r_p'}\cdot\overrightarrow{\omega}\right]\right]+2m\left[\overrightarrow{v_p'}\cdot\overrightarrow{\omega}\right]$$

or

$$\frac{d\vec{v_p}}{dt} = \frac{1}{m}\vec{F_{st}} - \vec{a_0} + \left[\vec{r_p} \cdot \vec{\omega}\right] + \left[\vec{\omega} \cdot \left[\vec{r_p} \cdot \vec{\omega}\right]\right] + 2\left[\vec{v_p} \cdot \vec{\omega}\right]$$

We write this equation in the projections on the axes of the coordinate system O'X'Y'Z':

$$\begin{cases} \dfrac{dV_{x'}}{dt} = \dfrac{1}{m}F_{stx'} + \dfrac{U_p^2}{r_p} \\ 0 = \dfrac{1}{m}F_{sty} - \dfrac{dU_p}{dt} - \dfrac{U_p V_p}{r_p} \\ 0 = \dfrac{1}{m}F_{stz} - \dfrac{dW_p}{dt} \end{cases}$$

or

$$\begin{cases} \dfrac{dV_p}{dt} = \dfrac{1}{m}F_{stx} + \dfrac{U_p^2}{r_p} \\ \dfrac{dU_p}{dt} = \dfrac{1}{m}F_{sty} - \dfrac{U_p V_p}{r_p} \\ \dfrac{dW_p}{dt} = \dfrac{1}{m}F_{stz} \end{cases} \quad (5)$$

We have the equation of motion of a particle in a rotating gas flow projected on the axis of the cylindrical coordinate system.

Substituting Eqs. (2) and (4) in (5) we obtain the system of equations of motion of the particle:

$$\begin{cases} \dfrac{dV_p}{dt} = \dfrac{18\mu}{\rho_p d_p^2}\left(V_g - V_p\right) + \dfrac{U_p^2}{r_p} \\ \dfrac{dU_p}{dt} = \dfrac{18\mu}{\rho_{\div} d_{\div}^2}\left(U_g - U_p\right)\dfrac{U_p V_p}{r_p} \\ \dfrac{dW_p}{dt} = \dfrac{18\mu}{\rho_p d_p^2}\left(W_g - W_p\right) \end{cases} \quad (6)$$

42.4 OUTPUT RELATIONSHIP BETWEEN THE GEOMETRICAL AND OPERATIONAL PARAMETERS

Formal analysis of relationships that define the motion of gas and solids in the scrubber. The analysis shows that the strict observance of similarity of movements in the devices of different sizes requires the preservation of four dimensionless complexes, such as

$$\mathrm{Re}_d = \frac{wD}{\nu}; F_r = \frac{w^2}{Dg}; A_r = \frac{\delta\rho_2}{D\rho_1}; \mathrm{Re}_\delta = \frac{\nu\delta}{\nu}$$

Not all of these systems are affecting the motion of dust. Experimentally found that the influence of the Froude number F_r negligible [10] and can be neglected. It is also clear that the effect of the Reynolds number for large values it is also insignificant. However, maintaining unchanged the remaining two complexes, still introduces significant difficulties in modeling devices.

On the other hand, there is no need for strict observance of the similarity in the trajectory of the particle in the apparatus. What is important is the end result – providing the necessary efficiency unit. To estimate the parameters that characterize the removal of particles of a given diameter, consider the approximate solution of the problem of the motion of a solid particle in a scrubber. A complete solution for a special case considered in the literature [11], this solution could be used to obtain the dependence of the simplified model of the flow.

For the three coordinates – radial, tangential and vertical equations of motion of a particle at a constant resistance can be written in the following form:

$$\frac{dw}{dt} - \frac{w_\varphi^2}{r} = -\alpha(w_r - v_r)$$

$$\frac{dw_z}{dt} \cong -\alpha(w_z - v_z)$$

$$\frac{dw_\varphi}{dt} + 2w_r\frac{w_\varphi}{r} - \alpha(w_\varphi - v_\varphi)$$

where α-factor resistance to motion of a particle, divided by its mass.

$$\alpha = \frac{\mu}{K\rho\delta^2}$$

K-factor, which takes into account the effect of particle shape (take K = 2).

The axial component of the velocity of gas and particles are the same, as follows from the equations of motion by neglecting gravity. Indeed, if the $\frac{dw_z}{dt} = \alpha(w_z - v_z)$, then taking $w_z - v_z = \Delta w_z$, we get $\frac{d\Delta w_z}{dt} = -\alpha\Delta w_z$ $\Delta W_z = \Delta w_{zo} e^{-dt}$.

If initially $w_z = v_z$ $(\Delta W_z = 0)$; $\Delta W_t = 0$ и $W_t = const$ projection speed:

$$w_\varphi = const \cdot \sqrt{r}$$

Valid law $W\varphi(r)$ may differ markedly from the accepted, but this is not essential. In this case, it only makes us enter into the calculation of average:

$$\left(\frac{w_\varphi^2}{r}\right)av$$

Under these simplifications, the first of the equations of motion is solved in quadrature. Indeed, for now,

$$\frac{dw_r}{dt} + aw_r = \frac{w_\varphi^2}{r}$$

then with the obvious boundary condition t = 0, yr = 0, we have:

$$w_r \cong \frac{1}{\alpha}\left(\frac{w_\varphi^2}{r}\right)av\left(1 - e^{-dt}\right)$$

The time during which the flow passes from the blade to the swirler exit from the apparatus as well:

$$t_1 = \frac{l}{w_{z\,av}}$$

On the other hand, knowing the law of the radial velocity, we can find the time during which the particle travels a distance of r_1 (the maximal distance from the wall) to the vessel wall (r_2).

$$r_2 - r_1 = \int_0 w_r dt = \frac{w_\varphi}{d_r} \int_0 \left(1 - e^{-dt}\right) dt$$

$$r_2 - r_1 = \frac{v_\varphi}{\alpha r_{av}} \left[t_1^l + \frac{1}{\alpha}\left(e^{-dt} - 1\right) \right]$$

Substituting in this equation is the limiting value of $t_1 = t_t$. we get:

$$\frac{\alpha r_{av}\left(r_2 - r_1\right)}{v_{\varphi av}^2} \geq \frac{l}{w_z} + \frac{1}{\alpha}\left(e^{\frac{\mu}{\rho\delta^2} * \frac{l}{w_z}} - 1 \right)$$

or

$$\frac{\mu}{K\rho\delta^2} \cdot \frac{r_1^2 - r_2^2}{2V_{\varphi\, av}^2} \cdot \frac{w_z}{l} \geq 1 + \frac{K\rho\delta^2}{\mu} \cdot \frac{w_z}{l}\left(e^{-\frac{\mu}{\rho\delta^2 K} * \frac{l}{w_t}} - 1 \right) \quad (7)$$

The presented approach is based on the known dependence and model of the flow, it's different in a number of studies approach is only in the details. However, further to allocate two sets, one of which characterizes the geometry of the device, and the other – operating data. The use of these systems simplifies the calculation and, most importantly, takes into account the influence of some key factors to the desired gas velocity and height of the apparatus $W_{Zav.}$

Dependence structure, Eq. (7) shows the feasibility of introducing two sets, one of which:

$$A = \frac{\mu l}{2\rho\delta^2 W_t K}$$

characterizes the effect of the flow regime and the particle diameter, and the other is a geometric characteristic of the device.

$$A_r = \frac{r_2}{l}\sqrt{\left(1 - r_1^{-2}\right)} ctg\beta \quad (8)$$

In Eq. (8) through r_1 marked relative internal radius apparatus:

$$r_1 = r_1 / r_2$$

and β_1 – the average angle of the flow at the exit of the guide apparatus,

$$tg\beta_1 = \frac{V_\varphi}{V_z}$$

Then Eq. (7) takes the simple form:

$$A_r \geq f(A)$$

where

$$f(A) = \sqrt{\frac{1}{A} + \frac{1}{2A^2}\left(l^{-2A} - 1\right)} \quad (9)$$

Expressed graphically in Fig. 3, this dependence allows to determine the minimum value of the mode parameter A_{min} for the scrubber with geometrical parameters of A_r. And must take $A > A_{min}$.

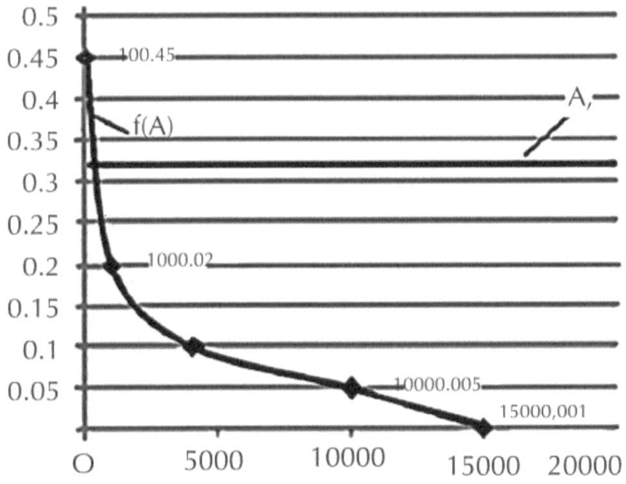

FIGURE 3 The relationship between the geometrical and operational complexes.

One of the important consequences of the resulting function is the relationship between the diameter of the dust particles and the axial velocity of the gas. With this machine, with data A_g, A_{min} = const and therefore,

$$w_z \delta^2 = const$$

This means that reducing the particle size axis (expenditure), the rate should be increased according to the dependence

$$\frac{w_z}{w_{zo}} = \left(\frac{\delta_o}{\delta}\right)^2$$

Unfortunately, a significant increase W_z permitted as this may lead to the capture of dust from the walls and ash. You can also change the twist angle β and the height of the flow system, without changing the axial velocity.

If the reduction of the particle diameter δ or increasing the size of the unit has increased the value of A, it must be modified accordingly A_g (using a graph Fig. 3), and the new value of A_r to find an angle β:

$$\frac{ctg\beta}{ctg\beta_0} = \left(\frac{A_g}{A_{go}}\right)$$

42.5 FINDINGS

1. Creating a mathematical model of the motion of a particle of dust in the swirling flow allowed us to estimate the influence of various factors on the collection efficiency of dust in the offices of the centrifugal type, and create a methodology to assess the effectiveness of scrubber.
2. Identified settlement complexes, one of which characterizes the geometry of the scrubber and the other operational parameters. The use of these systems and simplifies the calculation takes into account the influence of several key factors.
3. The developed method can be used in the calculation and design of gas cleaning devices, as constituent relations define the relationship between the technological characteristics of the dust collectors and their geometrical and operational parameters.

KEYWORDS

- **Centrifugal force**
- **Coefficient of resistance**
- **Coriolis acceleration**
- **Custodial complex**
- **Dynamic scrubber**
- **Geometric complex**
- **Trajectory**
- **Turbulent fluctuation**

REFERENCES

1. Pat. RF 2339435 (2008), Auto. See each other. USSR 1.340.410, 1987.
2. Barahtenko, G. M.; Idelchik, I. E. *Ind Sanitation Gas*, **1974**, *6*, 4–7.
3. Straus, V. *Industrial Cleaning Gases;* Chemistry: Moscow, 1981; 616 p.
4. Shilyaev, M. I. *Aerodynamics and Heat and Mass Transfer of gas-dispersion flow: studies Allowance;* Publishing House of Tomsk State Architect - Builds University: Tomsk, 2003; 272 p.
5. Lagutkin, M. G.; Tohti, D. A. *Sheep*, **2004**, *38*(1), 9–13.
6. Deutsch, M. E.; Filippov, G. A. *Gas Dynamics of Two-Phase Media;* Energy: Moscow, 1968; 424 p.
7. Starchenko, A. V.; Bells, A. M.; Burlutskiy E. S. *Thermophys. Aeromech.*, **1999**, *6*(1), 59–70.
8. Gupta, A.; Lilly, D.; Sayred N. *Swirling Flow* Trans. from English. Krashennikova, S. Y., Ed.; Mir: Moscow, 1987; 588 p.
9. Bezic, D. A. *Diss.kand.tehn. sciences, gos.inzh.-tehnol.akademiya Bryansk*, Bryansk, 2000; 150p.
10. Mizonov, V. E.; Blaschek, V.; Colin, R.; Greeks A. *Tohti*, **1994**, *28*(3), 277–280.
11. Litvinov, A. T. *J. Appl. Chem.*, **1971**, *44*(6), 1221–1231.

INDEX

C

D

E

F

I

K

L

M

N

O

P